Advances in Science, Technology & Innovation

IEREK Interdisciplinary Series for Sustainable Development

Advances in Science, Technology & Innovation (ASTI) is a series of peer-reviewed books based on important emerging research that redefines the current disciplinary boundaries in science, technology and innovation (STI) in order to develop integrated concepts for sustainable development. It not only discusses the progress made towards securing more resources, allocating smarter solutions, and rebalancing the relationship between nature and people, but also provides in-depth insights from comprehensive research that addresses the **17 sustainable development goals (SDGs)** as set out by the UN for 2030.

The series draws on the best research papers from various IEREK and other international conferences to promote the creation and development of viable solutions for a **sustainable future and a positive societal** transformation with the help of integrated and innovative science-based approaches. Including interdisciplinary contributions, it presents innovative approaches and highlights how they can best support both economic and sustainable development, through better use of data, more effective institutions, and global, local and individual action, for the welfare of all societies.

The series particularly features conceptual and empirical contributions from various interrelated fields of science, technology and innovation, with an emphasis on digital transformation, that focus on providing practical solutions to **ensure food, water and energy security to achieve the SDGs.** It also presents new case studies offering concrete examples of how to resolve sustainable urbanization and environmental issues in different regions of the world.

The series is intended for professionals in research and teaching, consultancies and industry, and government and international organizations. Published in collaboration with IEREK, the Springer ASTI series will acquaint readers with essential new studies in STI for sustainable development.

ASTI series has now been accepted for Scopus (September 2020). All content published in this series will start appearing on the Scopus site in early 2021.

Marzena Smol • Majeti Narasimha Vara Prasad •
Alexandros I. Stefanakis
Editors

Water in Circular Economy

Springer

Editors
Marzena Smol
Mineral and Energy Economy Research Institute
Polish Academy of Sciences
Krakow, Malopolskie, Poland

Majeti Narasimha Vara Prasad
School of Life Sciences
University of Hyderabad
(an Institution of Eminence)
Hyderabad, Telangana, India

Alexandros I. Stefanakis
School of Chemical and Environmental Engineering
Technical University of Crete
Chania, Greece

ISSN 2522-8714 ISSN 2522-8722 (electronic)
Advances in Science, Technology & Innovation
IEREK Interdisciplinary Series for Sustainable Development
ISBN 978-3-031-18164-1 ISBN 978-3-031-18165-8 (eBook)
https://doi.org/10.1007/978-3-031-18165-8

Cover image credit: Oksana Bessonova/shutterstock.com

This Springer imprint is published by the registered company Springer Nature Switzerland AG
The registered company address is: Gewerbestrasse 11, 6330 Cham, Switzerland

Introduction

Water is one of the most important resources in the world due to its key importance in human life and in the functioning of all industries. Water as a carrier of materials and energy is also a critical resource for a circular economy (CE), which aims to decouple economic growth and development from the consumption of finite raw materials. The CE dynamic offers a framework that is both resilient and restorative for generations to come and also a mechanism for transitioning growth into a positive trend for environment, economy and society. Currently, water is very unequally distributed and increasingly scarce as a result of urbanization and climate change. Therefore, actions to protect water resources, their sustainable management, as well as the re-use of water and water-based resources (such as materials and energy), are crucial for the CE, which aims to achieve resource efficiency, to reduce waste production and to improve environmental, economic and social sustainability.

Water sector is currently facing many challenges arising from the implementation of the CE model, which is the political priority of many countries and regions in the world. The current book provides the comprehensive knowledge on the importance of water resources in the CE model. The perspectives of environment, society and industry are analysed in the strategic areas of water resource management from the perspective of the CE implementation. They include the revision of global strategies towards resource efficiency, directions of activities and policies for integrated water management, smart cities or blue economy. The special interest is dedicated to the identification of key dimensions of circular water use—water as service, water as source of energy and water as carrier of materials—which should be integrated across all sectors. The book also presents an inventory of innovative solutions (technological, environmental, social) that can be implemented in water sector to accelerate the transformation from a linear to the CE model. An innovative approach in the book is also to present the scope of monitoring progress towards the CE in water sector, which includes the proposition of monitoring framework, inventory of the CE indicators and business models for the water sector and case studies in this area.

This book seeks to contribute to the implementation of the CE models through an inventory of multidisciplinary knowledge with high scientific and practical importance, which can contribute to improving the quality of the environment and achieving a balance between human activities and nature in the water sector. We invited research papers, critical reviews or case studies, which are relevant to the topic of the importance of water in the CE.

Marzena Smol

Contents

About the Editors

Marzena Smol is Associate Professor and Head of the Division of Biogenic Raw Materials in the Mineral and Energy Economy Research Institute, Polish Academy of Sciences (IGSMiE PAN, Poland). She holds a Master's degree in Environmental Protection and International Project Management, and Ph.D. and D.Sc. in Environmental Engineering. Expert in the field of biogenic raw materials management in the context of sustainable development (SD), circular economy (CE) and Green Deal Strategies. She participated in the development of the Roadmap Transformation towards circular economy in Poland. Holder of the Minister's Scholarship for Outstanding Young Scientists. She is author and co-author over 200 papers. She is holder of Springer Nature 2020 Highlights award in category "Water" for publication on CE in water and wastewater sector. Member of the Polish Young Academy of the Polish Academy of Sciences.

Majeti Narasimha Vara Prasad is Emeritus Professor, School of Life Sciences, University of Hyderabad, India. He has published over 219 papers and edited 34 books. He received doctorate in Botany from Lucknow University, India in 1979. Based on an independent study by Stanford University scientists in 2020, he figured in the top 2% of scientists from India, ranked number 1 in Environmental Sciences (116 in world). Prasad had several visiting assignments viz., University Quebec INRS-Eau, Canada, 1994; Department of Plant Physiology and Biochemistry, Jagiellonian University, Krakow, Poland, 1996; University of Coimbra, Portugal, 1999; Stockholm University, Institute of Botany, Sweden, 2000; University of Oulu, Oulu Finland, 2002; University of South Australia, Adelaide, Australia, 2005; Al-Farabi Kazakh National University, Department of Botany, Almaty, Kazakhstan, 2006; Ural Federal University, Ekaterinburg, Russia, 2007 to 2015; Ghent University Faculty of Bioscience Engineering, Belgium, 2011; Mahasarakham University, Thailand, 2013–14; University of Santa Cruz, Ilheus-Bahia, Brazil, 2015; Government of India deputation to Asian Institute of Technology, Thailand for one semester 2017.

Alexandros I. Stefanakis is Assistant Professor at Technical University of Crete (Greece). He received Ph.D. in Environmental Engineering at the Democritus University of Thrace. He is expert in ecological treatment of wastewater and sludge using natural systems. He has published over 15 papers in international peer-reviewed journals, international and national conference proceedings etc. His work includes numerous experiments with Constructed Wetland systems of different types and applications. He has participated in many national and European research projects. He has supervised and trained various undergraduate students. He has been awarded twice during his career for his research work. Finally, Dr. Stefanakis is also a practicing environmental engineer in Greece, dealing with the design of Constructed Wetlands facilities.

Water as Key Resource in Circular Economy

Impact of Climate Change on Water Status: Challenges and Emerging Solutions

Aiman Slimani, Fatima-Zahra Akensous, Khalid Oufdou, and Abdelilah Meddich

Abstract

Climate change effects on water, manifested by climate pattern fluctuations, have been exacerbating. Consequently, the globe is expected to witness even drier areas and water-related disasters by the end of the twenty-first century. Moreover, the intensification of water cycle is imposing a global socioeconomic impact, transforming agriculturally, industrially, and municipally water sector nexuses. Such consequences emerge as a challenging hurdle for institutions' ability to overpower failures in water sector management. Thus, the present paper underlines prominent findings surrounding climate change multifaceted effects on water status, stemming from selected case studies. The scope of research includes a thorough state-of-the-art analysis, tracked via desk research and comparative analysis. The main driving factors, impact of climate change on water resources, climate fluctuations, and precipitation intensity were most emphasized. Furthermore, agricultural, industrial, and municipal sectors have been highlighted as the major water-consuming fields. Finally, the applications of sustainable technological features, mainly artificial intelligence (AI), remote sensing, and biostimulants, were highlighted as effective tools for circular economy (CE) framework implementation. Both probable and discrepant results on projected water quantity and quality were noted and critically discussed. The paper concludes that developing prediction models and the enactment of the proposed CE framework go hand in hand.

Keywords

Water • Climate change • Circular Economy (CE)

A. Slimani (✉) · F.-Z. Akensous · A. Meddich
Laboratory of Agro-Food, Biotechnologies and Valorization of Plant Bioresources (Agrobioval), Faculty of Science Semlalia, Cadi Ayyad University, PO Box 2390 Marrakesh, Morocco
e-mail: slimani.aiman@gmail.com

F.-Z. Akensous
e-mail: fatimazahra.akensous@ced.uca.ma

A. Meddich
e-mail: a.meddich@uca.ma

A. Slimani · F.-Z. Akensous · A. Meddich
Center of Agrobiotechnology and Bioengineering, Research Unit Labelled CNRST (Centre AgroBiotech-URL-CNRST-05), "Physiology of Abiotic Stress" Team, Cadi Ayyad University, 40000 Marrakech, Morocco

A. Slimani · K. Oufdou
Laboratory of Microbial Biotechnologies, Agrosciences, and Environment (BioMagE), Labeled Research Unit-CNRST N°4, Faculty of Science Semlalia, Cadi Ayyad University, PO Box 2390 Marrakesh, Morocco
e-mail: oufdou@uca.ma

1 Introduction

Water, an excellent solvent, is the fundamental liquid on Earth and the most studied (Gleick, 1999; Zhou et al., 2019a, 2019b). In fact, this tiny molecule holds the key to development (Goswami & Bisht, 2017; Tiecco et al., 2019). It is found at the center of socioeconomics, energy and agriculture sectors, and ecosystems' balance (Hamidov & Helming, 2020). Nevertheless, during the last few decades, preserving water resources has become an urgent matter globally in the context of climate change (Liu et al., 2019). If no action is taken to address this problem, the situation will worsen, and an entire biosphere will collapse (Mello et al., 2020). Thus, to be able to maintain aquatic ecosystems, sustainability in water sector is crucial (Pires et al., 2017). Another problem arises as about 97% of water on Earth is saline. This water is impractical for irrigation, improper for drinking purpose, and unsuitable for industrial targets (Wijaya et al., 2021). However, the remaining 3% of freshwater are not fully available for use (Gray, 2017). Thus, tackling climate change is vital to save fresh water resources (Mello et al., 2020).

Climate change designates the long-term shifts in temperature and weather patterns (Werndl, 2016). Such

M. Smol et al. (eds.), *Water in Circular Economy*, Advances in Science, Technology & Innovation,
https://doi.org/10.1007/978-3-031-18165-8_1

alterations are incited by natural causes and anthropogenic action (Abbot & Marohasy, 2017). According to climatologists, temperature and precipitation variability, periods of drought, and heat wave frequencies are accentuating (Schilling et al., 2020). In addition, climate change and water nexus is mainly revealed by shifts in water patterns (Allan et al., 2020c). For instance, the vulnerability of freshwater resources is evidenced by observational records and climate models, fueled by global warming (Allan et al., 2020c). Climate models project that the twenty-first century will witness a significant rise in water-related complex phenomena (Asadieh & Krakauer, 2017). Prolonged intense droughts will further increase aridity (Liang et al., 2020). Moreover, futuristic water demand is indicating an increase of 55% by 2050 (Mohammad et al., n.d.). The extreme events adding to rising sea level are expected to exacerbate (Karegar et al., 2017). For all the aforementioned reasons, water is but the currently major concern globally.

Effectively managing water resources is gaining a compelling interest. Fortunately, a range of innovative approaches has emerged. In that sense, switching towards circular economy (CE) is gaining momentum (Ferronato et al., 2019). Achieving sustainable water use and attaining the United Nations' 17 sustainable development goals (SDG) go hand in hand (Shef & Wood, 2018). Hence, accommodating the twenty-first-century SDG at a global level requires a conceptualized outlook (Zhenmin & Espinosa, 2019). Shaping integration, planning, and cooperation, in respect of geography, jurisdiction, multidisciplinary, and the planning timelines duration is crucial (Chandra et al., 2021).

Fortunately, the application of innovative tools like artificial intelligence (AI) (Xiang et al., 2021), water remote sensing (Calera et al., 2017; Chawla et al., 2020), and biostimulants (Parađiković et al., 2019) can help make an effective decision towards sustainable water management. These eco-friendly means are able to provide important information to farmers, stakeholders, and water managers (Lin et al., 2018).

Thermodynamic tendencies are often cited as an explanation to the rise in atmospheric water vapor. Furthermore, precipitation and evaporation trends are conditioned by the Earth's energy balance. Though climatic variables like precipitation and evaporation are rising, the atmospheric adjusting leads to suppressing the increasing rates as a response to the greenhouse effect and absorbing aerosols. Taking precipitation as an important determinant in water cycle pattern alterations, the climatic variable trends are difficult to determine but are projected to upsurge, mainly in response to global warming. Such future predictions are likely to occur in ocean than on land surfaces. Consequently, flooding frequency will also depend on precipitation rising. Furthermore, the anthropogenic effect is further pushing extreme water cycle shifts due to extensive water use, which is projected to elevate water demand against an ever-growing global population (Allan et al., 2020a). Clearly, the gravest repercussions of climate change are already reflecting on water status. Hence, water organization concerns are rising and the implementation of adaptive measures is often obstructed. In addition, water policy installation is conditioned by both dynamic and contextual aspects, depending on developed and developing economies (Azhoni et al., 2018).

Therefore, the present paper intends to synthesize case studies and available data regarding the adverse impact of climate change on water. It also looks at promising CE tools that can effectively preserve and manage water resources.

2 Materials and Methods

The present paper was constructed based on a precise literature investigation. The investigation layout is presented in Fig. 1.

The literature investigation was organized into four steps. Step I consists of a state-of-the-art analysis, conducted through desk research. Step II concentrates on identifying knowledge gaps, highlighting differing and sometimes uncertain water future projections. Step III focuses on underlining findings that corroborate water crisis under climate change patterns, in addition to effective water managerial tools. Lastly, step IV represents an interpretative data explanation, closing in concluding and critical remarks.

In the first step, an objective revision of relevant data, available in the literature, was carried out through desk research methodology. The state-of-the-art analysis was built upon the review scheme of (Allan et al., 2020a, 2020b, 2020c) and (Azhoni et al., 2018), in order to screen the current knowledge on water and climate change as well as strategically deployed features to attenuate water-exacerbated issues. The overall procedure of treating selected study cases was achieved by combining the adopted approaches in these sources. The literature review analysis included peer-reviewed and selected publications available in databases like Springer, Elsevier, Google Scholar, Multidisciplinary Digital Publishing Institute (MDPI), BazTech, and company websites and reports provided by them (Smol, 2021; Smol & Koneczna, 2021). The used set of items during the whole process of research included "climate change", "climate change water", "climate change water crisis", "climate change water demand", "climate change water circular economy", "climate change water projections", "climate change water agriculture", "climate change water industry", and "climate change water municipal". The prominent results and findings are detailed in Sects. 3 and 4.

In the second and third steps of the investigation, discrepancies, variability, and uncertainties in regard to water

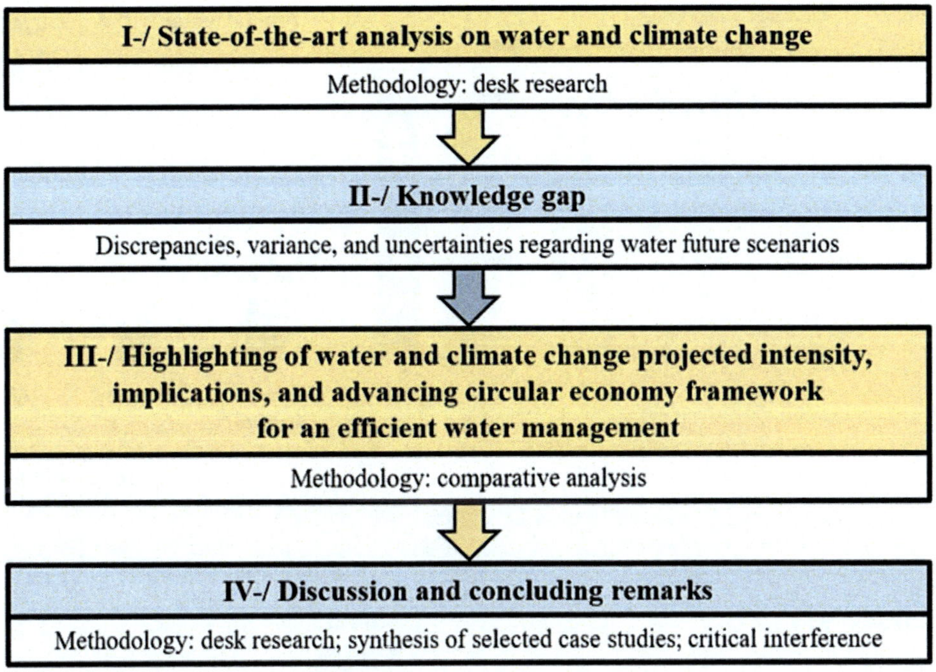

Fig. 1 The investigation layout (Smol, 2021; Smol & Koneczna, 2021)

future projections, focusing mainly on global circulation models (GCMs) generated data, were pointed out in a number of case studies. The findings were identified as knowledge gaps, which were discussed in Sect. 4 and highlighted in Sect. 5. A specific emphasis was put on climate fluctuations vis-à-vis water, precipitation patterns, and the impact of climate change on water resources. The culminating results are grouped in Tables 2, 3 and 4 and are discussed in Sect. 4. Furthermore, the implementation of water in vital sectors, such as agricultural, industrial, and municipal, in relation with water crisis, was summarized and interpreted in Sects. 4 and 5. Additional sources were relied upon to construct the main findings. Such sources include organization websites like the United Nations (UN), and scientific online publications like Our World in Data. Finally, the proposed circular economy framework, for a better management implementation in water sector, is developed from selected case studies grouped in Table 5.

The second step served as a bridge to the third step, which has the objective of underlining similarities and differences in the projected intensity of climate patterns, as well as futuristic water scenarios and the main implications, through comparative analysis.

The fourth and last step intended to critically discuss the synthesized studies, in comparison with corroborating and/or contradicting research cases. The overall methodology consists of synthesizing and interpreting the main findings. Furthermore, desk research concerning preceding study cases was effectuated, in order to conduct the comparative aspect. The culminating outcome of this step is mainly developed in Sects. 4 and 5.

3 Results

3.1 Water Consumption Worldwide and Implementation in Vital Sectors

Water consumption has been significantly important across the world. This is evident from the annual freshwater withdrawals, which represent the total water withdrawals from agricultural, industrial, and municipal sectors minus evaporation losses, measured in (m^3/year). The annual water withdrawals per geographic region in 2017 are shown in Fig. 2. According to Ritchie and Roser (2017), the total worldwide water withdrawals in 2017 is registered at 3894.02 billion m^3/year. Asia represents the largest water user, with a record of 2527.76 billion m^3/year, followed by North America (601.12 billion m^3/year) and Europe (292.18 billion m^3/year). Whereas Africa is ranked the 4th, with a record of 235.44 billion m^3/year, South America is ranked the 5th with a record of 211.21 billion m^3/year, and Oceania, is positioned as the least water user globally with a record of 26.31 billion m^3/year.

Water is significantly implemented in the agricultural, industrial, and municipal fields. The global annual water withdrawals per sector in 2017 are shown in Fig. 3. According to (Ritchie & Roser, 2017), the annual water withdrawals in 2017 reached a total of 3894.02 billion m^3/year. The agricultural sector represents the main water user, with an annual water withdrawals record of 2725.81 billion m^3/year, followed by the industrial (739.87 billion m^3/year) and municipal (428.34 billion m^3/year) sectors.

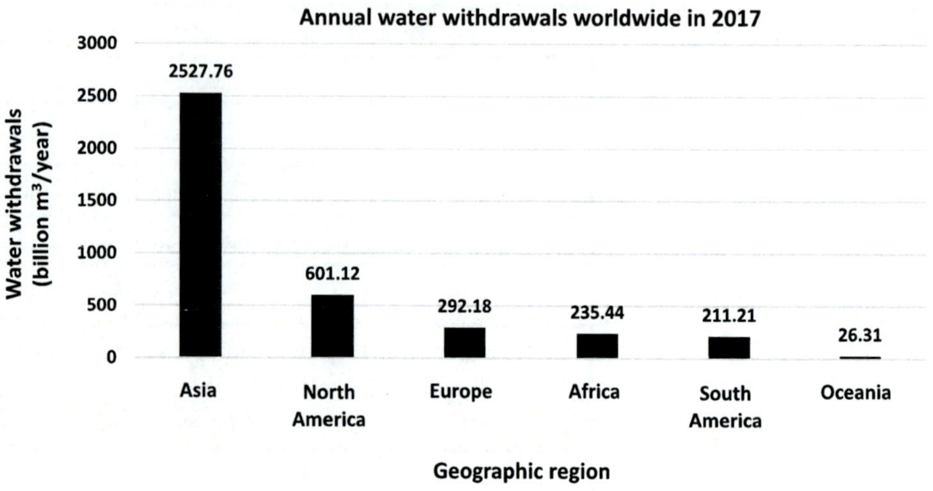

Fig. 2 Annual water withdrawals worldwide in 2017 (Ritchie & Roser, 2017)

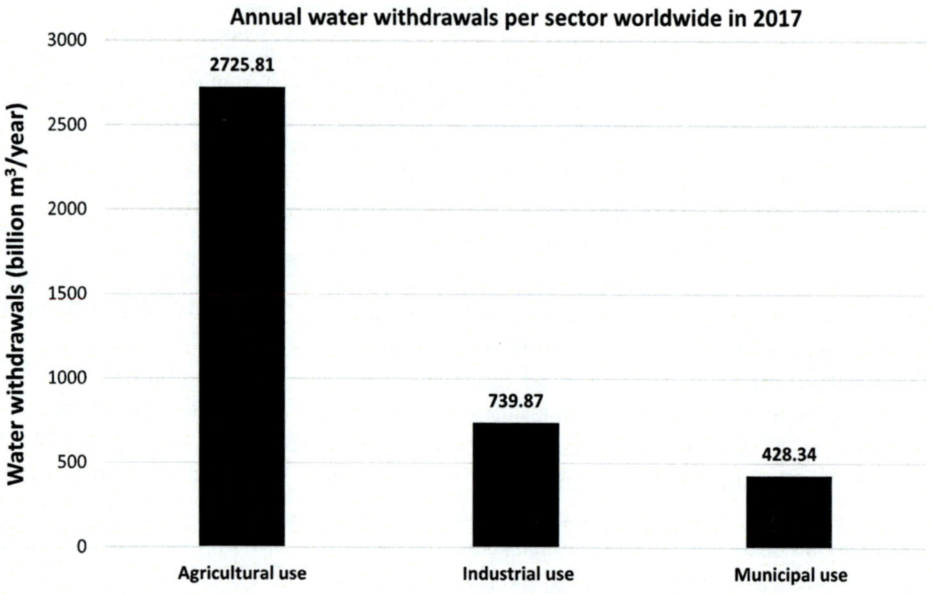

Fig. 3 Annual water withdrawals per sector worldwide in 2017 (Ritchie & Roser, 2017)

3.2 Water Crisis and Driving Factors

Water crisis has been a vexing issue globally due to several factors. The main driving factors and their direct impact on water are listed in Table 1. Climate change has been the most impactful factor on water, influencing precipitation patterns and leading to extreme weather events. Global population, which is currently at 7.9 billion in 2022, is already putting pressure on freshwater resources and causing increased water demand. Economic development, which highly relies on energy, is exacerbating water demand and leading to considerable pollution, whereas urbanization is putting pressure on urban water resources and their quality (Table 1).

3.3 Impact of Climate Change on Water Cycle

The impact of climate change on water is evidenced through climate pattern alterations (Zareian et al., 2017). Climate change is materialized in significant long-term alterations in temperature tendencies and weather patterns. Such alterations occur on a regional and/or global scale and are manifested through water cycle intensification (Allan et al., 2020a; Howe et al., 2019). Water cycle, also labeled hydrological cycle, governs water fluxes and the Earth atmosphere continuous circulation of water, which is essentially present in seas and oceans, rivers and lakes, ice sheets, and beneath Earth's surface (Allan et al., 2020b; Korenaga et al., 2017). This phenomenon is controlled by

Table 1 Major driving factors of water crisis and their impact

Driving factor	Impact
Climate change	Unprecedented major shifts in precipitation pathways, leading to geographic-dependent, uneven water distribution (Jiang et al., 2017; Szwed, 2019) Pronounced increase in drought and flooding events, conducting to deteriorated water quality due to contaminant concentration (droughts) and damaged water supply infrastructure (floods) (Miller & Hutchins, 2017; Shand et al., 2017)
Population growth	Pressure on freshwater resources due to extensive use and exploitation (Cramer et al., 2018; Koop & van Leeuwen, 2017) Increase in water demand and use (Ehsani et al., 2017; Tang, 2020)
Economic development	Higher water demand for energy-consuming sectors (Lee et al., 2017; Wijngaard et al., 2018) Heavier water pollution due to manufacturing and chemical processing (Inyinbor Adejumoke et al., 2018; Cai et al., 2020)
Urbanization	Elevated pressure on urban water resources due to population concentrating in cities (Hassan Rashid et al., 2018; Koop & van Leeuwen, 2017) Water quality deterioration due to pollution (Hassan Rashid et al., 2018; Miller & Hutchins, 2017)

natural factors like solar energy and volcanos. However, water cycle has been intensifying due to the anthropogenic factor, which is an inevitable consequence of the greenhouse gas emissions and aerosol releasings (Tang, 2020). One of the most ever-affected climatic variables is precipitation, which continuously undergoes modifications in quantity and distribution (Tabari, 2020). In this regard, precipitation as a crucial determinant of water delivery to Earth was chosen as the main climatic parameter of focus. Projected precipitation intensity and implications on water, by the end of the twenty-first century across different geographic areas, are grouped in Table 2. The Intergovernmental Panel on Climate Change (IPCC), which is the leading organization of the UN evaluating climate change, has come up with numeral statistical climate-modeling tools. Such means assist in quantifying possible effects of climate change on climatic variables (Zareian et al., 2017). In the present table, generated data by different downscaled GCMs indicate that precipitation fluctuates across regions and is predicted to mainly intensify. The intensification of precipitation patterns is expected to lead to extreme climatic events, such as prolonged droughts and flooding. Agriculture, economy, and infrastructure are among the main areas to be affected the most.

3.4 Impact of Climate Change on Water Resources

The impact of climate change on water resources is depicted through a preliminary evaluation of population exposure to water resources' stress, by 2050, on a global and regional scale (Table 3). The study in question highlights futuristic effects on water stress exposure through 4 Representative Concentration Pathways and 5 Shared Socio-economic Pathways, (4 RCPs and 5 SSPs, respectively), under the Coupled Model Intercomparison Project Phase 5 (CMIP5) climate models, on a regional and global scale. The focus was put on each RCP representing changes in 2040–2069 ("2050s"), relative to 1961–1990. As far as climate change spacial and seasonal pattern projections are concerned, water stress exposure in future is dominated by uncertainty. Hence, an important variability is recorded between the RCP and SSP models due to regional variation. The greater climate change impacts are registered under RCP8.5 in 2050. Furthermore, difference surrounds the absolute numbers of people subjected to escalating water resources' stress, varying from one RCP to another and between the 5 SSPs. Whereas under the highest RCP (RCP8.5) and with SSP2, it is predicted that (920–3400) million people will be subjected to elevated water resources' stress by 2050. Water stress exposure would decrease by 22–24% under RCP2.6 in 2050.

3.5 Impact of Climate Change on Water Availability and Quality

Global warming is predicted to further intensify water cycle's processes, mainly evaporation, evapotranspiration, and precipitation. Consequently, seasonal water availability is expected to increase, varying from dry to wet areas on global and regional levels (Tabari, 2020). Due to its pronounced impact on air temperature as well as hydrology, climate change has been severely affecting water quality. Furthermore, water temperature is predicted to rise owing to global warming. For instance, accessibility, seasonality, and variability of water present in rivers and glacier-fed river systems will be altered. In addition, a higher water temperature, combined with lower stream flow rates in summers, has been often shown to impair water quality (Hosseini et al., 2017).

Table 2 Projected precipitation intensity and implications on water by the end of the twenty-first century across different geographic areas

Geographic area	Global Circulation Model (GCM)	Downscaling models	Precipitation projected intensity	Implication	References
Cape Town and Maputo—South Africa Lagos—Sub-Saharan Africa Port Said—Northeast Africa	CORDEX multimodel: CCCMA, CNRM, GFDL, HADGEM, ICHEC, MIROC, MPI, and NCCN	CORDEX multimodel: RCA4, CCLM, HIRLAM, RACMO, REMO, and WRF regional climate models (RCMs)	*Cape town, Maputo, and Port Said* Mostly wet days and annual precipitation frequency and intensity drop *Lagos* Wet days drop, but annual precipitation frequency and intensity upsurge	Southern Africa's summer rainfall regions to possibly become drier Decline in agricultural production, impact on water resources, and food security threat	Abiodun et al. (2017)
Western Himalayas	LMDZ4 GCM	LMDZ4 GCM (stretchable for regional climate examination)	Slighter precipitation extremes' upsurge until the mid-twenty-first century, followed by a fast and significant rise afterwards	Upsurge in the orographic precipitation amplitude	Krishnan et al. (2019)
Middle East	WAS-CORDEX: CMIP5 GCMs: CNRM-CM5, EC-EARTH, GFDL-ESM2M, IPSL-CM5A-MR, MIROC5, and MPI-ESM-LR	WAS-CORDEX: SMHI, RCA4, and REMO RCMs	Shifts in the raining season Alteration in extreme precipitation Mean precipitation upsurge in autumn, against a pronounced decline in spring	Wetter autumn and drier spring Negative impact on rainfed crops like cereals Threat on yielding and food security Exacerbation of water management concerns	Tabari and Willems (2018)
Europe	EURO-CORDEX multimodel: CMIP5	EURO-CORDEX multimodel: various RCMs	Heavy precipitation upsurge over most European areas especially in autumn, winter, and spring	High-intensity precipitation events Possible river flooding Infrastructure damage	Nissen and Ulbrich (2017)
South Korea	CMIP5	Statistical Downscaling Model (SDQM), SD-DQM, and SD-QDM	Upsurge in extreme maximum daily precipitation frequency	Severity increase in the extreme precipitation events Impact on infrastructure	Sung et al. (2018)
Mediterranean climate zones of North Hemisphere	CMIP5	–	Winter precipitation decline Severe dry winters Precipitation drying	Water stress Impact on water resources Impact on crop production	Dong et al. (2021)

Table 3 Numbers of people (millions) exposed to increased water resources' stress under climate change by RCP and SSP: 2050 (Arnell & Lloyd-Hughes, 2014)

	SSP1	SSP2	SSP3	SSP4	SSP5
RCP8.5	1754 (802–2947)	2027 (919–3416)	2421 (1006–4201)	2105 (950–3519)	1810 (854–2981)
RCP6	1602 (759–2668)	1878 (807–3054)	2244 (924–3564)	1920 (809–3227)	1649 (803–2682)
RCP4.5	1514 (810–2845)	1794 (881–3239)	2157 (1037–3975)	1867 (884–3444)	1566 (854–2879)
RCP2.6	1330 (379–2997)	1575 (473–3434)	1887 (626–4088)	1656 (508–3481)	1375 (418–3033)

In this respect, the impact of climate change on water availability and quality is depicted through worldwide predicted hydrological cycle intensification, by the end of the twenty-first century, as shown in Table 4. Studies from the eight covered regions, namely Eastern Australia (Zhang et al., 2019), Eastern Africa (Hirpa et al., 2018), Southeast Europe (Idrizovic et al., 2020), South Asia (Nilawar & Waikar, 2019), East Africa (Näschen et al., 2019a), Western North America (Islam et al., 2017), Pacific Northwest of the Western USA (Clifton et al., 2018), and West Africa (Op de Hipt et al., 2018) mostly corroborate precipitation, rainfall, and evapotranspiration extremes, in addition to exacerbated extreme events (droughts and flooding). The projected scenarios mostly indicate surface water scarcity and water quality deterioration, in addition to seasonal fluctuations, impacting both livelihood and vital sectors like agriculture.

3.6 Circular Economy as an Efficient Water Management Framework

Circular economy (CE) represents an innovative approach that relies on circulation in the use, which can perfectly fit in water sector management (Smol et al., 2020). In order to further take CE framework, there exists a number of sustainable tools that can help achieve this goal (Sgroi et al., 2018; Voulvoulis, 2018). Such tools include artificial intelligence (AI), remote sensing, and biostimulants. Table 5 gathers considerable applications of these technological features in the three covered sectors. Enhancing agricultural production and lessening labor are considerably required. AI perfectly responds to these aspects as it helps with proper field and fresh water resources' management, thanks to real-time data monitoring and machine-based intelligent watering (Jha et al., 2019). AI can also help with assessing water quality on a spatio-temporal level, thanks to water quality indicator-based investigation and pollutant source identification. Therefore, AI can lead to improving water quality index (Wang et al., 2019). Remote sensing offers the advantage of operating as well as scheduling irrigation. Relying on multispectral imagery at a high spacial resolution, remote sensing can provide a better crop water use, based on timely given information (Calera et al., 2017). Furthermore, remote sensing represents an advantageous tool that can monitor water resources' parameters on a global and regional scale, speeding up decision-making processes and water management planning (Andres et al., 2018; Avtar et al., 2020). Finally, eco-friendly agricultural means, such as biostimulants, can help achieve improved water use efficiency and lowered pressure on water resources, which can limit water wastage (Del Buono, 2021).

4 Discussion

Water represents the most precious and ubiquitous mineral substance on Earth. Furthermore, it is associated with every aspect of development, including agricultural, industrial, and municipal.

Water contributes mainly to agriculture (Brown et al., 2018; Velasco-Muñoz et al., 2018). It is used for agricultural purposes like irrigation (Saccon, 2018) and livestock requirements (Gadomski, 2021). Hence, the sector is positioned as the largest consumer of water globally (Thevs et al., 2017). The vital implementation of water in industrialization is only second to agriculture (Ferrucci & Vocciante, 2021; Weber & Hogberg-Saunders, 2018). The industrial production process relies heavily on water for delivering products and equipment cooling (Pan et al., 2018). The municipal water use, ranked 3rd, is expected to even rise in the future (Bijl et al., 2018).

Globally, Asia is the largest water consumer, with most of the water use going to agriculture. Furthermore, population growth and developing economies are leading to extensive water consumption (Taniguchi et al., 2017). North America is positioned as the 2nd water consumer, which essentially stems from agricultural and industrial sectors (Chimner et al., 2017). In Europe, ranked the 3rd, water use is essentially municipal and domestic. Moreover, the sector's water demand is increasing due to demographic pressure (Ramsey et al., 2017). In Africa, ranked the 4th, water withdrawals are rather agricultural (Nhemachena et al., 2020). South America is positioned as the 5th water user worldwide, which mainly goes to agriculture. Industrial water use is also important due to fast economic growth and urbanization (Parker, 2018). Finally, Australia is commending the largest annual water withdrawn (15.95 billion m^3) from Oceania, which is essentially agricultural (Koech & Langat, 2018).

Climate change has been the major driving factor in water pattern alterations, leading to geography-dependent droughts and floods (Koop & van Leeuwen, 2017; Zhao et al., 2020), and conducting to water disparity and uneven distribution across the globe (Mpandeli et al., 2018). In addition, global warming is enhancing precipitation rates, which is further speeding up water cycle (Aleixo et al., 2019; Allan et al., 2020c).

Population growth and dynamism, which are rising rapidly, are contributing aspects to water crisis (Brown et al., 2019). On the other hand, the multi-scale economies and rapid economic development are exerting pressure on water resources. Though industrial societies tend to adapt by orienting production patterns towards less water consumption, the exposure to water risks is undeniable (Zhou et al., 2018). Economic growth and water pollution are associated. Industrialization generates important amounts of water

Table 4 Projected climate fluctuation impact on water by the end of the twenty-first century across different regions

Region	Water resource	Climate model	Variable	Predicted intensity	Impact	References
Eastern Australia	Manning River catchment	XAJ GCM	Temperature Rainfall Runoff Evapotranspiration Soil moisture	Temperature increase Seasonal rainfall fluctuations Runoff shifts Evapotranspiration shifts Soil moisture mostly decreasing trends	Important inter-seasonal changes in the future Endangered surface water supplies Drastic consequences for agriculture Land and water resource management issues	Zhang et al. (2019)
North-Western Kenya	Turkwel River basin	CMIP5	Precipitation Rainfall Temperature Evapotranspiration	Precipitation variability, mostly deficiency Scarce rainfall Wetter and warmer climate tendencies Increased evapotranspiration Surface water availability and groundwater supply decline	Long episodes of drought Water demand to grow three times Drier conditions Huge seasonal variation in rainfall	Hirpa et al. (2018)
Južna Morava, Serbia	Toplica River catchment	CMIP5	Temperature Precipitation Evapotranspiration Discharge	Global warming Seasonal upward and downward precipitation trends Potential evapotranspiration increase Water quantity to rise (winter) and drop (summer)	Non-uniformity of rainfall distributions Larger water deficit Elevated irrigation demand Snow cover melting Negative impact on agriculture and water quality	Idrizovic et al. (2020)
Maharashtra state, India	Purna River basin	MPI-CSC-REMO2009, SMHI-RCA4, and IITM-RegCM4-1 RCMs	Temperature Precipitation Stream-flow Sediment concentration	Considerable rise in temperature, particularly in the dry season Increased precipitation, especially in the wetter season Increased sediment concentration	Intense rainfall and flooding exacerbation Increase in water demand Pressure on water supply for both agriculture and drinking purposes Water quality deterioration Erosion and costly water treatment	Nilawar and Waikar (2019)

(continued)

Table 4 (continued)

Region	Water resource	Climate model	Variable	Predicted intensity	Impact	References
Morogoro region, South Central Tanzania	Tropical Catchment	CORDEX Africa GCM-RCM	Surface runoff Evapotranspiration Water yield	Surface runoff intensification Evapotranspiration decrease Water yield decrease Seasonality pattern alteration	Vulnerability of the catchment Seasonal-turned tributaries Difficulty in meeting high water requirements of crops like rice Groundwater recharge reinforcement	Näschen et al. (2019b)
Western North America	The Fraser River basin	CMIP5	Precipitation Air temperature Snowfall Snow water equivalent Snow cover	Precipitation fluctuations Air temperature increase Snow precipitation decrease Snow accumulation and covered area decline Snowmelt runoff yielding increase Snow cover loss	Surface albedo reduction Decreased water supply Affected regional weather	Islam et al. (2017)
Oregon, USA	Water resources of the Blue Mountains	CMIP3 and contributing GCMs	Stream temperature Snowpack persistence Peak flows Low flows	Warm streams to be more sensitive to warming temperature Mostly increased snowpack sensitivity Flooding magnitude increase Moderate streamflow decline	Forest mortality and wildfire risks Transition from snowmelt to rain-dominant basins Extreme hydrological event occurrences Seasonal flow fluctuations	Clifton et al. (2018)
Southern part of Burkina Faso, West Africa	A tropical catchment	CORDEX-Africa RCMs–GCMs ensemble	Temperature Precipitation Soil erosion	Temperature increase tendencies Precipitation pattern fluctuations (marked by uncertainty) Soil erosion pattern fluctuations (marked by uncertainty)	Impact on vegetation (zonation, properties and growth) Impact on rainfed agriculture	Op de Hipt et al. (2018)

pollutants (Sakamoto et al., 2019). As a consequence, industrial facilities tend to use freshwater in order to get rid of waste (Qadri & Bhat, 2020).

Moreover, urbanization trends are increasing, generating an imbalance between water availability and its demand in cities worldwide (Hassan Rashid et al., 2018). In a time span of seven decades (1950–2020), the global population residing in cities augmented from 0.8 to 4.4 billions, and is expected to attain 6.7 billon by 2050 (He et al., 2021). Consequently, city-resident populations are already facing situations where water demand surpasses its availability, which is affecting urban livelihood, environment, and socioeconomics (Niva et al., 2020).

To highlight the resulting water cycle intensification, a specific emphasis was put on precipitation as the prime hydrological variable through selected case studies.

Table 5 Technological features and their potential applications in water management

Sector	Technological feature	Application	References
Agricultural	Artificial intelligence (AI)	Irrigation scheduling Soil moisture sensing and estimation Sprinkler and drip irrigation Real-time data monitoring	Jha et al. (2019)
Agricultural	Remote sensing	Crop water requirement estimation Geo-hydrology and drainage conditions prevention Irrigation efficiency improvement Potential crop transpiration estimation	Calera et al. (2017)
Agricultural	Biostimulants	Water consumption improvement Inaccessible water reaching by plants' roots Water use efficiency enhancement Water shortage effective mitigation	Del Buono (2021)
Industrial	AI	Water quality assessment Industrial water-source pollutants' identification Abnormal wastewater discharge detection Discharging pollutant industries determination	Wang et al. (2019)
Industrial	Remote sensing	Groundwater rapid exploration Larger area coverage capacity Extent of surface water change observation Spatio-temporal scope capacity	Avtar et al. (2020)
Municipal	Remote sensing	Regional-scale precipitation monitoring Water quality monitoring Water budgets and soil moisture control Sectoral urban water demand modeling	Andres et al. (2018)

Downscaled GCM simulations, from Coordinated Regional Climate Downscaling Experiment (CORDEX), for the past (1986–2005) and future (2046–2065 and 2081–2100) time periods under RCP4.5 and RCP8.5 scenarios, have generally revealed a drop in wet days, upsurge in intense daily precipitation, and a decline in annual precipitation frequency and intensity over Cape Town, Maputo, and Port Said, whereas opposite trends were recorded in Lagos (Abiodun et al., 2017). These findings are in line with Pfahl et al. (2017), indicating through a physical scaling analysis that, for the South African region, futuristic extreme precipitation drop can partly be attributed to the dynamics of poleward circulation shifting. A decrease in futuristic precipitation over South Africa is also backed by Dosio et al. (2021).

Krishnan et al. (2019) investigated extreme precipitation trends over the Western Himalayas, using The Laboratoire de Meteorologie Dynamique Zoom General Circulation Model (LMDZ4 GCM). For a past reference of 1901–2005 and a future time period simulation (2006–2095) under RCP4.5, they indicated a significant rise from the mid-twenty-first century onwards. These findings were explained by the anthropogenic effect escalating western disturbances (WD), in addition to a strengthened positive East–West temperature gradient through the Tibetan Plateau. They also specified the subtropical westerly jet influence on the WD activity during winter–early spring, which is indeed corroborated by Hunt et al. (2018) and Midhuna et al. (2020).

The seasonal fluctuations in precipitation, projected in the Middle East over the South Asia domain WAS-CORDEX, were simulated for the current (1971–2000) and future (2070–2099) climate under RCP4.5 and RCP8.5 (Tabari & Willems, 2018). Recorded precipitation fluctuations were attributed to aridity. Furthermore, the authors indicated that the decline in spring precipitation could be due to storm tracks poleward shifting, weakening the Mediterranean storm track and decreasing the number of cyclones traversing the sea. This is in agreement with (Pfahl et al., 2017), who also indicated that the development in extreme precipitation dynamism can be partly influenced by poleward shifts over the subtropics.

Nissen and Ulbrich (2017), who investigated precipitation trends over Europe through EURO-CORDEX multimodel, for a referential time period of 1971–2000 and simulated future time periods (2021–2050 and 2071–2100) under RCP4.5 and RCP8.5, found that seasonal heavy precipitation is likely to further increase, with exceptions in summer. The authors stated an agreement that air temperatures' rise, induced by the greenhouse effect, which will lead to an increase in moisture-holding capacity, will possibly bring about an upsurge in seasonal extreme precipitation. These findings strongly align with those of (Rajczak & Schär, 2017), who recorded similar precipitation tendencies over Europe through the EURO-CORDEX multimodel, forced under RCP2.6, RCP4.5, and RCP8.5 scenarios.

Sung et al. (2018) have investigated extreme precipitation tendencies over South Korea through Coupled Model Intercomparison Project-Phase 5 (CMIP5) for a reference time period of 1980–2005, and three futuristic time periods (2011–2040, 2041–2070, and 2071–2100) under RCP4.5, registered future increase in maximum daily precipitation frequency. Similar findings were observed by Kim et al. (2021). Through CORDEX-East Asia Phase I, for a reference time period of 1981–2005, and a futuristic time period covering the mid-twenty-first century (2025–2049), the authors noted a rise in extreme precipitation over the Korean Peninsula. Generally, the results were linked to the extended East Asian summer monsoon, which generates South wind, favoring precipitation rise.

Precipitation tendencies over North Hemisphere were examined by Dong et al. (2021) through CMIP5 for historical (1962–2005) and future (2056–2099) time periods under RCP8.5. The findings pointed out a decrease in winter precipitation and drying. The authors related the future projections to alterations in the atmospheric circulation, incited by enhanced tropical rainfall but weaker slowdown Atlantic Meridional Overturning Circulation under global warming. While some authors attribute the precipitation decreasing tendency to seasonality and thermodynamic alterations, such as land–ocean warmth (Brogli et al., 2019), others associate it to atmospheric aerosol and the greenhouse effect (Tang et al., 2018).

Predictions indicate that water resources' stress will exacerbate, which will lead to an increased water demand and water supply imbalance, thus aggravating the ever-growing global population water issues (Petersen-Perlman et al., 2017). In this sense, Arnell and Lloyd-Hughes (2014) have led a preliminary evaluation of climate change impact regarding exposure to water resources' stress, on a global and regional scale, using 19 CMIP5 models. Groupings of RCP and Socio-Economic Futures (SSPs) were forced, in order to conduct estimations by 2050 (Table 3).

The authors have recorded greater SSPs than RCPs intra-variations regarding exposed population to water stress. RCPs variations were explained by climate policy impact on climate change. Take, for example, RCP2.6 related to an emission policy, with a target to achieve a 2 °C climate goal (van Vuuren et al., 2011). Hence, similar policies hold potential to decrease exposure to water scarcity by up to 530 million people, in comparison to RCP8.5. Furthermore, greater CMIP5 intra-model impact variations have been noted. For instance, under SSP2 and RCP8.5 scenarios, 920–3,500 million people are subjected to intensified water resources' stress. The authors attributed it to a variation in spacial and regional climate change patterns, especially in South and East Asia. Holden and Doshi (2019) corroborate that water resources' stress is already impacting ¼ of the global population, with regions like the Middle-East emerging as the topper.

In order to address water availability and quality exacerbating issues, a set of selected case studies were examined. Zhang et al. (2019) studied the effects of climate change on the availability of a Manning River catchment water resource, localized in Eastern Australia. The authors used the Xinanjiang (XAJ) model and relied on statistically downscaled climatic data under RCP8.5 scenario, for 1977–2016 as a past reference, and 2021–2060 and 2061–2100 as future time periods. Findings revealed that, while temperature will increase and soil moisture will generally decrease, seasonal rainfall, runoff, and evapotranspiration will exhibit fluctuating trends. The authors suggested that runoff fluctuations could be due to the theories of hydrological models and model calibrating criteria related to the used hydrological model. Furthermore, the authors stated XAJ model simulating limitations, which is backed by Zhou et al. (2019a, 2019b). Climate fluctuation trends due to climate change over Australia are also backed by Pearce et al. (2018).

Hirpa et al. (2018) modeled the responses of Turkwel River basin water resource to climate change, through CMIP5 and the Water Evaluation and Planning Version 21 (WEAP21) hydrological model. For a past reference of 1981–2005 and future time periods of 2021–2050 and 2071–2100 under RCP8.5 forced scenarios, the authors indicated climate variability due to global climate change, mostly characterized by important evapotranspiration, and precipitation and rainfall deficiency, leading to a drier environment. This is in accordance with the findings of Tan et al. (2020), which indicated that the Northeastern areas of Kenya are the most subjected to dryness, based on CMIP5 modeling under RCP4.5 and RCP8.5 forced scenarios, by the end of the twenty-first century. Furthermore, they emphasized uncertainty in future projections due to data and modeling limitations.

Idrizovic et al. (2020) studied the impact of climate change on the Toplica River water resource in Southern Serbia, based on data from RCM simulations generated from downscaling CMIP5. Under RCP4.5 and RCP8.5 forced scenarios, for a past reference of 1971–2000 and a future time period of 2021–2100, the authors recorded future global warming trends, potential evapotranspiration rise, seasonal precipitation extremes, and potential evapotranspiration increase. The findings were attributed to the greenhouse effect and climate change, whereas water discharge trends were linked to the Toplica River falling in the Danube River basin. The accentuation of global warmth trends is corroborated by Vukovic et al. (2018), who pointed out consequent temperature and precipitation pattern fluctuations.

Through MPI-CSC-REMO2009, SMHI-RCA4 and IITM-RegCM4-1 RCM models and under RCP4.5 and RCP8.5 scenarios, with 1980–2005 as past reference, and 2009–2031, 2032–2053, 2054–2075, and 2076–2099 as future time periods, Nilawar and Waikar (2019) attempted to assess climate change effects on streamflow and the concentration of sediments in the Purna river basin, located in India. According to the obtained results, temperature is expected to further rise, particularly in the dry season, whereas precipitation is projected to augment, especially during the wetter season. Similarly, Kumar et al. (2017) attempted to model the impact of climate change on the Upper Kharun Catchment, located in Chhattisgarh state in India, and recorded temperature and rainfall rising tendencies, in addition to evapotranspiration upward. Regarding precipitation in the Tibetan Plateau, the fluctuations can be attributed to monsoon tendencies and the influence of the subtropical jet (Hou et al., 2017).

Näschen et al., (2019a, 2019b) observed parameters like surface runoff and evapotranspiration tendencies over a tropical catchment localized in Tanzania, settling on a GCM–RCM combination from CORDEX Africa. The Authors recorded important results like the intensification of surface runoff, evapotranspiration drop, and water yield decrease by the end of the twenty-first century. They attributed the findings to the catchment concealing effects, the transforming of perennial tributaries of the Kilombero catchment to seasonal tributaries, and the conjugated effects of climate change plus land use and land cover change. This is in line with projections stating precipitation rise and augmenting water deficit by 2050 around Kilimanjaro, mainly due to the anthropogenic climate change factor (Said et al., 2019). Despite the observed variation in climate projections, the authors generally pointed out the importance of water management over Tanzania.

Researchers investigated the snow hydrology status over the Fraser River Basin (FRB) of British Columbia (BC), located in Canada, in the context of climate change (Islam et al., 2017). By using the CMIP5 and for a settled historical reference of 1980–2009 and future time period of 2040–2069, under RCP4.5 and RCP8.5 scenarios, the authors registered future precipitation fluctuating, which will lead to a decrease in snow accumulation and snow-covered areas by the 2050s. While the authors attributed the observed precipitation trends to the Pacific Ocean proximity, they associated snow precipitation to snowmelt season shifts. Furthermore, they stated the topography complexity and downscaling methods as a limitation for certainty in modeling. Comparable projected snowpack decline trends, during the twenty-first century, were recorded through CMIP5 modeling by Najafi et al. (2017). The observations were mainly linked to the anthropogenic effect.

Clifton et al. (2018) recorded more or less different climate trends in the semi-arid environment of the Blue Mountains, Oregon, situated in the United States. Based on downscaled contributing GCMs to CMIP3 assessment, through a past reference of 1977, and for future time periods of 2030–2059 and 2070–2099, the resulting data revealed projected temperature warming trends, which will lead to rain-dominant basins and warmer streams (up to 2 °C by 2080). These findings were explained not only by the geographical features but also by increasing warming trends, fueled by climate change. Accumulated snow melting over North America, which is influenced by air temperature and likely to lead to floods, is also backed by Tuttle et al. (2017).

Through CORDEX-Africa GCMs–RCMs modeling ensemble, climate change effects on a tropical catchment water resource, localized in Southwestern Burkina Faso were assessed (Op de Hipt et al., 2018). For historical (1971–2000) and future (2021–2050) time periods, and under RCP4.5 and RCP8.5 forced scenarios, predictions pointed out temperature increase but uncertainty regarding precipitation status. The authors attributed the uncertainties to uncorrected data overestimating precipitation, in addition to issues with correction biases. Moreover, such observations were even associated with future temperature status predictions. Those observations were further corroborated over the region (Op de Hipt et al., 2019).

Clearly, climate change is already impacting water resources in terms of quantity, quality, availability, and through precipitation variability (Miller & Hutchins, 2017). Thus, it is important to address this global phenomenon as a multifaceted and the most challenging matter of the twenty-first century. In this regard, circular economy (CE) represents an innovative model that forms a bridge, linking production and consumption (Marrucci et al., 2019). CE's primordial objective is to assure the sustainability of natural resources like water (Chojnacka et al., 2020). Hence, water crisis can be tackled through such frameworks (Fig. 4).

Artificial intelligence (AI) can play a major role in effectuating this transition (Jenny et al., 2020). AI, a set of smart machines powered by human skills, can be applied to enhance water quality, thanks to the faculty of detecting toxic contaminants (Gunda et al., 2019; Haghiabi et al., 2018). Furthermore, AI can allow the assessment and monitoring of water sanitary status, thus generating important data (Tiyasha et al., 2020). Another prominent role is the optimization of water use through sprinkler systems (Chukkapalli et al., 2020).

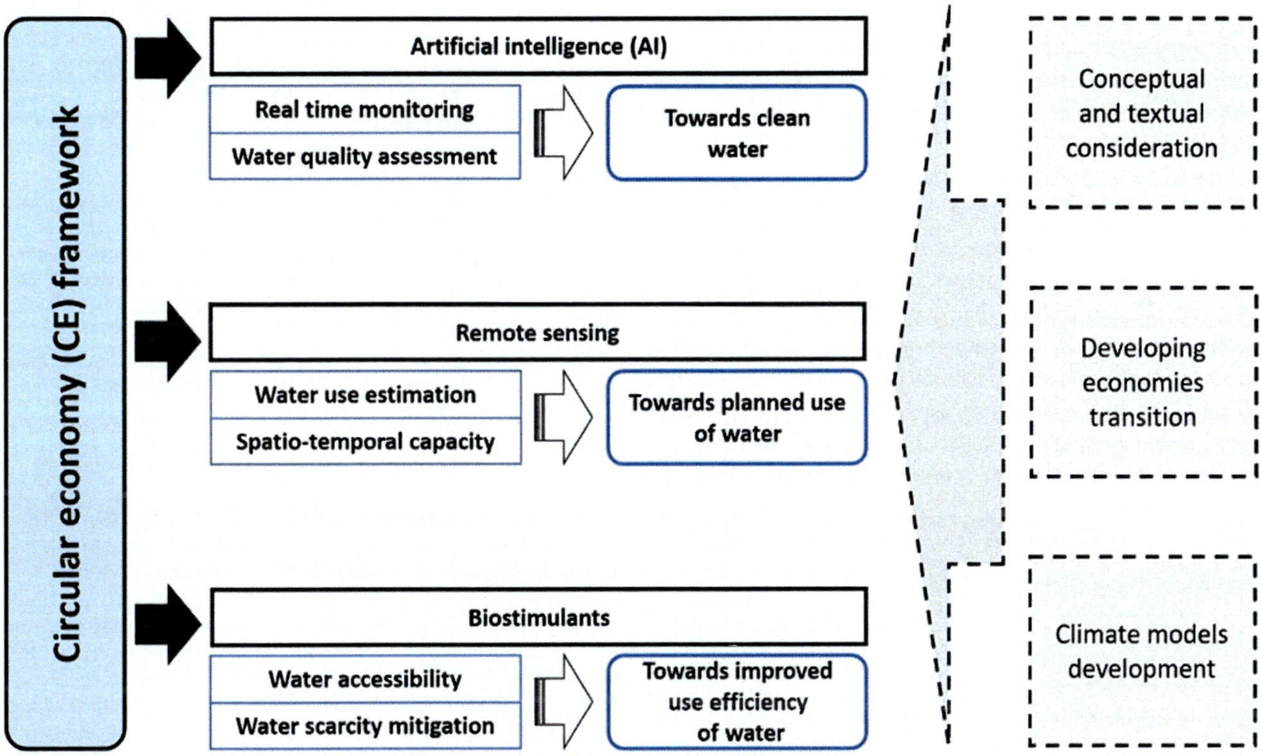

Fig. 4 CE framework proposal for a better management of water

Secondly, remote sensing has been a crucial means of analyzing water. Remote sensing forms a set of geospatial technological sensors that allow information acquisition from a considerable distance (Martos et al., 2021). This developed process helps gathering important information regarding hydrology, irrigation, and the assessment of drought among others (Alvino & Marino, 2017; Thakur et al., 2017; West et al., 2019). Further on, geographic information system (GIS), which is a category of database for geospatial data analysis, has been upgrading the convenience of remote sensing water-related data (Ramadas & Samantaray, 2018).

Finally, biostimulants, which are a range of beneficial microorganisms and organic amending products that benefit both the plant and soil, hold tremendous potential into improving water use efficiency (Rouphael & Colla, 2020). Besides the technological progress aiming towards irrigation monitoring, biostimulants can play a key role in augmenting water use efficiency, amid drastic climatic conditions (Akhtar et al., 2020). In fact, plants can establish advantageous associations with beneficial fungi and bacteria, which help them attract water. Thus, biostimulants can make water accessible for the root system to assimilate despite water scarcity (Ben Mrid et al., 2021).

5 Conclusions

Water is unequivocally the most important substance on Earth. It assures all forms of life, permits a better functioning of the environment, and is the engine to vital sectors like agriculture and economy. However, water crisis has been the most defining issue of the twenty-first century. We support that water crisis is the most direct magnitude of climate change, thus corroborate that tackling water scarcity-related issues is the most operative way to dealing with the exacerbated consequences of climate change. Climate patterns have been altering in an unprecedented manner. Thus, hydrological cycle patterns are transforming, leading to uneven distribution and unavailability of water on a global scale. Warming trends and sparse precipitation have also been ongoing scenarios for region-dependent, deteriorated water quality.

In this regard, the present paper attempts to shed light on the climate system's fluctuated patterns, amid climate change. In addition, the following adverse effects on water resources have been specifically highlighted. On the basis of gathered information, it is evident that the current water crisis trends are likely to intensify by the end of the twenty-first century. These unfortunate predictions have been a rising concern among the scientific community, climatologists, farmers, and stakeholders the most. Fortunately, emerging promising solutions have been gaining specific attention for their feasibility and efficient end-results. Such is the approach of circular economy (CE) implementation in water sector. Eco-friendly and cost-effective tools like artificial intelligence (AI), remote sensing, and biostimulants were underlined as constructive tools for an applicable and effective CE framework. Thus, management efficiency in water sector holds tremendous potential to be enacted through these means. We conclude that in order to attain this objective, a transition into sustainability in future water management policies can lead to effectively resolve water crisis across the globe. In addition, development of the available climate predicting models can help orient such operative mode towards success.

References

Abbot, J., & Marohasy, J. (2017). The application of machine learning for evaluating anthropogenic versus natural climate change. *GeoResJ, 14*, 36–46. https://doi.org/10.1016/j.grj.2017.08.001

Abiodun, B. J., Adegoke, J., Abatan, A. A., Ibe, C. A., Egbebiyi, T. S., Engelbrecht, F., & Pinto, I. (2017). Potential impacts of climate change on extreme precipitation over four African coastal cities. *Climatic Change, 143*(3–4), 399–413. https://doi.org/10.1007/s10584-017-2001-5

Akhtar, S. S., Amby, D. B., Hegelund, J. N., Fimognari, L., Großkinsky, D. K., Westergaard, J. C., Müller, R., Moelbak, L., Liu, F., & Roitsch, T. (2020). Bacillus licheniformis FMCH001 Increases Water Use Efficiency via Growth Stimulation in Both Normal and Drought Conditions. *Frontiers in Plant Science, 11*, 1–12. https://doi.org/10.3389/fpls.2020.00297

Aleixo, I., Norris, D., Hemerik, L., Barbosa, A., Prata, E., Costa, F., & Poorter, L. (2019). Amazonian rainforest tree mortality driven by climate and functional traits. *Nature Climate Change, 9*(5), 384–388. https://doi.org/10.1038/s41558-019-0458-0

Allan, R. P., Barlow, M., Byrne, M. P., Cherchi, A., Douville, H., Fowler, H. J., Gan, T. Y., Pendergrass, A. G., Rosenfeld, D., Swann, A. L. S., Wilcox, L. J., & Zolina, O. (2020a). Advances in understanding large-scale responses of the water cycle to climate change. In *Annals of the New York Academy of Sciences* (Vol. 1472, Issue 1, pp. 49–75). https://doi.org/10.1111/nyas.14337

Allan, R. P., Barlow, M., Byrne, M. P., Cherchi, A., Douville, H., Fowler, H. J., Gan, T. Y., Pendergrass, A. G., Rosenfeld, D., Swann, A. L. S., Wilcox, L. J., & Zolina, O. (2020b). Advances in understanding large-scale responses of the water cycle to climate change. *Annals of the New York Academy of Sciences, 1472*(1), 49–75. https://doi.org/10.1111/nyas.14337

Alvino, A., & Marino, S. (2017). Remote sensing for irrigation of horticultural crops. *Horticulturae, 3*(2), 40. https://doi.org/10.3390/horticulturae3020040

Andres, L., Boateng, K., Borja-Vega, C., & Thomas, E. (2018). A review of in-situ and remote sensing technologies to monitor water and sanitation interventions. *Water (Switzerland), 10*(6), 756. https://doi.org/10.3390/w10060756

Arnell, N. W., & Lloyd-Hughes, B. (2014). The global-scale impacts of climate change on water resources and flooding under new climate and socio-economic scenarios. *Climatic Change, 122*(1–2), 127–140. https://doi.org/10.1007/s10584-013-0948-4

Asadieh, B., & Krakauer, N. Y. (2017). Global change in streamflow extremes under climate change over the 21st century. *Hydrology and Earth System Sciences, 21*(11), 5863–5874.

Avtar, R., Komolafe, A. A., Kouser, A., Singh, D., Yunus, A. P., Dou, J., Kumar, P., Gupta, R. D., Johnson, B. A., Thu Minh, H. V., Aggarwal, A. K., & Kurniawan, T. A. (2020). Assessing sustainable development prospects through remote sensing: A review. *Remote Sensing Applications: Society and Environment, 20*, 100402. https://doi.org/10.1016/j.rsase.2020.100402

Azhoni, A., Jude, S., & Holman, I. (2018). Adapting to climate change by water management organisations: Enablers and barriers. *Journal of Hydrology, 559*, 736–748. https://doi.org/10.1016/j.jhydrol.2018.02.047

Ben Mrid, R., Benmrid, B., Hafsa, J., Boukcim, H., Sobeh, M., & Yasri, A. (2021). Secondary metabolites as biostimulant and bioprotectant agents: A review. *Science of the Total Environment, 777*, 146204. https://doi.org/10.1016/j.scitotenv.2021.146204

Bijl, D. L., Biemans, H., Bogaart, P. W., Dekker, S. C., Doelman, J. C., Stehfest, E., & van Vuuren, D. P. (2018). A global analysis of future water deficit based on different allocation mechanisms. *Water Resources Research, 54*(8), 5803–5824. https://doi.org/10.1029/2017WR021688

Brogli, R., Sørland, S. L., Kröner, N., & Schär, C. (2019). Causes of future Mediterranean precipitation decline depend on the season. *Environmental Research Letters, 14*(11), 114017. https://doi.org/10.1088/1748-9326/ab4438

Brown, J. J., Das, P., & Al-Saidi, M. (2018). Sustainable agriculture in the Arabian/Persian Gulf region utilizing marginal water resources: Making the best of a bad situation. *Sustainability (Switzerland), 10*(5), 1364. https://doi.org/10.3390/su10051364

Brown, T. C., Mahat, V., & Ramirez, J. A. (2019). Adaptation to future water shortages in the united states caused by population growth and climate change. *Earth's Future, 7*(3), 219–234. https://doi.org/10.1029/2018EF001091

Cai, H., Mei, Y., Chen, J., Wu, Z., Lan, L., & Zhu, D. (2020). An analysis of the relation between water pollution and economic growth in China by considering the contemporaneous correlation of water pollutants. *Journal of Cleaner Production, 276*, 122783. https://doi.org/10.1016/j.jclepro.2020.122783

Calera, A., Campos, I., Osann, A., D'Urso, G., & Menenti, M. (2017). Remote sensing for crop water management: From ET modelling to services for the end users. *Sensors, 17*(5), 1104. https://doi.org/10.3390/s17051104

Chandra, S., Chakrabortty, R., Roy, P., Chowdhuri, I., Das, B., Saha, A., & Shit, M. (2021). Changing climate and land use of 21st century in fluences soil erosion in India. *Gondwana Research, 94*, 164–185. https://doi.org/10.1016/j.gr.2021.02.021

Chawla, I., Karthikeyan, L., & Mishra, A. K. (2020). A review of remote sensing applications for water security: Quantity, quality, and extremes. *Journal of Hydrology, 585*, 124826. https://doi.org/10.1016/j.jhydrol.2020.124826

Chimner, R. A., Cooper, D. J., Wurster, F. C., & Rochefort, L. (2017). An overview of peatland restoration in North America: Where are we after 25 years? *Restoration Ecology, 25*(2), 283–292. https://doi.org/10.1111/rec.12434

Chojnacka, K., Moustakas, K., & Witek-Krowiak, A. (2020). Bio-based fertilizers: A practical approach towards circular economy. *Bioresource Technology, 295*, 122223. https://doi.org/10.1016/j.biortech.2019.122223

Chukkapalli, S. S. L., Mittal, S., Gupta, M., Abdelsalam, M., Joshi, A., Sandhu5, R., & Joshi, K. (2020). Ontologies and artificial intelligence systems for the cooperative smart farming ecosystem. *IEEE Access, 8*, 164045–164064.https://doi.org/10.1109/ACCESS.2020.3022763

Clifton, C. F., Day, K. T., Luce, C. H., Grant, G. E., Safeeq, M., Halofsky, J. E., & Staab, B. P. (2018). Effects of climate change on hydrology and water resources in the Blue Mountains, Oregon, USA. *Climate Services, 10*, 9–19. https://doi.org/10.1016/j.cliser.2018.03.001

Cramer, W., Guiot, J., Fader, M., Garrabou, J., Gattuso, J. P., Iglesias, A., Lange, M. A., Lionello, P., Llasat, M. C., Paz, S., Peñuelas, J., Snoussi, M., Toreti, A., Tsimplis, M. N., & Xoplaki, E. (2018). Climate change and interconnected risks to sustainable development in the Mediterranean. *Nature Climate Change, 8*(11), 972–980. https://doi.org/10.1038/s41558-018-0299-2

Del Buono, D. (2021). Can biostimulants be used to mitigate the effect of anthropogenic climate change on agriculture? It is time to respond. In *Science of the Total Environment* (Vol. 751, p. 141763). Elsevier B.V. https://doi.org/10.1016/j.scitotenv.2020.141763

Dong, L., Leung, L. R., Lu, J., & Song, F. (2021). Double-ITCZ as an emergent constraint for future precipitation over Mediterranean climate regions in the north hemisphere. *Geophysical Research Letters, 48*(3), 1–11. https://doi.org/10.1029/2020GL091569

Dosio, A., Jury, M. W., Almazroui, M., Ashfaq, M., Diallo, I., Engelbrecht, F. A., Klutse, N. A. B., Lennard, C., Pinto, I., Sylla, M. B., & Tamoffo, A. T. (2021). Projected future daily characteristics of African precipitation based on global (CMIP5, CMIP6) and regional (CORDEX, CORDEX-CORE) climate models. *Climate Dynamics, 57*(11–12), 3135–3158. https://doi.org/10.1007/s00382-021-05859-w

Ehsani, N., Vörösmarty, C. J., Fekete, B. M., & Stakhiv, E. Z. (2017). Reservoir operations under climate change: Storage capacity options to mitigate risk. *Journal of Hydrology, 555*, 435–446. https://doi.org/10.1016/j.jhydrol.2017.09.008

Ferronato, N., Rada, E. C., Gorritty Portillo, M. A., Cioca, L. I., Ragazzi, M., & Torretta, V. (2019). Introduction of the circular economy within developing regions: A comparative analysis of advantages and opportunities for waste valorization. *Journal of Environmental Management, 230*, 366–378. https://doi.org/10.1016/j.jenvman.2018.09.095

Ferrucci, A., & Vocciante, M. (2021). Improved management of water resources in process industry by accounting for fluctuations of water content in feed streams and products. *Journal of Water Process Engineering, 39*, 101870. https://doi.org/10.1016/j.jwpe.2020.101870

Gadomski, A. (2021). *Water in Biomechanical and Related Systems*.

Gleick, P. H. (1999). The human right to water. *Water Policy, 1*(1998), 487–503.

Goswami, K. B., & Bisht, P. S. (2017). The role of water resources in socio-economic. *Journal of Molecular Liquids, 5*, 1669–1674.

Gray, N. (2017). *Water Technology*.

Gunda, N. S. K., Gautam, S. H., & Mitra, S. K. (2019). Editors' choice —artificial intelligence based mobile application for water quality monitoring. *Journal of the Electrochemical Society, 166*(9), 1–6. https://doi.org/10.1149/2.0081909jes

Haghiabi, A. H., Nasrolahi, A. H., & Parsaie, A. (2018). Water quality prediction using machine learning methods. *Water Quality Research Journal, 53*(1), 3–13. https://doi.org/10.2166/wqrj.2018.025

Hamidov, A., & Helming, K. (2020). Sustainability considerations in water–energy–food nexus research in irrigated agriculture. *Sustainability, 12*(15), 6274.

Hannah Ritchie and Max Roser. (2017). "Water Use and Stress". Published online at OurWorldInData.org. Retrieved from: https://ourworldindata.org/water-use-stress [Online Resource]

Hassan Rashid, M. A. U., Manzoor, M. M., & Mukhtar, S. (2018). Urbanization and its effects on water resources: An exploratory analysis. *Asian Journal of Water, Environment and Pollution, 15* (1), 67–74. https://doi.org/10.3233/AJW-180007

He, C., Liu, Z., Wu, J., Pan, X., Fang, Z., Li, J., & Bryan, B. A. (2021). Future global urban water scarcity and potential solutions. *Nature Communications, 12*(1), 1–12. https://doi.org/10.1038/s41467-021-25026-3

Hirpa, F. A., Dyer, E., Hope, R., Olago, D. O., & Dadson, S. J. (2018). Finding sustainable water futures in data-sparse regions under climate change: Insights from the Turkwel River basin, Kenya. *Journal of Hydrology: Regional Studies, 19*, 124–135. https://doi.org/10.1016/j.ejrh.2018.08.005

Holden, E., & Doshi, V. (2019). Extreme water stress affects a quarter of the world's population, say experts. *The Gaurdian US, 6*.

Hosseini, N., Johnston, J., & Lindenschmidt, K. E. (2017). Impacts of climate change on the water quality of a regulated prairie river. *Water (Switzerland), 9*(3), 1–15. https://doi.org/10.3390/w9030199

Hou, J., D'Andrea, W. J., Wang, M., He, Y., & Liang, J. (2017). Influence of the Indian monsoon and the subtropical jet on climate change on the Tibetan Plateau since the late Pleistocene. *Quaternary Science Reviews, 163*, 84–94. https://doi.org/10.1016/j.quascirev.2017.03.013

Howe, P. D., Marlon, J. R., Mildenberger, M., & Shield, B. S. (2019). How will climate change shape climate opinion? *Environmental Research Letters, 14*(11), 1–18. https://doi.org/10.1088/1748-9326/ab466a

Hunt, K. M. R., Turner, A. G., & Shaffrey, L. C. (2018). The evolution, seasonality and impacts of western disturbances. *Quarterly Journal of the Royal Meteorological Society, 144*(710), 278–290. https://doi.org/10.1002/qj.3200

Idrizovic, D., Pocuca, V., Vujadinovic Mandic, M., Djurovic, N., Matovic, G., & Gregoric, E. (2020). Impact of climate change on water resource availability in a mountainous catchment: A case study of the Toplica River catchment. *Serbia. Journal of Hydrology, 587*, 124992. https://doi.org/10.1016/j.jhydrol.2020.124992

Inyinbor Adejumoke, A., Adebesin Babatunde, O., Oluyori Abimbola, P., Adelani Akande Tabitha, A., Dada Adewumi, O., & Oreofe Toyin, A. (2018). Water Pollution: Effects, Prevention, and Climatic Impact. In *Water Challenges of an Urbanizing World* (Vol. 33, pp. 33–47). InTech. https://doi.org/10.5772/intechopen.72018

Islam, S. ul, Déry, S. J., & Werner, A. T. (2017). Future climate change impacts on snow and water resources of the Fraser River Basin, British Columbia. *Journal of Hydrometeorology, 18*(2), 473–496. https://doi.org/10.1175/jhm-d-16-0012.1

Jenny, H., Wang, Y., Alonso, E. G., & Minguez, R. (2020). *Using artificial intelligence for smart water management systems, 143*.

Jha, K., Doshi, A., Patel, P., & Shah, M. (2019). A comprehensive review on automation in agriculture using artificial intelligence. *Artificial Intelligence in Agriculture, 2*, 1–12. https://doi.org/10.1016/j.aiia.2019.05.004

Jiang, R., Gan, T. Y., Xie, J., Wang, N., & Kuo, C. C. (2017). Historical and potential changes of precipitation and temperature of Alberta subjected to climate change impact: 1900–2100. *Theoretical and Applied Climatology, 127*(3–4), 725–739. https://doi.org/10.1007/s00704-015-1664-y

Karegar, M. A., Dixon, T. H., Malservisi, R., Kusche, J., & Simon, E. (2017). Nuisance flooding and relative sea-level rise: The importance of present-day land motion. *Scientific Reports, 7*(1), 1–9. https://doi.org/10.1038/s41598-017-11544-y

Kim, G., Cha, D. H., Park, C., Jin, C. S., Lee, D. K., Suh, M. S., Oh, S. G., Hong, S. Y., Ahn, J. B., Min, S. K., & Kang, H. S. (2021). Evaluation and projection of regional climate over East Asia in CORDEX-East Asia Phase I experiment. *Asia-Pacific Journal of Atmospheric Sciences, 57*(1), 119–134. https://doi.org/10.1007/s13143-020-00180-8

Koech, R., & Langat, P. (2018). Improving irrigation water use efficiency: A review of advances, challenges and opportunities in the Australian context. *Water (Switzerland), 10*(12), 1771. https://doi.org/10.3390/w10121771

Koop, S. H. A., & van Leeuwen, C. J. (2017). The challenges of water, waste and climate change in cities. *Environment, Development and Sustainability, 19*(2), 385–418. https://doi.org/10.1007/s10668-016-9760-4

Korenaga, J., Planavsky, N. J., & Evans, D. A. D. (2017). Global water cycle and the coevolution of the Earth's interior and surface environment. *Philosophical Transactions of the Royal Society A: Mathematical, Physical and Engineering Sciences, 375*(2094). https://doi.org/10.1098/rsta.2015.0393

Krishnan, R., Sabin, T. P., Madhura, R. K., Vellore, R. K., Mujumdar, M., Sanjay, J., Nayak, S., & Rajeevan, M. (2019). Non-monsoonal precipitation response over the Western Himalayas to climate change. *Climate Dynamics, 52*(7–8), 4091–4109. https://doi.org/10.1007/s00382-018-4357-2

Kumar, N., Tischbein, B., Kusche, J., Laux, P., Beg, M. K., & Bogardi, J. J. (2017). Impact of climate change on water resources of upper Kharun catchment in Chhattisgarh, India. *Journal of Hydrology: Regional Studies, 13*, 189–207. https://doi.org/10.1016/j.ejrh.2017.07.008

Lee, M., Keller, A. A., Chiang, P. C., Den, W., Wang, H., Hou, C. H., Wu, J., Wang, X., & Yan, J. (2017). Water-energy nexus for urban water systems: A comparative review on energy intensity and environmental impacts in relation to global water risks. *Applied Energy, 205*, 589–601. https://doi.org/10.1016/j.apenergy.2017.08.002

Liang, Y., Gillett, N. P., & Monahan, A. H. (2020). Climate model projections of 21st century global warming constrained using the observed warming trend. *Geophysical Research Letters, 47*(12), e2019GL086757. https://doi.org/10.1029/2019GL086757

Lin, Y.-P., Petway, J., Lien, W.-Y., & Settele, J. (2018). Blockchain with artificial intelligence to efficiently manage water use under climate change. *Environments, 5*(3), 34. https://doi.org/10.3390/environments5030034

Liu, X., Du, H., Zhang, Z., Crittenden, J., Michael, L. L., & Moreno-, J. (2019). Can virtual water trade save 1 water resources? *Water Research, 163*, 114848. https://doi.org/10.1016/j.watres.2019.07.015

Marrucci, L., Daddi, T., & Iraldo, F. (2019). The integration of circular economy with sustainable consumption and production tools: Systematic review and future research agenda. *Journal of Cleaner Production, 240*, 118268. https://doi.org/10.1016/j.jclepro.2019.118268

Martos, V., Ahmad, A., Cartujo, P., & Ordoñez, J. (2021). Ensuring agricultural sustainability through remote sensing in the era of agriculture 5.0. *Applied Sciences (Switzerland), 11*(13), 5911. https://doi.org/10.3390/app11135911

Mello, F. D. S., César, A., & Friaça, S. (2020). The end of life on Earth is not the end of the world: Converging to an estimate of life span of the biosphere ? *International Journal of Astrobiology, 19*(1), 25–42.

Midhuna, T. M., Kumar, P., & Dimri, A. P. (2020). A new Western Disturbance Index for the Indian winter monsoon. *Journal of Earth System Science, 129*(1), 1–14. https://doi.org/10.1007/s12040-019-1324-1

Miller, J. D., & Hutchins, M. (2017). The impacts of urbanisation and climate change on urban flooding and urban water quality: A review of the evidence concerning the United Kingdom. *Journal of Hydrology: Regional Studies, 12*, 345–362. https://doi.org/10.1016/j.ejrh.2017.06.006

Mohammad, S., Islam, F., & Karim, Z. (n.d.). *World ' s Demand for Food and Water : The Consequences of Climate Change*, 1–27.

Mpandeli, S., Naidoo, D., Mabhaudhi, T., Nhemachena, C., Nhamo, L., Liphadzi, S., Hlahla, S., & Modi, A. T. (2018). Climate change adaptation through the water-energy-food nexus in Southern Africa. *International Journal of Environmental Research and Public Health, 15*(10), 1–19. https://doi.org/10.3390/ijerph15102306

Najafi, M. R., Zwiers, F., & Gillett, N. (2017). Attribution of the observed spring snowpack decline in British Columbia to anthropogenic climate change. *Journal of Climate, 30*(11), 4113–4130. https://doi.org/10.1175/JCLI-D-16-0189.1

Näschen, K., Diekkrüger, B., Evers, M., Höllermann, B., Steinbach, S., & Thonfeld, F. (2019a). The impact of land use/land cover change (LULCC) on water resources in a tropical catchment in tanzania under different climate change scenarios. *Sustainability (Switzerland), 11*(24), 1–28. https://doi.org/10.3390/su11247083

Näschen, K., Diekkrüger, B., Evers, M., Höllermann, B., Steinbach, S., & Thonfeld, F. (2019b). The impact of land use/land cover change (LULCC) on water resources in a tropical catchment in Tanzania under different climate change scenarios. *Sustainability, 11*(24), 7083. https://doi.org/10.3390/su11247083

Nhemachena, C., Nhamo, L., Matchaya, G., Nhemachena, C. R., Muchara, B., Karuaihe, S. T., & Mpandeli, S. (2020). Climate change impacts on water and agriculture sectors in southern africa: Threats and opportunities for sustainable development. *Water (Switzerland), 12*(10), 1–17. https://doi.org/10.3390/w12102673

Nilawar, A. P., & Waikar, M. L. (2019). Impacts of climate change on streamflow and sediment concentration under RCP 4.5 and 8.5: A case study in Purna river basin. *India Science of the Total Environment, 650*, 2685–2696. https://doi.org/10.1016/j.scitotenv.2018.09.334

Nissen, K. M., & Ulbrich, U. (2017). Increasing frequencies and changing characteristics of heavy precipitation events threatening infrastructure in Europe under climate change. *Natural Hazards and Earth System Sciences, 17*(7), 1177–1190. https://doi.org/10.5194/nhess-17-1177-2017

Niva, V., Cai, J., Taka, M., Kummu, M., & Varis, O. (2020). China's sustainable water-energy-food nexus by 2030: Impacts of urbanization on sectoral water demand. *Journal of Cleaner Production, 251*, 119755. https://doi.org/10.1016/j.jclepro.2019.119755

Op de Hipt, F., Diekkrüger, B., Steup, G., Yira, Y., Hoffmann, T., & Rode, M. (2018). Modeling the impact of climate change on water resources and soil erosion in a tropical catchment in Burkina Faso, West Africa. *CATENA, 163*, 63–77. https://doi.org/10.1016/j.catena.2017.11.023

Op de Hipt, F., Diekkrüger, B., Steup, G., Yira, Y., Hoffmann, T., Rode, M., & Näschen, K. (2019). Modeling the effect of land use and climate change on water resources and soil erosion in a tropical West African catch-ment (Dano, Burkina Faso) using SHETRAN. *Science of the Total Environment, 653*, 431–445. https://doi.org/10.1016/j.scitotenv.2018.10.351

Pan, S.-Y., Snyder, S. W., Packman, A. I., Lin, Y. J., & Chiang, P.-C. (2018). Cooling water use in thermoelectric power generation and its associated challenges for addressing water-energy nexus. *Water-Energy Nexus, 1*(1), 26–41. https://doi.org/10.1016/j.wen.2018.04.002

Parađiković, N., Teklić, T., Zeljković, S., Lisjak, M., & Špoljarević, M. (2019). Biostimulants research in some horticultural plant species—

A review. In *Food and Energy Security* (Vol. 8, Issue 2, p. e00162). https://doi.org/10.1002/fes3.162

Parker, C. (2018). Energy transition in South America: Elite'S views in the mining sector, four cases under study. *Ambiente e Sociedade, 21*, 1–22. https://doi.org/10.1590/1809-4422asoc0089r1vu18l1ao

Pearce, T. D., Rodríguez, E. H., Fawcett, D., & Ford, J. D. (2018). How is Australia adapting to climate change based on a systematic review? *Sustainability (Switzerland), 10*(9), 3280. https://doi.org/10.3390/su10093280

Petersen-Perlman, J. D., Veilleux, J. C., & Wolf, A. T. (2017). International water conflict and cooperation: Challenges and opportunities. *Water International, 42*(2), 105–120. https://doi.org/10.1080/02508060.2017.1276041

Pfahl, S., O'Gorman, P. A., & Fischer, E. M. (2017). Understanding the regional pattern of projected future changes in extreme precipitation. *Nature Climate Change, 7*(6), 423–427. https://doi.org/10.1038/nclimate3287

Pires, A., Morato, J., Peixoto, H., Botero, V., Zuluaga, L., & Figueroa, A. (2017). Sustainability Assessment of indicators for integrated water resources management. *Science of the Total Environment, 578*, 139–147. https://doi.org/10.1016/j.scitotenv.2016.10.217

Qadri, H., & Bhat, R. A. (2020). Fresh water pollution dynamics and remediation. *Fresh Water Pollution Dynamics and Remediation*. https://doi.org/10.1007/978-981-13-8277-2

Rajczak, J., & Schär, C. (2017). Projections of future precipitation extremes over Europe: A Multimodel assessment of climate simulations. *Journal of Geophysical Research: Atmospheres, 122* (20), 10773–10800. https://doi.org/10.1002/2017JD027176

Ramadas, M., & Samantaray, A. K. (2018). Applications of remote sensing and GIS in water quality monitoring and remediation: A state-of-the-art review. *Energy, Environment, and Sustainability*, 225–246.https://doi.org/10.1007/978-981-10-7551-3_13

Ramsey, E., Berglund, E. Z., & Goyal, R. (2017). The impact of demographic factors, beliefs, and social influences on residential-water consumption and implications for non-price policies in urban India. *Water (Switzerland), 9*(11), 1–21. https://doi.org/10.3390/w9110844

Rouphael, Y., & Colla, G. (2020). Editorial: Biostimulants in agriculture. *Frontiers in Plant Science, 11*(February), 1–7. https://doi.org/10.3389/fpls.2020.00040

Saccon, P. (2018). Water for agriculture, irrigation management. *Applied Soil Ecology, 123*, 793–796. https://doi.org/10.1016/j.apsoil.2017.10.037

Said, M., Komakech, H. C., Munishi, L. K., & Muzuka, A. N. N. (2019). Evidence of climate change impacts on water, food and energy resources around Kilimanjaro. *Tanzania. Regional Environmental Change, 19*(8), 2521–2534. https://doi.org/10.1007/s10113-019-01568-7

Sakamoto, M., Ahmed, T., Begum, S., & Huq, H. (2019). Water pollution and the textile industry in Bangladesh: Flawed corporate practices or restrictive opportunities? *Sustainability (Switzerland), 11*(7), 1951. https://doi.org/10.3390/su11071951

Schilling, J., Hertig, E., Tramblay, Y., & Scheffran, J. (2020). Climate change vulnerability, water resources and social implications in North Africa. *Regional Environmental Change, 20*(1), 1–12. https://doi.org/10.1007/s10113-020-01597-7

Sgroi, M., Vagliasindi, F. G. A., & Roccaro, P. (2018). Feasibility, sustainability and circular economy concepts in water reuse. *Current Opinion in Environmental Science and Health, 2*, 20–25. https://doi.org/10.1016/j.coesh.2018.01.004

Shand, P., Grocke, S., Creeper, N. L., Baker, A. K., Fitzpatrick, R. W., & Love, A. J. (2017). Impacts of climate change, climate variability and management on soil and water quality in Wetlands of South Australia. *Procedia Earth and Planetary Science, 17*, 456–459. https://doi.org/10.1016/j.proeps.2016.12.115

Shef, J., & Wood, E. F. (2018). Satellite remote sensing for water resources management: Potential for supporting sustainable development in data-poor regions. *Water Resources Research, 54*(12), 9724–9758. https://doi.org/10.1029/2017WR022437

Smol, M. (2021). Inventory and comparison of performance indicators in circular economy roadmaps of the European Countries. *Circular Economy and Sustainability*, 1–28. https://doi.org/10.1007/s43615-021-00127-9

Smol, M., Adam, C., & Preisner, M. (2020). Circular economy model framework in the European water and wastewater sector. *Journal of Material Cycles and Waste Management, 22*(3), 682–697. https://doi.org/10.1007/s10163-019-00960-z

Smol, M., & Koneczna, R. (2021). Economic indicators in water and wastewater sector contributing to a circular economy (CE). *Resources, 10*(12), 129. https://doi.org/10.3390/resources10120129

Sung, J. H., Eum, H. I., Park, J., & Cho, J. (2018). Assessment of climate change impacts on extreme precipitation events: Applications of CMIP5 climate projections statistically downscaled over South Korea. *Advances in Meteorology, 2018*, 1–12. https://doi.org/10.1155/2018/4720523

Szwed, M. (2019). Variability of precipitation in Poland under climate change. *Theoretical and Applied Climatology, 135*(3–4), 1003–1015. https://doi.org/10.1007/s00704-018-2408-6

Tabari, H. (2020). Climate change impact on flood and extreme precipitation increases with water availability. *Scientific Reports, 10* (1), 1–10. https://doi.org/10.1038/s41598-020-70816-2

Tabari, H., & Willems, P. (2018). Seasonally varying footprint of climate change on precipitation in the Middle East. *Scientific Reports, 8*(1), 2–11. https://doi.org/10.1038/s41598-018-22795-8

Tan, G., Ayugi, B., Ngoma, H., & Ongoma, V. (2020). Projections of future meteorological drought events under representative concentration pathways (RCPs) of CMIP5 over Kenya, East Africa. *Atmospheric Research, 246*, 105112. https://doi.org/10.1016/j.atmosres.2020.105112

Tang, Q. (2020). Global change hydrology: Terrestrial water cycle and global change. *Science China Earth Sciences, 63*(3), 459–462. https://doi.org/10.1007/s11430-019-9559-9

Tang, T., Shindell, D., Samset, B. H., Boucher, O., Forster, P. M., Hodnebrog, Ø., Myhre, G., Sillmann, J., Voulgarakis, A., Andrews, T., Faluvegi, G., Fläschner, D., Iversen, T., Kasoar, M., Kharin, V., Kirkevåg, A., Lamarque, J. F., Olivié, D., Richardson, T., … Takemura, T. (2018). Dynamical response of Mediterranean precipitation to greenhouse gases and aerosols. *Atmospheric Chemistry and Physics, 18*(11), 8439–8452.https://doi.org/10.5194/acp-18-8439-2018

Taniguchi, M., Endo, A., Gurdak, J. J., & Swarzenski, P. (2017). Water-energy-food nexus in the Asia-pacific region. *Journal of Hydrology: Regional Studies, 11*, 1–8. https://doi.org/10.1016/j.ejrh.2017.06.004

Thakur, J. K., Singh, S. K., & Ekanthalu, V. S. (2017). Integrating remote sensing, geographic information systems and global positioning system techniques with hydrological modeling. *Applied Water Science, 7*(4), 1595–1608. https://doi.org/10.1007/s13201-016-0384-5

Thevs, N., Nurtazin, S., Beckmann, V., Salmyrzauli, R., & Khalil, A. (2017). Water consumption of agriculture and natural ecosystems along the Ili river in china and Kazakhstan. *Water (Switzerland), 9* (3), 207. https://doi.org/10.3390/w9030207

Tiecco, M., Cappellini, F., Nicoletti, F., Del, T., Germani, R., & Di, P. (2019). Role of the hydrogen bond donor component for a proper development of novel hydrophobic deep eutectic solvents. *Journal of Molecular Liquids, 281*, 423–430. https://doi.org/10.1016/j.molliq.2019.02.107

Tiyasha, T., & T. M., & Yaseen, Z. M. (2020). A survey on river water quality modelling using artificial intelligence models: 2000–2020. *Journal of Hydrology, 585*, 124670. https://doi.org/10.1016/j.jhydrol.2020.124670

Tuttle, S. E., Cho, E., Restrepo, P. J., Jia, X., Vuyovich, C. M., Cosh, M. H., & Jacobs, J. M. (2017). *Remote Sensing of Drivers of Spring Snowmelt Flooding in the North Central U.S.* (pp. 21–45). https://doi.org/10.1007/978-3-319-43744-6_2

van Vuuren, D. P., Stehfest, E., den Elzen, M. G. J., Kram, T., van Vliet, J., Deetman, S., Isaac, M., Goldewijk, K. K., Hof, A., Beltran, A. M., Oostenrijk, R., & van Ruijven, B. (2011). RCP2.6: Exploring the possibility to keep global mean temperature increase below 2°C. *Climatic Change, 109*(1), 95–116. https://doi.org/10.1007/s10584-011-0152-3

Velasco-Muñoz, J., Aznar-Sánchez, J., Belmonte-Ureña, L., & Román-Sánchez, I. (2018). Sustainable water use in agriculture: A review of worldwide research. *Sustainability, 10*(4), 1084. https://doi.org/10.3390/su10041084

Voulvoulis, N. (2018). Water reuse from a circular economy perspective and potential risks from an unregulated approach. *Current Opinion in Environmental Science and Health, 2*, 32–45. https://doi.org/10.1016/j.coesh.2018.01.005

Vukovic, A., Vujadinovic, M., Rendulic, S., Djurdjevic, V., Ruml, M., Babic, V., & Popovic, D. (2018). Global warming impact on climate change in Serbia for the period 1961–2100. *Thermal Science, 22*(6), 2267–2280. https://doi.org/10.2298/TSCI180411168V

Wang, P., Yao, J., Wang, G., Hao, F., Shrestha, S., Xue, B., Xie, G., & Peng, Y. (2019). Exploring the application of artificial intelligence technology for identification of water pollution characteristics and tracing the source of water quality pollutants. *Science of the Total Environment, 693*, 133440. https://doi.org/10.1016/j.scitotenv.2019.07.246

Weber, O., & Hogberg-Saunders, G. (2018). Water management and corporate social performance in the food and beverage industry. *Journal of Cleaner Production, 195*, 963–977. https://doi.org/10.1016/j.jclepro.2018.05.269

Werndl, C. (2016). On defining climate and climate change. *The British Journal for the Philosophy of Science, 67*, 337–364. https://doi.org/10.1093/bjps/axu048

West, H., Quinn, N., & Horswell, M. (2019). Remote sensing for drought monitoring & impact assessment: Progress, past challenges and future opportunities. *Remote Sensing of Environment, 232*, 111291. https://doi.org/10.1016/j.rse.2019.111291

Wijaya, D., Kusumadewi, R. A., Wijayanti, A., & Hadisoebroto, R. (2021). Saline water desalination with multilevel solar distillation. *IOP Conference Series: Earth and Environmental Science*, 012041. https://doi.org/10.1088/1755-1315/802/1/012041

Wijngaard, R. R., Biemans, H., Friedrich Lutz, A., Bhakta Shrestha, A., Wester, P., & Willem Immerzeel, W. (2018). Climate change vs. socio-economic development: Understanding the future South Asian water gap. *Hydrology and Earth System Sciences, 22*(12), 6297–6321. https://doi.org/10.5194/hess-22-6297-2018

Xiang, X., Li, Q., Khan, S., & Ibrahim, O. (2021). Urban water resource management for sustainable environment planning using artificial intelligence techniques. *Environmental Impact Assessment Review, 86*, 106515. https://doi.org/10.1016/j.eiar.2020.106515

Zareian, M. J., Eslamian, S., Gohari, A., & Adamowski, J. F. (2017). The effect of climate change on watershed water balance. In J. N. Furze, K. Swing, A. K. Gupta, R. H. McClatchey, & D. M. Reynolds (Eds.), *Mathematical Advances Towards Sustainable Environmental Systems*. Springer International Publishing. https://doi.org/10.1007/978-3-319-43901-3

Zhang, H., Wang, B., Li Liu, D., Zhang, M., Feng, P., Cheng, L., ... & Eamus, D. (2019). Impacts of future climate change on water resource availability of eastern Australia: A case study of the Manning River basin. *Journal of Hydrology, 543*, 1–46.

Zhao, Y., Weng, Z., Chen, H., & Yang, J. (2020). Analysis of the evolution of drought, flood, and drought-flood abrupt alternation events under climate change using the daily SWAP index. *Water (Switzerland), 12*(7), 1969. https://doi.org/10.3390/w12071969

Zhenmin, L., & Espinosa, P. (2019). Tackling climate change to accelerate sustainable development. *Nature Climate Change, 9*(7), 494–496.

Zhou, F., Hearne, Z., & Li, C. J. (2019a). Water—the greenest solvent overall. In *Current Opinion in Green and Sustainable Chemistry* (Vol. 18, pp. 118–123). https://doi.org/10.1016/j.cogsc.2019.05.004

Zhou, Q., Chen, L., Singh, V. P., Zhou, J., Chen, X., & Xiong, L. (2019b). Rainfall-runoff simulation in karst dominated areas based on a coupled conceptual hydrological model. *Journal of Hydrology, 573*, 524–533. https://doi.org/10.1016/j.jhydrol.2019.03.099

Zhou, Z., Liu, L., Zeng, H., & Chen, X. (2018). Does water disclosure cause a rise in corporate risk-taking?—Evidence from Chinese high water-risk industries. *Journal of Cleaner Production, 195*, 1313–1325. https://doi.org/10.1016/j.jclepro.2018.06.001

Integrated Water Management—Directions of Activities and Policies

Ewa Wiśniowska

Abstract

Integrated water management help countries deal with water resources and use. At present most countries have problems with the quality and availability of water resources. Some problems are because of natural circumstances, but many of them are the structural and institutional ones. Activities and policies in this area should take into consideration ecological sustainability, social equity, and economical effectiveness. The chapter presents the approaches of selected European, American, and African countries to water management, including European Union politics in water protection. Law requirements are presented, good practices and challenges have been indicated. Based on them some recommendations have been formulated.

Keywords

Water management • EU water framework directive • Clean water act • Good practices

1 Introduction

- Even in the middle ages, water was a crucial resource. Access to fresh water was essential in everyday life and during the wars. Rivers were used for communication purposes and trade. Nowadays, clean water resources are particularly relevant because of the significant increase in water demand. Global water use has increased six times since the 1920s. It continues to grow steadily at a rate of about 1% per year due to increasing population, economic development, and shifting consumption patterns (Water & Climate Change, 2020). At present, about half of the global population lives in areas that suffer water scarcity at least one month a year. Global water demand is at present 4,600 km^3, and it is estimated that, by 2050, it will increase up to 5,500–6,000 km^3 per year. A real challenge is the increase in the human population, and this increase is uneven. It is estimated that the most intensive growth of the human population will occur in Africa, which is one of the poorest areas with clearly visible problems with the drinking water supply. Followed by Asia, where the most overpopulation exists, e.g., in India and China (Boretti & Rosa, 2019). Four categories of countries can be indicated based on the susceptibility to water stress (based on (Falkenmark, 1997)):
- High-income countries with low water stress—this group of countries withdraw from using less than 10% of the renewable water resources within the country and have an annual per capita income of over 9000$. This type of country is Canada, and indeed they represent only 2% of the world's population. They have technical knowledge, financial resources, and a culture of environmental protection. They usually act as initiators for regional and global plans, cooperation, and actions.
- High-income countries with a middle or high water stress —this type of country, e.g., Spain, is at present focused on introducing a controlled water market to avoid water losses; good water management is crucial for their development, and they have financial resources to implement the organizational and technical solutions to support sustainable and efficient water management.
- Low-income countries with low water stress—these countries are the developing ones. They are rich in water resources. In these countries, the problems can be focused on intensive urbanization, including water supplies for the large cities and expanding water supply and sanitary systems quickly enough. As an example of this type of country, Brazil can be indicated.

E. Wiśniowska (✉)
Department of Sanitary Networks and Installations, Czestochowa University of Technology, 71 Dabrowskiego Str, Czestochowa, Poland
e-mail: ewa.wisniowska@pcz.pl

M. Smol et al. (eds.), *Water in Circular Economy*, Advances in Science, Technology & Innovation,
https://doi.org/10.1007/978-3-031-18165-8_2

- Low-income countries with medium or high water stress—have low water and financial resources, often simultaneously with a large population of citizens. There is a significant problem with providing access to drinking water for all citizens. These countries frequently need international aid and face problems connected with sanitary safety.

For each type countries, different strategies of water management are suitable. Estimated renewable water resources per capita by year in various regions of the world in 2014 are presented in Table 1.

The data given in Table 1 are ranked from the highest to the smallest one and compared to the average value. When we consider water resources in particular regions of the world, it is visible that in America, the resources are relatively high. Still, the real problem is in densely populated areas such as East and South Asia. Among the countries with the highest renewable water resources per capita are Colombia, Chile, and Peru (over 44,882 m^3/person/year).

Problems with water supply are also caused by other factors such as more and more common extreme weather conditions worldwide, changing natural landscapes, and damage to biodiversity and ecosystems. The worsening water supply problems in many areas result in people migrating in search of better living conditions; it will reshape population centers worldwide and lead to various conflicts (Water Management, 2040, 2021). To avoid conflicts based on water scarcity, it is necessary to prioritize common interest over the particular interest of individual corporations or regions. There is also a need to change paradigms, and economic growth should no longer be the primary measure of success (Water Management, 2040, 2021). The valuable tools for this are plans, directives, and activities, which allow for better water resources management. The application and development of such strategies were set in 1977 during the United Nations Water Conference in Argentina. The final document of the conference—the Mar del Plata Action Plan (MPAP)—was an essential step in developing integrated water management strategies. It was a framework concerning community water supplies and water for agriculture. The Mar del Plata Action Plan consisted of the recommendations and the resolutions. Among recommendations, assessment and systematic monitoring of water resources, efficient use of water to avoid its waste, and evaluation of large-scale water development projects and investments affecting the quality of the environment, including water, were described. Increased attention to the integrated planning of water management was also recommended, including the preparation, implementation, and evaluation of national water policies, institutional agreements, and legislation for a coordinated approach to water management. Public participation in the planning and decision-making was mentioned as an integral part of the water management processes. An important recommendation was also regional and international cooperation needed in the case of shared water. The document emphasized that effective legislation should be framed to promote the efficient and equitable use and protection of water and water-related ecosystems. Pricing and other economic incentives were recommended to be used to promote the efficient and fair use of water (MPAP, 1977). The Plan was only partially implemented, but most of the principles are valid today.

In 1992, Dublin Statement has included four principles that have started the understanding of the equitable and efficient management and appropriate use of water to be the problem that must be solved by involving the whole community. The Dublin Statement has stated that water is a finite and vulnerable resource. Management should be based on a participatory approach, involving users, planners, and policymakers at all levels, including the vital role of women. Water has been stated as a public good and has social and economic value in its competing uses.

At the 2000 Millennium Summit, the global community set the Millennium Development Goals (MDGs) for water and sanitation. There were eight goals focused on water that the United Nations member states have agreed to achieve by 2015. Among them, Goals 7 (To ensure environmental sustainability) and 8 (to develop a global partnership for

Table 1 Total Renewable Water Resources per capita by region (Arias et al., 2019)

Region	Total renewable water resources per capita, m^3/person/year
Latin America and the Caribbean	22,232
North America	16,004
Europe and Central Asia	7,855
East Asia and Pacific	4,522
Sub-Saharan Africa	3,987
South Asia	1,151
Middle East and North Africa	549
World Average	**5,921**

development) are well fitted to the globalization for environment protection. Goal 7 assumed the achievement of integrating the principles of sustainable development into country policies and programs, reversing a loss of environmental resources, and halving by 2015 the proportion of the population without sustainable access to safe drinking water and basic sanitation.

All the documents have formed the basis of integrated water resources management to deal with water crises worldwide.

2 Integrated Water Resources Management

Integrated Water Resources Management represents the holistic approach to water management and use. It "acknowledges the importance of the entire water cycle (…) and its interaction with other ecological systems in the natural ecosystem" (Manual 1, 2020).

Definition of integrated water resources management is given in Agenda 21 (Agenda 21, 1992) Chap. 18 document:

> Integrated water resources management is a process which promotes the coordinated development and management of water, land and related resources in order to maximize the resultant economic and social welfare in an equitable manner without compromising the sustainability of vital ecosystems.

Integrated water resources management is based on such vital concepts as integration, decentralization, participation, and sustainability, and its key policy goals are equity, ecological integrity, and efficiency (Meran et al., 2020). Equity means that water is a fundamental right of everybody, and as a result, everybody should have access to water of adequate quality and quantity. Ecological integrity means that water of the correct amount and quality should be persisted in the environment, used sustainably, to protect the rights of future generations to live in a clean environment. Efficiency means that water is not infinite, and because of this fact, it must be used at the maximum efficiency level. It has to be priced according to its economic value (Meran et al., 2020).

This approach also recognizes the diversified interests of various stakeholders. The intersectoral approach to water management requires the cooperation of different participants at different levels to simultaneously protect and conserve water resources with the uptake of water for many purposes, including domestic use, industry, and agriculture. The benefits of integrated water management are visible in many aspects: ecological, economic, social, cultural, and political—they concern the whole society and even humanity (Manual 1, 2020).

Success in integrated water resources management depends on planning, implementing, and monitoring the management process. To evaluate the effectiveness of integrated management, proper indicators should be applied. Based on the Manual of Stockholm International Water Institute (Manual 1, 2020), some critical indicators are listed in Table 2. It should be emphasized that in the area of law

Table 2 The most important indicators of successful implementation of integrated water resources management (based on Manual 1, 2020)

Evaluated area	Indicators of successful implementation
Policy laws, regulations, and involvement of the public	– Development of appropriate policy and legislative framework for integrated water resources management at national and sub-national levels – Stakeholder consultation and participation in management planning and implementation – Establishment of multisectoral water basin management body – Establishing a multi-level management framework – Decentralized to the lowest possible level decision-making process – Enabling information flow and transparency – Clear status of water entitlements – Education in environmental aspects including integrated water resources management – Public awareness on integrated water resources management – Training and building integrated resources management
Efficient water use and environmental sustainability	– Establishment of water demand management tools, use pricing of water as a tool to achieve sustainable water access – Development o water infrastructure – Development of new technologies – Good ecological quality of water resources, protection of the biodiversity – Support of natural functions and services of ecosystems – Rehabilitation and restoration of natural ecosystems – Monitoring of physical, chemical, and biological parameters of water and reduction of the contaminant level

and regulations, the tools and indicators are focused on achieving the participation of all stakeholders and making plans, which are then implemented, monitored, and corrected.

To transfer integrated water resources management to practice, it is recommended to use planning cycle which consists of (GWP, 2004; Meran et al., 2020):

1. Establishing status and overall goals: It can be mainly at a central level but with the participation of stakeholders. It is necessary to see the water problem at the national and first et al. at the international level. International cooperation is a crucial element. Federal programs and politics should meet the overall framework. The programs and politics implemented in neighborhood countries should be compatible.
2. Building commitment to reform: building and consolidating multistakeholder dialogue and agreements is one of the priority actions. The actions have to be built on the knowledge of the matter subject; proper tools should be chosen to implement the knowledge and to involve as the broad population as possible.
3. Analyzing gaps: analysis of the gaps in the framework is necessary in light of management functions.
4. Preparation of Strategy and Action plan: A portfolio of the actions and strategies should be created, e.g., as the road map. The actions and strategies should be set from the perspective of other international strategies and actions.
5. Building commitments to actions: is a crucial element for the progress, and it is based on the adaptation of action plans at the highest political levels. But it is necessary to remember that the entire stakeholders' acceptance is a crucial element in the implementation process. Means to finance are of great importance.
6. Implementation of Frameworks: appropriate tools on various levels are necessary to implement the frameworks, strategies, and plans; they should be chosen based on the knowledge and consultations. It is crucial to remember that financial means and appropriate infrastructure, including institutional, is an essential element of the implementation process.
7. Monitoring and evaluating process: it is necessary to monitor and evaluate the obtained outputs. An essential element of this process is to match descriptive indicators to the areas being assessed. Based on monitoring and evaluation, improvements in the actions and politics should be implemented. Monitoring and evaluation should be a continuous process.

Integrated management of water resources is complex because of many years of human-centered and particular interests-centered management systems. It is not easy to change the paradigms. Moreover, the system which undergoes management is highly complex. It acts at many levels (from governmental to individual households), and it is difficult to separate the problems originating from natural processes and the ones caused by industry and economy. The problem is that despite the integrated resource management, most limits and standards are regulated for individual contaminants or pollution sources. No legislation on cumulative effects is available. Management is also tricky because of the cumulative impacts of various sources on water resources. The critical problem is also how to change the public view on methods of contamination decrease (Schreier et al., 2014).

Applying the solutions is not easy because of the complexity and nonlinear nature of the stressors and pollution sources. It is almost impossible to establish a causal link between pollution and ecological effects. Therefore, holistic monitoring and management approaches are necessary (Bierioza et al., 2021).

This integrated approach is more and more visible in the legislation, politics, and directives of all world regions. It is evaluated, and now globalization in the management towards integrated management is well visible. It not only concerns more developed areas such as Europe or Northern America but also regions such as Africa.

3 European Policies for Water

Europe is where water resources protection should be of great interest. According to the European Environmental Agency Report in 2018, about 60% of its water bodies fail to achieve good ecological and chemical status. The two primary sources of the failure were hydromorphological pressure and diffuse pollution from agriculture (Bierioza et al., 2021; EEA, 2018).

Present European Policies on Water Management are based on "A Blueprint to Safeguard Europe's Water Resources," published by the European Commission in 2012 (A Blueprint, 2012). The document focuses on policy actions that should improve current water legislation application and the integration of water policy objectives with other policies. The aim is to build in member countries water policies effective in efficient water use and management. But A Blueprint has been preceded by other directives that form a combined system of legislation.

In the field of implementation, the following documents are essential: the Water Framework Directive (2000/60/EC), the Urban Wastewater Directive (91/271/EEC), the Bathing Water Directive (2006/7/EC), the Nitrates Directive (91/676/EEC), the Drinking Water Directive (98/83/EC and

EU 2020/2184), the Floods Directive (2007/60/EC), and the Marine Strategy Framework Directive (2008/56/EC). All directives mentioned above are complementary to the first one, referring to the Water Framework Directive.

The Water Framework Directive. The Directive required all Member States of the European Union to protect and improve the quality of rivers, lakes, groundwater, and transitional coastal waters, which means achieving good ecological and chemical status (both qualitative and quantitative) by the year 2015, at the latest by 2027. It prescribes the steps to reach. The good ecological status will be assessed according to physicochemical, chemical, biological, and hydromorphological quality criteria. It creates river basin districts—as natural geographical and hydrological units. River Basin Management Plans should be prepared and updated every 6 years to protect and improve the water environment. The Water Framework Directive is a flagship of European environmental policy. This directive was introduced to overcome the fragmentation of the European water policy. Implementation of such directive was, however, a great challenge. As the European Commission Fitness Check (European Commission Fitness Check, 2019) has revealed, its implementation is significantly delayed because of the problems with financing and lack of political will in member states.

The Urban Wastewater Directive. It concerns the collection, treatment, and discharge of urban wastewater and the treatment and release of sewage from specific industrial sectors. This regulation aims to "protect the environment from the adverse effects of urban wastewater discharges and discharges from the certain industrial sectors." Based on provisions of the Directive, urban agglomerations were created in the member states, with a population equivalent to over 2000 people. More advanced treatment should be applied in wastewater treatment plants with a population equivalent to more than 10,000 in sensitive areas. Sensitive areas mean "freshwater bodies, estuaries and coastal waters which are eutrophic or may become eutrophic if protective action is not taken." Under article 15 of the Directive, discharges from urban wastewater treatment plants have to be monitored to verify compliance with the emission control standards. On 13 July 2018, the European Commission published a Consultation on the Evaluation of the Urban Waste Water Treatment Directive to draw up the revision of the directive. The aim of the revision is to include circular economy rules in the wastewater treatment sector, take into account emerging pollutants of high toxicity and carcinogenicity, and implement new technologies and possibilities to treat wastewater. Also, energy self-sufficiency is essential in Urban Wastewater Directive revision.

The Bathing Water Directive. It was set to achieve safeguard public health and clean bath waters. The present directive is a revised version of standards from the 1970s. Based on the Directive Member States are obliged to monitor and access the bathing water for at least two parameters of fecal bacteria. They must inform the public about bathing waters' quality and create and develop bathing water profiles. The profiles should contain the data on the pollution sources and health risks connected with water use. Bathwater classification based on symbols was introduced in 2011. Evaluation of the implementation of the Directive was done in 2018–2019. The report was presented in march 2012. In October 2021, European Commission launched an online public consultation to gather views on the revision of the rules set in the Bathing Water Directive.

The Nitrates Directive. It was set to prevent the quality of water environment in member states of the European Union from nitrates from agricultural sources and to promote the use of good farming practices. The directive describes criteria based on which surface- and groundwaters had to be identified as affected by pollution. Based on the directions, nitrate-vulnerable zones have been identified as the ones in which the concentration of nitrates exceeded or was at risk of exceeding the value of 50 mg/L. Member states have to report nitrates concentrations in groundwaters and surface waters every 4 years. The reports also concern nitrate-vulnerable zones and eutrophication. Based on the Directive, farmers working in the nitrate-vulnerable zones have to implement action programs, and the implementation of the programs is also monitored and reported.

The Drinking Water Directive. 98/83/EEC Directive concerned the quality standards of drinking water. Not only set the standards but also obliged the member states to report their water quality to the European Commission and the public. In 2018, European Commission published a proposal for the revision of the directive to improve the quality of drinking water and ensure better access for the public to the data on drinking water quality, which was a response to the European Citizens' Initiative Right to Water. In 2020, new Drinking Water Directive (EU 2020/2184) replaced the 98/83/EEC Directive giving the member states 2 years to implement its requirements.

The Floods Directive. It concerns the assessment and management of flood risks. The member states, based on the data collected by the institutions and agencies, had to estimate the preliminary flood risk up to 2011. Based on these data, they were obliged to identify the areas at high risk of flood hazards and prepare flood hazards and risk maps based on models. Each Member State was also obliged to give Flood Risk Management Plans by December 2015.

The Marine Strategy Framework Directive. It aims to maintain Good Environmental Status in European Seas.

In addition to the directives mentioned above, which are established and implemented in a system of legislation, such documents as the Green Deal, campaigns, and politics illustrate well the changes in the approach to protecting the environment in the European Union towards integrated management of water resources.

European Union's efforts in water resources protection are summarized well in Green Deal from 2019 (The European Green Deal, 2019). I was approved in 2020. The European Deal's priorities are protecting biodiversity and ecosystems, reducing water protection, moving towards the circular economy, and ensuring sustainability. Green Deal is a crucial document in implementing the United Nations' Sustainable Development Goals. Its key objective is to provide a holistic policy framework to achieve carbon neutrality by 2050. It is a document of ambitious plans and the right directions. However, in European countries, it has faced some criticism. The importance of the criticism proves that, e.g., in Eastern Europe, it was perceived as potentially destructive for the European Union. However, the reservations were connected with the changes in promoted energy sources, as in these countries, coal is still a significant source of energy, not for water status.

On the other hand, such groups, e.g., Greenpeace, stated the strategy was not ambitious enough (Kelly, 2019). In the area more strictly connected with water pollution, the Green Deal adopted the "Zero pollution Action Plan" for all contaminants sources. It also contains ambitious plans for the protection of the environment against such pollutants as microplastics, pharmaceuticals, and chemicals.

The Green Deal is supported by several other documents, campaigns, and politics. Some are strictly involved in water resources management: European Citizens' Initiative Right to Water and the Farm to Work (part of the Green Deal).

European Citizens' Initiative Right to Water (Right2Water). It is a campaign to commit the European Union and the individual member states to implement the human right to water and sanitation. It is a nongovernmental organizations' initiative involving more than 120 NGOs. It was supported, among others, by German and Austrian trade unions. It is an impressive initiative among others because, in 2013, it was the first one that collected more than 1 million signatures. The initiative has set three main goals: guaranteed water and sanitation for all Europeans, no liberalization of water services, and global access to water and sanitation. It successfully affected European Commission's works.

Farm to work. It is a strategy mainly focused on food sustainability in Europe. But it also emphasizes the necessity of reducing the use of chemical pesticides and sustainable production by reducing fertilizers use. As a result, a decrease in water pollution should also occur. The time perspective of the Farm to Work strategy is 2030 to each reducing the fertilizer use by 20% and reducing nutrient loss by 50% should be achieved (Farm to Work, 2020).

European Union legislation at present fits well with integrated water management and involves all elements considered crucial in this type of approach. It is based not only on formal legislation but also on the citizens' initiatives. It involves all stakeholders and brings them together to achieve ambitious plans. But it should be emphasized that Europe is not homogenous socially, economically, and culturally. It is the area in which various interests, traditions, and approaches collide. It is well visible in the implementation process. Well visible border is between the countries of Old Europe and the new members of the European Union. The last one has low financial possibilities to achieve ambitious plans in the area of water protection. Analyzing the situation in these countries, it is also visible a significant improvement in the quality of the environment. Supporting environmental investments in the area of water protection by the European Union is a real chance for the poorer countries to achieve a good quality of water. As a conclusion, it can be said that European Union is strong by its diversity (it causes discussion and, as a result, also more thoughtful solutions), but simultaneously is also weak (conflicts of interests of individual countries are delaying the implementation of the solutions). The European approach to legislation changes is modern. It usually relies on replacing the older acts with new ones, not only on making improvements to the existing legislation. Because all solutions have to be approved by member states, the legislation is, however, not as flexible as it should be.

4 American Policies for Water

American approaches to integrated water management differ as North and Latin America differ significantly. That is why, for both continents, it should be described separately.

4.1 United States of America

Water policies in the United States have a long tradition that started in the nineteenth century. First policies (in the nineteenth and the beginning of the twentieth centuries) focused on maintaining the navigability of significant rivers and harbors. The next step in water policies in the USA was in the 1970s and institutionalized water environment protection, among others, by set Environmental Agency. It included, e.g., the National Environmental Policy Act (1970), the Clean Water Act (1972), and the Safe Drinking

Water Act (1974). The third step was connected with the formalization of federal policy to protect against damages and pollution caused by investments and projects. In the late 1980s, particular states became the creators of water policies. No federal but cross-border state collaboration paid an essential role in the water law system, e.g., in the Colorado River basin. Federal guidelines increasingly relied upon the partnership with state and local governments and private entities from the late 90 s. From the twentieth century to the present times, the national policies in water management still are characterized by collaboration with these partners with the leading role of individual states. An example of this kind of legislation is Water Resources Development Act, established in 1999. Visibly, this Act moved flood control efforts from large-scale investments to smaller-scale, local flood control (Reimer, 2012).

At present major water management and protection laws in the United States are Clean Water Act, Safe Drinking Water Act, National Environmental Policy Act, Endangered Species Act, Wild and Scenic Rivers Act, Federal Power Act, and Water Resources Development Act.

Clean Water Act (Federal Water Pollution Control Act)—it was originally enacted in 1948 and significantly revised in 1972. It is an essential law act that sets national standards for water quality to maintain the physical, chemical, and biological integrity of the waters in the US. The primary purpose of introducing national standards was ambitious—to make most US waters drinkable, swimmable, and fishable. The principles of the Act were the 1970s very modern. The Act consists of two major parts: provisions authorizing federal financial assistance for municipal sewage treatment plant constructions and other water quality improvement projects; the second are regulatory requirements concerning municipal and industrial discharges. At early on, emphasis was put on controlling discharges, and at present, more attention is paid to toxic pollutants discharges, which is similar to those in Europe (Copeland 2016).

Safe Drinking Water Act (An Act to amend the Public Health Service Act to assure that the public is provided with safe drinking water and other purposes)—it is the principal act that regulates drinking water standards at the national level for all public drinking water sources. It was established to ensure safe drinking water for the public. It applies to every public water system in the US. It includes, among others, mandatory requirements (e.g., maximum contaminant levels). Drinking water standards in the act include the following parameters: inorganic chemicals, organic chemicals, microorganisms, disinfectants, disinfectants by-products, and radionuclides. It allows the States to set the standards more strict than the federal ones and monitor and set the standards for contaminants not mentioned at the federal level. The system of drinking water protection has developed through the years, e.g., in 2011, Congress passed the amendment to the act based on which reduction of lead in drinking water was undertaken. In 2015, risk analysis was included to the system as the EPA was obliged to submit to Congress a strategic plan for assessing and managing risks connected with algal toxins in drinking water. In 2016, regulations helping small public water supply systems to meet the drinking water regulations were established (United States Grassroots Rural and Small Community Water Systems Assistance Act).

National Environmental Policy Act (National Environmental Policy Act of 1969)—is a US federal law that sets some federal agencies for assessment of environmental effects of the implemented regulations and plans. It is divided into three sections: thirst—which contains national environmental policies and goals, the second—it establishes provisions for federal agencies to enforce the policies and plans in environmental protection; and the third, which sets the Council of Environmental Quality (CEQ).

Endangered Species Act of 1973 prohibits both federal and private actions that can harm endangered species, including water species. Federal actions have to be approved by a special agency.

Wild and Scenic Rivers Act of 1968—it protects natural free-flowing conditions of designated rivers in the US.

The United States legislation in water management also relies on the balance of interests of many stakeholders. Contrary to the European Union legislation, it, however, rather uses amendments to the legal acts established in even 1900s, then replacing them with new legislation. It could be affected by tradition. In Anglo-Saxon countries, the precedential system is a cornerstone of justice. It probably affects also legislation. This system allows for long-term evaluation and improvements of the laws.

4.2 Canada

In Canada, the Federal Water Policy Act was established in 1987, and it is still in force. The Act regulates water resources management and balanced water use based on sustainable development rules. It defines two main goals: protection and enhancement of water resources and efficient management and use of water. The system is based on integrated planning, and the government applies an integrated approach to the planning and development of water resources. Cooperative, integrated planning is considered essential. New technologies of water and wastewater treatment, protection, and reuse are developed, and cooperation in implementing them is promoted. A helpful tool to encourage efficient water use is water pricing. The principle polluter pays is applied.

4.3 Latin America and the Caribbean

Latin America is commonly considered the area with enough water resources as it contains about 31% of the world's freshwater resources. Indeed Latin America also has large arid and semi-dry areas. It is estimated that in 2020 more than 160 million citizens of Latin America and the Caribbean do not have access to safe water (Latin America, 2021). Because of this common and sustainable access to water is an important issue also in this region.

In the 1980s, countries of Latin America underwent decentralization. As a result, water management responsibilities were transferred to the regions and provinces; most of them lacked the capacity required to operate and maintain the services effectively. In the 1990 decade in Latin America, a decrease in government provision of public goods and an increase in the private sector were observed in the water sector. Because of the shortcomings of such systems, many countries in the region have undertaken some actions to coordinate water policies, e.g., in Brazil, in 2000 National Water Agency was created as a part of the National Water Resources Management System. The agency is responsible for implementing the National Water Resources Policy implementation and implementing the principles of integrated water resources management. It is also responsible for the implementation and coordination of the National Water Law. A similar agency was set up in Peru in 2008. It aims to promote integrated water resources management. What is interesting in Cuba National Institute for Water Resources, with the mission of implementing and controlling the National Water Resources Policy, was established just in 1989. In Mexico Basin, authorities for more coordinated water management were established in the late 2010s. (Akhmouch, 2012).

Latin America is, at present, the arena of intensive changes in the water management approach. E.g., in 2018 in Argentina, Senate Bill No. 3563/18 would guarantee the right of access to drinking water to all inhabitants of Argentina was set. At the same time, in Mexico, a bill was proposed in the Senate that would modify the National Water Law to designate the environmental use of water as a priority and give it preference when granting water concessions or assignments (The National Laws Review, 2022).

5 Africa

In Africa, the need for sustainable and equal access to water is a crucial issue. The States have undertaken some activities to solve the problems connected with water resources management. One of the most important ones is the Guidelines on the Right to Water in Africa, established in 2019 (Guidelines on the Right, 2019). In the document approach based rights-based approach to water management is promoted. Equitable access to water has been categorized as the realization of human rights. The guidelines cover the States' obligation to respect, protect, promote, and fulfill the water right and mobilize resources to ensure the water right. Generally, the guidelines set the principles of equal access, non-discrimination, and non-retrogression in the context of human rights. What is essential in the context of the political situation in Africa is that the guidelines emphasize that States "shall refrain from using access to water as a political tool." States are also obliged to protect all water resources and to pursue, through an integrated water strategy, the realization of the right to water. While States are responsible for realizing the right to water, the guidelines outline accountability measures for infringements of this right, especially by companies and other particular groups of interests, by preventing corruption and guaranteeing sustainable environmental management.

As rightly observed by Professor Mdee (2020), in the policies mentioned above, African governments do not have efficient policies on the use of water in agriculture because the policies are focused on integrated water management and do not account for the real needs of the farmers. Whereas in Africa, patterns of use of water in agriculture are changing. Previously, there has been a focus on large irrigation schemes, and they, however, have failed. At present farmer, led-irrigation is more often not centered.

The African legislation and actions in integrated water management are of high importance. They both origin from the initiative of international organizations and as a result of individual country's operations. At present, compared to the rest of the world, the integrated water resources management in Africa does not fit the problems of this continent. It is because of the political and economic situation in the continent. Because of the climatic changes, Africa will face more severe problems than the ones it faces now. Effective and accepted solutions for Africa are of the highest importance.

6 Australian Policies for Integrated Water Management

As review directives, acts, and other policies on the protection of the environment, it is clearly visible that they all have started from traditional water management based on maximum allowable concentrations of pollutants, protection of individual elements of the environment against pollution, setting the agencies to ensure appropriate actions for the protection of the water, to multisectional, multidimensional and multisectoral (Meran et al., 2020).

Australia is one of the countries facing significant challenges in ensuring sustainable water access. A central government coordinates the water management system in Australia, but state and territory governments are responsible for managing their water resources. National law acts on water protection in Australia were consulted and set in collaboration with state and territory governments, agencies for the environment's safety, experts, and scientists, and widely discussed (Leadership in National Water Policy, 2022).

The main acts concerning the management of water in Australia are Water Efficiency Labelling and Standards Act 2005, Water Act 2007, Water Regulations 2007, Water Market Rules 2009, Water Charge Rules 2010.

Water Efficiency Labelling and Standards Act 2005. It is conceived to be a scheme to reduce demand for drinking water by informing consumers about the water efficiency of household appliances. It has been mandatory in the whole of Australia since 2006. This act supports achieving the main goals of water management.

Water Act 2007. This act is one of the most important concerning water management in Australia, establishing a national framework to manage its resources. It promotes the use and management of Murray–Darling Basin water resources to optimize social, economic, and environmental outcomes. Its goal is to protect and restore the ecological values of the Basin. Based on the Water Act, such as the Murray–Darling Basin Authority and Commonwealth Environmental Water Holder. It also sets requirements for the Murray–Darling Basin Plan. The Water Act also concerns informative aspects by setting the Bureau of Meteorology. It pays attention to the regulatory elements by involving the Australian Competition and Consumer Commission in developing and enforcing water charges and water market rules.

Water Regulations 2008. This is the legal act under the Water Act, which specifies the water information that certain organizations must give to the Bureau of Meteorology.

Water Market Rules 2009. The Water Act provides them and ensures irrigators can permanently transform their irrigation right into a statutory water access entitlement.

Water Charge Rules 2010. They provide transparency in charges and costs for customers of monopoly water infrastructure providers in Australia. Charges and costs of water in Australia are regulated and controlled by The Australian Competition and Consumer Commission.

7 Conclusions and Recommendations

Based on the review given above, the following conclusions and recommendations can be formulated:

1. Directives, policies, and other legal acts concerning water management gradually evolve from the control function (e.g., setting permissible concentrations of pollutants) or providing financial support for integrated management systems. This tendency is visible worldwide.
2. Legislative culture varies worldwide, but although there are several ways to reach the goals, they can be effective.
3. Integrated water resources management is the only alternative to protect the water environment and preserve them for future generations.
4. In the future, more attention should be paid to the effective implementation of the developed solutions, and assuring the participation of various groups of stakeholders is a good tool for it.

References

Agenda 21. Document of United Nations Conference, on Environment & Development, Rio de janerio, Brazil, 1992.

Arias, L., Rud, J. P., & Ruzzier, C. A. (2019). The regulation of public utilities of the future in Latin America and the Carraibean (LAC), Water and Sanitation Sector, Technical Note No IDB-TN-1678, Inter-American Development Bank.

A Blueprint to Safeguard Europe's Water Resources, Communication from the commission to the European Parliament, the Council and the European Economic and Social Committee of the Regions, Document 52012DC0673.

Bierioza, M. Z., Bol, R., & Glendell, M. (2021). What is the deal with the green deal: Will the new strategy help to improve European freshwater quality beyond the water framework directive? *Science of the Total Environment, 791*, 148080.

Boretti, A., & Rosa, L. (2019). Reassessing the projections of the world development report. *Clean Water, 2*, 15. https://doi.org/10.1038/s41545-019-0039-9

Copeland, C. (2016). Clean Water Act: A summary of the Law, Congressional Research Service.

Council Directive 91/271/EEC of 21 May 1991 concerning urban waste-water treatment, *OJ L 135, 30.5.1991*, pp. 40–52.

Directive 2008/56/EC of the European Parliament and of the Council of 17 June 2008 establishing a framework for community action in the field of marine environmental policy, *OJ L 164, 25.6.2008*, pp. 19–40.

Directive 2007/60/EC of the European Parliament and of the Council of 23 October 2007 on the assessment and management of flood risks, *OJ L 288, 6.11.2007*, pp. 27–34.

Directive 2000/60/EC of the European Parliament and of the Council of 23 October 2000 establishing a framework for Community action in the field of water policy, *OJ L 327, 22.12.2000*, pp. 1–73.

Directive 2006/7/EC of the European Parliament and of the Council of 15 February 2006 concerning the management of bathing water quality and repealing Directive 76/160/EEC, *OJ L 64, 4.3.2006*, pp. 37–51.

Directive (EU) 2020/2184 of the European Parliament and of the Council of 16 December 2020 on the quality of water intended for human consumption, *OJ L 435, 23.12.2020.*

Council Directive 98/83/EC of 3 November 1998 on the quality of water intended for human consumption, *OJ L 330, 5.12.1998, p. 32–54.*

EEA. European Environmental Agency. European Waters. Assessment of Status and Pressures, 2018, 7.EEA Report, Luxembourg (2018), 1–90, cited after: (Bierioza et al., 2021).

European Commission Fitness Check of the Water Framework Directive and the Floods Directive. SWD (2019) 439 Final, Brussels, 2019, 1–184.

Falkenmark, M. (1997). Mar del Plata Anniversary Seminar, Analytical summary, Stockholm International Water Institute, Stockholm.

Farm to Work (2020), https://ec.europa.eu/food/horizontal-topics/farm-fork-strategy_en

Guidelines on the Right to Water in Africa, African Commission on Human and Peoples' Rights, 2019.

GWP. (2004). Catalyzing change: A handbook for developing integrated water resources management (IWRM) and water efficiency strategies. Technical advisory committee (TAC) background paper no. 5. Stockholm, Sweden: Global Water Partnership., cited after: Meran et al. 2020.

Kelly, E. (2019). EU Council meets to debate Green Deal grand plan for tackling climate change. *Science Business.*

Latin America looks to achieve descent water management, 2021, https://www.caf.com/en/currently/news/2021/11/latin-america-looks-to-achieve-decent-water-management/

Leadership in national water policy, Australian Government, Department of Agriculture, Water and the Environment, 2022, https://www.awe.gov.au/water/policy/policy

Manual 1. Principles and practices of Integrated Water Resources Management, Stockholm International Water Institute, Stockholm, 2020.

Mdee, A. (2020). Why water policies in Africa need to change, https://medium.com/university-of-leeds/why-water-policies-in-africa-need-to-change-ab65f4cb3813

Meran, G., Siehlow, M., & von Hirschhausen, C. (2020). Integrated Water Resource Management: principles and Applications, Chapter 3 in: The Economics of Water, Rules and Institutions.

MPAP. (1977). Mar del Plata Action Plan, United Nations.

Reimer, A. (2012). U.S. Water Policy: Trends and Future Directions, National Agricultural & Rural Development Policy Center.

Schreier, H., Kurian, M., & Ardakanian, R. (2014). Integrated Water resources Management: a practical solution to address complexity by employing the nexus approach, Institute for Integrated Management of Material Fluxes and of Resources, Dresden.

The European Green Deal. Communication from the commission to the European Parliament, the European Council, the Council, the European Economic and Social Committee and the Committee of the Regions The European Green Deal (2019), Document 52019DC0640.

The National Laws Review, 2022, https://www.natlawreview.com/article/latin-american-environmental-regulatory-tracker-september-16-october-15-2018.

Water and Climate Change. The United Nations World Water Development Report 2020, UNESCO 2020.

Water Management 2040. Future scenarios. Kemira, 2021.

Circular Water Management in Smart Cities

Joanna Bąk

Abstract

The ongoing climate change forces the need for more and more attentive water management in the city. On the other hand, citizens expect solutions that make their lives easier, or increase their comfort. The answer to these needs seems to be the introduction of circular water management with the use of solutions characteristic for smart cities. Therefore, the aim of the article is to present the framework and perspectives for implementing circular water management in smart cities. At the beginning, the definitions of the circular economy and smart city were discussed. The differences in the linear and circular approaches to water management in cities were also presented. The study presented the stages of the product (water) life cycle as part of a circular model together with a discussion of various circular water management practices used in a smart city, with particular emphasis on innovative solutions in this area. A SWOT analysis was conducted to assess the prospects for the implementation of circular water management in smart cities. The conclusions summed up the framework of circular water management in smart cities and perspectives for its implementation. On the basis of the analysis, suggested directions of activities for the cities of the future were indicated.

Keywords

Circular economy • Water management • Water resources • Conserving water • Rainwater • Drinking water

J. Bąk (✉)
Faculty of Environmental Engineering and Energy, Cracow University of Technology, Kraków, Poland
e-mail: joanna.bak@pk.edu.pl

1 Introduction

Both to survive and improve the quality of life, it is necessary to make continuous efforts to innovations (Lee & Trimi, 2018). The world needs innovation and further progress, and this development should meet the needs of the present generation, "without compromising the ability of future generations to meet their own needs", according to (United Nations, 1987). This mainly applies to areas that are particularly sensitive to the ongoing climate change. They include, among others, water management in cities (Interdisciplinary advisory team for the climate crisis at the President of the Polish Academy of Sciences, 2020). Cities especially need it, because in the face of increasing urbanization, the life of the inhabitants becomes more and more onerous, and the water supply may not be guaranteed in the short term, which can be confirmed, inter alia, by the list of cities most at risk of running out of water (Chapman, 2019).

Trends such as the circular economy and the construction of smart cities respond to the needs of today's world. The development of smart cities is treated as the preferred response to the urbanization challenges of the twenty-first century (Praharaj & Han, 2019). Their combination may generate innovative solutions with environmental benefits, allowing cities to adapt to new climatic conditions.

The number of articles on issues related to the implementation of circular water management with the use of solutions typical of smart cities is still insufficient. There are many definitions of the concept of smart city, and water management can be treated in many dimensions. Moreover, the circular economy model in the water and sewage sector can be presented in more detail. There is also a lack of lists of various available solutions enabling the implementation of circular water management in smart cities. The author of the work compared various definitions of water management and smart city. He proposed a more detailed model of the circular economy in the water and sewage sector, as well as presented a list of solutions characteristic of smart cities

M. Smol et al. (eds.), *Water in Circular Economy*, Advances in Science, Technology & Innovation,
https://doi.org/10.1007/978-3-031-18165-8_3

enabling the implementation of transition towards circularity. The aim of the article is to present the framework and perspectives for the implementation and development of circular water management in smart cities. The scope of the work includes a discussion on the definition of a smart city and water management, presentation of the differences between linear and circular economy and an overview of the stages of the water life cycle along with the presentation of solutions for the implementation of circular water management in smart cities. Moreover, perspectives for the implementation of circular water management in smart cities of the future were analyzed in terms of the existing opportunities, as well as possible risks.

This study on water management relates mainly to water supply (especially to rational use of water resources) and to rainwater management in smart cities.

2 Basic Definitions and Scope of the Analysis

Each analysis should be preceded by insightful considerations regarding its scope, and thus also the definitions concerning it. Particular attention should be paid to this if the definitions are not unambiguous and fully consistent, or when the issue can be treated in many dimensions. This is the case with the concept of smart city. There are over 120 different definitions of this concept, and as a concept—smart city is not a matter of agreement (Camero et al., 2019; Juma & Shaalan, 2020). There are also studies analyzing, i.e., various definitions of the concept of smart city, such as the work (Lim et al., 2019), which identifies the results of the development of smart cities. One of the areas of interest under consideration is conceptualization, i.e., the definition of a smart city (Lim et al., 2019). There are also questions about the difference between a sustainable city and a smart city, as well as recommendations to use the term "smart sustainable city" as more adequate (Ahvenniemi et al., 2017). On the other hand, the concept of water management in relation to cities can be treated in many dimensions and may refer to various scopes. The issues of water supply to the population, the risk of flooding, but also the management of blue-green infrastructure and the use of rainwater can be considered (rainwater management).

Smart city has many definitions, not necessarily exactly overlapping with each other. In scientific articles and studies, it can be found numerous considerations on the definition of the concept of "smart city" (Camero et al., 2019; Juma & Shaalan, 2020; The Committee of Digital and Knowledge-Based Cities of UCLG, 2017). One of the more traditional definitions of smart city defines it as a city that uses information and communication technologies (ICT) to increase the interactivity and efficiency of critical infrastructure, its elements and public services, making residents more aware of them (The Committee of Digital and Knowledge-Based Cities of UCLG, 2017). On the other hand, the definition referred to as standard by (Juma & Shaalan, 2020), defines a smart city as a city aspiring to provide a better version of life through active involvement of residents (based on the use of data power and useful application of energy, technology, resources, and waste). Due to the fact that the concept of smart cities is an evolving concept, according to EIP-SC European Innovation Partnership on Smart Cities and Communities, smart cities can be described as "systems of people who interact and use flows of energy, materials, services and financing to catalyze sustainable economic development, resilience, and a high quality of life" (The Committee of Digital and Knowledge-Based Cities of UCLG, 2017). These interactions and flows are defined by the adjective "smart" thanks to the strategic use of ICT infrastructure and services as part of a transparent urban planning and management process that responds to the social and economic needs of society (The Committee of Digital and Knowledge-Based Cities of UCLG, 2017).

The concept of a smart city also has various dimensions. They include Smart Economy, Smart Governance, Smart Living, Smart Mobility, Smart People and Smart Environment (Sharif & Pokharel, 2022).

A definition closely related to water management in a city seems to be an explanation that defines smart city as settlement units in which ICT is used to make the standard infrastructure more durable, practical, and secure (Juma & Shaalan, 2020). It is also worth mentioning a broader definition in which there is wise management of natural resources, and thus also water management. The term "smart" can be used for a city that invests in human and social capital and communication infrastructure to promote sustainable economic development and high quality of life, including above mentioned wise management of natural resources (through participatory government) (Bielawska, 2019; Fundación Telefónica, 2011; The Committee of Digital and Knowledge-based Cities of UCLG, 2012).

Water management is defined by goals and tasks to be implemented. It covers the management of strategic natural resources (surface and groundwater) necessary for human life as well as for the functioning of the economy, leading to the strategic goal of achieving and maintaining good water status (Zaleski et al., 2020).

Water management in cities can be treated in many ways. As surface water management in terms of water quality protection or flood risk minimization, as water and sewage management (for water supply and sewage disposal) or rainwater management in the city. Various scopes of activities and tasks of water management are also adopted or considered. It is worth adding here that currently an integrated approach to water management is indicated as important and necessary (Nachlik et al., 2020). As early as

1977 at the United Nations Conference on Water in Mar de Plata, Integrated Water Resource Management was the recommended approach for the inclusion of multiple competing uses of water resources (Rahaman & Varis, 2005). In practice, the concept of integrated management still does not meet the needs (Nachlik et al., 2020) and it is necessary to identify future challenges for its further implementation (Rahaman & Varis, 2005).

The tasks of water management include satisfying the broadly understood water needs, water safety, economic use of rivers, and protection of water resources by ensuring their good condition (Zaleski et al., 2020). The section of water management (as a section of government administration), in accordance with Polish legislation (Act on government administration departments, 1997), covers matters of shaping, protection, and rational use of water resources, maintenance of inland surface waters together with technical infrastructure, as well as defining the principles and conditions of collective water supply for human consumption and collective sewage disposal, but also issues related to flood protection, maintenance of inland waterways, drainage and others. On the other hand, the authors of water resource use efficiency rankings in Polish cities, the Water City Index, divide the urban water policy into four areas (Ćmielewski et al., 2021). As part of managing water resources in the city, they distinguish such categories as life (e.g., drinking water), threats (floods, droughts, and sectoral losses), economy and business (water for industry and others) as well as culture and inhabitants (incl. landscape and education) (Ćmielewski et al., 2021).

3 Differences in the Linear and Circular Approach to Water Management

Limited availability and depletion of natural resources lead to a departure from the traditional linear model of the economy. The process of continuous depletion of natural resources, which is related to human lifestyle, cannot last forever (Cobo et al., 2018), it cannot be allowed to happen. It should also be added that water is one of the natural resources. However, along with the development of civilization, it began to be treated as an economic value and a factor of production (Thier, 2016). The circular economy (CE) is the answer to the imperfection of the linear model of the economy. Its aim is to overcome the linear pattern (take—make—dispose) of production and consumption and propose a circular system in which the value of products, materials, and resources is kept in the economy as long as possible (Merli et al., 2018). On this occasion, it is also worth mentioning the product design model called "cradle to cradle". It takes into account all aspects of product life, and the name shows the essence of the circular economy (gozwpraktyce.pl).

In work (Zvimba, 2019), one of the figures presents a linear approach to water and wastewater business cycle. In this approach, successive processes of water purification, water distribution, water consumption, sewage collection, and sewage treatment were distinguished. At the beginning, water is taken from water resources, and finally, treated sewage is discharged to the receiving body. The author of the work (Zvimba, 2019) also takes into account the formation of sludge for landfilling (from the WTP and WWTP). A document from the US Environmental Protection Agency is available providing guidelines for the location of water supply intakes downstream from municipal wastewater treatment plants (US Environmental Protection Agency, 1978). It confirms the existence of concepts already in the past assuming the location of the abstraction of water from the river downstream from treated wastewater discharge. With this sequence, the effluent treatment efficiency must be very high. Such concepts can also be compared to a kind of circular system in the field of water and wastewater management. The circular model for the water sector distinguishes water treatment preceded by a technology project, water distribution, and consumption, as well as sewage collection, as well as water recycling and recovery ending in the recovery of the product (which is water) (Zvimba, 2019). Of course, the circular approach includes supplementing the cycle with raw water and discharging waste (Zvimba, 2019). Figure 1 proposes a circular economy model for the water and sewage sector, which includes inter alia, use of greywater (thanks to the use of integrated sanitary utensils and installation of greywater in the building) and rainwater on the property. The starting point for the proposed model was the circular approach to the water and sewage business cycle in the work (Zvimba, 2019).

4 Stages of the Water Life Cycle as Part of a Circular Economy Model

The stages that can be distinguished in the circular economy model for the water and sewage sector are crucial due to the answer to the question which of them and how can be realized (or implemented) together with the concept of smart cities.

The discussion of the individual stages of the circular economy model should begin with water resources. Typically surface and groundwater are used as a water source. In regions with a deficit of this type of water—sea water is used after costly desalination processes. Rainwater and sewage (after appropriate treatment processes) can also be used as a water source. The source of drinking water can also be water

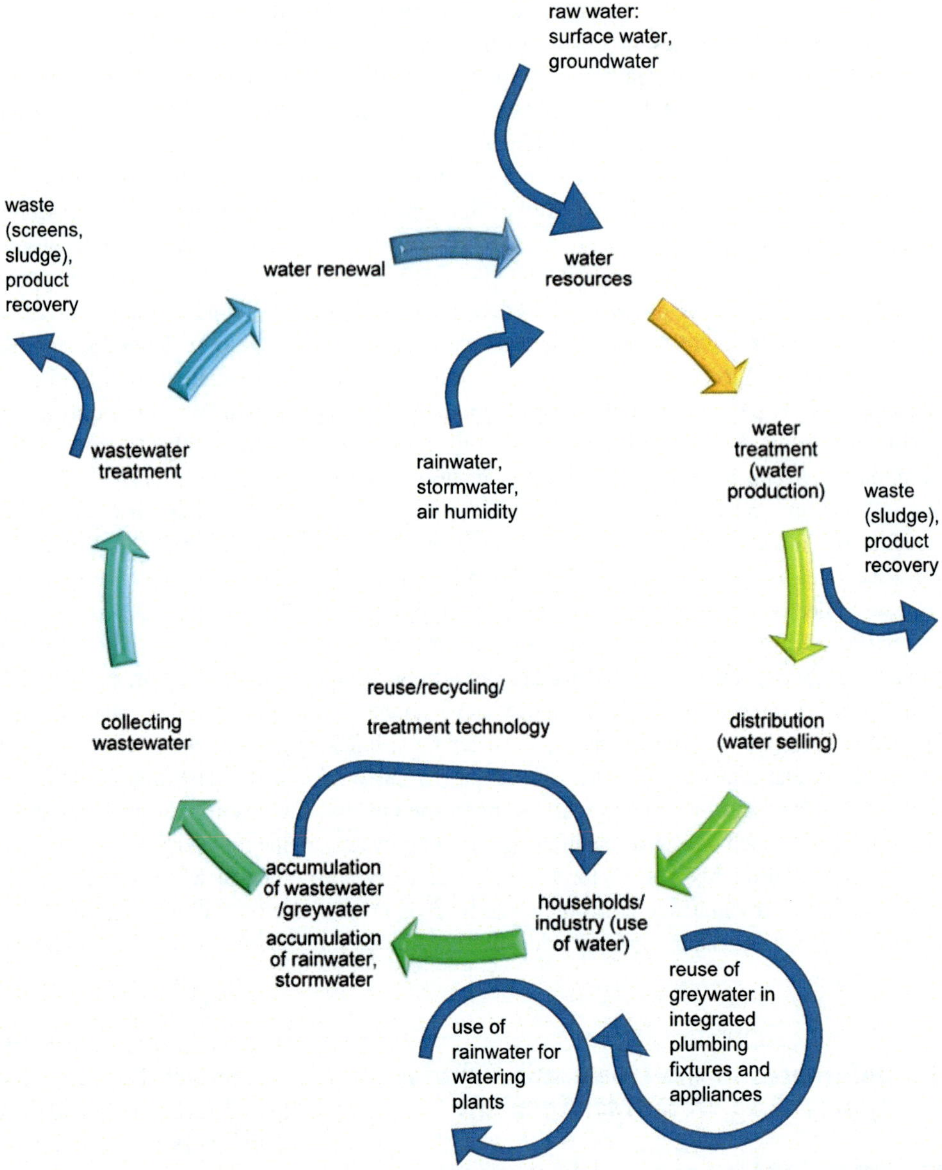

Fig. 1 Circular economy model for water and wastewater sector

from air humidity (www.watergen.com). An element of intelligent circular water management is the diversification of water sources to maintain reliable and resilient water supply (Zvimba, 2019).

As a first step, after taking into account water resources, water treatment should be mentioned. There are treatment systems of varying degrees of complexity, but assuming a circular cycle of the raw material, which is water, technologically advanced systems will prevail. It should also be added that the selection of unit processes depends on the quality of the raw material, so the processes of rainwater, groundwater, and surface water treatment will vary. If the use of a mixture of water from various sources is considered—it should be preceded by tests and a detailed selection of processes adapted to the quality of the raw material. The water-wastewater business cycle from work (Zvimba, 2019) also includes the technology design stage (before water treatment). As far as the stage of water purification is concerned, the development of methods for monitoring the parameters of raw and purified water quality and the operation of water treatment facilities and devices, as well as collecting and transferring data for system checking and control, is already at an advanced stage. It should be added that due to the high cost of monitoring, there is a lack of data on water quality for the proper management of this resource (Silva et al., 2022). A number of physical sensors have been available for a long time for online monitoring of water quality data used in control strategies (Ratnaweera & Fettig, 2015). Already in the past, the so-called software sensors based on multivariate regression analysis, artificial neural network models, or fuzzy logic models, were designed (Ratnaweera & Fettig, 2015). The main new technologies under development are based on optical or electrochemical sensors (Silva et al., 2022). However, further development in this area and further improvements are desirable. In the scientific literature, it can be found indications for the need to improve physical sensors or further develop software sensors (Ratnaweera & Fettig, 2015), as well as the necessity to subject new methods of water quality monitoring to exhaustive checks in terms of accuracy and repeatability before being released for use (Silva et al., 2022). They should particularly concern the efficient operation of equipment or the rationalization of chemical dosing, as well as energy consumption. The generation of waste/residue (water treatment sludges) is also related to the water treatment processes (Ahmad et al., 2016). There is a field here for the implementation of the principles of circular economy related to smart solutions for the recovery of materials or energy from waste. An example of a solution consistent with the principles of circular economy may be the innovative approach presented in (Górka et al., 2022), which allows the use of water and sewage sludge and their combined processing with a positive effect on biogas production.

The next stage includes transport, storage, and distribution of water to recipients. In terms of this stage, first of all, attention should be paid to the quick detection of leaks and failures, maintaining appropriate water quality in the system, as well as monitoring aimed at detecting unusual parameters of the system's operation and allowing the analysis and elimination of certain unfavorable events on the basis of observation. Mention should also be made of demand management and supply system management related to the initial stages of the cycle (Zvimba, 2019).

The next, one of the most important stages, is the use of water—both in the industrial and municipal sectors. Global water consumption is predicted to increase significantly, including for industry, which is one of its main users (OECD, 2012; Willet et al., 2019). In industry, water should remain in use as long as it is fit to fulfill its purpose reliably and safely. Smart solutions can be used in this area to check and control the operation of a water-using system. There is scope for industry to achieve large changes in the water cycle that can have a significant impact on total water use (Andrews et al., 2011). The potential in this respect depends mainly on the type of industry. Benchmarked data from various industry sectors show that some industries have the potential to reduce consumption by around 50% (Andrews et al., 2011). In turn, in households, it is extremely important to encourage consumers to use water efficiently. Intelligent solutions focus here mainly on collecting data on water consumption in homes and informing users about it, as exemplified by the activities carried out in Milton Keynes in Great Britain (www.mksmart.org). This is to accomplish two main tasks—to control possible leaks and to encourage the minimization of water consumption. There are also intelligent solutions that enable customers to reduce their water consumption, such as intelligent household appliances (www.home-connect.com).

Before collecting wastewater from the city area, there is also the stage of storing wastewater (including rainwater, greywater), which is then sent for treatment or pre-treatment on site and reused. Greywater, including used water, e.g., from washbasins, can be directed to flushing toilets after pre-treatment. Similarly, rainwater, after initial treatment, can also be used to flush toilets. Such solutions require dual installation. They are used not only in housing construction, but also in public utility buildings—incl. in hotels (www.greenwatersolutions.pl) or sports facilities. For example, in Poland, rainwater is used to flush toilets at stadiums in Warsaw, Gdansk, and Wroclaw (Kryszyńska et al., 2019, polsatplusarenagdansk.pl, www.pgenarodowy.pl). It should be added that in industrial plants, when closed circuits are used, used water is also collected and treated. Noteworthy are also solutions in the form of integrated sanitary utensils—urinals integrated with the washbasin or washbasins integrated with the flushing tanks. In these solutions, the used water from hand washing is reused to flush the toilet or urinal. There are also rainwater collection systems in buildings and on properties—directly at the point of rainfall. This water is then stored and used, for example, for watering plants (i.e., for supplying groundwater). Rainwater, after collection, is also sometimes directed, instead of to the sewage system, to blue-green infrastructure facilities in the city (e.g., to rain gardens). Smart solutions may be used here in the future, among others, in order to avoid overflow of rain gardens (in case of heavy rainfall) or excessive water retention—as this may result in the development of mosquitoes.

The last steps include collecting sewage and rainwater and directing them to a sewage treatment plant, and then treatment. After wastewater treatment, water renewal processes are applied. In wastewater treatment and water renewal, as with water purification, intelligent solutions can be used to monitor, check, and control systems, allowing for minimizing the use of chemicals and energy. In the case of generated waste—the recovery of raw materials is extremely important.

5 Circular Water Management Projects and Practices in Smart City

5.1 xRs Approaches

The transformation of the European economy into a circular economy model requires more sustainable practices in resource and waste management (Commission of European Communities, 2018; Smol et al., 2020). In practice, waste management already uses methods of prevention such as 2Rs (reduce and reuse), 3Rs (reduce, reuse, and recycling), or 4Rs (reduce, reuse, recycling, and recovery) (Smol et al., 2020). The International Water Association (IWA) has developed the 5Rs approach to water management, enriching these 4Rs with an action such as restore (WBCSD, 2017). The idea of 5R also appears in the context of managing urban water resources—recover, reduce, recycle, resource, and reuse (Zhao et al., 2021). There was also a proposal to extend the aforementioned xRs techniques in the scope of the water and sewage sector by two aspects: reclamation (removal) and rethink (Smol et al., 2020). The implementation of these activities will enable the introduction of the CE idea in a smart city.

5.2 Practices and Projects in Smart City

The first and most important practice is the implementation of Smart Water Management (SWM) in cities. SWM uses information and communication technology (ICT) with real-time data and responses as an integral part of addressing water management challenges (www.iwra.org). It requires integration of systems and a set of measures to monitor, control and regulate the use and quality of water resources, as well as the maintenance of related equipment (www.digiteum.com). A wide range of hardware and software instruments connecting people to water systems include sensors, smart meters and monitoring concentrators, data processing and visualization tools, GIS and satellite maps, actuators, and web and mobile controls (www.digiteum.com, www.iwra.org). Individual countries are at different stages of advancement in implementing SWM. The leaders in this area include among others Singapore and the Netherlands (www.xylem.com). International Water Resources Association (IWRA) is involved in a project to promote the continued implementation of SWM (www.iwra.org).

Among the activities carried out for the implementation of circular water management in smart cities, one can distinguish such basic directions as limiting water losses through leaks, minimizing water consumption and effective water management, including rainwater management and control of illegal discharges to sewage systems (datasmart.ash.harvard.edu, hub.beesmart.city). Numerous solutions are already available in these areas, such as a water leakage detection system using IoT-enabled sensors to locate and assess water leaks in real time, or a system using a network of sensors to monitor municipal sewage systems (hub.beesmart.city). Other solutions are in the research phase—like the system described by Figueiredo et al. (2021) being a digital solution for monitoring and controlling the water distribution system in real time. There is also a practice of constant access of consumers to data on their water consumption thanks to the Internet of Thing IoT technology (hub.beesmart.city, www.mksmart.org).

For buildings, mention should be made of the Building Management Systems (BMSs) used, also known as Building Control Systems (BCSs). These are building automation systems that enable monitoring and management of all devices and systems in the building, as well as its surroundings, and transforming the building into a Smart House facility. Within the functional area of the discussed tool, i.e., the BMS safety system, there is a flood protection subsystem. Based on information from the sensor, this subsystem signals the appearance of water on the floor, enabling quick action to reduce material losses. Within another functional area, there is a subsystem for monitoring energy and utilities consumption, which provides users with access to information on historical, current, and forecast water consumption, which allows for the identification of potential areas for changes (Moskwa & Dechnik, 2020). It is worth adding that some anti-flood devices have the option of being connected to central building control devices (www.instalacjebudowlane.pl).

Greywater recovery systems are also used within the buildings. An example is an installation used in a hotel facility in Zakopane, Poland. The system is equipped with smart technology, allowing for remote supervision and control by specialists (www.greenwatersolutions.pl).

It should also be mentioned that numerous studies are conducted on monitoring and controlling rainwater harvesting systems, the examples of which are the works: (Bhosale et al., 2020; Oberascher et al., 2021; Xu et al., 2020). There are also solutions available on the market in this respect, an example of which can be the system described on (www.raingrid.com).

Research on the development of a smart pipe is also carried out. The result of the project is to be the development of a system based on data readings from sensors for continuous monitoring of pipelines (www.shmsystem.pl).

Equipment dedicated to households is extremely important. Single device is low—efficient, but with the effect of scale—a lot can be achieved. In addition, starting with smaller solutions can help engage residents and encourage further action in this regard. Among household appliances, smart washing machines and dishwashers must be mentioned.

Smart washing machines are technologically advanced devices with Wi-Fi support. They allow for remote starting, pausing, and scheduling washing, as well as receiving notifications about finished work or device failure. Thanks to the use of sensors, technology based on fuzzy logic and other solutions, they allow for automatic dosing of detergents, selection of the amount of water, or setting the washing time. They also take into account the level of soiling of the laundry, water hardness, the type of detergent or the type of load in the drum and its quantity (Spernacka, 2016, www.bosh-home.pl, www.samsung.com). These functions allow to save water, energy, and detergents in households (while maintaining the expected results in the form of effectively washed fabrics).

Dishwashers belong to the same group of appliances as the aforementioned washing machines. Smart devices from this group can select their settings or the appropriate program depending on the type of dishes, their quantity and degree of soiling-saving water and energy. Thanks to automation, they can change, among others temperature and speed of washing adjusting them to the load (www.home-connect.com, www.pocket-lint.com).

An interesting project that has received funding from the European Union for development is a smart shower. The innovative shower panel includes an intelligent drain cover that allows water and energy to be recycled. Water is taken from the shower floor, treated with filters and UV radiation, and then mixed with hot tap water to maintain the desired temperature. The solution can be used in place of a standard shower (cordis.europa.eu).

A useful solution also seems to be a water management system in the household (in terms of security) consisting of a base and sensors. The base checks water pipes for leaks, detects increased consumption, automatically cuts off the water in the event of a threat, and sends information to a smartphone. The sensors will inform about too high or too low air humidity and activate an alarm if there is a risk of water-related losses (www.hansgrohe.com, www.red-dot.org).

It is also worth mentioning the sanitary flushing cistern for irregularly used buildings such as hotels, schools, or hospitals. This solution, which can be operated through a smartphone application and connection to the building management system, prevents water stagnation in the installation and is reliable when changing the water in the installation. The temperature and volumetric flow rate sensor minimizes water consumption (www.geberit.pl).

6 Perspectives for the Implementation of Circular Water Management in Smart Cities of the Future

A SWOT analysis was carried out to assess the perspectives for the implementation of circular water management in smart cities of the future. Based on the analysis, the strengths and weaknesses as well as the existing opportunities and threats were listed. The results are presented in Table 1.

There are many strengths of implementing and developing circular water management in smart cities worth mentioning, but one of the most important is the possibility of saving water. It generates financial benefits related to reduced water consumption, mentioned among the factors driving water-smart CE by work (Salminen et al., 2022). It should be added that these solutions are usually good for the environment, because according to (Smol et al., 2021) the transformation towards CE is a response to the global environmental crisis. Among the weaknesses, there are undoubtedly the investment and operating costs related to recycling. According to (Andrews et al., 2011), in the case of industry, often the economics determine the viablity of recycling. For the implementation of circular water management with the use of smart city solutions, many potential opportunities can be mentioned, although the most important are the existing and adopted policies, programs, and regulations. CE regulatory policy packages and instruments play a key role in supporting the desired transition towards circularity (Salminen, 2022). The role of universities, which create opportunities to implement CE both through research and education, is also of considerable importance (Salas et al., 2021). The drivers also include the availability of public finance for the implementation of CE (Saliminen et al., 2022), which can also be taken as an opportunity. In turn, for threats, apart from war or uncertain geopolitical situation and economic crisis, it is necessary to mention possible cyber-physical attacks (Oberascher et al., 2021), which is related to solutions for smart cities. (Liu et al., 2021) also points to the lack of awareness of city dwellers about environmental issues and the benefits of CE. An important threat is the issue that points out work (Salminen et al., 2022). It highlights the risks that CE may pose to the environment (water-related ecosystems). In applications such as use of secondary water for irrigation, contaminants may pose a risk to the quality of surface and groundwater (Salminen et al., 2022).

A detailed analysis of drivers and barriers (cultural, regulatory, and market) for water-smart CE identified by the Finnish interviewees representing actors who develop, use or both develop and use solutions for a water-smart CE can be found in work (Salminen et al., 2022). Analyzing the SWOT method used for the CE implementation in the field of raw materials recovery sector (Smol et al., 2021), some analogies can also be found to circular water management in smart cities. Also (Liu et al., 2021) compiled (based on the literature) a preliminary list of challenges and barriers to CE implementation in SWM.

Table 1 SWOT analysis of implementing and developing circular water economy in a smart city—own elaboration based on own analysis and literature

Strengths	*Weaknesses*
• allows to save raw materials/resources (water, but also energy, chemicals), mainly through their efficient use and avoidance of losses • actions and activities (related, inter alia, to the operation) are facilitated by the use of new technologies/techniques • provides faster information and therefore allows for faster (earlier) decision-making and reaction • the solutions are usually good for the environment and climate • the operation is usually intuitive and very simple • the collected data from different areas facilitate the integration of water management	• sometimes there is a need for training in the use of hardware and software instruments • dependence on energy supplies (including battery power supply) and Internet access • the necessity to invest in resources (purchase or modernization)—investment costs • operating costs (infrastructure maintenance) • high costs of wastewater treatment (to the drinking water quality) • high energy demand for wastewater treatment (to the drinking water quality) • difficulties in implementing solutions to existing buildings, existing infrastructure • lack of technical standards and regulations (in some areas)
Opportunities	*Threats*
• existing and future policies, strategies, and programs may facilitate the implementation of CE in a smart city (Salminen et al., 2022) • new legal regulations • the ongoing climate change (may increase the demand and accelerate the implementation of CE, and smart city solutions) • public education, raising awareness of society (Salas et al., 2021) • further development of IoT, ICT, • minimizing components, increasing the capacity and efficiency of system components, other improvements • seeking savings by consumers/ enterprises • ongoing research and projects (Salas et al., 2021) • subsidies (Salminen et al., 2022) • competitions promoting the CE idea in a smart city • rating systems (e.g., buildings, fixtures) • environmental management systems introduced • use of renewable energy • existing good practices that may be followed	• war • unstable geopolitical situation • economic crisis • hacking attacks, breaking into systems, cyber-physical attacks (Ma, 2021; Oberascher et al., 2021) • unauthorized use of obtained data, information leakages (Ma, 2021) • residents' concerns about their over-monitoring • lack of social acceptance for recovered water (Liu et al., 2021) • technological problems in obtaining the required degree of wastewater treatment (to the quality of drinking water) • lack of awareness in the society regarding the need to implement CE (Liu et al., 2021), and smart city solutions • lack of funds (due to redirection of them to other priority objectives) • climate change—hindering activities (drought and torrential rainfall) • CE impact on the environment (Salminen et al., 2022)

7 Conclusions

The circular water management implementation and development stage with the use of solutions characteristic of smart cities was analyzed. It is different all over the world. On the basis of the work carried out, the following conclusions can be drawn:

- currently, there are many interesting and innovative solutions that allow the implementation of the assumptions of the circular economy in smart cities
- the implementation of circular water management solutions with the use of solutions characteristic of smart cities has both strengths and weaknesses
- particular attention should be paid to mitigating weaknesses and developing solutions and mechanisms to eliminate the impact of threats on the process of implementing circular water management in smart cities.

The paper proposes a more detailed model of the CWM in smart cities, including, inter alia, inclusion of water from air humidity. Currently, it seems crucial to educate the society to raise awareness of the need to introduce solutions such as CE, and the idea of a smart city. At the same time, further works should be carried out on the technical and technological development of the existing solutions. Legal and economic solutions should also be developed to facilitate the financing and maintenance of such solutions.

References

Ahmad, T., Ahmad, K., & Alam, M. (2016). Characterization of water treatment plant's sludge and its safe disposal options. *Procedia Environmental Sciences, 35*, 950–955, ISSN 1878-0296,https://doi.org/10.1016/j.proenv.2016.07.088

Ahvenniemi, H., Huovila, A., Pinto-Seppä, I., & Airaksinen, M. (2017). What are the differences between sustainable and smart cities?, *Cities, 60*, Part A, pp. 234–245, ISSN 0264-2751,https://doi.org/10.1016/j.cities.2016.09.009

Andrews, M., Berardo, P., & Foster, D. (2011). The sustainable industrial water cycle—a review of the economics and approach. *Water Supply, 11* (1), 67–77. https://doi.org/10.2166/ws.2011.010

Bhosale, M., Ukhalkar, P., & Kadam, M. (2020). Smart rain water harvesting system (SRWHS) using IoT enabled smart kit for better tracking and management in Pune City. *Allana Inst of Management Sciences, Pune, 10*(1), 1–3.

Bielawska, M. (2019). Warsztaty z zakresu ochrony środowiska w kontekście. *Human Smart Cities*. Available at: https://www.popt.gov.pl/media/58944/Ochrona_srodowiska_i_ekologia.pdf. Accessed 19 April 2022.

Camero, A., & Alba, E. (2019). Smart City and information technology: A review. *Cities, 93*, 84–94, ISSN 0264–2751, https://doi.org/10.1016/j.cities.2019.04.014.Chapman; W. (2019). 10 Cities Most At Risk of Running Out of Water, U.S. News & World Report, Cities, https://www.usnews.com/news/cities/slideshows/10-cities-most-at-risk-of-running-out-of-water. Accessed 19 July 2022

Cobo, S., Dominguez-Ramos, A., & Irabien, A. (2018). From linear to circular integrated waste management systems: A review of methodological approaches. *Resources, Conservation and Recycling, 135*, 279–295, ISSN 0921–3449, https://doi.org/10.1016/j.resconrec.2017.08.003

Commission of European Communities. (2018). Communication No. 29, 2018. Monitoring framework for the circular economy (COM no. 29, 2018).

Ćmielewski, M., Głowcki, J., Hausner, J., Kudłacz, M., Kutek, K., & Ramm, K. (2021). Water City Index 2021. Ranking efektywności wykorzystania zasobów wody w polskich miastach, 2021, Fundacja GAP, Open Eyes Economy Summit, Uniwersytet Ekonomiczny w Krakowie, Arcadis sp. z o.o., https://oees.pl/water-city-index-2021. Accessed 19 April 2022.

cordis.europa.eu/article/id/422217-smart-shower-aims-to-get-europe-cleaner-in-a-greener-way. Accessed 29 April 2022.

datasmart.ash.harvard.edu/news/article/come-drought-or-high-water-728. Accessed 29 April 2022.

Figueiredo, I., Esteves, P., & Cabrita, P. (2021). Water wise—A digital water solution for smart cities and water management entities. *Procedia Computer Science, 181*, 897–904, ISSN 1877–0509, https://doi.org/10.1016/j.procs.2021.01.245

Fundación Telefónica. (2011). Smart cities, a first step towards the Internet of Things, 2011. Available at: https://www.telefonica.com/en/communication-room/blog/presentation-of-the-report-smart-cities-a-first-step-towards-the-internet-of-things. Accessed 19 April 2022.

Górka, J., Cimochowicz-Rybicka, M., & Poproch, D. (2022). Sludge management at the Kraków-Płaszów WWTP—case study. *Sustainability, 14*, 7982. https://doi.org/10.3390/su14137982

gozwpraktyce.pl/slownik/cradle-to-cradle. Accessed 23 April 2022, Accessed 29 April 2022.

hub.beesmart.city/en/solutions/the-need-of-smart-cities-to-get-smart-about-water. Accessed 29 April 2022.

Interdisciplinary advisory team for the climate crisis at the President of the Polish Academy of Sciences, Communication 01/2020 on climate change and water management in Poland (2020), http://eko.org.pl/index_news.php?dzial=2&kat=20&art=2419. Accessed 19 July 2022.

Juma, M., & Shaalan, K. (2020). Cyberphysical systems in the smart city: challenges and future trends for strategic research. In: Intelligent Data-Centric Systems, Swarm Intelligence for Resource Management in Internet of Things, Academic Press, pp. 65–85, ISBN 9780128182871, https://doi.org/10.1016/B978-0-12-818287-1.00008-5.1.2

Kryszyńska, H. (ed.), Wiełądek K., Wolańska K., Świerkula E. (2019) Poradnik dla zarządców nieruchomości—Eko-lokator, Fundacja Instytut na rzecz Ekorozwoju.

Lee, S. M., & Trimi, S. (2018). Innovation for creating a smart future. *Journal of Innovation & Knowledge, 3*, 1–8.

Lim Y., Edelenbos, J., & Gianoli, A. (2019). Identifying the results of smart city development: Findings from systematic literature review. *Cities, 95*, 102397, ISSN 0264-2751.https://doi.org/10.1016/j.cities.2019.102397

Liu, Q., Yang, L., & Yang, M. (2021). Digitalisation forWater sustainability: Barriers to implementing circular economy in SmartWaterManagement. *Sustainability, 13*, 11868. https://doi.org/10.3390/su132111868

Ma, C. (2021). Smart city and cyber-security; technologies used, leading challenges and future recommendations. *Energy Reports, 7*, 7999–8012, ISSN 2352–4847, https://doi.org/10.1016/j.egyr.2021.08.124

Merli, R., Preziosi, M.,& Acampora, A. (2018). How do scholars approach the circular economy? A systematic literature review. *Journal of Cleaner Production, 178*, 703-722, ISSN 0959-6526https://doi.org/10.1016/j.jclepro.2017.12.112

Moskwa, S., & Dechnik, M. (2020). BMS—system zarządzania budynkami. Na czym polega system automatyki BMS?, MuratorPlus. Available at: https://www.muratorplus.pl/technika/ochrona-osob-i-mienia/bms-systemy-zarzadzania-budynkami-aa-w96W-ZJt3-Axkd.html

Nachlik, E., Januchta-Szostak, A., Kundzewicz, Z., Okruszko, T., Ramm, K., Rosiek, K., Wawer, R., & Zaleski, J. (2020). Alert wodny 8, Zintegrowane podejście w gospodarowaniu wodą, GAP, Open Eyes Economy Summit, 2020. Available at: https://oees.pl/alerty-eksperckie/?doing_wp_cron=1650263610.7226409912109375000000. Accessed 15 April 2022

Oberascher, M., Dastgir, A., Li, J., Hesarkazzazi, S., Hajibabaei, M., Rauch, W., & Sitzenfrei R. (2021). Revealing the challenges of smart rainwater harvesting for integrated and digital resilience of urban water infrastructure. *Water, 13*, 1902, https://doi.org/10.3390/w13141902

OECD. (2012). OECD Environmental Outlook to 2050, OECD Publishing.

Praharaj, S., & Han, H. (2019). Cutting through the clutter of smart city definitions: A reading into the smart city perceptions in India, City. *Culture and Society, 18*, 100289, ISSN 1877-9166,https://doi.org/10.1016/j.ccs.2019.05.005

polsatplusarenagdansk.pl/odpowiedzialnosc-spoleczna/Zielony-stadion,a,3036. Accessed 29 April 2022.

Rahaman, M. M., & Varis, O. (2005). Integrated water resources management: Evolution, prospects and future challenges, Sustainability: Science. *Practice and Policy, 1*(1), 15–21. https://doi.org/10.1080/15487733.2005.11907961

Ratnaweera, H., & Fettig, J. (2015). State of the art of online monitoring and control of the coagulation process. *Water, 7*(11), 6574–6597. https://doi.org/10.3390/w7116574

Salas, D. A., Criollo, P., & Ramirez, A. D. (2021). The role of higher education institutions in the implementation of circular economy in Latin America. *Sustainability, 13*(17), 9805. https://doi.org/10.3390/su13179805

Salminen, J., Määttä, K., Haimi, H., Maidell, M., Karjalainen, A., Noro, K., Koskiaho, J., Tikkanen, S., & Pohjola, J. (2022). Water-smart circular economy—conceptualisation, transitional policy instruments and stakeholder perception. *Journal of Cleaner Production, 334*, 130065, ISSN 0959–6526, https://doi.org/10.1016/j.jclepro.2021.130065

Sharif, R. A., & Pokharel, S. (2022). Smart city dimensions and associated risks: review of literature. *Sustainable Cities and Society, 77*, 103542, ISSN 2210-6707,https://doi.org/10.1016/j.scs.2021.103542

Silva, G. M. E., Campos, D. F., Brasil, J. A. T., Tremblay, M., Mendiondo, E. M., & Ghiglieno, F. (2022). Advances in technological research for online and in situ water quality monitoring—A review. *Sustainability, 14*(9), 5059. https://doi.org/10.3390/su14095059

Smol, M., Christian, A., & Preisner, M. (2020). Circular economy model framework in the European water and wastewater sector. *Journal of Material Cycles and Waste Management, 22*, 682–697. https://doi.org/10.1007/s10163-019-00960-z

Smol, M., Marcinek, P., & Koda, E. (2021). Drivers and barriers for a circular economy (CE) implementation in Poland—A case study of raw materials recovery sector. *Energies, 14*, 2219, https://doi.org/10.3390/en140822. Accessed 29 April 2022.

Spernacka, M. (2016). Pralki z funkcjami smart—popularne modele. Available at: https://agdlab.pl/artykuly/Pralki.z.funkcjami.smart-popularne.modele,9427. Accessed 29 April 2022.

The Committee of Digital and Knowledge-based Cities of UCLG. (2012). Smart cities study: International study on the situation of ICT, innovation and Knowledge in cities Bilbao.

The Committee of Digital and Knowledge-Based Cities of UCLG chaired by Juan Mari Aburto, Mayor of the City of Bilbao. (2017). Smart cities study 2017. International study on the situation and development of ICT, innovation and knowledge in cities, Bilbao.

Thier, A. (2016). Woda jako zasób naturalny oraz surowiec w ujęciu teorii ekonomii. *Hydrological Cycle in the Natural Environment and Its Implications, Aura, 7–8*, 3–5. https://doi.org/10.15199/2.2016.7-8.1

United Nations. (1987). *Report of the World Commission on Environment and Development: Our Common Future, tzw* (p. 1987). Oxford University Press.

United States Environmental Protection Agency. (1978). Guidance for Planning the Location of Water Supply Intakes Downstream from Municipal Wastewater Treatment Facilities, pp. 1–239.

Ustawa z dnia 4 września. (1997). r. o działach administracji rządowej (Act on government administration departments), Dz.U. 1997 nr 141 poz. 943.

WBCSD. (2017). Business guide to circular water management: spotlight on reduce, reuse and recycle, Switzerland. Available at: https://docs.wbcsd.org/2017/06/WBCSD_Business_Guide_Circular_Water_Management.pdf. Accessed 29 April 2022.

Willet, J., Wetser, K., Vreeburg, J., & Rijnaarts, H. H. M. (2019). Review of methods to assess sustainability of industrial water use. *Water Resources and Industry, 21*, 100110, ISSN 2212-3717, https://doi.org/10.1016/j.wri.2019.100110

www.bosch-home.pl/produkty/pranie-i-suszenie/pralki. Accessed 29 April 2022.

www.digiteum.com/smart-water-management-iot. Accessed 29 April 2022.

www.geberit.pl/planowanie-instalacja/wiedza/higiena-wody-pitnej. Accessed 29 April 2022.

www.greenwatersolutions.pl/aktualnosci/systemy-szarej-wody-w-hotelarstwie. Accessed 29 April 2022.

www.hansgrohe.com/smartliving/pontos/products. Accessed 29 April 2022.

www.home-connect.com/global/smart-home-appliances/dishwashers#/Togglebox=10160507-10160303-1. Accessed 29 April 2022.

www.home-connect.com/global/smart-home-appliances/washers. Accessed 19 April 2022.

www.instalacjebudowlane.pl/5328-25-77-urzadzenia-przeciwzalewowe-kessel-.html. Accessed 29 April 2022.

www.iwra.org/swm. Accessed 29 April 2022.

www.mksmart.org/water. Accessed 29 April 2022.

www.pgenarodowy.pl/aktualnosci/2612,ekologiczny-stadion-narodowy. Accessed 29 April 2022.

www.pocket-lint.com/smart-home/buyers-guides/146405-best-smart-dishwashers. Accessed 29 April 2022.

www.raingrid.com/argon. Accessed 29 April 2022.

www.red-dot.org/project/pontos-45295. Accessed 29 April 2022.

www.samsung.com/us/washers. Accessed 29 April 2022.

www.shmsystem.pl/smart_pipe. Accessed 29 April 2022.

www.watergen.com/commercial. Accessed 29 April 2022.

www.xylem.com/pl-pl/making-waves/water-utilities-news/5-leaders-in-smart-water-management. Accessed 29 April 2022.

Xu, W. D., Fletcher, T. D., Burns, M. J., & Cherqui, F. (2020). Real time control of rainwater harvesting systems: The benefits of increasing rainfall forecast window. *Water Resources Research, 56*, e2020WR027856, https://doi.org/10.1029/2020WR027856

Zaleski, J., Balcerowicz, M., Dolny, J., Godyń, I., Kozyra, J., Kundzewicz, Z., Majewski, W., Pierzgalski, E., & Wawer, R. (2020). Alert wodny 9, Zarządzanie gospodarką wodną w Polsce, GAP, Open Eyes Economy Summit, 2020. Available at: https://oees.pl/alerty-eksperckie/?doing_wp_cron=1650263610.7226409912109375000000. Accessed 15 April 2022.

Zhou, L., Wang, H., Zhang, Z., et al. (2021). Novel perspective for urban water resource management: 5R generation. *Frontiers of Environmental Science & Engineering, 15*, 16. https://doi.org/10.1007/s11783-020-1308-z

Zvimba, J. N. (2019). Circular economy model for water and wastewater management, UJ—Montipilier workshop: New frontiers in separation processes & membrane development. *Water Research Comission*. Available at: http://www.unesco-simev.org/wp-content/uploads/4-Circular-economy-water-and-ww_John-Zvimba.pdf. Accessed 29 April 2022.

The Influence of Mineral Parameters on the Geochemistry of Heavy Metals in Bottom Sediments

Małgorzata Wojtkowska

Abstract

As a result of increasing environmental pollution, elevated concentrations of heavy metals, including those toxic to aquatic biocenosis, are observed in surface water reservoirs. Heavy metals are introduced to water systems in dissolved and solid form, the final link of their reception is bottom sediments. Accumulation takes place here, which is a specific barrier protecting living organisms from excessive impact of these pollutants. Metals can be present in water in dissolved, colloidal and suspended forms. Depending on many physical and chemical factors, they may remain in the dissolved phase or transition to the suspended phase. It depends on the type of metal, its solubility in water, ability to adsorb on solid particles, complexity, pH and pE of water, temperature, and the ionic strength and chemical composition of the water. Most metals do not stay dissolved in water for a long time, but they undergo numerous processes leading to their deposition in bottom sediments. If aerobic conditions prevail in the overlying water layer, mobile forms associated with carbonates and hydrated iron and manganese oxides predominate in the sediments. Under strongly anaerobic conditions, there are mainly stable forms of metals bound in the form of organo-sulphide and aluminosilicate bonds. The forms of heavy metal binding in bottom sediments depend on their structure and chemical composition, and they remain in dynamic equilibrium with the water phase.

Keywords

Heavy metals • Water • Bottom sediments

M. Wojtkowska (✉)
Warsaw University of Technology, Warsaw, Poland
e-mail: malgorzata.wojtkowska@pw.edu.pl

List of Abbreviations

DOC	Dissolved organic carbon
COD	Chemical oxygen demand
TOC	Total organic carbon
J + LK	Hydrated metal ions and bound into labile complexes
IK	Inert complexes
ISO	International Organization for Standardization
DPASV	Differential pulse inversion voltammetry
I	Ion-exchange fraction
II	Carbonate fraction
III	Associated with iron and manganese oxide fraction
IV	Organic fraction
V	Residual fraction

1 Introduction

Inland surface waters, due to their geographical location, geological substrate, size, water movement, their physical and chemical properties and biological life, belong to very diverse natural environments. Water quality in surface water bodies is significantly affected by human activity. Surface waters are the main receiver of municipal and industrial sewage, surface run-off from agricultural areas and transport routes. The amount of inflowing pollutants, including heavy metals, and the management carried out in the catchment area determine water quality and its appropriate purity class in a given water body or watercourse (Preisner et al., 2020; Wojtkowska, 2020).

There are micropollutants in the aquatic environment, which include heavy metals also known as trace elements. Metals in unpolluted surface water are found in very low concentrations. Heavy metals enter surface waters as a result

M. Smol et al. (eds.), *Water in Circular Economy*, Advances in Science, Technology & Innovation,
https://doi.org/10.1007/978-3-031-18165-8_4

of natural erosion processes, with precipitation rich in dust containing significant amounts of metals, with surface run-off from agricultural areas and from industrialized areas containing elevated metal concentrations. The presence of metals in surface water, which is a potential source of drinking water supply for the population, makes it unsuitable for this purpose (Bain et al., 2020; Zhang & Davison, 2000). Heavy metals in the aquatic environment bioaccumulate in living organisms. The harmfulness of such micropollutants depends on the type of element, the chemical composition of the water, and the type of aquatic organisms inhabiting the water environment (Aderinola et al., 2009). It is now known that to determine the impact of metals on aquatic biocoenoses it is not sufficient to determine their total concentrations, which does not inform about the bioavailability of metals and the resulting risk to living organisms. The bioavailability of metals depends on the forms in which metals occur in a given element of the environment, and speciation studies are used to determine these forms (Wojtkowska & Bogacki, 2012).

Knowledge of the forms in which metals occur in the aquatic environment makes it possible to assess their migration from bottom sediments to water and to analyse their transformation and binding processes. Dissolved forms of metals constantly undergo chemical exchange with suspended solids and bottom sediments, which disturbs their equilibrium in the water–sediment system (Markich et al., 2001; Wojtkowska et al., 2016).

Accumulation of metals in sediment depends on its mineral-organic composition as well as on the physical and chemical conditions at the water–sediment interface. The granulometric composition of the sediment is also important (Wojtkowska & Matula, 2016). It is assumed that fine fractions adsorb metals more strongly and are more easily transported by flowing water.

Studies on bottom sediments, using speciation analysis, taking into account the varied nature of the collected sediments and their stratification, allow characterization of metal combinations in individual fractions and specific chemical compounds. Determination of metal forms facilitates prediction of their bioavailability and the results obtained provide a basis for risk assessment due to the presence of trace elements in the aquatic environment (Olaniran et al., 2013; Slaveykova & Wilkinson, 2005).

At present, it is necessary to conduct research on metal forms, especially in the least recognized area, which is waters and bottom sediments of surface water bodies. At the same time, control of the chemical composition (macro- and micropollutants) of sediments allows us to describe the mechanisms of migration of toxic pollutants in the aquatic environment. A comprehensive assessment of the lability of metals, taking into account the transformation of their forms and the influence of physical and chemical parameters of water and bottom sediments on their migration, is the basis for risk assessment (Olutona et al., 2012; Safoura et al., 2015).

The aim of this study was to assess the degree of migration and accumulation of selected metals in surface waters as a function of selected macroelements present in the waterbed and bottom sediments of a reservoir. In order to achieve this goal, it was necessary to determine the level of contamination of water and bottom sediments with metals, and to determine the forms of occurrence of the metals in the waterbed and bottom sediments.

2 Materials and Methods

Water and bottom sediment samples for the description of heavy metal forms were taken from Lake Zegrzyńskie. This is the eighth largest water retention reservoir in Poland. The reservoir is fed by rivers: 42% by the Narew and 57% by the Bug. One of the main functions of the Zegrzyński Reservoir since 1986 is the supply of drinking water to the northern districts of Warsaw. Other tasks that Lake Zegrzyński performs are regulating the flow of water in the Vistula River through the Żerań Canal, flood protection, power generation, shipping and tourism, and leisure and recreation.

Ten measurement sites were selected in the reservoir, from which water samples were taken from the depth of 0.5 m and at the bottom, as well as samples of bottom sediments. Water analysis included the following parameters: temperature, radiation, colour, turbidity, pH, conductivity, dissolved substances, suspended solids, alkalinity, calcium and magnesium, chlorides, sulphates, dissolved oxygen, phosphorus, ammonium nitrogen, nitrate nitrogen, Kjeldahl nitrogen and Chemical Oxygen Demand (COD). The following tests were carried out in samples of bottom sediments: hydration, dry mass, organic substances, carbonates, pH in water extract, total phosphorus and nitrogen after mineralization. All determinations were made in accordance with current ISO standards.

Metals (Pb, Cd, Cu, Zn, Ni) were determined in water and sediment samples by atomic absorption spectroscopy. The following metal forms were determined in water samples: hydrated metal ions and bound into labile complexes (J + LK) and inert complexes (IK) using the differential pulse inversion voltammetry (DPASV) method. Nickel determination was performed by cathodic adsorption inversion voltammetry. The total heavy metal content in the suspension phase and in the bottom sediment samples was determined after wet digestion with a mixture of concentrated acids (HNO_3, $HClO_4$) in a digestion block.

Tessier's sequential scheme was used to determine the forms of metals present in bottom sediments (Tessier et al., 1979). In this scheme, five metal fractions were distinguished: ion exchange (I), carbonate (II), fraction associated with iron and manganese oxides (III), organic fraction (IV) and residual fraction (V).

Heavy metal content was determined in suspension and bottom sediment samples after mineralization and after sequential extraction by flame atomic absorption spectrometry.

3 Results

The qualitative and quantitative composition of water and bottom sediments was assessed to determine the migration and accumulation of metals in the aquatic environment. The chemical composition of water was variable in time and varied over the study area. The values of such indicators as suspended solids, COD, nitrogen and phosphorus compounds, as well as metals, with the exception of nickel and iron, in extreme cases reached very high concentrations, exceeding the values specified in the legislation in force. From the data obtained, it is clear that there is some differentiation in the values of chemical indicators in the vertical section, between the surface layer and the water bottom layer. The bottom layer of water was characterized by relatively higher values of such quality indicators as suspended solids, COD, forms of nitrogen and phosphorus compounds, and heavy metals. Other indicators remain at similar levels. In the case of oxygen, higher concentrations are found in the surface layer of lake water.

The study also revealed quite specific horizontal spatial distribution of values of the analysed chemical indicators of water. Mean values of pH were at a very similar level at all measurement sites. In the case of dissolved oxygen, there were considerable differences especially between measurement sites selected at the river inflow, in the main stream of the reservoir, in the stagnant area and at the lake outflow. The greatest differentiation was observed for DOC, dissolved phosphate and total phosphorus.

Analysis of the surface layer of the bottom sediments showed seasonal variability of the chemical composition in the annual time section and considerable spatial variability (measurement sites) in the area of Lake Zegrzyńskie. On the basis of the obtained data, a seasonal distribution of mean values of the percentage share (%) of particular chemical indicators of the quality of the examined bottom sediments at the measurement sites was determined (Table 1).

The analysed bottom sediments showed a near-neutral reaction varying from 6.8 in spring to 7.6 in autumn. The percentages of the determined chemical parameters of bottom sediments showed a similar spatial distribution for the spring and summer periods. In these periods, the highest percentages of chemical parameters such as carbonates, iron and manganese oxides, total organic carbon, total nitrogen and total phosphorus were found in the chemical composition of bottom sediments. The maximum values were recorded at sites in the stagnant part of the channel, while the minimum values were found in the sediments of the river mouth in summer. In autumn, relatively high variability was observed in the content of carbonates in bottom sediments, with the maximum at the sites in stagnant areas and the minimum at inflows and outflows. The content of iron and manganese oxides was less differentiated, with only a higher proportion occurring in winter at the sites in the central part of the reservoir. Some differentiation was observed in the case of percentages of organic matter in bottom sediments, especially in autumn in the whole area of the reservoir. Percentages of total nitrogen and phosphorus in bottom sediments at all sites were quite similar in spring and distinctly different in summer. In summer, relatively high concentrations of total nitrogen were recorded at sites in the main basin of the reservoir. The contribution of total phosphorus, on the other hand, was differentiated in autumn and equalled and was very low in winter.

A seasonal analysis of the chemical composition of the bottom sediments showed that the sediments deposited in the stagnant part generally showed a higher percentage contribution of the analysed chemical indicators, reaching maximum values most frequently in spring and summer. Generally, during spring and summer there were quite large variations in the percentage contribution of most chemical components of bottom sediments.

The analysis of heavy metal concentrations in water showed particularly significant differentiation in horizontal spatial distribution. The course of this distribution shows a

Table 1 Percentage of chemical indicators in bottom sediments (%)

Season/indicator	Wiosna	Lato	Jesień	Zima
pH	6,8	7,0	7,6	6,95
HCO_3^-	4,81	14,27	11,70	10,68
Oxides Fe i Mn	17,42	15,46	5,31	8,04
TOC	12,08	19,82	9,54	6,63
Nog	2,4+	3,58	0,15	0,13
Pog	2,89	0,56	7,15	0,19

Table 2 Seasonal changes in concentrations of dissolved metals in water layers [μgZn/l]

Layer	Zn		Cd		Cu		Ni		Pb	
	Surface	Bottom	Surface	Bottom	Surface	Bottom	Surface	Bottom	Surface	Bottom
Spring	28.4	51.0	1.38	3.67	16.2	35.9	9.3	34.6	4.1	7.1
Summer	39.2	50.4	8.13	15.14	15.4	19.1	53.2	80.5	14.5	18.7
Autumn	37.7	104.1	7.51	18.17	13.5	29.8	32.0	65.5	16.3	16.3
Winter	34.2	61.8	0.40	1.72	7.95	27.8	32.2	140.3	8.5	12.7

tendency of increasing metal concentrations, especially of zinc and lead, in the central, stagnant zone and a clear differentiation of concentrations between the surface and bottom layers of the lake water. The five analysed metals showed mainly a tendency to remain in higher concentrations in the supra-bottom water layer. The smallest differentiation in concentrations between the surface and bottom layers was observed for copper.

The contents of heavy metals underwent seasonal changes in the annual time section and were differentiated in the vertical and horizontal spatial arrangement of the water tone of Lake Zegrzyńskie. The data presented in Table 2 show the image of the seasonal distribution of average values of dissolved heavy metal concentrations in individual water layers.

From the data presented above, some general trends in the behaviour of dissolved forms of heavy metals in the water depth were apparent. Higher concentrations of these metals were found in the supra-bottom layer of the water column. The occurrence of maximum and minimum values of metals in both layers of the water tone in the seasonal pattern was quite different:

- in the surface layer: maximum values occurred in summer (Zn, Cd, Ni), spring (Cu) or autumn (Pb); minimum values occurred in winter (Cd and Cu) and spring (Zn, Pb, Ni);
- in the superficial layer: maximum values occurred in autumn (Zn, Cd, Pb), spring (Cu) or winter (Ni); minimum values occurred in winter (Zn and Cu) and spring (Cd, Pb, Ni).

The calculated mean values of concentrations of each metal for each season made it possible to determine the general trends of changes in the concentration level of each metal in lake bottom sediments over the annual cycle.

In terms of content in bottom sediments, the studied heavy metals behaved differently in the annual time section. Similarities were found for zinc and nickel and cadmium and copper concentrations. In the case of zinc and nickel, there were differences in values of concentrations in bottom sediments, whereas the variation of concentrations in seasons was identical (Table 3).

Cadmium and copper differed in their concentrations in bottom sediments, while the distribution of concentrations during the seasons was quite similar: winter > summer > spring > autumn. The seasonal distribution of lead concentrations in bottom sediments followed the order: spring > summer > autumn > winter. In terms of concentration levels in the bottom sediments of lake, the heavy metals studied can be ranked as follows: Zn > Cu > Pb > Ni > Cd.

The application of speciation analysis made it possible to differentiate the forms of the studied metals occurring in surface waters. Dissolved metal species determined in water samples include hydrated ions and labile complexes, neutral complexes and insoluble species bound to suspensions and colloid (Fig. 1).

Zinc in the waterbed was bounded mainly in suspended form and constituted 98–99% of its total content. In the surface layer, zinc was present in 36.46% and in the supra-bottom layer in 63.54% of the total amount of this metal in the waterbed.

The percentages of J + LK and inert complexes IK in each season were as follows:

	Spring	Summer	Autumn	Winter
J + LK [%]	40.0	44.6	45.6	48.1
IK [%]	60.0	55.4	54.4	51.8

Table 3 Seasonal changes in concentration metals in sediments [mg/kg d.m.]

	Zn	Cd	Cu	Ni	Pb
Spring	98.9	11.28	58.47	24.75	44.70
Summer	102.37	9.25	61.99	33.70	43.34
Autumn	76.30	12.94	39.92	19.48	37.92
Winter	65.87	26.72	64.25	17.59	29.47

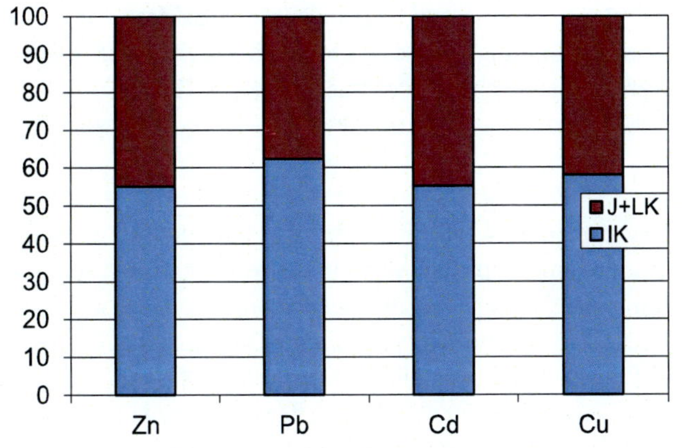

Fig. 1 Percentage of J + LK and IK fractions of heavy metals in water

The data presented show that the proportion of hydrated ions and labile zinc complexes had an increasing trend over the annual time section, from 40.0% in the spring season to 48.1% in the winter season.

The percentages of cadmium in suspended form were of the order of 96–99%. The content of cadmium in the surface layer of water constituted 25.17%, and in the upper layer 74.83% of the total amount of this metal in water.

Mean values of concentrations of dissolved forms of cadmium varied in seasons and measurement sites. The average percentages of J + LK and IK in spring, summer, autumn and winter were

J + LK:	41.7	41.8	45.3	42.7
IK:	58.3	58.2	54.7	57.3

The above data show that the percentage of J + LK cadmium in the water was in the three seasons (spring, summer and winter) at a similar level of about 42%, in the autumn it was higher (45.3%).

Suspended nickel percentages were in the range of 96–99% of the total metal content in water. The content of nickel (dissolved and suspended forms) in the surface layer was 34.59%, in the superficial layer 65.41% of the total amount of this metal in water. Generally, the proportion of dissolved nickel forms in the water of Lake Zegrzyńskie did not exceed 4%. Mean values of dissolved nickel form concentrations were seasonally and spatially differentiated at measurement sites. The average percentages of J + LK and IK in spring, summer, autumn and winter were

J + LK	37.7	41.3	46.3	38.6
IK	62.3	58.7	53.7	61.4

The percentages of the J + LK forms of nickel in the water body were similar in spring and winter and significantly increased in summer and autumn (maximum values).

The percentages of suspended lead ranged from 93.8% to 99.4% and varied seasonally and spatially from 0.6% to 6.2%. The content of lead in dissolved and undissolved form in the surface layer was 26.06% and in the supra-bottom layer 73.94% of the total amount of Pb in water. The share of dissolved forms of lead in the water depth was slightly above 5%. Mean values of dissolved forms of lead concentrations varied seasonally and spatially at the measurement sites. The average percentages of J + LK and IK lead in spring, summer, autumn and winter were

J + LK	32.8	33.3	44.5	35.2
IK	67.2	66.7	55.5	64.8

The percentages of the J + LK forms of lead in the water tone were at similar levels in spring and summer and maximum in autumn.

The total content of copper (dissolved and suspended forms) in the water in the surface layer was 29.83% and in the upper layer was 70.17%.

From the above data, it can be seen that the proportion of dissolved forms of copper in the water table was at a similar level in both layers. The level was relatively higher compared to the other heavy metals studied. Mean values of dissolved forms of copper concentrations were seasonally and spatially differentiated at the measurement sites.

The obtained concentrations showed a low content of dissolved forms of zinc in the water depth up to 1.7% in the bottom layer of the water. The greatest differences in these active forms were observed for cadmium, which reached up to 8.6% in the surface layer and only 1.6% in the bottom layer, and for lead, which reached the highest values in the bottom layer of water (5.2%) and the lowest in the surface layer (1.6%). Copper and nickel were characterized by equal proportions of dissolved forms in both water layers, reaching for Cu about 7% and for Ni 3.6%, respectively. This indicates a very different behaviour of the studied metals in the water depth, resulting from the different properties of the elements in aqueous solution of different composition.

The studied metals showed varying susceptibility to binding with basic chemical components of bottom sediments, with this property changing to a greater or lesser extent in the area of bottom sediment deposition. Quite significant differences were also found in the values of heavy metal percentages in individual seasons of the annual time section (Fig. 2).

The affinity of the studied heavy metals towards specific fractions of the bottom sediments varied and was metal specific.

Zn

The percentage of zinc was highest in the third fraction, which means that this metal was most strongly bound by hydrated iron and manganese oxides. In this fraction was on

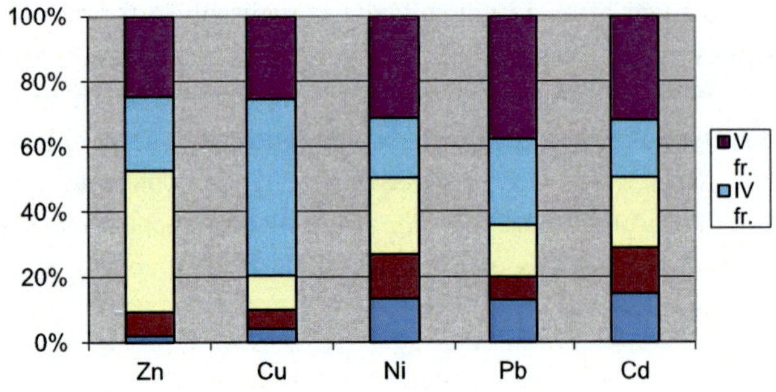

Fig. 2 Percentage of metals in bottom sediments

average bound in 43.2%, ranging from 31.1% to 52.2%. Zinc was the least bound by the ion-exchange fraction. The percentage in this fraction was on average 2.0% with a range of 1.1%–3.9%. On average 22.8% of zinc was bound to organic matter, 24.7% to residual fraction and 7.3% to carbonates. The different bottom sediment fractions in terms of their zinc-binding capacity can be ranked as follows:

$$\text{III} > \text{V} > \text{IV} > \text{II} > \text{I}$$

Considering the proportion of zinc in the five fractions, it can be concluded that the level of Zn accumulation in bottom sediments depends mainly on their content of iron and manganese oxides, natural minerals and organic matter. The study on seasonal character of zinc speciation, in general, confirmed the leading role of fraction III in Zn fixation in bottom sediments. In the autumn–winter season, the contribution of organic matter to zinc fixation in bottom sediments was significant and the percentages were, respectively, 47.8 and 45.9%.

Cd

The dominant fraction of cadmium in bottom sediments was the residual fraction. The percentage of cadmium in this fraction was on average 31.9%, ranging from 23.0 to 36.3%. In fractions I and II, cadmium was bound to a similar extent (14.9% and 14.0%). The role of hydrated iron and manganese oxides was less important with an average percentage of 21.7%. The share of cadmium in fraction IV (17.5%) indicates a relatively small contribution of organic matter to cadmium binding. However, in sediments rich in organic matter, the percentage of cadmium in fraction IV increased to 37.5%. The individual fractions of bottom sediment in terms of their ability to bind cadmium can be ranked as follows: V > III > IV > II = I. The structure of cadmium binding in individual bottom sediment fractions, as presented above, changes seasonally. In spring and summer season, fraction I (ion-exchangeable) becomes dominant, while cadmium contribution to fraction V (residual) at that time was only 13.1%. In the winter season, the role of hydrated iron and manganese oxides in the cadmium-binding process clearly increased (25.9% on average) at the expense of fractions I and II.

Ni

The percentages of nickel in the five bottom sediment fractions form a pattern very similar to that of cadmium. The dominant fraction was the residual fraction, in which Ni percentage was on average 31.9%. The percentage of Ni in fractions I and II was almost identical (13%), in fraction III the average was 23.6%. The different fractions of bottom sediment in terms of nickel-binding capacity can be ranked as follows: V > III > IV > II = I. Accumulation of nickel in bottom sediments, like cadmium, depends on the basic chemical constituents of the sediment. The results indicate that the pattern given above does not change significantly between seasons. In each season, Ni showed the highest affinity for the mineral fraction (more than 35%). In the summer and the winter seasons, the proportion of the ion exchange and the carbonate fraction of Ni increased significantly to 21%.

Cu

The percentages of copper in the five bottom sediment fractions were characterized by the absolute dominance of the organic and residual fractions. The mean copper percentage in fraction IV was 54.1% and in fraction V 25.5%. The percentages of copper in fractions I and II were similar (4.1 and 5.8%). In terms of percentage contribution, the copper fractions in the bottom sediments could be ranked in the following order: IV > > V > III > II > I. The analysis of the data indicates that the accumulation of copper in bottom sediments will depend primarily on the organic matter content. The minerals forming the residual fraction are also a significant factor. In the spring season, the residual fraction (V) was dominant—on average 43%. The remaining seasons were dominated by the percentage of organic copper fraction (above 57%) in bottom sediments.

Pb

The highest percentage of Pb was found in fraction V of bottom sediments. This share was on average 37.7%. There was also a significant percentage in the organic fraction. In terms of the ability to bind lead, the different fractions could be ranked in the following order: V > IV > III > I > II. Thus, the dominant factors in the accumulation of this metal in bottom sediments are minerals forming the residual fraction and organic matter, and to a lesser extent hydrated iron and manganese oxides and the ion-exchange fraction. The pattern given above changes with the seasons.

In summer, the proportion of fractions III, IV, V and partly I was quite similar, while in winter fraction IV dominated.

Based on the speciation analysis, the dominant fraction in the bottom sediments was.

- the residual fraction and hydrated Fe and Mn oxides for Pb, Cd and Ni;
- the organic and residual fractions for Cu;
- the fraction of hydrated iron and manganese oxides for Zn.

The proportion in the fractions changed strongly with temperature and ion composition in the water.

4 Discussion

Heavy metals are present in the water body as total dissolved forms and in suspension. The concentrations of dissolved forms of metals show high variability in the annual cycle. The highest concentrations of total dissolved forms of cadmium and copper occurred in spring. Maximum concentrations occurred in this period in river waters. The site situated in the stagnant section is a receiver of all pollutants flowing into the rivers. The highest concentrations of dissolved forms of zinc and nickel were found in autumn at the mouth of rivers. The highest concentrations of lead in the water depth were recorded in summer at the sites situated in the central part.

The mean percentage concentration of J + LK of the studied metals in relation to total dissolved forms did not vary over the reservoir area. The percentage of labile forms of all metals was quite high. The results obtained were consistent with the studies of Capadaglio et al. (1990), who obtained the proportion of J + LK forms in the range of 30–50% in Pacific waters. Observations of seasonal changes in labile forms of metals showed that the highest J + LK concentrations occurred in winter for zinc (61.5%) and cadmium (50.2%), in spring for copper (45.6%), and in autumn for lead (45.1%) and nickel (46.5%).

Such a high proportion of ionic and labile complexes in the studied waters may be due to anthropogenic pollution flowing into the reservoir with river waters. Due to a lack of publications determining concentrations of hydrated ions and labile complexes in surface waters, especially in dammed reservoirs, it is not possible to compare the obtained results with literature data.

The data presented above show that in winter the concentration of inert complexes (IK) of zinc and cadmium was lower than that of chemically active J + LK. Hydrated ions and labile complexes are highly toxic forms and are present in elevated concentrations which are dangerous to aquatic organisms. At low pH values and insufficient water oxygenation, conversion of inert complexes to bioavailable metal forms occurs. In reservoirs heavily laden with organic matter, with increasing concentration of total organic carbon (TOC), decrease of J + KL concentration of analysed heavy metals is observed in favour of increase of IK concentration as a result of immobilization of metals from suspension form. Such a phenomenon is observed in the period from spring to autumn, when an intensive inflow of pollutants occurs.

The distribution of metal content in the vertical profile of the aqueous phase is similar for all elements. The total dissolved forms of metals were found in elevated concentrations in the supra-bottom layer of water. The highest content of dissolved forms in the upper layer of lake water was found for nickel and zinc–ca. 35%. Zinc and cadmium were more strongly bound to suspended matter in the surface water layer than in the upper water layer. The remaining metals (copper, nickel and lead) were found mainly in suspension in the superficial layer of water. The strong mixing occurring in the reservoir enhances the remobilization of metals from the bottom sediments and enriches the super-bottom water layer in heavy metals. Depending on prevailing temperature conditions and chemical and physical adsorption processes, metals can be bound in suspension or remain in dissolved form.

During the ice cover there is an increase in the remobilization of metals from the surface sediment layer into the overlying water layer. This is related to the occurrence of oxygen differentiation in the vertical profile. The amount of metals released depends on their properties and form of association in the sediment. During this period, lead and cadmium, and to a lesser extent zinc, nickel and copper are strongly bound by the solid suspended phase in the overlying water layer. Metals can be bound by easily lifted colloidal particles (Zahrim & Khairunis, 2019). The increased remobilization during winter was probably the result of chemical reduction of metal forms contained in the sediments due to turbulent mixing in the overlying water layer and reduced transport of metals to the surface water layers. Similar conclusions were obtained in a study by Zhang et al. (2014).

The total concentrations of dissolved heavy metal forms in the surface and supra-bottom layers are negligible compared to the metal forms bound in the suspended phase, often accounting for more than 95% of the metal present in the vertical profile of the water body. In the present study, it was observed that with small changes in pH there was an increase in the adsorption fraction of metals. The activity of metal cations is proportional to the concentration of H^+, and the activity of anions is inversely proportional to the concentration of H^+ ions. Under the pH conditions determined by the study, the increase in the suspended adsorption fraction ranked the metals in the following order in the superdense layer: Cu < Pb < Ni < Zn < Cd and in the surface layer: Cd < Cu < Ni < Pb < Zn. In a dammed reservoir, where apart from natural water movements caused by winds and natural flow there are variable kinetic conditions of water resulting from dam operation, which periodically favours sedimentation and then the role of heavy metals migration is limited only to their removal from the water phase and deposition in the form of bottom sediments (Namieśnik & Rabajczyk, 2010; Wojtkowska, 2020).

In my own research, it was noted that in the aqueous layer zinc is more strongly bound by hydrated iron oxides than copper, and this process takes place at pH = 7–7.5, in which zinc and copper salts are hydrolysed (Dojlido, 1995). The highest concentrations of copper in surface water were observed in spring and autumn, when the content of organic matter including humus compounds was the highest. This was confirmed with the studies of Szarek-Gwiazda and Mazurkiewicz-Boroń (2002), who observed a strong association of copper with humic substances. The copper content associated with the suspended matter accounted for about 93% of the total copper concentration in the aqueous phase. The concentration of this metal in natural waters depends on the acid–base nature of the environment. Copper and zinc in dissolved form occur as Cu^{2+} and Zn^{2+} ions and in the form of copper anions $[Cu(CO_3)_2]^{2-}$, $[Cu(OH)_3]^{3-}$ and $[Cu(OH)_4]^{2-}$ and the same of zinc. Metal anions can co-precipitate with particles of magnesium hydroxide $Mg(OH)_2$, which has a positive surface electric charge. Heavy metal cations co-precipitate with $CaCO_3$ molecules with a negative electrical charge. This process depends on the pH of water and its hardness. As noted by Li et al. (2013), the pH value of heavy metal removal decreases with increasing water hardness.

The conducted research showed that the predominant form of the analysed metals in the water depth at all sites was the suspended form. Low concentrations of dissolved forms of heavy metals, caused by strong adsorption on suspensions and co-precipitation or precipitation in the form of hardly soluble forms into bottom sediments, may increase due to inflow of pollutants and change of thermo-oxygen conditions in the supra-bottom water layer.

In the studies carried out, the influence of the study periods on changes in the concentration of dissolved and suspended forms of metals in the water depth was observed. The lowest concentrations of dissolved forms for copper and cadmium, in contrast to suspended forms, occurred in winter under ice cover. No enhanced remobilization of dissolved forms from bottom sediments was observed for these metals, or the remobilized forms were strongly bound in suspension. After the ice cap receded, cadmium concentration in the surface layer of water increased slightly, while a significant increase in cadmium concentration was recorded in the supra-sedimentary layer of water. At this time, there is an increased inflow of water to the lake from surface run-off and rainfall. During the study period, maximum cadmium concentrations in the water depths occurred in spring and autumn (April and September). The concentration of dissolved forms of copper increased after the ice had receded, reaching its highest values in the water in April.

In the winter period, the concentration of cadmium and copper in the surface layer of sediments was the highest during the whole study period. Such variability in Cd and Cu concentrations confirms Segura's theory (Segura et al. 2006) that the process of sorption and desorption of cadmium and copper is controlled by the concentration of organic compounds both in the aqueous phase and in bottom sediments. In winter, when there is a smaller inflow of organic compounds, the contribution of these metals to the water depth is small. The obtained analytical results confirmed with the studies conducted by Cottenie (1993). Sedimented organic matter in bottom sediments, especially macromolecular humic compounds, which are an important fraction of organic matter and mostly poorly soluble in acidic environments, permanently binds cadmium and copper during winter depositing them in bottom sediments. The process of desorption from bottom sediments is also limited by the concentration of chloride ions, as stated by Flower in his work (Flower et al., 2017). In studies conducted on Zegrzyński Lake, limited migration of cadmium from sediments at low chloride ion concentrations was observed. Increase of salinity, which is caused mainly by concentration of chloride and sulphate ions causes fast exchange of ions (cations) between dissolved and bound forms of cadmium. The binding capacity of Cd ions is closely dependent on pH and dissolved oxygen concentration in the aqueous phase. Street has shown that at high oxygen concentrations, cadmium in alkaline media precipitates as $CdCO_3$ or $Cd_3(PO_4)_2$. The results of the conducted studies showed a negative correlation between cadmium concentration and the concentration of total phosphorus and organic carbon in bottom sediments. Such a correlation is observed especially in winter, when organic matter decomposes under anaerobic conditions. Saturation of bottom sediments with hydrogen sulphide leads to the appearance of insoluble sulphides of Cd and other metals. At the same time, the phosphorus bound in the sediment is released:

$$3H_2S + Cd_3(PO_4)_2 = 3\,CdS + 2H_3PO_4$$

At times when there is an increase in the flow of oxygenated waters (spring), cadmium may be released from the bottom sediments into the waterbed in dissolved form.

Lowering the pH of the water environment leads to dissolution of carbonates, hydroxides and desorption of metal cations. This phenomenon is observed in winter for lead, nickel and zinc. During the period of ice cover occurrence in the Zegrzyński Lake, the concentration of dissolved forms of these metals in the aqueous phase was the highest, while the concentration of suspended forms and bound forms in the bottom sediments of the studied heavy metals was the lowest.

All the analysed metals were, to varying degrees, associated with fractions determined in bottom sediments. The proportion of metals in individual fractions depended to a large extent on the period during which the studies were carried out and on the prevailing conditions. The main factors determining accumulation of metals in individual fractions were the macroelements of bottom sediments and aerobic conditions of supra-sedimentary water.

5 Conclusions

Concentrations of heavy metals in surface water and bottom sediments are influenced primarily by the chemical composition and quality of inflowing water from rivers and surface run-off. The occurrence of elevated concentrations of heavy metals in bottom sediments is related to the anthropogenic pollution of the water body. As a result of lowered water flow velocity, which transports pollutants in solid and dissolved forms, sedimentation of the carried material deteriorates the conditions in the upper water layer. The influence of organic matter is largely responsible for the formation of metal complexes occurring in different chemical equilibria. The adsorption and colloidal fraction of sediment can cause migration of heavy metals from the sediment to the water. The adsorption process depends on the type of ions, co-reaction with iron and manganese oxides, formation of hydrocomplexes and interaction with organic matter. The phenomenon of adsorption, with a sufficient amount of exchangeable ions, e.g. Ca^{2+} and Mg^{2+} with Cd^{2+} or Pb^{2+}, enables retention and immobilization of heavy metals in bottom sediments. Desorption occurs at low pH as a result of exchange of metal cations with H $ions^{+}$. In a reducing environment, the solubility of various metals increases, including that of Fe and Mn oxides. Metals bound to the fraction of hydrated iron and manganese oxides under reducing conditions can be released to the water depth in dissolved form. In a strongly reducing environment, a competitive reaction to the desorption process is the reaction of binding metals in the form of sulphides: CuS, ZnS, PbS and CdS, which as insoluble compounds are accumulated in the organic and residual fraction of bottom sediments. All analysed heavy metals undergo continuous migration in water–sediment system. The phenomena of precipitation and dissolution, adsorption and desorption, and complexation and decomposition depend on the pH and dissolved oxygen concentration in the supernatant layer of water. These parameters are the most important factors that determine the transfer of metals from the solid-to-the liquid phase and it is possible to study their migration as a function of pH and dissolved oxygen. The percentage of metals in the bottom sediment fractions (speciation) is characteristic of the individual heavy metals. It is determined by the transformations the trace elements undergo. In contaminated sediments, cadmium, nickel, lead, copper and zinc were characterized by different percentages in individual fractions. The binding of the analysed heavy metals in the individual fractions of bottom sediments depended on the chemical properties of the metals, the conditions in the water environment and the study period.

References

Aderinola, O. J., Clarke, E. O., Olarinmoye, O. M., Kusemiju, V., & Anatekhai, M. A. (2009). Heavy metals in surface water sediments, fish and perwinkles of Lagos Logoon. *American-Eurasian Journal of Agricultural and Environmental Sciences, 5*(5), 609–617.

Bain, R., Johnston, R., & Slaymaker, T. (2020). Drinking water quality and the SDGs-npj Clean. *Water, 3*, 1–3.

Capadaglio, G., Coale, K. H., & Bruland, K. W. (1990). Lead speciation in surface waters of the eastern North Pacific. *Marine Chemistry, 29*, 221–233.

Cottenie, A. (1993). Laboratory of Analytical and Agrochemistry, State University of Ghent, Belgium

Dojlido, J.R. (1995). Chemia wód powierzchniowych. (Surface water chemistry). Wydawnictwo Ekonomia i Środowisko. Białystok.

Flower, H., Rains, M., Lewis, D., Zhang, J.-Z., & Price, R. (2017). Saltwater intrusion as potential driver of phosphorus release from limestone bedrock in a coastal aquifer. *Estuarine, Coastal and Shelf Science, 184*(5), 166–176.

Li H., Shi A., Li M., & Zhang X. (2013). Effect of pH, temperature, dissolved oxygen, and flow rate of overlying water on heavy metals release from storm sewer sediments, Hindawi Publishing Corporation. *Journal of Chemistry*, 2013, ID 434012, 11. https://doi.org/10.1155/2013/434012 https://doi.org/10.1155/2013/434012.

Markich, S. J., Brown, P. L., Batley, G. E., Apte, S. C., & Stauber, J. L. (2001). Incorporating metal speciation and bioavailability into water quality guidelines for protecting aquatic ecosystems. *Australasian Journal of Ecotoxicology, 7*, 109–122.

Namieśnik, J., & Rabajczyk, A. (2010). The speciation and physico-chemical forms of metals in surface waters and sediments. *Chemical Speciation & Bioavailability, 22*(1), 1–24. https://doi.org/10.3184/095422910X12632119406391

Olaniran, A. O., Balgobind, A., & Pillay, B. (2013). Bioavailability of heavy metals in soil: impact on microbial biodegradation of organic compounds and possible improvement strategies. *International Journal of Molecular Sciences, 14*, 10197–10228. https://doi.org/10.3390/ijms140510

Olutona, G. O., Aribisala, O. G., & Akintunde, E. A. (2012). A study of chemical speciation of metals in aquatic bottom sediment of Aiba reservoir, Iwo, Nigeria. *African Journal of Environmental Science and Technology, 6*(8), 312–321. https://doi.org/10.5897/AJEST11.227

Preisner, M., Neverova-Dziopak, E., & Kowalewski, Z. (2020). An analytical review of different approaches to wastewater discharge standards with particular emphasis on nutrients. *Environmental Management, 66*, 694–708. https://doi.org/10.1007/s00267-020-01344-y.

Safoura, J., Hessam, H. A., Gholamalizadeh, A. A., & Jaber, S. (2015). Fractionation of heavy metals in bottom sediments in Chahnimeh 1, Zabol, Iran. *Environmental Monitoring and Assessment, 187*, 340. https://doi.org/10.1007/s10661-015-4510-x

Segura, R., Arancibia, V., Zúñiga, M. C., & Pastén, P. (2006). Distribution of copper, zinc, lead and cadmium concentrations in stream sediments from the Mapocho River in Santiago, Chile. *Journal of Geochemical Exploration, 91*(1–3), 71–80.

Slaveykova, V. I., & Wilkinson, K. J. (2005). Predicting the bioavailability of metalcomplexea: Critical review of the biotic ligand model. *Environmental Management, 2*, 9–24.

Szarek-Gwiazda, E., & Mazurkiewicz-Boroń, G. (2002). Deposition of copper in the eutrophic, submontane dobczyce dam reservoir (Southern Poland)–Role of speciation. *Water, Air, and Soil Pollution, 140*, 203–218.

Tessier, A., Campbell, P. G. C., & Bisson, M. (1979). Sequential extraction procedure for the speciation of particulate trace metals. *Analytical Chemistry, 51*, 7.

Wojtkowska, M., & Bogacki, J. (2012). Use of speciation analysis for monitoring heavy metals in the bottom sediments of the Utrata River. *Ochrona Środowiska, 34*(4), 43–46.

Wojtkowska, M., & Matula, M. (2016). Analysis of heavy metals in selected particular granulometric fractions of bottom sediments. *Rocznik Ochrona Srodowiska, 18*, 667–680.

Wojtkowska, M. (2020). Heavy metal pollution of water and bottom sediments of the Wilanowskie Lake. *Desalination and Water Treatment, 191*, 110–117. https://doi.org/10.5004/dwt.2020.25178

Wojtkowska, M., Bogacki, J., & Witeska, A. (2016). Assessment of the hazard posed by metal forms in water and sediments. *Science of the Total Environment, 551*, 387–392. https://doi.org/10.1016/j.scitotenv.2016.01.073

Zahrim A.Y., & Khairunis, I. (2019). Nanoparticles enhanced coagulation of biologically digested Leachate. In *Nanotechnology in water and wastewater treatment* (pp. 205–241). https://doi.org/10.1016/B978-0-12-813902-8.00011-3

Zhang, C., Yu, Z., Zeng, G., Jiang, M., Yang, Z., Cui, F., Zhu, M., Shen, L., & Hu, L. (2014). Effects of sediment geochemical properties on heavy metal bioavailability. *Environment International, 73*, 270–281. https://doi.org/10.1016/j.envint.2014.08.010

Zhang, H., & Davison, W. (2000). Direct in situ measurements of labile inorganic and organically bound metal species in synthetic solutions and natural waters using diffusive gradients in thin films. *Analytical Chemistry, 72*, 4447–4457. https://doi.org/10.1021/ac0004097

Sustainable Water Use in Agriculture—Circular Economy Approach

Majeti Narasimha Vara Prasad

Abstract

Irrigation is one of the most important sustainable applications in agriculture. Nutrient-rich soil and adequate soil moisture are the most essential requirements for agricultural productivity. Some areas are rain fed while others have the provision of irrigation either through canals or ground water. Innovation for reducing the freshwater footprint in agriculture such as the adoption of low-cost and energy-saving suitable irrigation techniques like solar, drip, sprinkler, controlled tube irrigation, and fertigation will reduce the water wastage in agriculture. Constructed ponds will facilitate groundwater recharge by rainwater harvesting. Soils represent an important part of the ecosystem. They control the water flow, the recharge of ground water, the nutrient supply of the plants, and can reduce the number of pollutants or protect the ground and drinking water from them. Pesticides from agricultural industry, greenhouse gases from the atmosphere, and wastewater from contaminated sites are reflected with many other parameters in the chemical and microbiological situation of soils which is of primary importance for the plant quality. Therefore, the knowledge of the situation below the soil surface is an important task for the agricultural and industrial researches. Lysimeters represent a practical tool to get a view in the deeper soils. The level of the lysimeter vessel surface flushes with the surface of the surrounding area. They simulate the natural relation between soil, atmosphere, and plants. The dimension of the soil monoliths can vary from small (less than 0,5 m^2 surface and not deeper than 1 m) to big types (2 m^2 surface and up to 3.5 m depth). An overview of the various requirements for sustainable water uses in agriculture that contributes to cross-discipline discussions and understanding are reviewed in this chapter. Irrigation water use trends for sustainable agriculture in selected parts of the globe are cited as examples.

Keywords

Agricultural water management • Best management practice • Cropland • Drip systems • Integrated water management • Irrigation • Lysimeter • Sprinklers • Sustainable development • Water use efficiency

1 Introduction

Several variables play a significant role for sustainable agriculture. Irrigation is one of the most important sustainable applications in agriculture (Lampridi, et al., 2019; Mpanga & Idowu, 2021; Muñoz-García & Hernández-Callejo, 2022) (Fig. 1). Nutrient-rich soil and adequate soil moisture are the most essential requirements for agriculture. Some areas are rain fed while others have the provision of irrigation either through canals or use of ground water. Innovation for reducing the freshwater footprint in agriculture such as the adoption of low-cost and energy-saving suitable irrigation techniques like solar, drip, sprinkler, controlled tube irrigation, and fertigation will reduce the water wastage in agriculture. Wastewater irrigation in India is gaining considerable significance (Minhas et al., 2022). The water availability and irrigation challenges caused by global climate crisis require scientific approach and technological solutions for Water-Energy-Food (WEF) nexus (Saray et al., 2022).

Constructed ponds will facilitate groundwater recharge by rainwater harvesting. Soils represent an important part of the ecosystem. They control the water flow, the recharge of ground water, the nutrient supply of the plants, and can reduce the number of pollutants or protect the ground and drinking water from them. Pesticides from agricultural industry, greenhouse gases from the atmosphere, and

M. N. V. Prasad (✉)
School of Life Sciences, University of Hyderabad (An Institution of Eminence), Hyderabad, Telangana 500 046, India
e-mail: mnvsl@uohyd.ac.in

M. Smol et al. (eds.), *Water in Circular Economy*, Advances in Science, Technology & Innovation,
https://doi.org/10.1007/978-3-031-18165-8_5

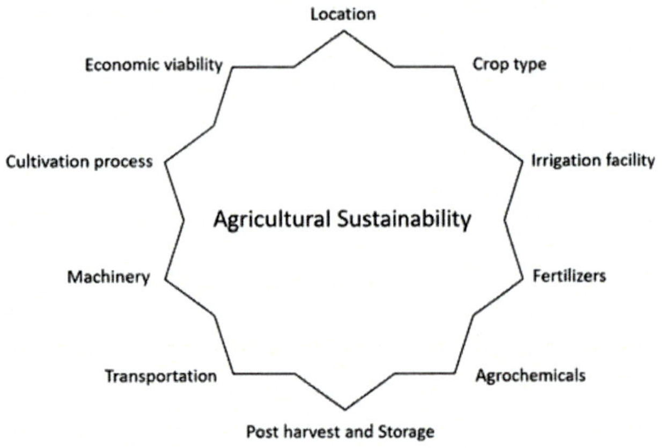

Fig. 1 Factors responsible for sustainable agriculture

wastewater from contaminated sites are reflected with many other parameters in the chemical and microbiological situation of soils which is of primary importance for the plant quality.

2 Hydrological Cycle

Hydrological cycle depicting the main water resources is crucial aspect of the agroecosystem (Fig. 2). Knowledge of the situation below the soil surface is an important task for the agricultural and industrial researches. Lysimeters represent a practical tool to get a view in the deeper soils. The level of the lysimeter vessel surface flushes with the surface of the surrounding area. They simulate the natural relation between soil, atmosphere, and plants. The dimension of the soil monoliths can vary from small (less than $0.5m^2$ surface and not deeper than 1 m) to big types (2 m^2 surface and up to 3.5 m depth).

Water use efficiency is essential in semiarid regions in order to achieve sustainability. For this purpose, the use of compact removable weighing lysimeters can be a suitable tool for determining the adjusted water amounts that a crop is consuming along the irrigation scheduling. The use of these devices can help farmers to determine the crop water requirements and to improve the irrigation system efficiency. A weighing lysimeter provides a direct method to measure the crop evapotranspiration (ETC) through the mass balance of an isolated soil volume.

An overview of the various requirements for sustainable water uses in agriculture that contributes to cross-discipline discussions and understanding is reviewed in this chapter. Irrigation water use trends for sustainable agriculture in selected parts of the globe are cited as examples.

According to Hegde (2012), 81% of India's fresh water is used for irrigation, 13% for industry, and 6% for domestic use. Human water consumption is expected to increase by up to 26% by 2025. This data demonstrates the poor use of water resources, particularly in agriculture. As a result, there is a critical need to develop strategies to increase the efficiency of freshwater utilization in agriculture. Agriculture is the backbone and most significant gift to human life not only in India, but also across the world. However, the bulk of the world's cropland suffers from a severe lack of irrigation

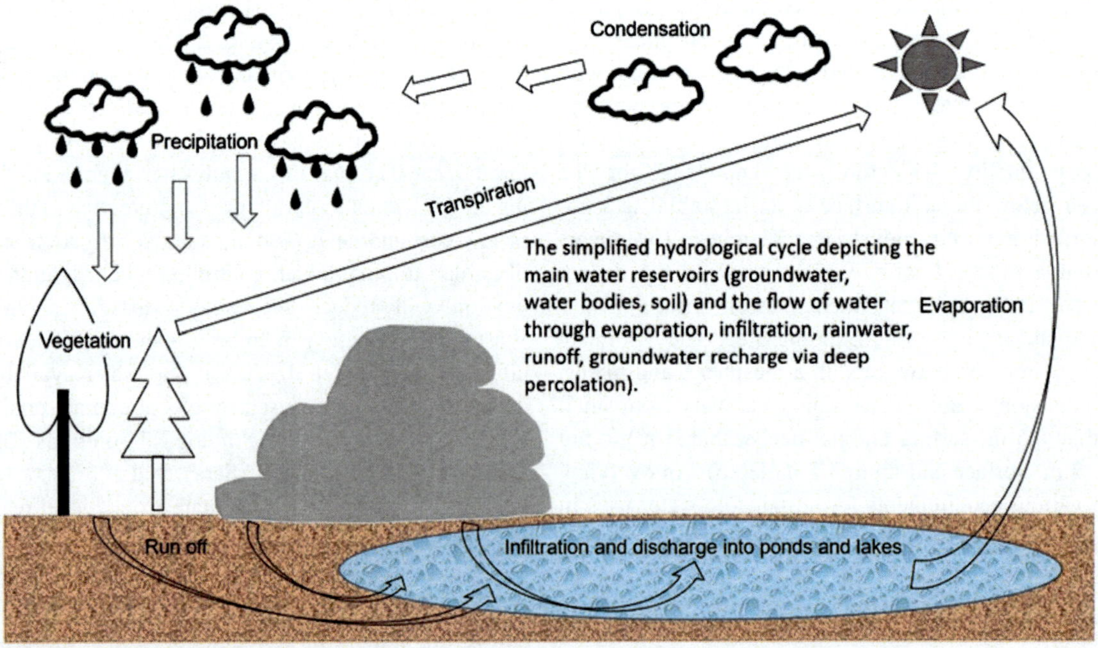

Fig. 2 The simplified hydrological cycle depicting the main water reservoirs (groundwater, water bodies, soil) and the flow of water through evaporation, infiltration, rainwater, runoff, groundwater recharge via deep percolation

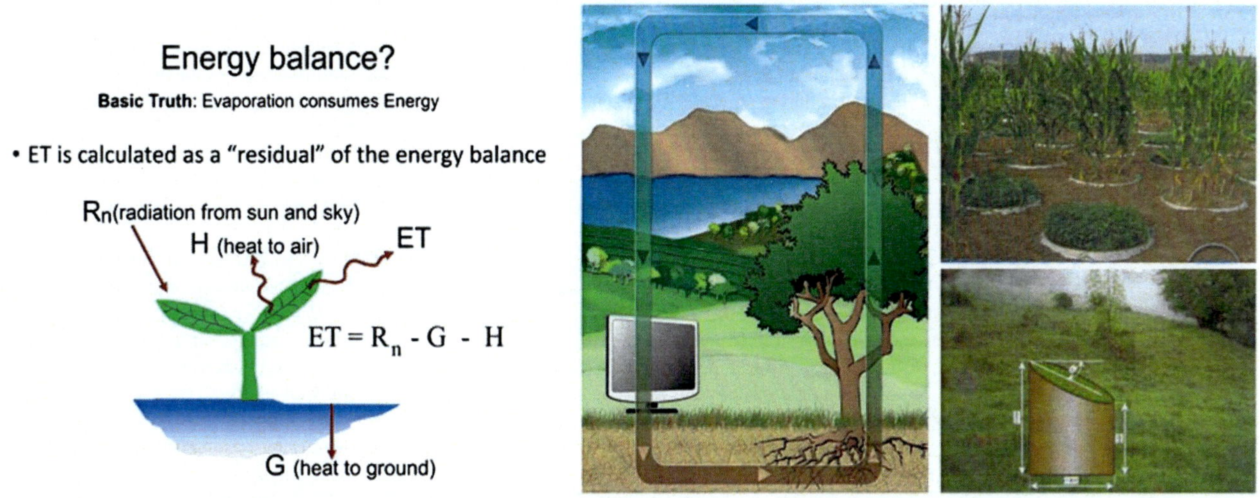

Fig. 3 Lysimeter simulates natural relation between soil–plant–atmosphere representing the link between lab and real-time studies (*Courtesy* Prof. Ralph Meissner, Germany)

water. Drip irrigation systems have been created by researchers in such a scenario to minimize water use in arid locations. This is due to the lack of standards and systematic approaches for utilizing water and power in a positive manner. As a result, farmers' overheads for conventional drip irrigation have increased, and they must personally visit and monitor the farms on a regular basis. In order to optimize irrigation for agriculture under climate change (Chartzoulakis & Bertaki, 2015; Nyika, 2022), different instruments are employed. For example, Lysimeter for measuring the water and solute fluxes with high-precision (Meissner et al., 2010) (Figs. 3 and 4, Table1), solar-powered submersible pump system without the use of batteries in agriculture (Bhosale, 2022) remote sensing tools (Bretreger et al., 2022).

Water use efficiency is essential in semiarid regions in order to achieve sustainability. For this purpose, the use of compact removable weighing lysimeters can be a suitable tool for determining the adjusted water amounts that a crop is consuming along the irrigation scheduling. The use of these devices can help farmers to determine the crop water requirements and to improve the irrigation system efficiency. A weighing lysimeter provides a direct method to measure the crop evapotranspiration (ETC) through the mass balance of an isolated soil volume (Meissner et al., 2010; Meissne & Rupp Hand Haselow, 2020).

Fig. 4 There has been a gradual rise in publications using lysimeters for a variety of hydrological applications. *Data source* www.sciencedirect.com

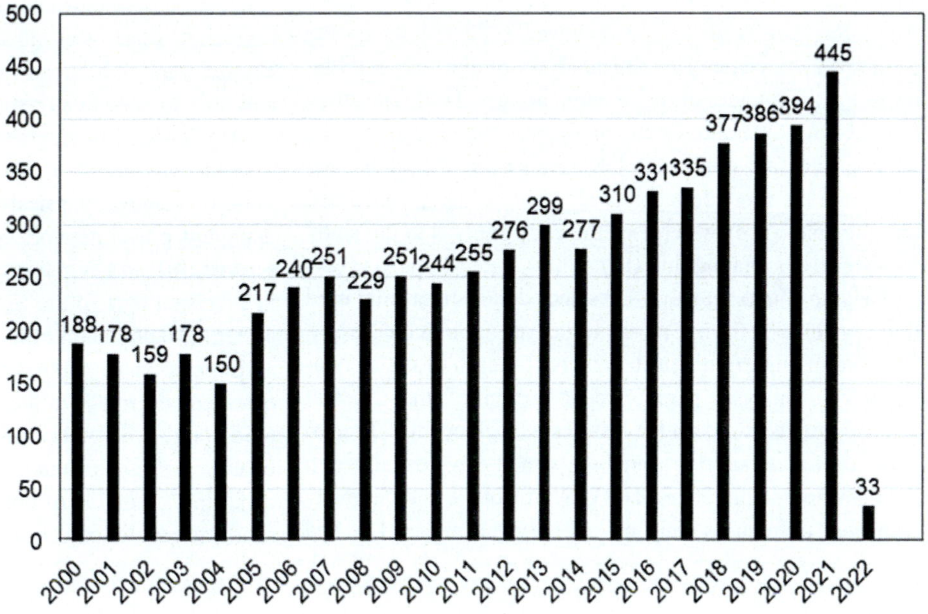

Table 1 Lysimeter applications (A–Z)

1. Anion–cation leaching	18. Nitrate leaching
2. Behavior of heavy metals in floodplain sites	19. Nitrogen use efficiency
3. CO2 release of peat soils	20. Nitrous oxide emissions
4. Diffuse pollution of water resources	21. Nutrient leaching from different soil types
5. Dissolved organic carbon investigation in wetlands	22. Nutrient transport from agricultural systems
6. Drought and irrigation on the fate of nitrogen	23. Performance of different genotypes under field
7. Environmental fate and behavior of agrochemicals	24. Phosphorus in soil-leaching—transfer
8. Evapotranspiration in different agroclimatic regions	25. Rainfall of interception and canopy storage
9. Fulvic acid and humic acid in degraded lands	26. Salinity impacts on drainage
10. Groundwater seepage	27. Snowpack dynamics
11. Herbicide transport in macroporous soils	28. Soil CO_2 efflux
12. Hydrological studies	29. Solute transport
13. Interception loss from forest	30. Surface impoundments and landfills
14. Leaching and reclamation studies	31. Surface hydrology
15. Measurement of dew and fog	32. Pesticide leaching
16. Mineral fertilizers use efficiency	33. Toxicity tests
17. Modeling the water balance and monitoring fertilizer losses due to drainage	34. Tracer release in melting snow
	35. Transport of [14c] benazolin and bromide
	36. Transport of organic compounds in macroporous soil
	37. Water dynamics
	38. Water use by intensively cultivated agri-silvicultural crops

3 Bibliographic Resume of Sustainable Water Use in Agriculture (SWUA)

The analysis of the number of articles published during 1993–2017 proves that research on SWUA has attracted increased attention throughout the years. In particular, the number of articles published (A) grew from 4 in 1993 to 267 in 2017. Figure 5 shows how the number of articles published has grown exponentially within this thematic field. Aznar-Sánchez et al. (2018) reviewed 1022 articles on the economics of water use, and analysis of the productivity, impact, and structure of the research is carried out. The study found USA, Australia, China, Spain, and Netherlands are the most productive countries; UK, France, Netherlands, and China are the most collaborative countries. They concluded that the analysis of water use efficiency evolves to the study of water use sustainability (Fig. 5).

Velasco-Muñoz et al. (2018) selected most frequently used five keywords by the five most productive countries on SWUA. These terms signify the main hotspots for each of these countries. This analysis permits identification of the different research trends based on the features of each country. In general, all of the countries shared the same keywords, although there are some differences. The U.S., Australia, and Germany displayed differences only on one keyword. The U.S. and Germany had the same keywords, while Australia included the name of the country among the most used keywords and did not include water supply. For China and India, the most repeated keyword was the name of its country itself. China is the only country that included water resources and the country that gave less importance to water management and sustainability. Chinese works were focused on water availability and water use efficiency. India was the only country that did not include irrigation, but it was focused on the sustainable management of water use and its efficiency (Table 2).

Figure 6 has been created in order to compare the publication trends of articles versing on water as a general topic and articles specifically studying SWUA.

Applications of Internet of Things (IoT) have increased over the past several years (Table 3). The Internet of Things aims to connect physical objects (Things) to the Internet using low power network connectivity that is embedded with sensors. IoT and Wireless Sensor Networks (WSN) are used in precision agriculture to create a smart farm management system (Madhumathi et al., 2022).

Experts have estimated that IoT devices will see an exponential growth in the coming years with an average of 6.58 connected devices per person. Radio frequency identification (RFID) and its adoption in the agriculture supply chain in the context of both traceability and management of products revolutionized the IoT.

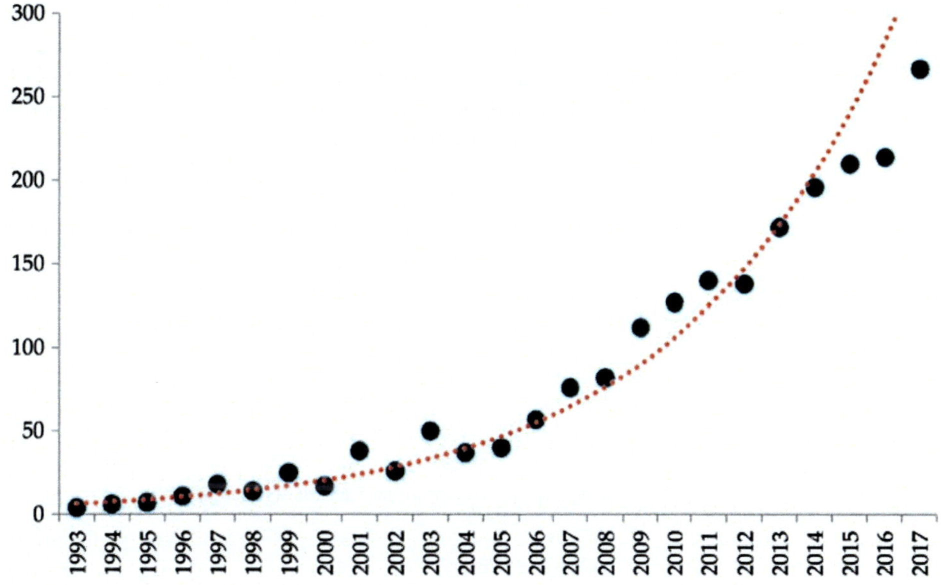

Fig. 5 Trend in the number of articles in the broad area of sustainable water use in agriculture (with MDPI open-access license)

Table 2 Main keywords by country (with MDPI open-access license)

Country	Keyword1	Keyword2	Keyword3	Keyword4	Keyword5
China	China	Water supply	Water resources	Water use efficiency	Irrigation
The United States	Water supply	Irrigation	Water use	Sustainability	Water management
Australia	Water use	Australia	Irrigation	Sustainability	Water management
India	India	Sustainability	Water management	Water use	Water use efficiency

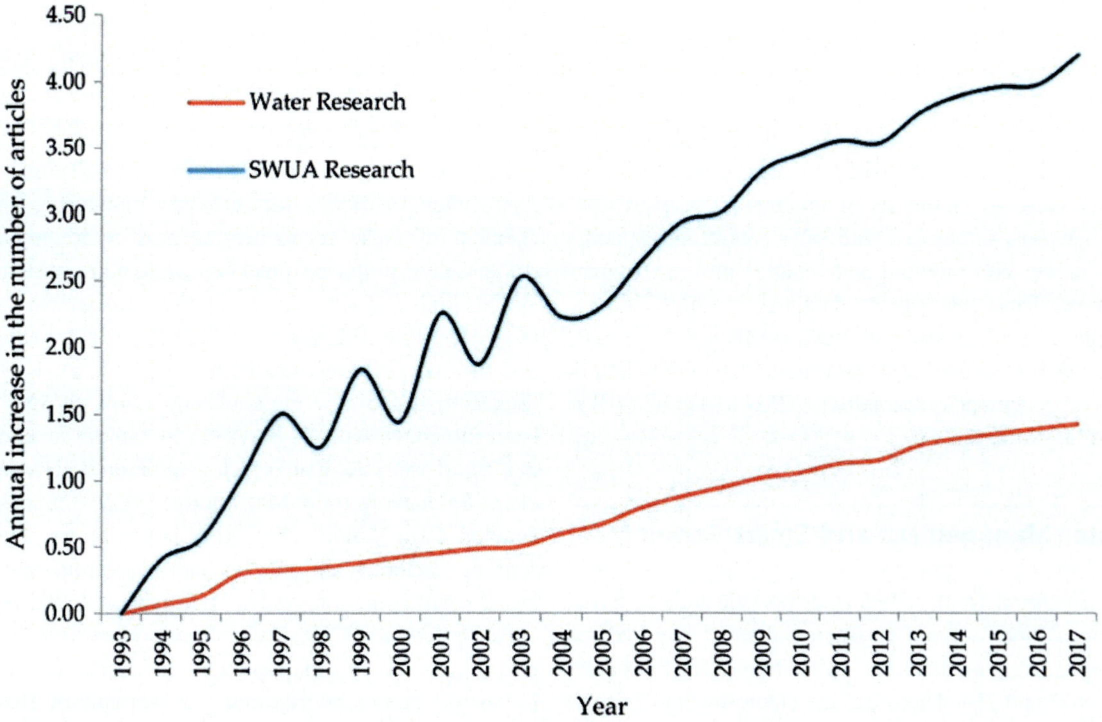

Fig. 6 Comparative trends in the number of articles on water as a general topic versus sustainable water use in agriculture (SWUA) (with MDPI open-access license)

Table 3 Chonological review of literature on wireless sensor networks in agriculture for smart water management

The use of the Multiple Access Time Division allows more efficient transmission of data into WSN (wireless sensor network) irrigation systems	Nesa Sudha et al. (2011)
Developed wireless sensor network architecture to monitor horticulture crops. It is based on the Berkeley Medium Access Control (B-MAC) protocol, and it takes into account various components such as base station, gateway, soil mote, water mote, and environmental conditions, and it interconnects the properties in order to achieve better throughput and reduce delays	Lopez et al. (2011)
A self-contained agricultural precision system based on central pivot irrigation systems was established	Dong et al. (2013)
Prototype of WSN sensor networks for precision agriculture	Jao et al. (2013)
Automated zone-specific irrigation with wireless sensor/actuator network and adaptable decision support	Goumopoulos et al. (2014)
Review of wireless sensors and network applications in agriculture	Abbasi et al. (2014)
WSN sensors for precision agriculture which boosts field productivity by lowering agricultural	Kodali et al. (2014)
WSN sensor networks for precision agriculture	Kassim et al. (2014)
Automated irrigation system using a wireless sensor network and GPRS module	Gutierrez et al. (2014)
Designing a wireless sensors network for monitoring and predicting droughts	Gaddam et al. (2014)
Introduced a novel approach for effective sensor node Efficient Dynamic Deployment Approach of Sensor Nodes (EDSNDA), deployment in wireless networks	Li et al. (2015)
Smart humidity sensors for precision agriculture:	Imam et al. (2015)
The lifetime property of Wireless sensor networks (WSNs) metaheuristic algorithm that operates in three phases: transition, evaluation, and determination	Tsai et al. (2016)
Explored frequency reuse, which involves reusing a frequency for industrial, scientific, and medical (ISM) applications	Hassan et al. (2016)
A sensor array system is used to measure soil temperature and moisture so that watering requirements could be predicted in real time. Such an approach helped in developing smartphone application for irrigation scheduling of cotton crop	Vellidis et al. (2016)
WSN sensor networks for precision agriculture	Deepika and Rajapirian (2016)
Smart irrigation system using wireless sensor	Gajendran et al. (2017)

Wireless sensor networks (WSN) (Grace et al., 2015) are capable of sensing conditions of products. These are distributed networks of sensors with self-networking capabilities, low power consumption and requires minimal human intervention. These also allow data transmission using a wide range of wireless standards such as Bluetooth (Smart), Wi-Fi, ZigBee, near-field communication (NFC) (Krishna & Biradar, 2022, Kumar & Choudhury, 2022; Li et al., 2015; Loiskandl & Nolz, 2021).

4 Water Management and Smart Farming

On an agricultural front, IoT devices provide precise information on a wide range of parameters that are required for enhancing farming methods and cultivation of fresh produce (Figs. 7, 8, 9 and 10). These include environmental factors, growth conditions, soil, farming equipment (Nathan, 2015), greenhouse production environment (Ji-chun Zhao et al., 2010), water irrigation, pest, and fertilizers (TongKe, 2013). Application of WSNs for real-time monitoring and management of these parameters is now increasing. Several studies have proposed precision agriculture systems using WSN (Abd El-kader and Mohammad El-Basioni 2013) for crop cultivation (Mafuta et al., 2013) for automating irrigation methods, environment monitoring (Zhu et al. 2011) and (Sakthipriya, 2014) for monitoring crops. WSN has also been integrated with GPS system to capture location information of the area that is being monitored (Hwang et al., 2010; Satyanarayana & Mazaruddin, 2013). Smart-network-enabled farm would offer great potential for automating farming methods and further enhance productivity. IoT-based smart irrigation systems (Garc´ıa et al. 2020), climate smart agriculture (CSA) technologies (Hussain et al., 2022), and smart water management platform enabled with IoT-based precision irrigation for agriculture (Kamienski et al. 2019) are widely applied in different parts of the world (Mpanga & Idowu, 2021; Muñoz-García &

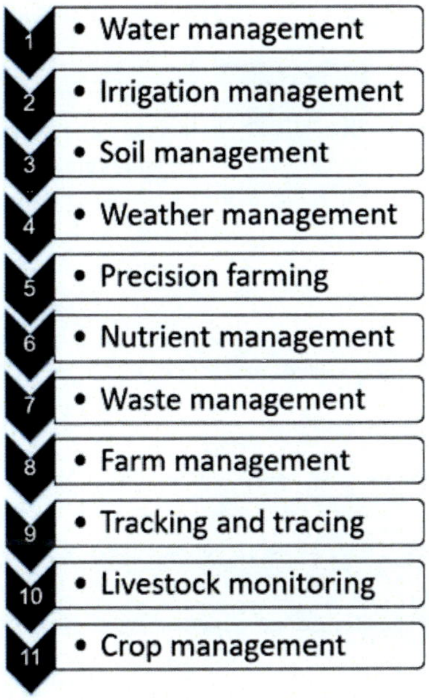

Fig. 7 Role of Internet of things (IoT) in agriculture

Hernández-Callejo, 2022; Phasinam et al. 2022; Phillips & Encel 2022; Ruiz-Canales & García, 2021; Ouda & Zohry 2022).

Reference be made to Fig. 11, for various farm innovations for reducing the freshwater footprint in agriculture for sustainable water use.

Circular economy approach in the water and wastewater sector has been reviewed recently (Smol, 2022). Developments in agriculture technology using Internet of things, advancements, and challenges of Internet of things in smart agriculture are also highlighted recently (Sinha & Dhanalakshmi, 2022; Sucharitha & Sai, 2022).

According to Velasco-Muñoz et al. (2018), the number of articles published during 1993–2017 proves that research on Sustainable Water Use in Agriculture (SWUA) has attracted increased attention throughout the years. Sustainability of water use in agriculture is a line of research that has gained importance worldwide. A bibliometric analysis was presented by Velasco-Muñoz et al. (2018) taking a sample 2084 articles. Results indicate exponential growth in the

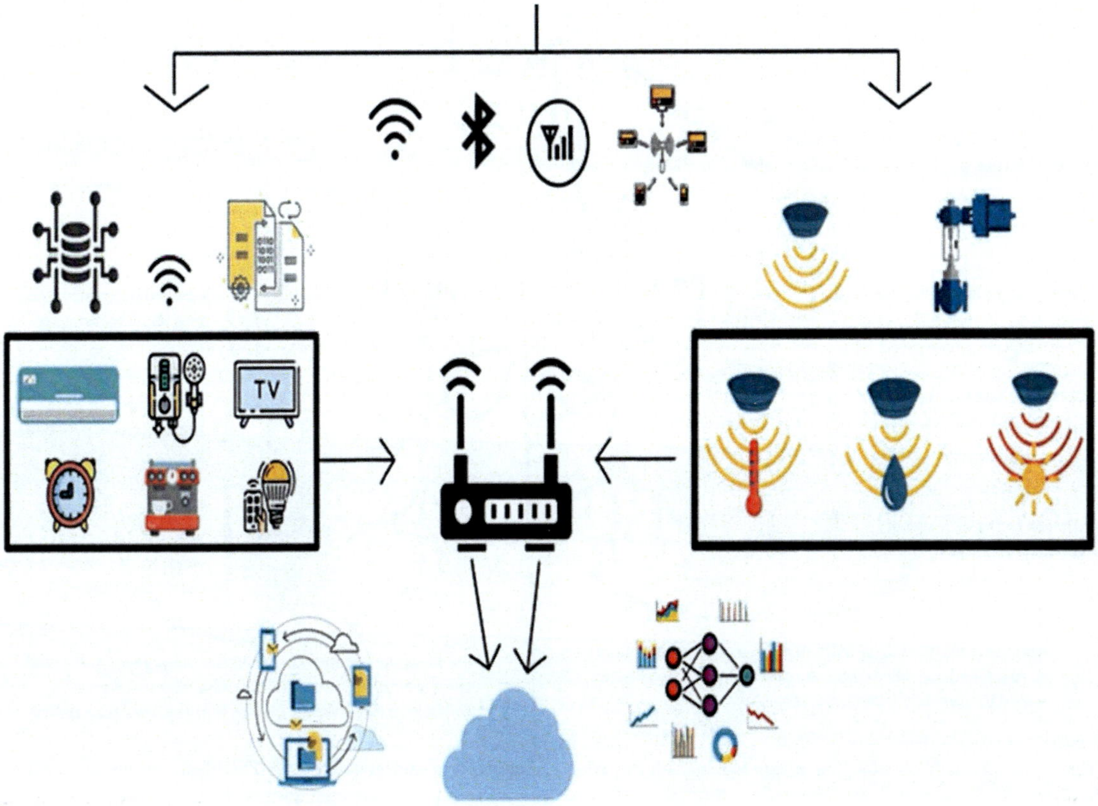

Fig. 8 Diversity of IoT devices

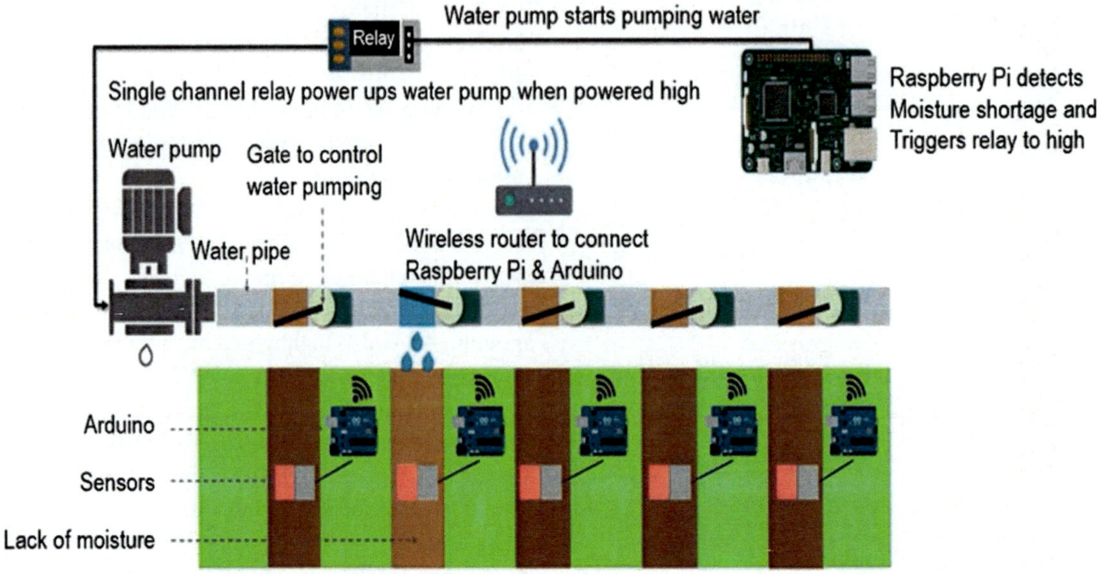

Fig. 9 Smart irrigation system and its key components

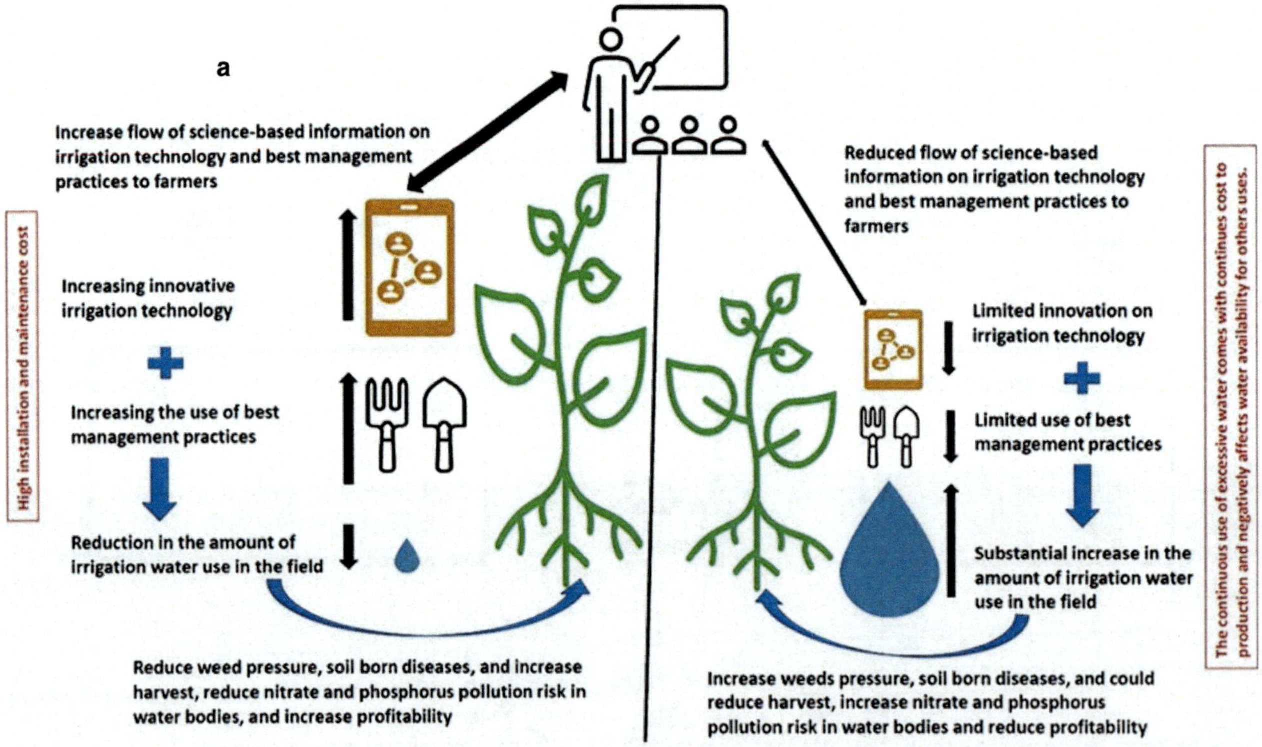

Fig. 10 An ideal approach on how the integrated use of irrigation technology, and best management practices through effective science-based extension and education information flow, could influence irrigation water management and sustainable water use positively **a** and negatively. **b** The black arrows show increase (arrow pointing up) or decrease (arrow pointing down) and wide (wider = more information flow) and narrow (narrow = limited information flow) (with permission from Mpanga & Idowu, 2021)

Fig. 11 Pictorial depictions of various farm innovations for reducing the freshwater footprint in agriculture (Dubey et al., 2021). The adoption of suitable irrigation techniques like solar, drip, sprinkler, controlled tube irrigation, and even fertigation will reduce the water wastage in agriculture whereas constructed ponds will facilitate groundwater recharge by rainwater harvesting. Figures **a–j** show innovative water conservation methods. Figure annotations: **a** solar pumps for irrigation in rice–wheat cropping system, **b** plastic mulching for tomato cultivation, **c** canal irrigation, **d** solar irrigation in rice field, **e** sprinkler irrigation in vegetable field, **f** constructed pond for rainwater harvesting, **g** sprinkler irrigation in mustard field, **h** plastic mulching for chillies, **i** solar irrigation in agroforestry system, **j** controlled tube irrigation **k** mixed irrigation in wheat field (with permission from Dubey et al., 2021)

number of articles published per year, with research in this field having acquired a global scale. Environmental Science and Agricultural and Biological Sciences are the main categories. Three journals—Agricultural Water Management, Water Resources Management, and Nongye Gongcheng Xuebao Agricultural Engineering published most of the articles. China, the U.S., Australia, India, and Germany produced the most research. The three institutions that published the most articles were all Chinese (Chinese Academy of Sciences, China Agricultural University and Northwest A&F University). Further, Velasco-Muñoz et al. (2018) mentioned that the most cited authors were Ridoutt, Hoekstra, and Zhang. The keywords most frequently used include: water use, irrigation, water management, water supply, and sustainability. A network map shows three clusters that focus on the environmental, agronomic, and management aspects. The findings of this study can assist researchers in this field by providing an overview of research on the sustainability of hybrid resources.

5 Micro-Irrigation for Sustainable Agriculture—The Case of India

The agriculture sector is the largest consumer of water in India. In India, farmers face the challenge of water scarcity. The ongoing water crisis has affected nearly 600 million people and is expected to worsen. The country's population is touted to increase to 1.6 billion by 2050. It accounts for approximately 90% of 761,000 billion liters of annual freshwater withdrawals in the country. Per capita consumption of water in agriculture sector ranges from 4913 to 5800 kiloliter per capita per year. Climate change too has aggravated water scarcity concerns. The impact of climate change is much more evident in Indian agriculture, where around 85% farmers are small and marginal and 60% agriculture is dependent upon the vagaries of monsoon. The role of irrigation, therefore, takes the front seat.

In the age of climate change and water scarcity, micro-irrigation can help increase crop yield and decrease water, fertilizer, and labor requirements. The continued irrigation through traditional practices since the introduction of Green Revolution in the 1960s, however, has begun to show its multitudinous ill effects on groundwater quality and height, water logging, soil salinity, soil health, crop productivity, partial factor productivity, and cost economics of farm practices. This is where micro-irrigation assumes significance. Micro-irrigation can increase yields and decrease water, fertilizer, and labor requirements. By applying water directly to the root zone, the practice reduces loss of water through conveyance, runoff, deep percolation, and evaporation.

Micro-irrigation gained prevalence when the Indian Parliament was rocked with issue of farmer suicides. Sensing the significance and probable benefits of the process to double the farmers' income along with agricultural sustainability and environmental quality, the Union government launched a comprehensive flagship program called Pradhan Mantri Krishi Sinchai Yojana or "more crop per drop".

Under the program, financial assistance of up to 55% is available for small and marginal farmers and 45% for other farmers for adoption of micro-irrigation systems. The funding pattern between the Union Government's and the State Government's share since November 2015 has been 60:40 for all states except the Northeast and the Himalayan states, for which the funding pattern is 90:10.

These losses are unavoidable in traditional irrigation practices; micro-irrigation, through its water-saving approach, has paved the way for higher water use efficiency of around 75–95%. Another resource-saving practice possible through micro-irrigation is fertigation, which comprises combining water and fertilizer application through irrigation. Fertigation results in balanced nutrient application, reduced fertilizer requirement of around 7–42% (thus, saving expenditure cost incurred by farmer), higher nutrient uptake, and nutrient use efficiency. It is quite apparent that in the present scenario, vertical expansion of agricultural lands is not possible. Therefore, in order to increase the yield and productivity, we have to focus on degraded and waste lands.

Micro-irrigation provides this opportunity. A national-level survey undertaken for the Union Government showed that farmers were able to bring 519.43 ha of degraded land under cultivation through the technique. It also helped use of saline water for irrigation without causing salinity or osmotic stress to plants (Khorsandi, 2022) (see chapter, Figs. 13 and 14). The practice needs wide-scale adaptation in India, particularly in the Indo-Gangetic plains where the soil salinity is high. Israel can be a good example —a desert nation with water scarcity has become a water surplus nation because it adapted micro-irrigation practises, especially drip irrigation that saves almost three-fourths of the water used for irrigation done through open canals. Drip irrigation is the most effective practice with water use efficiency of around 85–90%.

Significant electricity savings—on an average 30.5%—have been estimated and high fertilizer-use efficiency reported, resulting in an average consumption reduction of 28.5%, according to a Federation of Indian Chambers of Commerce and Industry report.

Another advantage is maintenance of optimum soil moisture conditions that help increase overall productivity and profitability. Across various studies, it has been found that the adoption of micro-irrigation systems helped boost the yield of fruit as well as vegetable crops.

The productivity for fruit crops increased 42.3% and that of vegetable crops by 52.8%. There was an average reduction of 31.9% in irrigation cost thorough higher water use efficiency. Another gain has been the adaptation of diverse cropping patterns.

It is quite evident that importance of micro-irrigation to achieve sustainability in Indian agriculture cannot be neglected. But it's a long way ahead and requires extensive demonstrations, training, and awareness programs to bring Indian farming community abreast with micro-irrigation practices (Kakar 2020).

Velasco-Muñoz et al. (2018) used the 20 most frequently keywords in articles on SWUA for the period of 1993–2017 (Table 4). The period was divided into 5-year sub-periods, which show the evolution of keywords. The values that appear in the table refer to the position that keywords occupied in each sub-period compared with the total number of keywords of the sample, the number of articles in which it appeared [R (A)], and the repetition percentage (%). The 20 most frequently used keywords included terms related to water (water use, water management, water supply, water resources, water use efficiency, water conservation, groundwater and groundwater resources); sustainability (sustainability, sustainable development, and climate change); and crops and agricultural practices (irrigation, agriculture, crops, irrigation system, crop yield, crop production, *Triticum aestivum* evapotranspiration, and *Zea mays*).

Table 4 Evolution of the main keywords (with MDPI open-access license)

Keywords	1993–2017		1993–1997		1998–2002		2003–2007		2008–2012		2003–2017	
	A	%	R(A)	%	R(A)	%	R(A)	%	R(A)	%	R(A)	%
Water use	697	33.45	9(6)	13.04	3(29)	24.17	3(89)	34.23	2(255)	42.57	3 (318)	30.03
Irrigation	691	33.16	1(11)	23.91	1(35)	29.17	1(104)	40.00	4(207)	34.56	1(334)	31.54
Water management	668	32.05	4(8)	17.39	2(33)	27.50	2(97)	37.31	3(216)	36.06	4(314)	29.65
Water supply	661	31.72	310)	21.74	7(13)	10.83	8(50)	19.23	1(263)	43.91	2(325)	30.69
Sustainability	572	27.45	15(4)	8.70	5(21)	17.50	4(74)	28.46	5(191)	31.89	6(282)	26.63
Water resources	547	26.25	7(7)	15.22	5(21)	17.50	5(68)	26.15	7(152)	25.38	5(299)	28.23
Water use efficiency	477	22.89	12(5)	10.87	9(9)	7.50	9(49)	18.85	6(160)	26.71	7(251)	23.70
Sustainable development	475	22.79	4(8)	17.39	4(24)	20.00	7(67)	25.77	8(134)	22.37	8(242)	22.85
Agriculture	342	16.41	21(3)	6.52	12(8)	6.67	12(39)	15.00	11(81)	13.52	9(211)	19.92
Water conservation	311	14.92	21(3)	6.52	51(3)	2.50	24(22)	8.46	9(106)	17.70	10(177)	16.71
Crops	281	13.48	0	0.00	25(5)	4.17	14(34)	13.08	13(79)	13.19	12(161)	15.20
Ground water	279	13.39	33(2)	4.35	9(9)	7.50	11(41)	13.46	12(81)	13.52	14(141)	13.31
Irrigation system	267	12.81	0	0.00	19(6)	5.00	30(18)	6.92	10(93)	15.53	13(150)	14.16
Crop yield	228	10.94	72(1)	2.17	19(6)	5.00	15(27)	10.38	15(72)	12.02	16(122)	11.52
Crop production	209	10.03	0	0.00	43(4)	3.33	29 (19)	7.31	17(69)	11.52	17(118)	11.14
Triticum aestivum	197	9.45	33(2)	4.35	16(7)	5.83	13(38)	14.62	18(68)	11.35	26 (80)	7.55
Climate change	192	9.21	0	0.00	51(3)	2.50	81(7)	2.69	21(57)	9.52	15(125)	11.80
Evapotranspiration	176	8.45	0	0.00	43(4)	3.33	25 (21)	8.08	30(46)	7.68	18(105)	9.92
Groundwater resources	175	8.40	33(2)	4.35	25(5)	4.17	30(18)	6.92	23(56)	9.35	22(94)	8.88
Zea mays	171	8.21	72(1)	2.17	43(4)	3.33	21(23)	8.85	22(57)	9.52	25(86)	8.12

A = the annual number of total articles; R = ranking position

India is a country which depends on monsoon, hence floods are annual feature due to lack of appropriate drainage system and reservoirs (Fig. 12). Rain water storage in wells and pumping for drip irrigation in dry/rain-fed agroecosystem is a feasible practice by farmers (Fig. 13).

Agriculture causes diffused water pollution (Mancuso et al., 2021). Nature-Based Solutions (NBS) are suitable and sustainable to apply for environmental issues related to diffuse water pollution from agriculture for nitrogen removal and recovery. NBSs can help with wastewater treatment and management, water storage, drinking water provision, groundwater recharge, coastal protection, and agricultural practice, ensuring the provision of ecosystem services and a wide range of other environmental benefits (Fig. 14).

Fig. 12 Urban floods in India and many countries are an annual feature due to lack of appropriate drainage system. With the help of engineered drainage system and reservoirs, rain water could be used for irrigation. See Fig. 13 for drip irrigation using water stored in well

Fig. 13 Rain water storage in well and pumping for drip irrigation in dray/rain-fed agroecosystem **a** Well for storage of rain water **b–e** Water pumping devise fitted with filter f and **g** Drip irrigation to annual crops **h** drip irrigation to perennial crops, e.g., Mango orchard

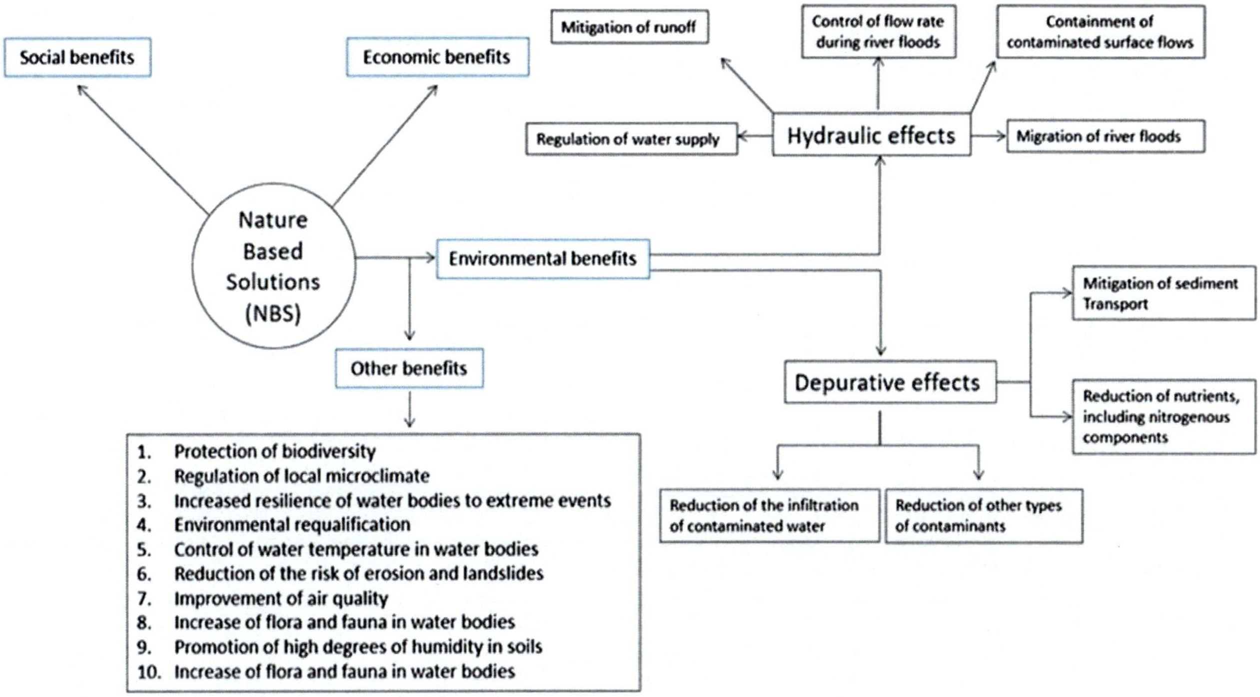

Fig. 14 Main effects attributable to the implementation of NBS (redrawn from Mancuso et al., 2021)

Acknowledgements Thanks are due to Elsevier for according permission to reproduce Figures 10 and 11; MDPI Figs. 5 and 6 and Tables 2 and 4.

References

Abbasi A. Z., Islam, N., & Shaikh, Z. A. (2014). A review of wireless sensors and network applications in agriculture. *Computer Standards & Interfaces, 36*(2), 263–270.

Abd El-kader, S. M., & Mohammad El-Basioni, B. M. (2013). Precision farming solution in Egypt using the wireless sensor network technology. *Egyptian Informatics Journal, 14*(3), 221–233.

Aznar-Sánchez, J. A., Belmonte-Ureña, L. J., Velasco-Muñoz, J. F., & Manzano-Agugliaro, F. (2018). Economic analysis of sustainable water use: A review of worldwide research. *Journal of Cleaner Production, 198*, 1120–1132.

Bhosale, S. K. (2022). Development of a solar-powered submersible pump system without the use of batteries in agriculture. *Indonesian Journal of Educational Research and Technology, 2*(1), 57–64.

Bretreger, D., Yeo, I. Y., & Hancock, G. (2022). Quantifying irrigation water use with remote sensing: Soil water deficit modelling with uncertain soil parameters. *Agricultural Water Management, 260*, 107299.

Chartzoulakis, K., & Bertaki, M. (2015). Sustainable water management in agriculture under climate change. *Agriculture and Agricultural Science Procedia, 4*, 88–98.

Deepika, G., & Rajapirian, P. (2016). Wireless sensor network in precision agriculture: a survey. In *Proceedings of the 2016 International Conference on IEEE Emerging Trends in Engineering, Technology and Science (ICETETS)*, Pudukkottai, India.

Dong, X., Vuran, M. C., & Irmak, S. (2013). Autonomous precision agriculture through integration of wireless underground sensor networks with center pivot irrigation systems. *Ad Hoc Networks, 11*(7), 1975–1987.

Dubey, P. K., Singh, A., Chaurasia, R., Pandey, K. K., Bundela, A. K., Dubey, R. K., & Abhilash, P. C. (2021). Planet friendly agriculture: Farming for people and the planet. *Current Research in Environmental Sustainability, 3*, 100041.

Gaddam, A., Al-Hrooby, M., & Esmael, W. F. (2014). Designing a wireless sensors network for monitoring and predicting droughts. In *Proceedings of the 8th International Conference on Sensing Technology*, Liverpool, UK.

Gajendran, E., Prabhu, S. B., & Pradeep, M. (2017). An analysis of smart irrigation system using wireless sensor. *Multidisciplinary Journal of Scientific Research & Education, 3*(3), 230–234.

Garc'ıa L., L. Parra, J. M. Jimenez, J. Lloret, & Lorenz, P. (2020). IoT-based smart irrigation systems: An overview on the recent trends on sensors and IoT systems for irrigation in precision agriculture. *Sensors, 20*(4), 1042.

Goumopoulos, C., O'Flynn, B., & Kameas, A. (2014). Automated zone-specific irrigation with wireless sensor/actuator network and adaptable decision support. *Computers and Electronics in Agriculture, 105*, 20–33.

Grace, K. V., Kharim, S., & Sivasakthi, P. (2015). Wireless sensor based control system in agriculture field. In *Proceedings of the Global Conference on Communication Technologies(GCCT)*, Uckalay, India.

Gutierrez, J., Villa-Medina, J. F., Nieto-Garibay, A., & Porta-Gandara, M. A. (2014). Automated irrigation system using a wireless sensor network and GPRS module. *IEEE Transactions on Instrumentation and Measurement, 63*(1), 166–176.

Hassan, F., Roy, A., & Saxena, N. (2016). Convergence of WSN and cognitive cellular network using maximum frequency reuse. *IET Communications, 11*, 664–672.

Hegde, N. G. (2012). Water scarcity and security in India. In *Proceedings of the 2012 BAIF-Indian Science Congress* (pp. 1–19). Kolkata, India.

Hussain, S., Amin, A., Mubeen, M., Khaliq, T., Shahid, M., Hammad, H. M., Sultana, S. R., Awais, M., Murtaza, B., Amjad, M., & Fahad, S. (2022). Climate smart agriculture (CSA) technologies. In *Building climate resilience in agriculture* (pp. 319–338). Springer, Cham.

Hwang, J., Shin, C., & Yoe, H. (2010). Study on an agricultural environment monitoring server system using wireless sensor networks. *Sensors, 10*(12), 11189–11211, Dec.

Imam, S. A., Choudhary, A., & Sachan, V. K. (2015). Design issues for wireless sensor networks and smart humidity sensors for precision agriculture: a review. In *Proceedings of the 2015 International Conference on Soft Computing Techniques and Implementations-(ICSCTI)*, Faridabad, India.

Jao, J., Sun, B., & Wu, K. (2013). A prototype wireless sensor network for precision agriculture. In *Proceedings of the IEEE33rd International Conference on Distributed Computing Systems Workshops*, Philadelphia, PA, USA.

Kakar, R. (2020). Micro-irrigation: The way ahead for sustainable agriculture. https://www.downtoearth.org.in/blog/agriculture/micro-irrigation-the-way-ahead-forsustainable-a griculture-73153. Accessed on 3 Feb 2022.

Kamienski, C., Soininen, J. P., Taumberger, M., Dantas, R., Toscano, A., Salmon Cinotti, T., Filev Maia, R., & Torre, N. A. (2019). Smart water management platform: IoT-based precision irrigation for agriculture. *Sensors., 19*(2), 276.

Kassim M. R. M., Mat, I., & Harun, A. N. (2014). Wireless sensor network in precision agriculture application. In *Proceedings of the 2014 International Conference on Computer, Information and Telecommunication Systems (CITS)*, Jeju, Republic of Korea.

Khorsandi, F. (2022). Haloculture: Strategy for sustainable utilization of saline land and water resources. *Iranian Journal of Earth Sciences, 8*(2), 164–172.

Kodali R. K., Rawat, N., & Boppana, L. (2014). WSN sensors for precision agriculture. In *Proceedings of the 2014 Region 10 Symposium*, Kuala Lumpur, Malaysia.

Krishna, G., & Biradar, C. (2022). Geo-big data in digital augmentation and accelerating sustainable agroecosystems. In *Data science in agriculture and natural resource management* (pp. 221–242). Springer, Singapore.

Kumar, D., & Choudhury, U. (2022). Agriculture-IoT-based sprinkler system for water and fertilizer conservation and management. In *Design and development of efficient energy systems* (pp. 229–244).

Lampridi, M. G., Sørensen, C. G., & Bochtis, D. (2019). Agricultural sustainability: A review of concepts and methods. *Sustainability, 11* (18), 5120.

Li, Y., Gao, W., Wu, C., & Wang, Y. (2015). Deployment of sensors in WSN: An efficient approach based on dynamic programming. *Chinese Journal of Electronics, 24*(1), 33–36.

Loiskandl, W., & Nolz, R. (2021). Requirements for sustainable irrigated agriculture. *Agronomy, 2021*(11), 306. https://doi.org/10.3390/agronomy11020306

Lopez, J. A. A. J., Garcia-Sanchez, F., Soto, A., Iborra, F.-S., & Garcia-Haro, J. (2011). Design and validation of a wireless sensor network architecture for precision horticulture applications. *Precision Agriculture, 12*(2), 280–295.

Madhumathi, R., Arumuganathan, T., & Shruthi, R. (2022). Internet of things in precision agriculture: A survey on sensing mechanisms, potential applications, and challenges. In *Intelligent sustainable systems* (pp. 539–553). Springer, Singapore.

Mafuta, M., Zennaro, M., Bagula, A., Ault, G., Gombachika, H., & Chadza, T. (2013). Successful deployment of a wireless sensor network for precision agriculture in Malawi. *International Journal of Distributed Sensor Networks, 2013*, 1–13.

Mancuso, G., Bencresciuto, G. F., Lavrnić, S., & Toscano, A. (2021). Diffuse water pollution from agriculture: A review of nature-based solutions for nitrogen removal and recovery. *Water, 13*(14), 1893.

Meissner, R., & Rupp Hand Haselow, L. (2020). Use of lysimeters for monitoring soil water balance parameters and nutrient leaching. In M. N. V. Prasad & M. Pietrzykowski (Eds.), *2020* (pp. 171–205). Elsevier.

Meissner, R., Prasad, M.N.V., Du Laing, G., & Rinklebe, J. (2010). Lysimeter application for measuring the water and solute fluxes with high precision. *Current Science*, *99*(5).

Minhas, P. S., Saha, J. K., Dotaniya, M. L., Sarkar, A., & Saha, M. (2022). Wastewater irrigation in India: Current status, impacts and response options. *Science of the Total Environment, 808*, 152001.

Mpanga, I. K., & Idowu, O. J. (2021). A decade of irrigation water use trends in Southwest USA: The role of irrigation technology, best management practices, and outreach education programs. *Agricultural Water Management, 243*, 106438.

Muñoz-García, M. A., & Hernández-Callejo, L. (2022). Photovoltaics and electrification in agriculture. *Agronomy, 12*(1), 44.

Nathan. (2015). IoT and Agriculture: How the Internet of Things is Changing Agricultural Operations, *Davra Networks*, 26-Aug.

Nesa, S. M., Valarmathi, M. L., & Babu, A. S. (2011). Energy efficient data transmission in automatic irrigation system using wireless sensor networks. *Computers and Electronics in Agriculture, 78*(2), 215–221.

Nyika, J.M. (2022). Climate change situation in Kenya and measures towards adaptive management in the water sector. In *Research anthology on environmental and societal impacts of climate change* (pp. 1857–1872). IGI Global.

Ouda, S., & Zohry, A.E.H. (2022). Water-smart practices to manage water scarcity. In *Climate-smart agriculture* (pp. 3–26). Springer, Cham.

Phasinam, K., Kassanuk, T., Shinde, P. P., Thakar, C. M., Sharma, D. K., Mohiddin, M., & Rahmani, A. W. (2022). Application of IoT and cloud computing in automation of agriculture irrigation. *Journal of Food Quality, 2022.*

Phillips, P., & Encel, K. (2022). Measuring sustainable development goal 6. In *The Routledge handbook of sport and sustainable development* (pp. 158–168). Routledge.

Ruiz-Canales, A., & García, M.F.-V. (2021). Sustainable applications in agriculture. *Sustainability, 2021*(13), 4136. https://doi.org/10.3390/su13084136

Sakthipriya, N. (2014). An effective method for crop monitoring using wireless sensor network. *Middle-East Journal of Scientific Research, 20*(9), 1127–1132.

Saray, M. H., Baubekova, A., Gohari, A., Eslamian, S. S., Klove, B., & Haghighi, A. T. (2022). Optimization of water-energy-food nexus considering CO2 emissions from cropland: A case study in northwest Iran. *Applied Energy, 307*, 118236.

Satyanarayana, G. V., & Mazaruddin, S. D. (2013). Wireless sensor based remote monitoring system for agriculture using Zigbee and GPS. In *Conference on Advances in Communication and Control Systems* (pp. 110–114).

Sinha, B. B., & Dhanalakshmi, R. (2022). Recent advancements and challenges of Internet of Things in smart agriculture: A survey. *Future Generation Computer Systems, 126*, 169–184.

Smol, M. (2022). Circular economy approach in the water and wastewater sector. In *Circular economy and sustainability* (pp. 1–19). Elsevier.

Sucharitha, G., & Sai, M. M. (2022). *Developments in agriculture technology using internet of things. in internet of things and its applications* (pp. 341–360). Springer, Cham.

TongKe, F. (2013). Smart agriculture based on cloud computing and IOT. *Journal of Convergence Information Technology*, *8*(2), 210–216, Jan.

Tsai, C.-W., Hong, T.-P., & Shiu, G.-N. (2016). Metaheuristics for the lifetime of WSN: A review. *IEEE Sensors Journal, 16*(9), 2812–2831.

Velasco-Muñoz, J. F., Aznar-Sánchez, J. A., Belmonte-Ureña, L. J., & Román-Sánchez, I. M. (2018). Sustainable water use in agriculture: A review of worldwide research. *Sustainability*, *10*, 1084. https://doi.org/10.3390/su10041084.

Vellidis, G., Liakos, V., Andreis, J. H., Perry, C. D., Porter, W. M., Barnes, E. M., Morgan, K. T., Fraisse, C., & Migliaccio, K. W. (2016). Development and assessment of a smartphone application for irrigation scheduling in cotton. *Computers and Electronics in Agriculture, 127*, 249–259.

Zhao, J.-C., Zhang, J.-F., Feng, Y., & Guo, J.-X. (2010) The study and application of the IOT technology in agriculture (pp. 462–465).

Zhu, Y., Song, J., & Dong, F. (2011). Applications of wireless sensor network in the agriculture environment monitoring. *Procedia Eng.*, *16*, 608–614.

Innovative Solutions in the Transition to Circular Economy in Water Sector

Advanced Treatment Technologies in Removal of Pollutants from Water and Wastewater

Santhoshi Chitthaluri, Revanth Mamidala, Kiruthika Eswari Velmaiel, Vineeth Manthapuri, Kasadi Naveen, and P. RajaSekhar

Abstract

The water and wastewater treatment and its management is a significant element in a circular economic purview since a number of industries depend on water and also on the recovered resources out of wastewater. Because limited access to water or these resources can limit production capacities and profits for the industries. Also, the discharge of large quantities of effluents, containing hazardous pollutants from agricultural, municipal, and industrial sources is leading to the contamination of the water supplies to a great extent. This paper communicates insights on the recent advanced water and wastewater treatment technologies through a comprehensive review. Also, it highlights the benefits and challenges faced by these removal mechanisms providing scope for future research. Through this work, adsorption can be identified as a simple, sustainable, cost-effective, and environmentally friendly technique for water and wastewater treatment compared to other technologies discussed. Nevertheless, there is a dire need for future research and development focusing on evaluating hybrid treatment systems, large-scale practicality, and techno-economic assessment of the systems to understand the feasibility of these technologies for a broad range of applications.

Keywords

Wastewater treatment • Membrane technology • Microalgae • Microbial fuel cell • Adsorption • Advanced oxidation process • Contaminants

S. Chitthaluri (✉)
Environmental Engineering, Department of Civil Engineering, National Institute of Technology Warangal, Hanamkonda, Telangana State, India
e-mail: schittha@student.nitw.ac.in

R. Mamidala · P. RajaSekhar
Water Resources Engineering, Department of Civil Engineering, Osmania University, Hyderabad, India
e-mail: prscivilou@osmania.ac.in

K. E. Velmaiel · V. Manthapuri
Environmental and Water Resource Engineering Division, Department of Civil Engineering, IIT Madras, Chennai, India
e-mail: mvineeth@vt.edu

K. Naveen
Water Resources Engineering, Department of Civil Engineering, KU Leuven, Leuven, Belgium

1 Introduction

Exponentially increasing population in the global level has put all the developmental activities in a nation at risk especially the two major problems faced are resource scarcity and environmental pollution (Tan et al., 2021). By 2025, it is estimated that greater than 3 billion people worldwide will experience water shortages (Jafarinejad, 2021; Bin Abdullah, 2005). The problems associated with the pollution in water if left unchecked lead to greater harm to the public health on large scale. And the increasing challenges associated with the decreasing water availability and rapid industrialization activities, in turn, with more and more wide range of contaminants coming into existence are causing a huge burden on the conventional water and wastewater treatment processes (Armah et al., 2021). The effectiveness of these conventional treatment processes in removing the chemical and microbial contaminants has gotten limited over the last two decades. The presence and persistence of emerging contaminants and their transformed forms are a matter of serious health and environmental concerns (Zhou & Smith, 2001).

Water occupies about 71% of the earth's surface and it constitutes to 65% of our bodies. Water and wastewater treatment plants are critical to ensuring public health as they strive towards maintaining sanitation and hygiene. We use water for a range of purposes like cooking, gardening, washing, etc. and it gets ended up as wastewater (Jayalekshmi et al., 2021). Water as it gets contaminated poses threat

M. Smol et al. (eds.), *Water in Circular Economy*, Advances in Science, Technology & Innovation,
https://doi.org/10.1007/978-3-031-18165-8_6

to the health and environment and subsequently loses its aesthetic and economic value (Jayalekshmi et al., 2021; Sonune & Ghate, 2004). It is estimated that around 70–80% of the total water supplies gets converted to wastewater and released into the natural water streams. The wastewater contains suspended solids, minerals, nutrients, pathogenic bacteria, toxic metals, micro-organisms, etc. Proper treatment of wastewater involves many different processes, involving physical, chemical and biological treatment methods (Arthurson, 2008) to remove solids, organic matter and nutrients from wastewater. Preliminary, primary, secondary and tertiary treatments are the stages involved, in general.

Firstly, the increased knowledge about the consequences from water pollution and the desire of public to have accessibility to water of better quality prompted the implementation of a great deal of regulations enforced getting stricter, by mounting the range of contaminants being regulated and reducing the limits of contaminant levels. Secondly, the continuous decrease in the water availability, its resources, and the rapid population growth rates and industrial evolution made the reuse of the municipal and industrial wastewaters and the recovery of potential pollutants used in industrial processes more vital. This is more critical in the arid or semiarid regions as the import of potable water and irrigation water is carried out at great expense (Zhou & Smith, 2001).

The various human activities and increasing industrialization have created new contaminants in wastewater called emerging contaminants (EC). In the view of growing concern over the contamination of water resources by the release of more toxic contaminants, advanced treatment technologies have established to eliminate diverse potentially toxic compounds which possibly will not be successfully separated by traditional treatment processes. Also, the advances in the manufacturing industries and the associated market growth in the advanced treatment systems have profited the system as a whole in its versatility and costs at industrial scales (Zhou & Smith, 2001). The present study initially briefs on the conventional water and wastewater treatment systems and then details on the working principles, efficiencies, and limitations of various advanced modern water and wastewater treatment technologies, so as to resolve new challenges and to have a better use of economical resources.

2 Conventional Water and Wastewater Treatment Systems

This section briefly details on conventional water and wastewater treatment systems.

2.1 Conventional Water Treatment

Majority of the diseases, greater than 80% of them, which affect humankind are waterborne and a large number of these diseases can be prevented by processing water used for domestic purposes through basic treatment. Therefore, the call for treatment of water for supply to communities, whether it is from surface water or groundwater sources, holds great importance. In particular, when the same source is being utilized for various purposes, such as navigation, irrigation, and domestic, the concern for water quality for drinking purposes can be alleviated only through satisfactory treatment using suitable processes (Vigneswaran et al., 2009).

A drinking water treatment plant can be varied as follows:

Filtration plant: This is usually used to treat surface water.

Disinfection plant: This is used for high-quality water source to make sure that water does not contain pathogens.

Softening plant: This is generally used to treat groundwater.

Typical filtration plant is illustrated in Fig. 1; the basic motive behind its designed is to get rid of turbidity, colour

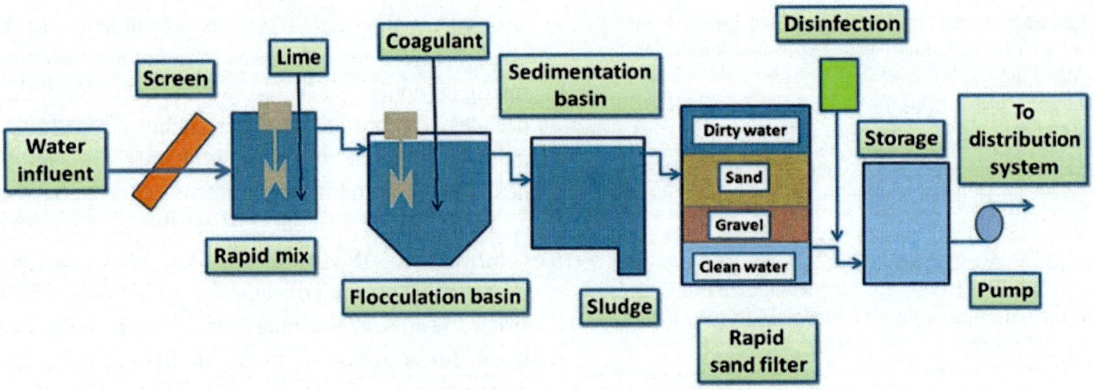

Fig. 1 Filtration treatment plant. *Source* from Shmeis (2018)

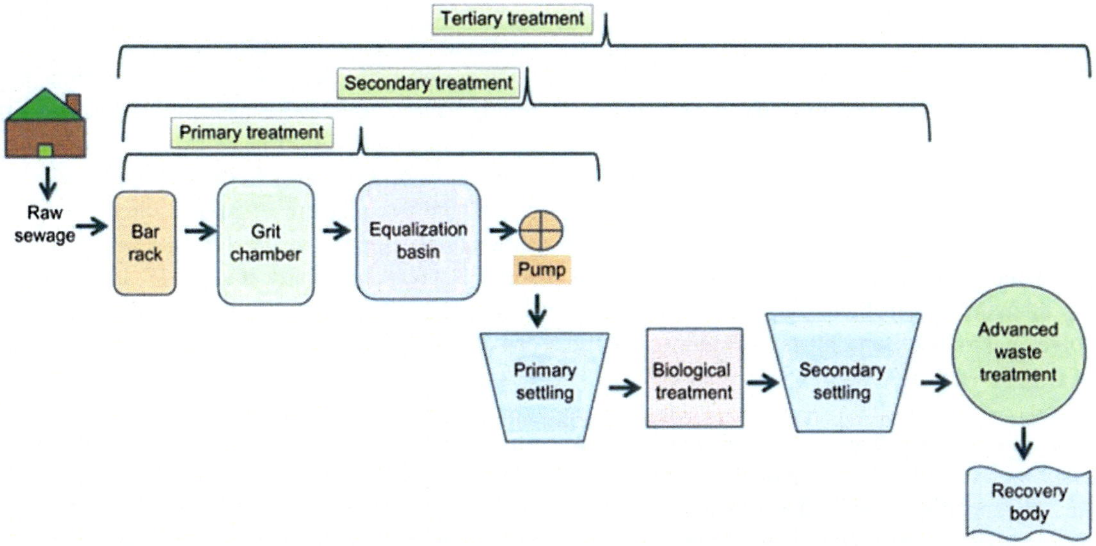

Fig. 2 Conventional wastewater treatment system. *Source* from Shmeis (2018)

and odours, in addition to bacteria and other contaminants. A filtration plant, in general, involves the following stages (Shmeis, 2018):

Rapid mixing: This is the units into which the chemicals (like alum) are added and mixing is done to rapidly disperse them throughout the inflow water.

Flocculation: Slow mixing is carried out to make them come close to each other and in turn to neutralize the charge of the particles and subsequently form large flocs that further possibly can readily be settled.

Sedimentation: During this stage, the flocs due to its increased weight can get settled at the bottom of the tank.

Filtration: The settled water which is clear water on top is allowed to pass through a series of filters with altering compositions (like gravel, sand and charcoal) and pore sizes to facilitate the removal of fine particles that were not settled, for example, chemicals, dust, bacteria, parasites and viruses.

Disinfection: To reduce or to kill a number of pathogens, chemicals are added to the filtered water in this stage.

The same operational units are employed in softening plants, as of the filtration plants but different chemicals are used to facilitate the water hardness removal.

2.2 Conventional Wastewater Treatment

A typical conventional wastewater treatment comprises a combination of physical, chemical and biological operations and processes to eliminate solids, organic matter and, occasionally, nutrients from wastewater. In general, the terms used to describe different degrees of treatment, in order to increasing treatment level, are preliminary, primary, secondary and tertiary and/or advanced wastewater treatment (Saravanathamizhan & Perarasu, 2021). Figure 2 illustrates a conventional wastewater treatment system.

Preliminary and Primary Treatment: The removal of floating materials like papers, pieces of wood, rags, heavily in settleable inorganic solids, dead animals, etc. Also involves removal of oils, grease, etc., from the sewage. Screening, comminutors, detritus tank, skimming tanks and floatation unit are the range of units involved in this stage of treatment. The floating matter removal is carried out by screening units. Grit chamber is used for elimination of grit and sand. The oil and grease components are eliminated by skimming tanks (Amina, 2017). BOD of the wastewater reduces by 15–30% by this stage of treatment (Jayalekshmi et al., 2021).

Primary treatment contributes to removal of incoming biochemical oxygen demand of about 5–50%, total suspended solids of about 50–70%, and grease and oils of about 65%. This stage involves the removal of organic and inorganic solids by sedimentation and floatation. Though doesn't remove colloidal and dissolved contaminants, this greatly associates in removal of organic nitrogen and phosphorus, and heavy metals associated with solids acting as the pre-application treatment of wastewater in many industrialized countries (Jayalekshmi et al., 2021).

Primary sludge is usually removed from the bottom of tanks to a central well, where it is pumped to sludge-processing units while scum as well is separated from the surface of the tank either by mechanical means or by using water jets to sludge-processing units (Ghodeif, 2013; Jayalekshmi et al., 2021).

Secondary Treatment: About 85% of BOD_5 is removed, but does not eliminate significant amount of nitrogen, phosphorous or heavy metals, nor does it completely remove pathogenic bacteria and viruses. The organic matter of colloidal nature which passes through the primary clarifiers has to be further treated. Filtration and activated sludge process are the two different types of treatment methods used here, which involves intermittent sand filters, trickling filters and contact beds (Fujie et al., 1983).

Advanced or Tertiary Treatment: Contaminants that are not separated by secondary are removed in this stage, resulting in a better quality water that is ready for either reuse, recycle or to be discharged into the environment. This stage involves the removal of refractory substances and compounds such as nitrogen and phosphorous (Metcalf et al., 1991).

3 Recent Developments in Water and Wastewater Treatment Technologies

Before studying the treatment technologies it is very important to understand the sources of contamination and characteristics of the water to be treated, so that the technology to treat it can be decided accordingly.

Figure 3 shows the sources and flow of emerging contaminants in the environment.

There are two chief sources of contaminants in wastewater: (i) natural, like weathering of rocks, volcanic activity and soil erosion and (ii) dispersion of mineral contaminants through waste disposal sites, anthropogenic activities, urban runoff, the manufacture of printed circuit boards, mining, agricultural activities, textile dyes, the electroplating of metal surfaces, etc. (Chowdhary et al., 2018; Launay et al., 2016).

Wastewater generated from industries, household sector and agriculture contains altering quantities of toxic inorganic (excessive nutrients and heavy metals) and organic (polyaromatic hydrocarbons, pigments, etc.) contaminants that pose serious environmental and health risks (Ariffin et al., 2017; Rashed, 2013; Roccaro et al., 2013; Zheng et al., 2013). Among the metalloids and heavy metals (potentially toxic elements or PTEs), PTEs are of trace elements category with a density >4 ± 1 g cm^{-3}. They comprise mercury (Hg), copper (Cu), cadmium (Cd), tin (Sn), zinc (Zn), iron (Fe), silver (Ag), lead (Pb), manganese (Mn), cobalt (Co), chromium (Cr), arsenic (As), nickel (Ni) and aluminium (Al) (Hussain et al., 2021; Mohammed et al., 2011). Due to their solubility, persistence and higher mobility, wastewater with these PTEs if not treated effectively and released into freshwater bodies could lead to various environmental and health effects. These PTEs are generally consumed by crops and aquatic organisms, and enter the human food chain, thus adversely affecting human health (Akpor et al., 2014; Barakat, 2011; Harvey et al, 2015).

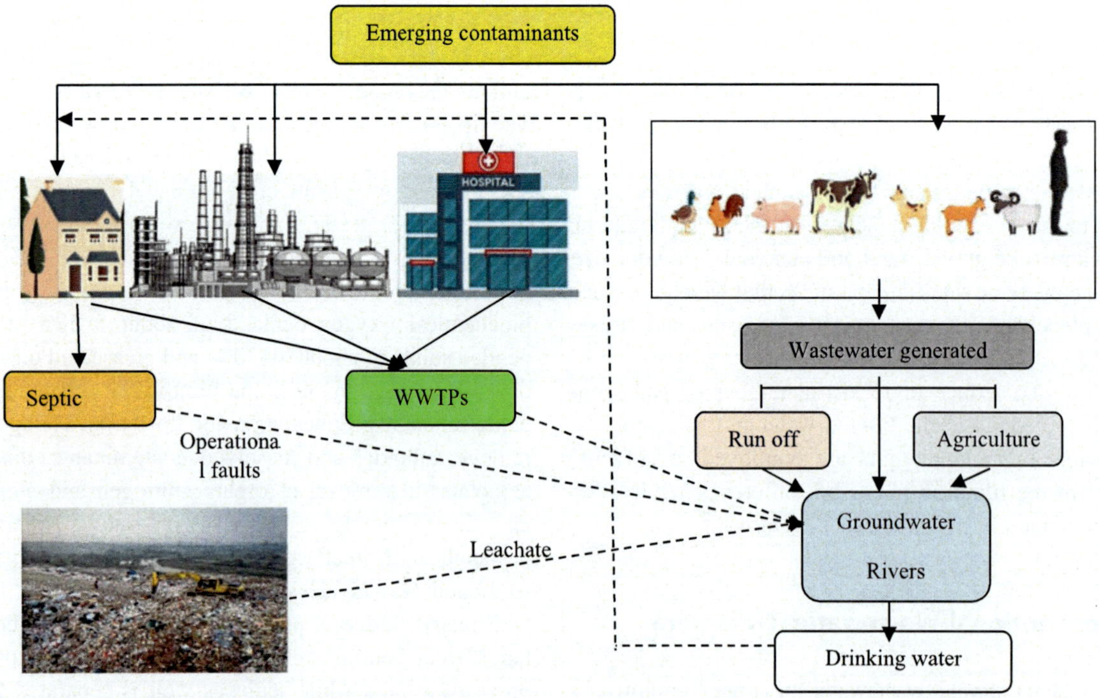

Fig. 3 Sources and the flow of ECs in the environment (Shahid et al., 2021)

Apart from anthropogenic and natural sources, there are two important types of wastewater pollutants, they are organic and inorganic pollutants. Organic pollutants include phenols, pesticides, herbicides, dyes, petroleum, oils, fats, biphenyls, proteins, medicines and starches, while inorganic pollutants contain PTEs, excess nutrients and chemical fertilizers (Ali, 2008; Alssgeer et al., 2018).

It is crucial to characterize the water to determine key wastewater characteristics, since the treatment technology to be preferred depends on the nature of the wastewater to be treated (Crini & Lichtfouse, 2019).

Wastewater characteristics: The water containing contaminants majorly due to human use is termed as wastewater. The chemical quality of wastewater is identified by the inorganic and organic components of wastewater.

The parameters generally considered when measuring the chemical characteristics of wastewater are pH, alkalinity, volatile solids (VS), total solids (TS), chemical oxygen demand, biochemical oxygen demand, total nitrogen, total phosphorus, etc. Table 1 presents various wastewater parameters and their definitions.

Table 2 summarizes the raw wastewater characteristics of a few industries. To mitigate the environmental and health hazards created by the wastewater generated from such various industries, numerous technologies are used that are based on different intensities of treatments, and processes, for instance, membrane filtration, chemical precipitation, reverse osmosis, solvent extraction, adsorption, oxidation, etc. (Bolan, 2008; Caicedo et al., 2019; Choudhary et al., 2020; Lehr et al., 2005).

This chapter specifically focuses on membrane technologies, microalgal treatment, microbial fuel cells, adsorption techniques and advanced oxidation processes.

3.1 Membrane Filtration Technologies

Membrane filtration can be largely defined as a separation process that uses semi-permeable membrane to split the feed

Table 1 Wastewater parameters and their definitions

Parameters	Definition
Total solids	This is the organic and inorganic matter; suspended and dissolved solids; settleable and volatile solid content of wastewater (Armah et al., 2021)
Chemical oxygen demand (COD)	COD is measured as the quantity of oxygen required to stabilize the carbonaceous organic matter chemically (Shmeis, 2018)
Biochemical oxygen demand (BOD)	This is the oxygen quantity required by micro-organisms for the organic matter degradation under aerobic conditions. BOD is an indication of the degree of contamination (Armah et al., 2021)
Total nitrogen (TN) and total phosphorus (TP)	TN is the combination of both the organic and inorganic nitrogen, and ammonia in wastewater, it exists as nitrite, nitrate, ammonium and organic dissolved compounds such as urea, amino acids and organic nitrogen composites. Phosphorus is present as phosphates such as orthophosphates, condensed phosphates, etc. (Islam et al., 2019)
Metals	Even as low concentrations as 1–3 mg/L of metals in wastewater is toxic since they are non-biodegradable and they do accumulate eventually (Al-Saydeh et al., 2017; Baysal et al., 2013). Even though a few metals which are crucial to human, plants and animals may still be tolerated in minimal quantities, for example, zinc, copper, but if found beyond the limits can be exceptionally toxic
Pharmaceutical compounds	The pharmaceutical compounds that have been recently found in the wastewater are antibiotics, anti-cancer agents, anticonvulsants, contrast agents, beta-blockers, lipid-regulators, antidepressants, hormones and analgesics. Since their long-term effect on human and aquatic habitats is unknown, they are considered as a part of emerging pollutants in wastewater (Deblonde et al., 2011)
Bacteria and viruses	Viral and bacterial infections from waterborne outbreaks are usually connected with environments associated with the discharge of wastewater. Enteric viruses such as rotaviruses, noroviruses, enteroviruses, astroviruses, bocaviruses, hepatitis-A virus, papillomaviruses, hepatitis-E virus, human polyomaviruses, Aichi virus, a plant virus called pepper mild mottle virus and enteric bacteria such as faecal coliforms, bacteriophages and E-coli are found in wastewater, and the full particulars of their occurrence and concentration in untreated and treated wastewater by continents are reviewed by Farkas et al. (2020)

Table 2 Raw industrial wastewater chracteristics (Chang et al., 2008; Fazal et al., 2018; Enitan et al., 2015; Munter & Lundin, 2000; Armah et al., 2021)

Parameters/industry	Abattoir	Brewery	Cane sugar	Coke oven	Textile	Oil refinery	Tannery
BOD_5, mg/L	476–3850	1609.34–3980.61	350–2750	510–1360	50–500	100–500	1000–2000
COD, mg/L	935–6600	1096.41–8926.08	1000–4340	930–3120	250–8000	150–800	2000–4000
TSS, mg/L	750–4400	530.67–3728.02	760–800	19–3330	100–700	130–600	2000–3000
pH	6.85–8.19	4.6–7.3	5–6.5	6.8–8.2	5.6–9	2–6	11–12

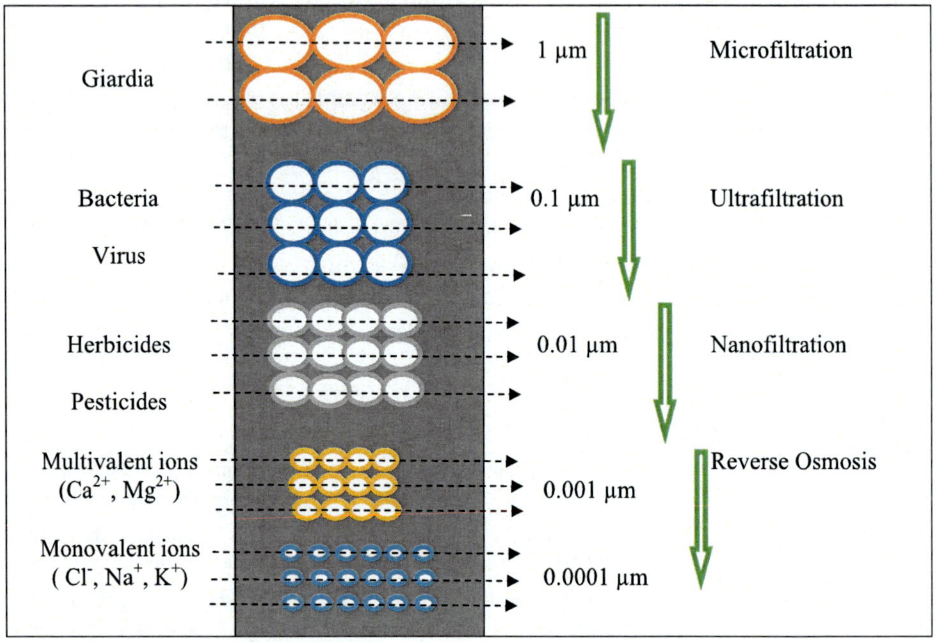

Fig. 4 Sieving of emerging contaminants through membranes with different pore sizes. Adapted from Sengupta et al. (2021)

stream into two portions: (a) the material passing through the membranes known as permeate and the species being left behind called retentate (Mallevialle et al., 1996).

Based on the range of the size of the permeating species, the corresponding employed driving forces, the rejection mechanisms, composition of membranes and its chemical structure, and the geometry of construction, all these constituents define the kind of membrane filtration established (Li et al., 2018). Successful use of membrane processes depends on the proper selection of membrane material (Fig. 4). Inorganic membranes have high chemical and temperature resistance but their restricted use is due to brittleness and expense whereas the organic polymers remain the most widely used commercial membrane materials. They have a thin active polymeric layer coated over microporous support providing mechanistic strength required for their durability while having high chemical resistance and water permeability. Polyamides, polysulfones, polypropylene and acetates are the typical polymers used as active layers (Zhou & Smith, 2001; Mallevialle et al., 1996).

Figure 5 illustrates the ideal membrane characteristics. In the selection of a membrane process, as through a practical application point of view, one of the most significant properties of membrane is the pore size or molecular weight cut-off (MWCO), which specifies the maximum molecular weight of a solute to be rejected (Zhou & Smith, 2001).

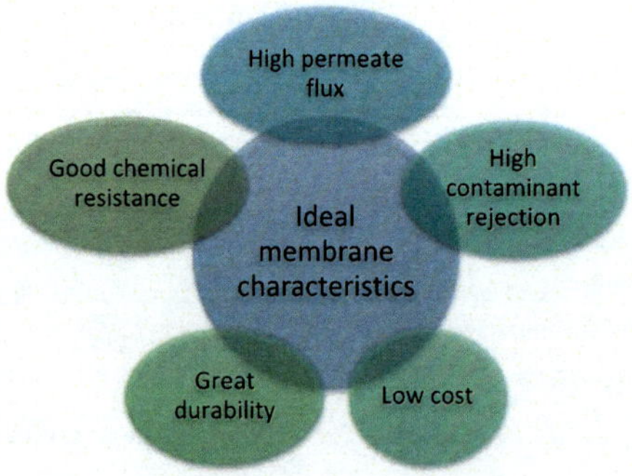

Fig. 5 Ideal membrane characteristics (Zhou & Smith, 2001)

Classification of membrane filtration processes: Table 3 summarizes classification based on characteristics of common membrane filtration processes.

Table 3 Main characteristics of common membrane filtration processes. Adapted from Zhou and Smith (2001)

Membrane	Separation size (μm)	Typical transmembrane pressure, ΔP (MPa)	Main mechanisms	Permeate flux
Reverse osmosis (RO)	<0.001	5–8	Diffusion+exclusion	Low
Nanofiltration (NF)	0.001–0.00	0.5–1.5	Diffusion+exclusion	Medium
Ultrafiltration (UF)	0.003–0.1	0.05–0.5	Sieving	High
Microfiltration (MF)	>0.05	0.03–0.3	Sieving	High

Microfiltration requires a relatively low TMP and the large pore sizes of MF limit the removal of suspended solids, protozoan, virus, cysts, bacteria and, on lesser extent, organic colloids within the region (Pabby et al., 2008).

As a pre-treatment for membrane bioreactor and desalination applications, UF is increasingly receiving recognition. UF and MF both use physical sieving as a separation mechanism.

In operation, UF with a specific MWCO is impermeable to compounds with molecular weights beyond the MWCO and has established a 3–6 log removal of chlorine-resistant colloids, protozoan cysts, coliform bacteria and viruses. As a pre-treatment to RO, the usage of MF and UF has seen an increasing demand at the industrial scales. Both MF and UF can be used as pre-treatment strategies for NF and RO processes, while applied as a post-treatment to chemical precipitation of pH adjustment, organic chemical removal, metal, hardness and phosphorous removal (Zhou & Smith, 2001).

NF is eminent by means of the elimination of magnesium and calcium ions consequential in water softening, and of no sodium ions addition all through the filtration process, when judged against ion exchange units. It doesn't require chemical treatment additionally so as to remove hardness (Nageswara, 2014).

NF has a restricted application in the industry owing to the pore size of the membrane, which is restricted to nanopore size. RO and UF are favoured in view of the fact that they can cover the UF range efficiently devoid of the cost constraints of NF due to high initial, operating and maintenance costs (Mohammad et al., 2007). The membranes have to be changed depending on the amount of total dissolved solids treated, ahead of the actual life which amplifies the NF cost.

RO is frequently referred to as a tight membrane and has been extensively used in brackish and wastewater treatment. The high external pressure in this process is a result of hypertonic feed and is typically more than the osmotic pressure which is applied to retain dissolved solute, and avert and permit the permeation of solvent at a MWCO of about 100 Da through diffusion mechanism (Obotey et al., 2020). It is highly effective in desalination systems.

The driving force being the concentration gradient, the separation and concentration in forward osmosis (FO) occur as the concentrated solution (e.g. salts such as NaCl) draw water from a less concentrated feed solution. The use of FO operates at ambient conditions, therefore irreversible fouling is little (Armah et al., 2021).

An MBR is a blending of membrane process (MF and UF) and a biological wastewater treatment using activated sludge. In current years, the MBR has been extensively used for municipal and industrial wastewater treatment. Latest publications point towards the feasibility of using an MBR to remove ECs (Besha et al., 2017; Ji et al., 2020; Prado et al., 2017).

In current MBRs, membrane modules are placed above the aeration tube, inside the bioreactor vessel or aerobic tank (Fig. 6a) (Cirja et al., 2008), while in first-generation MBRs, the membrane modules are placed outside the aerobic tank, and the aeration system is used to supply oxygen to the micro-organisms, which in turn are responsible for the biodegradation (Goswami et al., 2018) as shown in Fig. 6. Compared to other configurations, the current MBRs consume lower energy and subsequently lower EC biodegradation is seen (Su et al., 2020). The first-generation MBRs are costlier, hence generally used in small-scale high-strength applications since they can treat concentrations of higher mixed liquor suspended solids (MLSS). The latter is compact and requires easy maintenance, cleaning and module replacement (Wang & Menon, 2011).

A hybrid MBR with NF is used for the treatment of the textile wastewater streams to remove toxic dyes and other substances, at pilot scale which was studied by Li et al. (2020), where this MBR-NF process was found to reduce the footprint of the system by 13.7% compared to existing process.

Mass transport can lead to the accumulation, adsorption or attachment of materials onto the surfaces of membranes and (or) within pores of membranes, leading to reduction in permeate flux with time, and this occurrence is known as membrane fouling. Five principal fouling mechanisms seen are (i) concentration polarization, (ii) cake formation, (iii) organic adsorption, (iv) inorganic precipitation and (v) biological fouling. Every one of these mechanisms has diverse effects on recovery of flux from backwash.

Unless getting dissolved by cleaning agents in backwash water, fouling caused by small colloids and salt precipitates in the pores of membrane is generally considered as an irreversible process. This whole process leads to declination

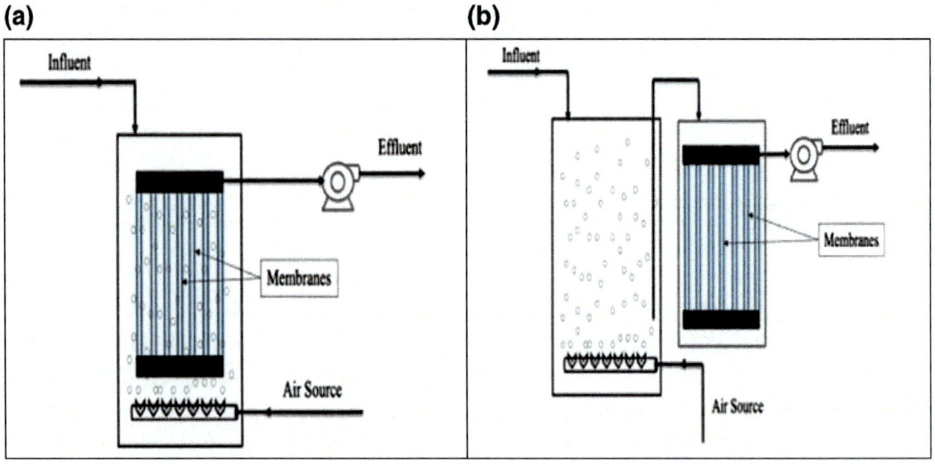

Fig. 6 **a** Current MBRs; **b** first-generation MBRs. *Source* from Sengupta et al. (2021)

of the flux rate in the long term. On the contrary, reduced flux rates owing to the growth of a surface cake are mainly reversible. Bio-fouling is usually accompanied first by adsorption of macromolecules to condition the membrane surface for the adhesion of micro-organisms, and then by microbial multiplication (Baker et al., 1995; Zhou & Smith, 2001).

The chemical modification of the membrane surface, physical improvement of the module geometric configuration by promoting hydrodynamic shear such as dean vortex and using rotational discs, the use of proper cleaning techniques, adjustment of operating conditions, addition of powdered activated carbon to form a more porous surface cake and the pre-treatment of influents are a few fouling control strategies. The effectiveness of these strategies seems to be specific to each site. In a few applications, integrated approaches by coupling satisfactory pre-treatment with selection of the proper membrane module and operating conditions are crucial to alleviate fouling problems efficiently and economically (Zhou & Smith, 2001).

Various advantages and disadvantages of membrane technology are mentioned in Fig. 7.

Clean technology, ability to replace conventional processes, energy saving (in most cases), flexibility in system design and ability to produce high-quality products are the major promoting aspects, for the domestic and industrial application of MT like water supply, pharmaceutical, chemical, biotechnological, food, beverages, metallurgy and in various other separation processes (Armah et al., 2021; Zheng et al., 2015).

3.2 Microalgal Wastewater Treatment

Population dynamics and the increased industrial set-ups have induced a disproportion in the water-resource equation. The demand for water in production sector, agricultural sector and commercial services, and the domestic use of water have exceeded the supply capacity of potable water sources (Naidoo et al., 2016). Functional wastewater

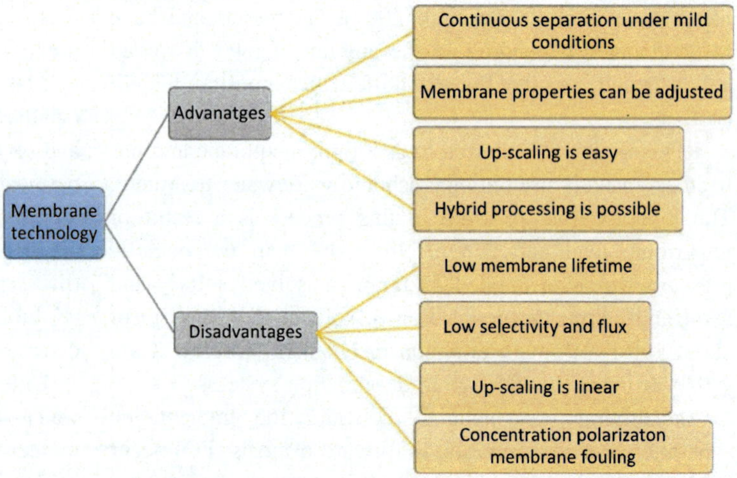

Fig. 7 Membrane technologies advantages and disadvantages. Adapted from Burggraaf and Cot (1996)

treatment plants (WWTP) have demonstrated to be greatly challenging to operate in terms of chemical input and energy. Even though the fundamental stages of treatment are primary, secondary and tertiary, the effluent from these plants contributes to secondary pollution as they are incapable of meeting the green-drop guidelines (Naidoo et al., 2016). The use of microalgae-based WWT systems has received intense examination in the research community, and in synergy with industry, various WWT technologies and strategies have been developed to tackle specific needs in the sector (Chalivendra, 2014).

Microalgae are the microscopic organisms discovered in freshwater and marine ecologies. They are monocellular species that stay alive independently or in clusters. Microalgae are efficient and perform the photosynthesis by consuming the greenhouse gas CO_2 and growing photoautotrophically. They are identified to provide nearly half of the atmospheric oxygen. Microalgae are known to grow profusely in small quantities of water and nutrients when compared to plants growing on land. Production of 1 kg of soy requires 2204 L of water, whereas 1 kg of microalgal biomass requires only 333 L of water. Moreover, microalgae can also grow in industrial wastewater. They can utilize the excess nutrients present in it, thereby allowing safe disposal of the wastewater. The microalgal biodiversity is enormous and they symbolize an unexploited reserve (Mondal & Khan, 2021).

In aquatic habitats, algae and bacteria have a symbiotic relationship (Fig. 8). Algae aid aerobic bacterial oxidation of organic substances, by offering oxygen through photosynthesis and utilize carbon dioxide and nutrients that are produced through aerobic bacterial oxidation, for their development. The majority of nitrogen in algal cell bound to proteins which compose to 45–60% of dry weight and phosphorous is vital for the synthesis of phosphate esters, nucleic acids and phospholipids (Jayalekshmi et al., 2021). Phosphorus sedimentation, hydrogen sulphur and ammonia elimination are resulted by escalating pH and dissolved oxygen concentration. High pH in algal ponds also promotes disinfection or pathogen destruction (Chaiprasert, 2011).

For specific microalgae strains, standard culture media are optimized and are subsequently modified to cultivate many other strains which are then used as templates to define wastewater characteristics (Armah et al., 2021).

The algal wastewater treatment systems have turned out to be a feasible inexpensive substitute to more complicated and costly treatment systems, mainly for domestic wastewater. Algae are competent of eradicating toxic material, for example, zinc, selenium and arsenic from the aquatic ecosystem by accumulating those substances within them. Algal biomass recovered from these ponds is typically utilized as a phosphorus and nitrogen supplement in agriculture, and they could as well be fermented to generate energy from methane. Therefore, microalgal biomass generated signifies a carbon sink, and consequently alleviates the negative effect of CO_2 by photosynthetic sequestration of this greenhouse gas (Armah et al., 2021; Mambo, 2014).

Various research teams (Fagan et al., 2016; Gimeno et al., 2016; Hom-Diaz et al., 2017; Xiong et al., 2016) reported the presence of emerging contaminants in wastewater and the probable detrimental effects they can cause on the environment and living organisms. EC removal using pure microalgae strains has been proven to be effective. However, microalgae-based EC removal technologies have not received appreciable attention in the global research community. The advocacy for employing microalgae to sequester wastewater nutrients as a treatment option has fascinated global acceptance. Nevertheless, there is scepticism in employing wastewaters for microalgal cultivation to produce biomass and bio-products. This is principally because of the certainty that wastewaters are of a wide range of sources and therefore have a broad range of properties whose stability is in question.

Apart from the capability of microalgae to sequester NH_3-N, NO_3-N and PO_4^{3-}, microalgae also eliminates heavy metals in addition to organic carbon from wastewater, while

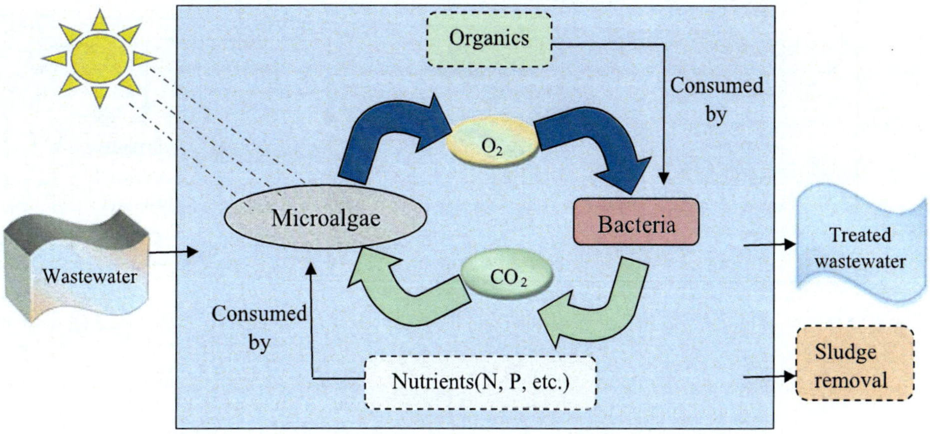

Fig. 8 Schematic diagram of symbiotic interactions in algal-bacteria system (Al-Jabri et al., 2020)

Table 4 Microalgal nutrient removal efficiencies in diverse wastewaters

Type of wastewater	Strain/microalgae used	Total nitrogen (mg/L)	TN removed (%)	Total phosphorous (mg/L)	TP removed (%)	COD reduction (%)	Biomass productivity (g/L/day)	References
Domestic wastewater	Chlorella pyrenoidosa	47.36	95	3.22	81	78	0.09	Dahmani et al. (2016)
Poultry wastewater	Scenedesmus obliquus	122.9	97.1	27.9	99.3	96	0.08–0.13	Oliveira et al. (2018)
Sewage wastewater	Scenedesmus abundans	197	95	286	83	–	–	Lekshmi et al. (2015)
Agro-industry wastewater	Microalgal consortia	1570	49	154	70	–		Singh et al. (2011)
	Algal consortia	44	36	88	13	46	–	de Godos et al. (2010)
	Scenedesmus obliquus	44	34	88	65	42	–	de Godos et al. (2010)
Potato processing wastewater	Chlorella sorokiniana	3.4	95	0.4	81	84.8	0.0188	Hernández et al. (2013)
Secondary pig manure	Chlorella sorokiniana	3.2	83	5	58	62.3	0.0263	Hernández et al. (2013)

avoiding secondary pollution. Nevertheless, prior research has showed that microalgae can seldom grow in undiluted wastewater due to high concentrations of ammonium and other compounds often present in wastewater. Diverse microalgae species show different growth indices in each wastewater treatment application. It is consequently vital to select appropriate microalgal strain to treat a given wastewater source. Table 4 summarizes a few microalgal nutrient removal efficiencies in diverse wastewaters treated. A schematic of the microalgal bioremediation of wastewater and potential applications of the produced biomass is shown in Fig. 9.

There are different types of microalgal wastewater treatment systems like open and closed microalgal systems which are detailed as follows:

Open ponds: Waste stabilization ponds/open ponds (Fig. 10) are widely used small communities for WWT around the world. These ponds are simple to operate, low cost and provide effective treatment in the view of pathogen and organic carbon removal. However, phosphorus removal in waste stabilization ponds is generally low, often around 15–50%.

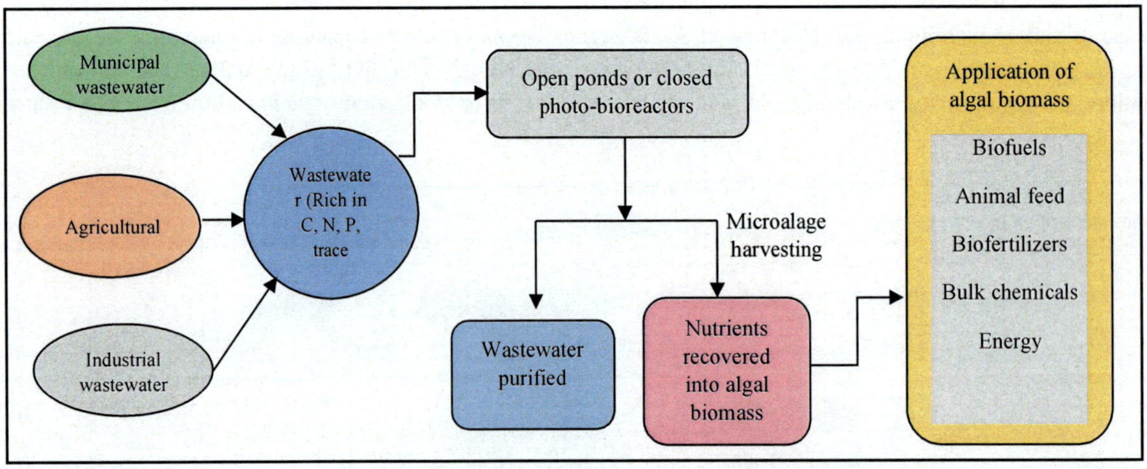

Fig. 9 Microalgal bioremediation of wastewater and the prospective applications of corresponding produced biomass (Adapted from Liu & Hong, 2021)

Fig. 10 Microalgal open pond systems (*Source* from Mambo et al., 2014; Cowan et al., 2016; Powell, 2009)

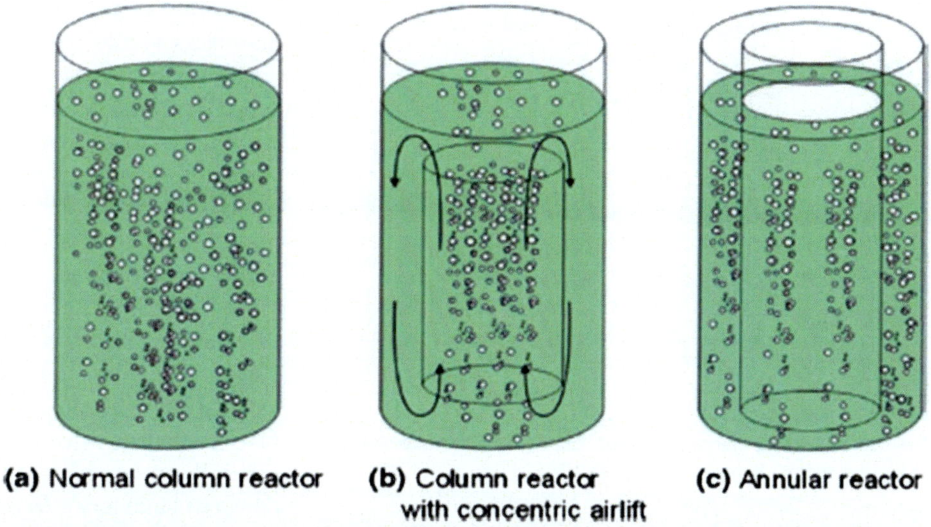

Fig. 11 Bubble column reactors (*Source* from Armah et al., 2021; Płaczek et al., 2017)

Closed photobioreactor systems: They (Fig. 11) are characterized by (i) lower risk of contamination, (ii) minimized evaporation losses, which is a severe concern in open systems and (iii) ample control over the operational variables resulting in efficient photosynthetic performances. Yet, closed systems are more costly, as they must be built using transparent materials, and are more complex to operate and tough to scale up.

The production of microalgae biomass across the globe was reported to be approximately 9000 tonnes per year with an approximate production cost of $20 to $200 per kilogram (Wang et al., 2016). Figure 12 summarizes a few benefits and challenges associated with microalgal treatment systems.

3.3 Microbial Fuel Cells (MFC) for Wastewater Treatment

Electricity generation coupled with wastewater treatment is possible with the use of MFCs that use bacteria as biocatalyst for conversion of chemical energy into electricity (Zhang et al., 2008).

MFC contains an anaerobic anode chamber, a cathode chamber and a proton-exchange membrane (PEM) that only permits the transfer of H^+ proton from the anode chamber to the cathode chamber (Fig. 13). Transfer of electrons from central metabolic system of bacteria to gain energy to the anode acts as the final electron acceptor in MFC. The electron is now conducted across an external circuit to the

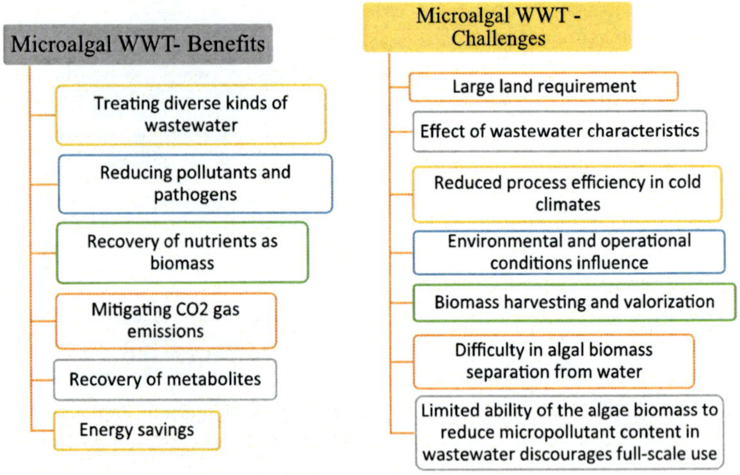

Fig. 12 Benefits and challenges of microalgal WWT (Lavrinovičs & Juhna, 2018; Molinuevo-Salces et al., 2019)

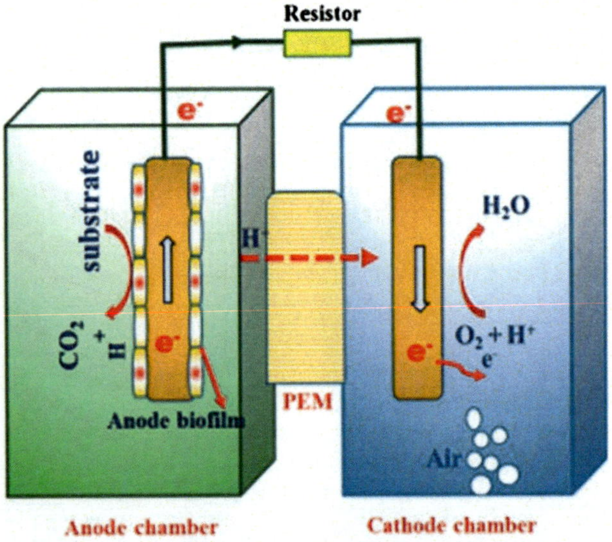

Fig. 13 Schematic illustration of the double-chamber microbial fuel cell. *Source* from Rahmani et al. (2022)

cathode where they form water by combination of O_2 and H^+ ions (Mahesh et al., 2021). Bacteria in the anode chamber act as catalysts for the breakdown of organic matter producing electrons and protons (Kumbhar et al., 2021). Some heavy metals that are good electronic acceptors have strong reducibility for MFC, which produces electricity. Hence, metals can be electrochemically reduced and are obtained from the cathode surface. Heavy metals that can be treated by this method are Cr (VI), Cu(II), Ag(I), Mn(II) and Cd(II) (Kumbhar et al., 2021).

Degradation of heavy metals by MFCs is studied separately in two types (Chen et al., 2018).

Single-chambered MFCs: One-chambered MFCs have both anode and cathode in the same compartment. This type of MFC lacks ion-exchange membrane.

Two-chambered MFCs: It is made up of two separate compartments made up of acrylic glass, polycarbonate, bottles, plexiglass, etc., which holds two electrodes well separated by proton-exchange membrane.

Inability to reduce all metal ions is one of the major drawbacks of MFC. MFC reduces a majority of heavy metal ions such as mercury, vanadium, cobalt, chromium and copper but not zinc, nickel, lead and cadmium, since only when it is thermodynamically complimentary, reduction occurs. Therefore, metals can act as direct electron acceptors exclusive of any external power supply, when metal ions have positive standard potential. Yet, some metal ions having redox potential lower than the anode potential cannot be reduced as they cannot spontaneously accept electrons from the cathode. Table 5 shows the electrode potential for reduction of a few heavy metals. As a result, they need an external power supply to transfer the electrons to the cathode for reduction (Cai et al., 2016). Also, the bio-toxicity of certain heavy metals deteriorates the performance of MFC due to its inhibitory effects on the metabolism and activity of the micro-organisms (Kumbhar et al., 2021).

MFC technologies show potential yet a novel strategy in WWT, considering that the treatment process itself turns out to be a technique to capture energy in the form of electricity or hydrogen gas, rather than being a net consumer of electrical energy. Some advantages and disadvantages of MFCs are summarized in Fig. 14.

3.4 Adsorption

Adsorption is a mass transfer process which involves the accumulation of substances at the interface of two phases, such as liquid–liquid, gas–liquid, gas–solid or liquid–solid interface (Fig. 15). The material being adsorbed is the adsorbate and the adsorbing matter is termed the adsorbent.

Table 5 Electrode potential for reduction of heavy metals by MFC (Cecconet et al., 2018; Kumbhar et al., 2021)

Pollutant	Equation of reduction	Electrode potential (V)
Cd (II)	$Cd^{2+} + 2e^{-} \rightarrow Cd$	−0.4
Mn (VII)	$MnO^{4-} + 4H^{+} + 3e^{-} \rightarrow MnO_2 + 2H_2O$	+1.7
Ag (I)	$Ag^{+} + e^{-} \rightarrow Ag$	+7.99
Cr (VI)	$Cr_2O_7{}^{2-} + 14H^{+} + 6e^{-} \rightarrow 2Cr^{3+} + 7H_2O$	+1.33
Cu (II)	$Cu^{2+} + 2e^{-} \rightarrow Cu$	+0.337

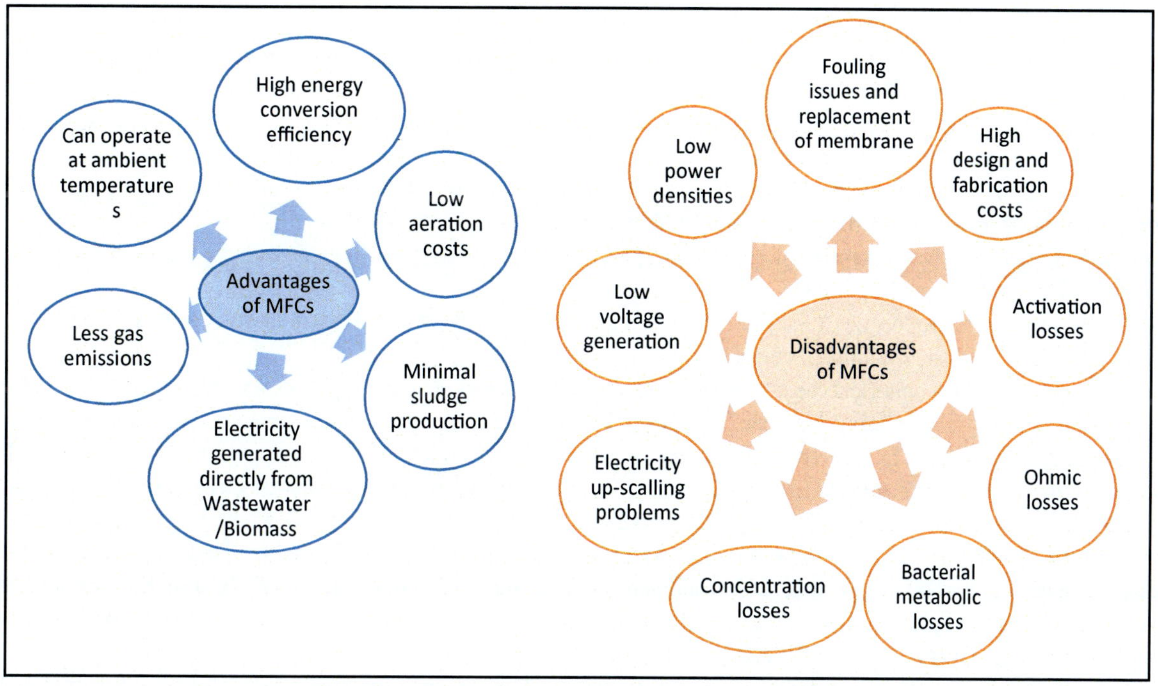

Fig. 14 Advantages and disadvantages of MFCs (Quach-Cu et al., 2018)

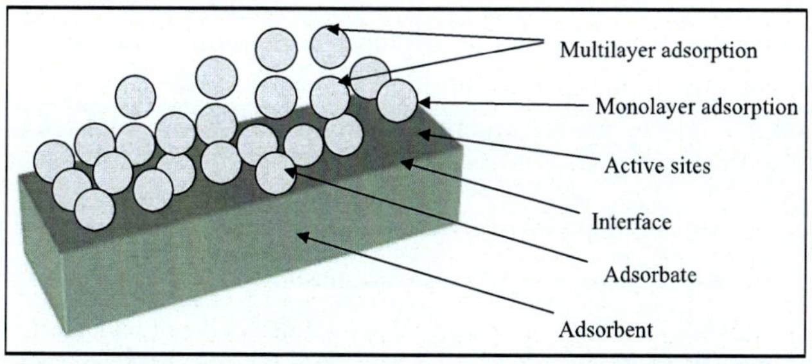

Fig. 15 General mechanism of adsorption to remove contaminants

The properties of adsorbates and adsorbents largely depend upon their constituents (De Gisi et al., 2016). If the interaction between them is of physical nature, then it is physisorption. Here, the interactions are van der Waals forces of attraction and since they are weak the process is reversible. Conversely, if the attraction forces between adsorbed molecules and the solid surface are due to chemical bonding, it is called chemisorptions, which occurs only as a monolayer and, furthermore, substances chemisorbed on solid surface are hardly removed because of stronger forces at stake. Under favourable conditions, both processes can occur simultaneously or alternatively. Physical adsorption results in a decrease in free energy and entropy of the adsorption system and, thus, it is exothermic (Cooney, 1998).

In a solid–liquid system, adsorption results in the removal of solutes from solution and their accumulation at solid surface. The solute remaining in the solution reaches a dynamic equilibrium with that adsorbed on the solid phase. The amount of adsorbate that can be taken up by an adsorbent as a function of both temperature and concentration of adsorbate, and the process, at constant temperature, can be described by an adsorption isotherm according to the general Eq. (1):

$$q_t = C_0 - C_t * V/m \ [mg/g] \quad (1)$$

where q_t (mg/g) is the amount of adsorbate per mass unit of adsorbent at time t; V is the volume of the solution (l); C_0 and C_t (mg/l) are the initial and at time t concentration of adsorbate, respectively, and m is the mass of adsorbent (g) (Cooney, 1998).

Adsorption was advantageous over other physico-chemical methods, as solid adsorbent has the potential of most efficiently removing the inorganic and organic pollutants such as hydrocarbons, pesticides, fertilizers, pharmaceuticals compounds, oils and phenols.

Studies on different types of adsorbents in use: Many organic and inorganic materials and minerals are commonly used as adsorbents, each having different adsorption capacities for specific pollutants removal from wastewater (Wang & Peng, 2010).

Activated carbon is one of the most extensively used adsorbents. It is very efficient due to its high porosity which is created by treating carbon high temperature (1300 °C) and steam in the presence of inorganic salts with or without oxygen. Granulated activated carbon and powdered activated carbon are two widely used forms of carbon, based on the particle size, and as macroporous (≥ 50 nm), mesoporous (2–50 nm) and microporous (2 − >0.8 nm) based on the pore size. The activated alumina consists mainly of aluminium oxide (Al_2O_3) spherical beads, highly porous with increased surface area ranging between 345 and 415 m^2/g (Sharma & Sharma, 2021). The adsorptive property of zeolite is due to honeycomb structure and cavities.

Zeolites have the surface area of 1–20 m^2/g and synthetic zeolites may be developed by hydrothermal processes in a temperature range of 90–100 °C, an autoclave followed by ion exchange with certain cations such as Ca^{2+}, Na^+, Li^+, K^+ and NH^{4+} (Sharma & Sharma, 2021). Titania (TiO2)-based nano-composite materials attract researcher's attention in materials science because of the broad possibilities of property modification (Rashid et al., 2021; Tatarchuk et al., 2020).

Activated carbon has several properties that make it particularly suitable for the purpose such as the high specific surface, the affinity with numerous compounds in addition to its ease of regeneration. The efficiency relies on the nature of the carbon particles, the raw material using which it is prepared, pore size, surface area, density and hardness. In spite of huge use, the general idea is to trim down the use of activated carbon due to its high costs. Hence, scientific world is looking at low-cost adsorbents as sustainable alternatives for wastewater treatment (De Gisi et al., 2016).

In order to replace the commercial activated carbon, a suitable non-conventional low-cost adsorbent should (i) be efficient to remove many and different contaminants, (ii) have high rate of adsorption and adsorption capacity and (iii) have high selectivity for different concentrations (Grassi et al., 2012). Different types of adsorbents used in wastewater treatment are categorized in Table 6 and with a few benefits and challenges of adsorption being summarized in Table 7.

3.5 Advanced Oxidation Processes

Advanced oxidation processes (AOPs) are a set of chemical treatment procedures intended to remove organic and inorganic materials in wastewater by oxidation through reactions with highly reactive oxygen species like hydroxyl radical (OH), ozone (O_3), hydrogen peroxide (H_2O_2) and ultraviolet (UV) light. Their application becomes indispensable since many of the organic compounds present in industrial water are resistant to conventional treatment.

During the AOP treatment of wastewater, hydroxyl radicals or sulphate radical are generated in sufficient quantity to remove refractory organic matters, traceable organic contaminants (such as aromatics, pesticides, petroleum constituents and volatile organic compounds) or certain inorganic pollutants or to increase wastewater biodegradability as a pre-treatment prior to following biological

Table 6 Different types of adsorbents for wastewater treatment (Rashid et al., 2021)

Low-cost adsorbents	Nano-adsorbents	Bio-sorbents
• Natural adsorbents (such as clay, chitin, peat, zeolite, moss, wood, coal, etc.) • Agricultural waste (vegetable and fruit peel like lemon peel, orange peel, banana peel, wheat bran, coconut, rice husk, pulse seed coat, sugarcane bagasse, soybean hulls, rice straws, rice bran, lotus leaves, pumpkin seeds, hazelnut shells, almond shells, walnut shells, apricot shells, etc.) • Industrial waste (like red mud, coffee waste, bagasse ash, fly ash, palm oil ash, etc.)	• Carbon-based nanomaterials (carbon nanomaterials and three-dimensional graphene) • Metal oxide based (Fe_3O_4, CuO, ZnO, their composites, etc.) • Nanocomposites (spinal ferrite nanoparticles as well as their nanocomposites, etc.) • Boron nitride material	They are derivative of biological sources or biological means. They tend to attract or bind to pollutants due to particular functional groups present on their surface • Enzyme based • Microorganism based (such as algal biochar, waste-based bio-sorbent, modified lemon leaf, ZrO_2-modified diatoms, fungal and bacterial biomass)

Table 7 Benefits and challenges of adsorption (Jayalekshmi et al., 2021)

Adsorption	
Benefits	Challenges
• Simple operation • Resource recovered by using waste material in their generation • No additional nutrient requirement • Low quantity of sludge • High efficiency • Regeneration possible • No increase in COD of water • Can remove contaminants even in dilute concentration	• Cost and stability is a concern • No regeneration can be done in some cases • Performance depends on adsorbent • Decrease in binding sites' number (fouling) • Application is still confined to lab scale; hence assets and liabilities on industrial scale are yet to be explored • Poor understanding and reluctance in general to adopt new technologies

Table 8 Redox potential of commonly used oxidants in water. *Sourced* from Zhou and Smith (2001)

Oxidative species	Redox potential (V)
Hydroxyl radical	2.80
Ozone	2.07
Hydrogen peroxide	1.78
Permanganate	1.69
Chlorine dioxide	1.56
Chlorine	1.36
Oxygen	1.23

treatment (Saravanathamizhan & Perarasu, 2021). Table 8 summarizes redox potential of commonly used oxidants in water treatment.

Types of AOPs: A range of AOP techniques such as photocatalytic oxidation, ozonation and Fenton process have been used as a pre-treatment technique in the water and wastewater treatment. They are detailed in the further sections.

Photocatalytic oxidation: Photocatalysis is an AOP that uses an ultraviolet light in the presence of catalyst and completely decomposes the organic pollutant. This is an emerging technology as it uses no chemical and is the major benefit over the other processes, as only light radiation is used (Zhang et al., 2019).

$$\text{Organic pollutant} + O_2 \xrightarrow{\text{Catalyst}} + h\nu \quad CO_2 + H_2O \qquad (2)$$

Ozonation: Ozonation is a complex oxidation process that takes place through the addition of ozone and is responsible for significantly increasing wastewater biodegradability (Fig. 16). Various studies have reported that the removal of personal care and pharmaceutical products is highly efficient using ozonation. Nevertheless, there is no denying on the reality that ozone has a short half-life and if its concentration exceeds a definite level, which is close to 23%, it is

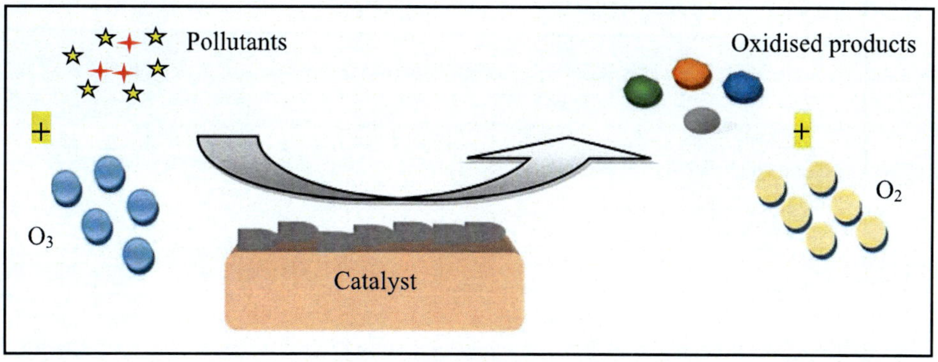

Fig. 16 Schematic illustration of catalytic ozonation

considered a threat (Ahmed et al., 2021; Shen et al., 2019). Generally, ozone cannot be stored and generated from oxygen for the ready use in the water treatment plants. Ozone may react with water and produce hydroxyl radicals that have a higher oxidation potential.

In ozonation, under the influence of pH, two reactions are considered (Rashid et al., 2021):

Direct pathway: pH < 4 reaction occurs between dissolved compounds in water and molecular ozone.
Indirect pathway: pH > 10 reactions occur between hydroxyl radicals produced by ozone breakdown and dissolved compounds.

Ozone can be used with other technologies like H_2O_2, UV, ultrasound, and others, to increase the colour or organic matter removal efficiency (Sicardi, 2020).

Fenton oxidation process: Fenton oxidation process is where the ferrous ions react with the hydrogen peroxide in generating hydroxyl radicals. This process is well known to be effective in destroying the toxicity of organic compounds present in wastewater. However, some of the apparent restrictions of this process are the small range of pH and the risk and cost associated with transporting, handling and storing the reagents. A good number of reports have also established the effectiveness of this approach in reducing heavy metal contents like copper, chromium, aluminium and a few others (Chen et al., 2020).

The reactions that involve Fenton's reagent with any organic compound (RH) are shown below:

$$H_2O_2 + Fe^{2+} \rightarrow Fe^{3+} + OH^- + OH^* \quad (3)$$

$$RH + OH^* \rightarrow H_2O + R^* \quad (4)$$

$$R^* + Fe^{3+} \rightarrow R^+ + Fe^{2+} \quad (5)$$

$$R + H_2O \rightarrow ROH + H^+ \quad (6)$$

Hydrogen peroxide generates hydroxyl radicals upon interaction with the iron ions, as H_2O_2 is not good enough for dye decolourization alone at normal conditions (Sicardi, 2020).

Biodegradability index (BI) indicator is the proportion of biologically degradable organic matter to that of the total organic matter present, BI is the ratio of BOD_5 to COD. When this ratio is greater than 0.5, then the effluent is easily biodegradable. When this ratio is within the range of 0.4–0.5, the effluent is average biodegradable and within the range of 0.2–0.4 the effluent is slowly biodegradable. In order to make such valuations, respirometry tests were conducted that measure the rate of carbon dioxide production or oxygen consumption of an organism or organic system (Saravanathamizhan & Perarasu, 2021).

In order to improve the biodegradability of landfill leachate, Fenton treatment (Fe^{2+}/H_2O_2) and different ozone-based AOPs (O_3, O_3/OH and O_3/H_2O_2) were carried out as pre-treatment techniques. Fenton process improves the BI from 0.01 to 0.15 and ozone with H_2O_2 improves the BI from 0.01 to 0.24 (Cortez et al., 2011). Some of the advantages and disadvantages of AOPs are presented in Fig. 17.

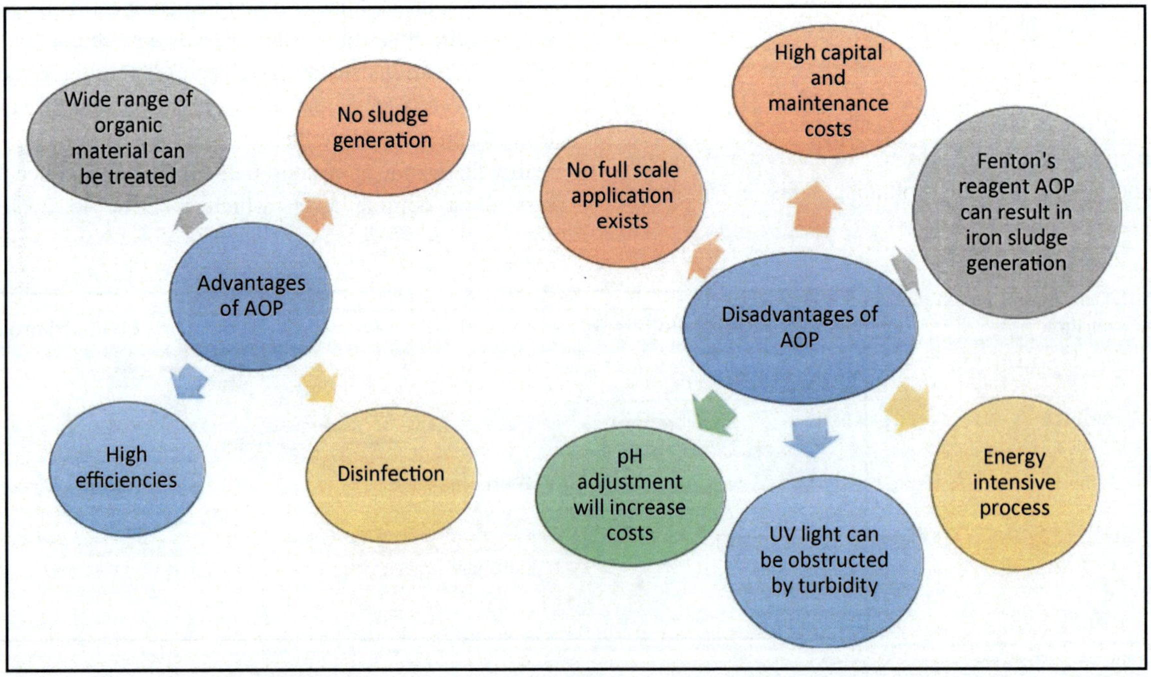

Fig. 17 Advantages and disadvantages of AOP

4 Conclusions and Outlook

Municipal wastewater and industrial effluents are the major sources of pollution of water bodies with varying contaminants. With increasing urbanization and industrialization, contaminants of various kinds are polluting the water resources, necessitating the need to develop technologies with wider applicability and efficiencies. Four groups of advanced water and wastewater treatment methods with their working principles, corresponding advantages, disadvantages and applications are discussed and reviewed broadly in this chapter. Membrane technology is resulting in high efficiencies but the durability being questioned makes them uneconomical. The microalgal systems are fascinating and are highly economical but their greater dependency on the climatic conditions is a huge concern. Microbial fuel cell is an attractive treatment method but the scalability and economic aspects are great challenges. Advanced oxidation processes, though being effective, operation costs are very high. Adsorption is an emerging technology with simple operation mechanism giving high efficiencies, specifically in removing emerging contaminants, which is still yet be explored to a large extent, like in techno-economical aspects.

Though many recent technologies like membrane filtration and adsorption are providing treatment results meeting standards of disposal to a great extent, furthermore advanced treatment technologies and hybrid systems are urgently necessary to be explored, which could promise high-quality treated water with reduced inputs to treatment in terms of resources or energy invested, in turn, to strengthen the public and environmental health, economy and also industrial production processes. Majority of the studies published involved findings based on indoor laboratory-scale experiments. Recently, studies on pilot scale are gradually increasing, but further detailed studies are required to be conducted in pilot scale and on large scale, to understand the treatment efficiencies on a long-term perspective under fluctuating environments and wastewater characteristics, consequently with economic considerations involved, to be explored in line.

References

Ahmed, S. F., Mofijur, M., Nuzhat, S., Chowdhury, A. T., Rafa, N., Uddin, M. A., Inayat, A., Mahlia, T. M. I., Ong, H. C., Chia, W. Y., & Show, P. L. (2021). Recent developments in physical, biological, chemical, and hybrid treatment techniques for removing emerging contaminants from wastewater. *Journal of Hazardous Materials, 416*, 125912. https://doi.org/10.1016/j.jhazmat.2021.125912.

Akpor, O. B., Otohinoyi, D. A., Olaolu, D. T., & Aderiye, B. I. (2014). Pollutants in wastewater effluents: Impacts and remediation processes. *International Journal of Environmental Research and Earth Science, 3*(3), 050–059.

Ali, M. (2008). Assessment of some water quality characteristics and determination of some heavy metals in Lake Manzala, Egypt. *Egyptian Journal of Aquatic Biology and Fisheries, 12*(2):133–154. https://doi.org/10.21608/ejabf.2008.1998.

Al-Jabri, H., Das, P., Khan, S., Thaher, M., & AbdulQuadir, M. (2020). Treatment of wastewaters by microalgae and the potential applications of the produced biomass—A review. *Water, 13*(1), 27. https://doi.org/10.3390/w13010027.

Al-Saydeh, S. A., El-Naas, M. H., & Zaidi, S. J. (2017). Copper removal from industrial wastewater: A comprehensive review. *Journal of Industrial and Engineering Chemistry, 56*, 35–44. https://doi.org/10.1016/j.jiec.2017.07.026.

Alssgeer, H. M. A., Gasim, M. B., Hanafiah, M. M., Abdulhadi, E. R. A., & Azid, A. (2018). GIS-based analysis of water quality deterioration in the Nerus River, Kuala Terengganu, Malaysia. *Desalination Water Treat, 112*, 334–343. https://doi.org/10.5004/dwt.2018.22335.

Amina, T. (2017). Wastewater Screening & Classification of screens (complete list) wastewater treatment. *Environmental Engineering, 17*, 2017. https://engineeringcivil.org/articles/environmental-engineering/wastewater-screening-classification-screens-complete-list-wastewater-treatment/.

Ariffin, N., Abdullah, M. M. A. B., Zainol, M. R. R. M. A., Murshed, M. F., Faris, M. A., & Bayuaji, R. (2017). Review on adsorption of heavy metal in wastewater by using geopolymer. In *MATEC Web of Conferences* (Vol. 97, p. 01023). EDP Sciences. https://doi.org/10.1051/matecconf/20179701023.

Armah, E. K., Chetty, M., Adedeji, J. A., Kukwa, D. T., Mutsvene, B., Shabangu, K. P., & Bakare, B. F. (2021). Emerging trends in wastewater treatment technologies: The current perspective. In *Promising techniques for wastewater treatment and water quality assessment* (Vol. 1, p. 71). https://doi.org/10.5772/intechopen.93898.

Arthurson, V. (2008). Proper sanitization of sewage sludge: A critical issue for a sustainable society. *Applied and Environmental Microbiology, 74*(17), 5267–5275. https://doi.org/10.1128/AEM.00438-08.

Baker, J., Stephenson, T., Dard, S., & Côté, P. (1995). Characterisation of fouling of nanofiltration membranes used to treat surface waters. *Environmental Technology, 16*(10), 977–985. https://doi.org/10.1080/09593331608616335.

Barakat, M. A. (2011). New trends in removing heavy metals from industrial wastewater. *Arabian Journal of Chemistry, 4*(4), 361–377. https://doi.org/10.1016/j.arabjc.2010.07.019.

Baysal, A., Ozbek, N., & Akman, S. (2013). Determination of trace metals in waste water and their removal processes. In *Waste water-treatment technologies and recent analytical developments* (pp. 145–171). https://doi.org/10.5772/52025.

Besha, A. T., Gebreyohannes, A. Y., Tufa, R. A., Bekele, D. N., Curcio, E., & Giorno, L. (2017). Removal of emerging micropollutants by activated sludge process and membrane bioreactors and the effects of micropollutants on membrane fouling: A review. *Journal of Environmental Chemical Engineering, 5*(3), 2395–2414. https://doi.org/10.1016/j.jece.2017.04.027.

Bin Abdullah, K. (2005). Use of water and land for food security and environmental sustainability. *Water Wheel*. https://doi.org/10.1002/ird.254.

Bolan, N. S. (2008). Water encyclopedia: Domestic, municipal, and industrial water supply and waste disposal. *Journal of Environmental Quality, 37*(3), 1299. https://doi.org/10.2134/jeq2008.0002br.

Burggraaf, A. J., & Cot, L. (Eds.). (1996). Fundamentals of inorganic membrane science and technology. Elsevier.

Cai, W. F., Geng, D. L., & Wang, Y. H. (2016). Assessment of cathode materials for Ni (ii) reduction in microbial electrolysis cells. *RSC*

Advances, 6(38), 31732–31738. https://doi.org/10.1039/C6RA02082H.

Caicedo, C., Rosenwinkel, K. H., Exner, M., Verstraete, W., Suchenwirth, R., Hartemann, P., & Nogueira, R. (2019). Legionella occurrence in municipal and industrial wastewater treatment plants and risks of reclaimed wastewater reuse. *Water Research, 149*, 21–34. https://doi.org/10.1016/j.watres.2018.10.080.

Cecconet, D., Callegari, A., & Capodaglio, A. G. (2018). Bioelectrochemical systems for removal of selected metals and perchlorate from groundwater: A review. *Energies, 11*(10), 2643. https://doi.org/10.3390/en11102643.

Chaiprasert, P. (2011). Biogas production from agricultural wastes in Thailand. *Journal of Sustainable Energy and Environment*. 63–65. (Spec. Issue).

Chalivendra, S. (2014). Bioremediation of wastewater using microalgae. University of Dayton. http://rave.ohiolink.edu/etdc/view?acc_num=dayton1418994496.

Chang, E. E., Hsing, H. J., Chiang, P. C., Chen, M. Y., & Shyng, J. Y. (2008). The chemical and biological characteristics of coke-oven wastewater by ozonation. *Journal of Hazardous Materials, 156*(1–3), 560–567. https://doi.org/10.1016/j.jhazmat.2007.12.106.

Chen, Y. J., Fan, T. Y., Wang, L. P., Cheng, T. W., Chen, S. S., Yuan, M. H., & Cheng, S. (2020). Application of Fenton method for the removal of organic matter in sewage sludge at room temperature. *Sustainability, 12*(4), 1518. https://doi.org/10.3390/su12041518.

Chen, Z., Lu, X., Yin, R., Luo, Y., Mai, H., Zhang, N., Xiong, J., Zhang, H., Tang, J., & Luo, D. (2018). Research on treatment of wastewater containing heavy metal by microbial fuel cell. In *IOP Conference Series: Earth and Environmental Science* (Vol. 121, No. 3, p. 032022). IOP Publishing. https://doi.org/10.1088/1755-1315/121/3/032022.

Choudhary, M., Peter, C. N., Shukla, S. K., Govender, P. P., Joshi, G. M., & Wang, R. (2020). Environmental issues: a challenge for wastewater treatment. In *Green materials for wastewater treatment* (pp. 1–12). Springer, Cham. https://doi.org/10.1007/978-3-030-17724-9_1.

Chowdhary, P., Raj, A., & Bharagava, R. N. (2018). Environmental pollution and health hazards from distillery wastewater and treatment approaches to combat the environmental threats: A review. *Chemosphere, 194*, 229–246. https://doi.org/10.1016/j.chemosphere.2017.11.163.

Cirja, M., Ivashechkin, P., Schäffer, A., & Corvini, P. F. (2008). Factors affecting the removal of organic micropollutants from wastewater in conventional treatment plants (CTP) and membrane bioreactors (MBR). *Reviews in Environmental Science and Bio/technology, 7* (1), 61–78. https://doi.org/10.1007/s11157-007-9121-8.

Cooney, D. O. (1998). *Adsorption design for wastewater treatment* (pp 182). CRC Press.

Cortez, S., Teixeira, P., Oliveira, R., & Mota, M. (2011). Evaluation of Fenton and ozone-based advanced oxidation processes as mature landfill leachate pre-treatments. *Journal of Environmental Management, 92*(3), 749–755. https://doi.org/10.1016/j.jenvman.2010.10.035.

Cowan, A. K., Mambo, P. M., Westensee, D. K., & Render, D. S. (2016). Evaluation of integrated algae pond systems for municipal wastewater treatment: the Belmont valley WWTW pilot-scale IAPS case study. WRC Report, (TT 649/15). http://www.wrc.org.za/wp-content/uploads/mdocs/TT%20649-15.pdf.

Crini, G., & Lichtfouse, E. (2019). Advantages and disadvantages of techniques used for wastewater treatment. *Environmental Chemistry Letters, 17*(1), 145–155. https://doi.org/10.1007/s10311-018-0785-9.

Dahmani, S., Zerrouki, D., Ramanna, L., Rawat, I., & Bux, F. (2016). Cultivation of Chlorella pyrenoidosa in outdoor open raceway pond using domestic wastewater as medium in arid desert region. *Bioresource Technology, 219*, 749–752. https://doi.org/10.1016/j.biortech.2016.08.019.

De Gisi, S., Lofrano, G., Grassi, M., & Notarnicola, M. (2016). Characteristics and adsorption capacities of low-cost sorbents for wastewater treatment: A review. *Sustainable Materials and Technologies, 9*, 10–40. https://doi.org/10.1016/j.susmat.2016.06.002.

de Godos, I., Vargas, V. A., Blanco, S., González, M. C. G., Soto, R., García-Encina, P. A., Becares, E., & Muñoz, R. (2010). A comparative evaluation of microalgae for the degradation of piggery wastewater under photosynthetic oxygenation. *Bioresource Technology, 101*(14), 5150–5158. https://doi.org/10.1016/j.biortech.2010.02.010.

Deblonde, T., Cossu-Leguille, C., & Hartemann, P. (2011). Emerging pollutants in wastewater: A review of the literature. *International Journal of Hygiene and Environmental Health, 214*(6), 442–448. https://doi.org/10.1016/j.ijheh.2011.08.002.

Enitan, A. M., Adeyemo, J., Kumari, S. K., Swalaha, F. M., & Bux, F. (2015). Characterization of brewery wastewater composition. *International Journal of Environmental and Ecological Engineering*. https://doi.org/10.51415/10321/2992.

Fagan, R., McCormack, D. E., Dionysiou, D. D., & Pillai, S. C. (2016). A review of solar and visible light active TiO2 photocatalysis for treating bacteria, cyanotoxins and contaminants of emerging concern. *Materials Science in Semiconductor Processing, 42*, 2–14. https://doi.org/10.1016/j.mssp.2015.07.052.

Farkas, K., Walker, D. I., Adriaenssens, E. M., McDonald, J. E., Hillary, L. S., Malham, S. K., & Jones, D. L. (2020). Viral indicators for tracking domestic wastewater contamination in the aquatic environment. *Water Research, 181*, 115926. https://doi.org/10.1016/j.watres.2020.115926.

Fazal, T., Mushtaq, A., Rehman, F., Khan, A. U., Rashid, N., Farooq, W., Rehman, M. S. U., & Xu, J. (2018). Bioremediation of textile wastewater and successive biodiesel production using microalgae. *Renewable and Sustainable Energy Reviews, 82*, 3107–3126. https://doi.org/10.1016/j.rser.2017.10.029.

Fujie, K., Bravo, H. E., & Kubota, H. (1983). Operational design and power economy of a rotating biological contactor. *Water Research, 17*(9), 1153–1162. https://doi.org/10.1016/0043-1354(83)90056-8.

Ghodeif, K. (2013). Baseline assessment study for wastewater treatment plant for Al Gozayyera Village, West Kantara City, Ismailia Governorate, Egypt. In *Network of demonstration activities for sustainable integrated wastewater treatment and reuse in the Mediterranean*: Cairo, Egypt. https://doi.org/10.13140/RG.2.2.34897.63844.

Gimeno, O., García-Araya, J. F., Beltrán, F. J., Rivas, F. J., & Espejo, A. (2016). Removal of emerging contaminants from a primary effluent of municipal wastewater by means of sequential biological degradation-solar photocatalytic oxidation processes. *Chemical Engineering Journal, 290*, 12–20. https://doi.org/10.1016/j.cej.2016.01.022.

Goswami, L., Kumar, R. V., Borah, S. N., Manikandan, N. A., Pakshirajan, K., & Pugazhenthi, G. (2018). Membrane bioreactor and integrated membrane bioreactor systems for micropollutant removal from wastewater: A review. *Journal of Water Process Engineering, 26*, 314–328. https://doi.org/10.1016/j.jwpe.2018.10.024.

Grassi, M., Kaykioglu, G., Belgiorno, V., & Lofrano, G. (2012). Removal of emerging contaminants from water and wastewater by adsorption process. In *Emerging compounds removal from wastewater* (pp. 15–37). Springer, Dordrecht. https://doi.org/10.1007/978-94-007-3916-1_2.

Harvey, P. J., Handley, H. K., & Taylor, M. P. (2015). Identification of the sources of metal (lead) contamination in drinking waters in north-eastern Tasmania using lead isotopic compositions. *Environmental Science and Pollution Research, 22*(16), 12276–12288. https://doi.org/10.1007/s11356-015-4349-2.

Hernández, D., Riaño, B., Coca, M., & García-González, M. C. (2013). Treatment of agro-industrial wastewater using microalgae–bacteria consortium combined with anaerobic digestion of the produced biomass. *Bioresource Technology, 135*, 598–603. https://doi.org/10.1016/j.biortech.2012.09.029.

Hom-Diaz, A., Jaén-Gil, A., Bello-Laserna, I., Rodríguez-Mozaz, S., Vicent, T., Barceló, D., & Blánquez, P. (2017). Performance of a microalgal photobioreactor treating toilet wastewater: Pharmaceutically active compound removal and biomass harvesting. *Science of the Total Environment, 592*, 1–11. https://doi.org/10.1016/j.scitotenv.2017.02.224.

Hussain, M. M., Wang, J., Bibi, I., Shahid, M., Niazi, N. K., Iqbal, J., Mian, I. A., Shaheen, S. M., Bashir, S., Shah, N. S., & Hina, K. (2021). Arsenic speciation and biotransformation pathways in the aquatic ecosystem: The significance of algae. *Journal of Hazardous Materials, 403*, 124027. https://doi.org/10.1016/j.jhazmat.2020.124027.

Islam, M. M. M., Shafi, S., Bandh, S. A., & Shameem, N. (2019). Impact of environmental changes and human activities on bacterial diversity of lakes. In *Freshwater microbiology* (pp. 105–136). Academic Press. https://doi.org/10.1016/B978-0-12-817495-1.00003-7.

Jafarinejad, S. (2021). Forward osmosis membrane technology for nutrient removal/recovery from wastewater: Recent advances, proposed designs, and future directions. *Chemosphere, 263*, 128116. https://doi.org/10.1016/j.chemosphere.2020.128116.

Jayalekshmi, S. J., Biju, M., Somarajan, J., Muhammad Ajas, P. E., & Sunny, D. (2021). Wastewater treatment technologies: A review. *International Journal of Engineering Research & Technology, ICART–2021, 09*(09). https://www.ijert.org/wastewater-treatment-technologies-a-review.

Ji, J., Kakade, A., Yu, Z., Khan, A., Liu, P., & Li, X. (2020). Anaerobic membrane bioreactors for treatment of emerging contaminants: A review. *Journal of Environmental Management, 270*, 110913. https://doi.org/10.1016/j.jenvman.2020.110913.

Kumbhar, P., Savla, N., Banerjee, S., Mathuriya, A. S., Sarkar, A., Khilari, S., Jadhav, D. A., & Pandit, S. (2021). Microbial electrochemical heavy metal removal: Fundamental to the recent development. In *Wastewater treatment* (pp. 521–542). Elsevier. https://doi.org/10.1016/B978-0-12-821881-5.00026-X.

Launay, M. A., Dittmer, U., & Steinmetz, H. (2016). Organic micropollutants discharged by combined sewer overflows–characterisation of pollutant sources and stormwater-related processes. *Water Research, 104*, 82–92. https://doi.org/10.1016/j.watres.2016.07.068.

Lavrinovičs, A., & Juhna, T. (2018). Review on challenges and limitations for algae-based wastewater treatment. *Construction Science, 20*, 17–25. https://doi.org/10.2478/cons-2017-0003.

Lehr, J. H., Keeley, J., & Lehr, J. (2005). *Domestic, municipal, and industrial water supply and waste disposal; Wiley Interscience: Hoboken*. Wiley Interscience Hoboken.

Lekshmi, B., Joseph, R. S., Jose, A., Abinandan, S., & Shanthakumar, S. (2015). Studies on reduction of inorganic pollutants from wastewater by Chlorella pyrenoidosa and Scenedesmus abundans. *Alexandria Engineering Journal, 54*(4), 1291–1296. https://doi.org/10.1016/j.aej.2015.09.013.

Li, C., Feng, G., Song, C., Zhong, G., Tao, P., Wang, T., & Shao, M. (2018). Improved oil removal ability by the integrated electrocoagulation (EC)-carbon membrane coupling with electrochemical anodic oxidation (CM/EAO) system. *Colloids and Surfaces a: Physicochemical and Engineering Aspects, 559*, 305–313. https://doi.org/10.1016/j.colsurfa.2018.09.043.

Li, K., Liu, Q., Fang, F., Wu, X., Xin, J., Sun, S., Wei, Y., Ruan, R., Chen, P., Wang, Y., & Addy, M. (2020). Influence of nanofiltration concentrate recirculation on performance and economic feasibility of a pilot-scale membrane bioreactor-nanofiltration hybrid process for textile wastewater treatment with high water recovery. *Journal of Cleaner Production, 261*, 121067. https://doi.org/10.1016/j.jclepro.2020.121067.

Liu, X. Y., & Hong, Y. (2021). Microalgae-based wastewater treatment and recovery with biomass and value-added products: A brief review. *Current Pollution Reports, 7*(2), 227–245. https://doi.org/10.1007/s40726-021-00184-6.

Mahesh, R., Panda, S. K., Das, M., Yashavanth, P. R., Dhull, S., Negi, B. B., Jakhwal, P., & Maiti, S. K. (2021). Advances in biotechnological tools for bioremediation of wastewater using bacterial–algal symbiotic system. In *Wastewater Treatment* (pp. 385–411). Elsevier. https://doi.org/10.1016/B978-0-12-821881-5.00019-2.

Mallevialle, J., Odendaal, P. E., & Wiesner, M. R. (Eds.). (1996). Water treatment membrane processes. American Water Works Association.

Mambo, M. P. (2014). Integrated algae pond systems for the treatment of municipal wastewater. Doctoral dissertation, Rhodes University.

Mambo, P. M., Westensee, D. K., Render, D. S., & Cowan, A. K. (2014). Operation of an integrated algae pond system for the treatment of municipal sewage: A South African case study. *Water Science and Technology, 69*(12), 2554–2561. https://doi.org/10.2166/wst.2014.187.

Metcalf, L., Eddy, H. P., & Tchobanoglous, G. (1991). *Wastewater engineering: Treatment, disposal, and reuse* (Vol. 4). McGraw-Hill.

Mohammad, A. W., Hilal, N., Al-Zoubib, H., Darwish, N. A., & Ali, N. (2007). Modelling the effects of nanofiltration membrane properties on system cost assessment for desalination applications. *Desalination, 206*(1–3), 215–225. https://doi.org/10.1016/j.desal.2006.02.068.

Mohammed, A. S., Kapri, A., & Goel, R. (2011). Heavy metal pollution: source, impact, and remedies. In *Biomanagement of metal-contaminated soils* (pp. 1–28). Springer, Dordrecht. https://doi.org/10.1007/978-94-007-1914-9_1

Molinuevo-Salces, B., Riaño, B., Hernández, D., & Cruz García-González, M. (2019). Microalgae and wastewater treatment: Advantages and disadvantages. In *Microalgae biotechnology for development of biofuel and wastewater treatment* (pp. 505–533). Springer, Singapore.

Mondal, M., & Khan, A. A. (2021). Immobilized microalgae for removing industrial pollutants: A greener technique. In *Wastewater treatment* (pp. 367–384). Elsevier. https://doi.org/10.1016/B978-0-12-821881-5.00018-0.

Munter, R., & Lundin, L. C. (2000). Industrial wastewater treatment. In *Sustainable water management in the Baltic Sea Basin, Book II, water use and management*. http://www.balticuniv.uu.se/index.php/boll-online-library/831-swm-2-water-use-and-management.

Nageswara, L. (2014). A Nano technological methodology for treatment of Waste water. *International Journal of ChemTech Research, 6*(4), 2529–2533.

Naidoo, S., Rodda, N., Stenstrom, T. A., Schmidt, S., Dent, M., Bux, F., Hanke, N., Buckley, C. A., & Fennemore, C. (2016). Water security in South Africa: Perceptions on public expectations and municipal obligations, governance and water re-use. *Water SA, 42*(3), 456–465. https://doi.org/10.4314/wsa.v42i3.11.

Obotey Ezugbe, E., & Rathilal, S. (2020). Membrane technologies in wastewater treatment: A review. *Membranes, 10*(5), 89. https://doi.org/10.3390/membranes10050089.

Oliveira, A. C., Barata, A., Batista, A. P., & Gouveia, L. (2018). Scenedesmus obliquus in poultry wastewater bioremediation. *Environmental Technology*. https://doi.org/10.1080/09593330.2018.1488003.

Pabby, A. K., Rizvi, S. S., & Requena, A. M. S. (2008). Handbook of membrane separations: Chemical, pharmaceutical, food, and biotechnological applications. *CRC Press*. https://doi.org/10.1201/b18319.

Płaczek, M., Patyna, A., & Witczak, S. (2017). Technical evaluation of photobioreactors for microalgae cultivation. In *E3S web of conferences* (Vol. 19, p. 02032). EDP Sciences. https://doi.org/10.1051/e3sconf/20171902032.

Powell, N. (2009). Biological phosphorus removal by microalgae in waste stabilisation ponds: A thesis presented in partial fulfilment of the requirements for the degree of Doctor of Philosophy in Environmental Engineering at Massey University, Palmerston North, New Zealand. Doctoral dissertation, Massey University. http://hdl.handle.net/10179/1128.

Prado, M., Borea, L., Cesaro, A., Liu, H., Naddeo, V., Belgiorno, V., & Ballesteros, F., Jr. (2017). Removal of emerging contaminant and fouling control in membrane bioreactors by combined ozonation and sonolysis. *International Biodeterioration & Biodegradation, 119*, 577–586. https://doi.org/10.1016/j.ibiod.2016.10.044.

Quach-Cu, J., Herrera-Lynch, B., Marciniak, C., Adams, S., Simmerman, A., & Reinke, R. A. (2018). The effect of primary, secondary, and tertiary wastewater treatment processes on antibiotic resistance gene (ARG) concentrations in solid and dissolved wastewater fractions. *Water, 10*(1), 37. https://doi.org/10.3390/w10010037.

Rahmani, A. R., Navidjouy, N., Rahimnejad, M., Alizadeh, S., Samarghandi, M. R., & Nematollahi, D. (2022). Effect of different concentrations of substrate in microbial fuel cells toward bioenergy recovery and simultaneous wastewater treatment. *Environmental Technology, 43*(1), 1–9. https://doi.org/10.1080/09593330.2020.1772374.

Rashed, M. N. (2013). Adsorption technique for the removal of organic pollutants from water and wastewater. *Organic Pollutants-Monitoring, Risk and Treatment, 7*, 167–194. https://doi.org/10.5772/54048.

Rashid, R., Shafiq, I., Akhter, P., Iqbal, M. J., & Hussain, M. (2021). A state-of-the-art review on wastewater treatment techniques: The effectiveness of adsorption method. *Environmental Science and Pollution Research, 28*(8), 9050–9066. https://doi.org/10.1007/s11356-021-12395-x.

Roccaro, P., Sgroi, M., & Vagliasindi, F. G. (2013). Removal of xenobiotic compounds from wastewater for environment protection: Treatment processes and costs. *Chemical Engineering Transactions, 32*. https://doi.org/10.3303/CET1332085.

Saravanathamizhan, R., & Perarasu, V. T. (2021). Improvement of biodegradability index of industrial wastewater using different pretreatment techniques. In *Wastewater treatment* (pp. 103–136). Elsevier. https://doi.org/10.1016/B978-0-12-821881-5.00006-4.

Sengupta, A., Jebur, M., Kamaz, M., & Wickramasinghe, S. R. (2021). Removal of emerging contaminants from wastewater streams using membrane bioreactors: a review. *Membranes, 12*(1), 60. https://doi.org/10.3390/membranes12010060.

Shahid, M. K., Kashif, A., Fuwad, A., & Choi, Y. (2021). Current advances in treatment technologies for removal of emerging contaminants from water–A critical review. *Coordination Chemistry Reviews, 442*, 213993. https://doi.org/10.1016/j.ccr.2021.213993.

Sharma, V. P., & Sharma, P. (2021). Environmental contaminants: treatment, threats, toxicity, and tools for sustainability. In *Wastewater treatment* (pp. 93–102). Elsevier. https://doi.org/10.1016/B978-0-12-821881-5.00005-2.

Shen, J., Ding, T., & Zhang, M. (2019). Analytical techniques and challenges for removal of pharmaceuticals and personal care products in water. In *Pharmaceuticals and personal care products: Waste management and treatment technology* (pp. 239–257). Butterworth-Heinemann. https://doi.org/10.1016/B978-0-12-816189-0.00010-X.

Shmeis, R. M. A. (2018). Water chemistry and microbiology. In *Comprehensive analytical chemistry* (Vol. 81, pp. 1–56). Elsevier. https://doi.org/10.1016/bs.coac.2018.02.001.

Sicardi, S. (2020). Treatment of wastewater from textile dyeing by ozonization.

Singh, M., Reynolds, D. L., & Das, K. C. (2011). Microalgal system for treatment of effluent from poultry litter anaerobic digestion. *Bioresource Technology, 102*(23), 10841–10848. https://doi.org/10.1016/j.biortech.2011.09.037.

Sonune, A., & Ghate, R. (2004). Developments in wastewater treatment methods. *Desalination, 167*, 55–63. https://doi.org/10.1016/j.desal.2004.06.113.

Su, F., Liang, Y., Liu, G., Mota Filho, C. R., Hu, C., & Qu, J. (2020). Enhancement of anti-fouling and contaminant removal in an electro-membrane bioreactor: Significance of electrocoagulation and electric field. *Separation and Purification Technology, 248*, 117077. https://doi.org/10.1016/j.seppur.2020.117077.

Tan, X., Xie, G. J., Nie, W. B., Xing, D. F., Liu, B. F., Ding, J., & Ren, N. Q. (2021). High value-added biomaterials recovery from granular sludge based wastewater treatment process. *Resources, Conservation and Recycling, 169*, 105481. https://doi.org/10.1016/j.resconrec.2021.105481.

Tatarchuk, T., Mironyuk, I., Kotsyubynsky, V., Shyichuk, A., Myslin, M., & Boychuk, V. (2020). Structure, morphology and adsorption properties of titania shell immobilized onto cobalt ferrite nanoparticle core. *Journal of Molecular Liquids, 297*, 111757. https://doi.org/10.1016/j.molliq.2019.111757.

Vigneswaran, S., Ngo, H. H., Visvanathan, C., & Sundaravadivel, M. (2009). Conventional water treatment technologies. *Wastewater Recycling, Reuse, and Reclamation, II*, 36.

Wang, L. K., & Menon, R. (2011). Treatment of industrial effluents, municipal wastes, and potable water by membrane bioreactors. In *Membrane and desalination technologies* (pp. 201–236). Humana Press, Totowa, NJ. https://doi.org/10.1007/978-1-59745-278-6_5.

Wang, S., & Peng, Y. (2010). Natural zeolites as effective adsorbents in water and wastewater treatment. *Chemical Engineering Journal, 156*(1), 11–24. https://doi.org/10.1016/j.cej.2009.10.029.

Wang, Y., Ho, S. H., Cheng, C. L., Guo, W. Q., Nagarajan, D., Ren, N. Q., Lee, D. J., & Chang, J. S. (2016). Perspectives on the feasibility of using microalgae for industrial wastewater treatment. *Bioresource Technology, 222*, 485–497. https://doi.org/10.1016/j.biortech.2016.09.106.

Xiong, J. Q., Kurade, M. B., Abou-Shanab, R. A., Ji, M. K., Choi, J., Kim, J. O., & Jeon, B. H. (2016). Biodegradation of carbamazepine using freshwater microalgae Chlamydomonas mexicana and Scenedesmus obliquus and the determination of its metabolic fate. *Bioresource Technology, 205*, 183–190. https://doi.org/10.1016/j.biortech.2016.01.038.

Zhang, M. H., Dong, H., Zhao, L., Wang, D. X., & Meng, D. (2019). A review on Fenton process for organic wastewater treatment based on optimization perspective. *Science of the Total Environment, 670*, 110–121. https://doi.org/10.1016/j.scitotenv.2019.03.180.

Zhang, T., Cui, C., Chen, S., Yang, H., & Shen, P. (2008). The direct electrocatalysis of Escherichia coli through electroactivated excretion in microbial fuel cell. *Electrochemistry Communications, 10* (2), 293–297. https://doi.org/10.1016/j.elecom.2007.12.009.

Zheng, C., Zhao, L., Zhou, X., Fu, Z., & Li, A. (2013). Treatment technologies for organic wastewater. *Water Treatment, 11*, 250–286. https://doi.org/10.5772/52665.

Zheng, X., Zhang, Z., Yu, D., Chen, X., Cheng, R., Min, S., Wang, J., Xiao, Q., & Wang, J. (2015). Overview of membrane technology applications for industrial wastewater treatment in China to increase water supply. *Resources, Conservation and Recycling, 105*, 1–10. https://doi.org/10.1016/j.resconrec.2015.09.012.

Zhou, H., & Smith, D. W. (2001). Advanced technologies in water and wastewater treatment. *Canadian Journal of Civil Engineering, 28* (S1), 49–66. https://doi.org/10.1139/s02-020.

The Role of Water Recycling in Building a Circular Economy in the Textile Industry

Marta Gmurek and Lucyna Bilińska

Abstract

The lifestyle of modern society is deeply connected to massive industrialization and the production of goods. One of the trends of the modern age is fast fashion. For the last decades, customers of fashion concerns have been used to buying new clothing every single season. But, we must reconsider the motivation of buying a brand new pair of blue jeans and start to bother that producing every pair of jeans needs between 7000 and 10,000 L of fresh water? Textile production is one of the most water-consuming industrial branches, which undoubtedly poses an environmental threat. This chapter describes the most relevant issues related to textile wastewater problems. The most serious critical points were stressed by discussing the consumption of water at each level of textile goods production and the most serious sources of environmental pollution during textiles production. At the same time, this chapter provides the most beneficial circular economy solutions for textile wastewater reuse, such as heat recovery or the use of greywater for hygiene purposes. But the main and most promising path was found to be wastewater recycling and closing the water loop during the textile production processes. By giving not only the data from the literature but also practical industrial examples, the authors showed effective methods for textile wastewater treatment in the service of recycling. The technical issues and economics of these treatments concluded their applicability. Finally, the chapter includes the topic of minimal water parameters that need to be kept for recycling and the system for ensuring the proper final quality of textiles.

Keywords

Textile wastewater • Textile wastewater treatment • Advanced oxidation processes • Bio-treatment • Wastewater recycling

1 Introduction

As far as colossal textile production benefits the living condition, it poses a severe environmental threat. According to Islam and co-authors, even 10% of global greenhouse gas emissions are from textiles manufacturing (Islam, 2020). Moreover, textile waste occupies 5% of landfill space, and 20% of overall freshwater contamination originates in textile production sources (Islam, 2020). Many toxic substances can be found in gases exhausted by textile factories. Besides the apparent pollutants, such as carbon dioxide, carbon monoxide, or sulfur oxides, textile processing can release formaldehyde, acids, softeners, benzene-derived compounds, and other volatile compounds (Islam, 2020). Correspondingly, the water pollution caused by the textile industry can also be harmful. Metals, chlorides, sulfates, phosphates, non-biodegradable dyes, formaldehyde-based dye fixing agents, hydrocarbon-based softeners, and other toxic compounds disturb living conditions in water bodies. At the same time, they are a source of genotoxicity for water organisms and cause health hazards for people using polluted water for living purposes (Bhatia et al., 2018; Mirza et al., 2020). The new group of textile pollutants is microplastic. The European Commission estimates that 0.5 million tons of microplastic are emitted into water bodies due to clothing washing. Figure 1 shows the significant environmental impacts of the textile industry.

M. Gmurek (✉) · L. Bilińska
Faculty of Process and Environmental Engineering, Lodz University of Technology, Wolczanska 213, 90-924 Lodz, Poland
e-mail: marta.gmurek@p.lodz.pl

L. Bilińska
e-mail: lucyna.bilinska@p.lodz.pl

L. Bilińska
Bilinski Factory of Colour, Mickiewicza 29, 95-050 Konstantynow Lodzki, Poland

M. Smol et al. (eds.), *Water in Circular Economy*, Advances in Science, Technology & Innovation,
https://doi.org/10.1007/978-3-031-18165-8_7

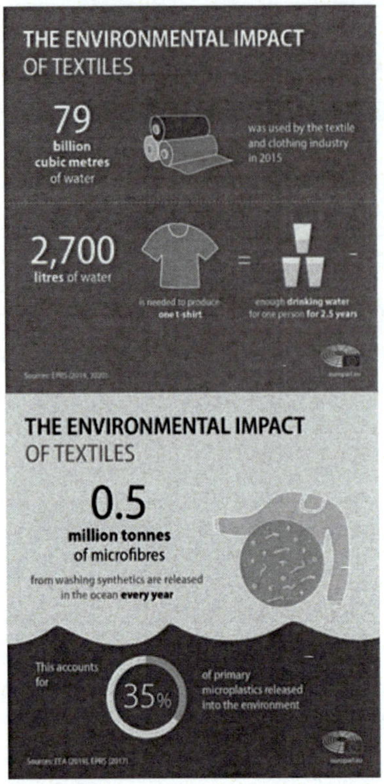

Fig. 1 The main environmental impacts of the textile industry adapted from (European Parliament, 2021)

Unfortunately, the abovementioned environmental impacts can escalate in industrial areas. China and the EU are the most important world textiles markets. According to the World Trade Statistical Review 2021 released by the World Trade Organization, China had a 43.5% share in the world textile market in 2020. The second-largest was the EU (18.1%), then India (4.2%) and Turkey (3.3%) (World Trade Organization, 2020). These regions are significantly exposed to textile-originated pollution. The EU recognized this branch as the fourth highest-pressure category for the use of primary raw materials and water, after food, housing, and transport. Therefore, the EU governance overtook the active management of minimizing the environmental risk and proposed circular water management as one of the explicit topics within the Textiles Green Deal Program (European Commission, 2020).

This chapter tries to deal with water circularity in the textile industry by giving hints on the wastewater characteristics and the possibilities of treatment and water reuse. Two main areas were discussed within the scope. The first was textile wastewater characterization and giving the specific groups of pollutants that originated in textile production. The second was a discussion on treatment methods, which can be dedicated to effective textile waste removal from water. Furthermore, a debate on water recycling demands and the conclusion of wastewater reuse in industrial practice found a place within the scope.

2 Textile Wastewater Characteristic

Considering that textile production is a multistep operational process, it is visible that many entire dissent types of pollutants can occur in wastewater originating in it. Spinning, weaving, knitting, and subsequent wet processing are some of the following stages of textile production (Kalliala & Talvenmaa, 2000). But thousands of assistant agents used in chemical processing make it the most environmentally harmful. The other crucial aspect is water consumption, and the cubic meters of highly polluted water are produced in textile manufacturing. A high quantity of 100–300 L per kilogram of fresh water is used in dyeing operations, and this value oscillates between 5 L/kg for some wovens and more than 900 L/kg for finishing felted fabric. As a result, a textile plant generates about 1000–3000 m^3 of wastewater after processing about 12–20 tonnes of textiles per day (Ghaly et al., 2013).

Figure 2 shows how complex and divaricated can be the chemical textiles processing. Processes differ in terms of the type of textile material. The highest number of operations is needed to process the woven fabric, which is caused by the necessity of warp desizing. Because knitted has no warp, the step of desizing is not present in this processing, and this treatment is shorter and less operational. The nonwoven fabrics need less demanding processing. Mainly because they are produced for technical purposes, and they do not need to fulfill stringent quality standards required in the garment segment. It should be kept in mind that each textile material needs a specific technological condition and many different chemical agents to operate it. Therefore, every single stage of textile processing brings a particular composition of pollutants into the wastewater. The types of potential contaminants and their environmental risk depending on textile material type are shown in Fig. 3.

Figure 3 shows that despite both the processes and the pollutants diversification, some impurities are common for all types of textile production. These can be characterized as dyes, organics, and microplastic. The first group, dyes, seems to be the most obvious one and, at the same time, the most studied. Numerous papers were devoted to color removal (Paździor et al., 2019), and at the same time, the possible ecotoxicological impacts were proven for dyes (Al-Tohamy et al., 2022). The second group of textile pollutants can be specified as organics. This group covers many assistant agents, such as resins, waxes, softeners, organic

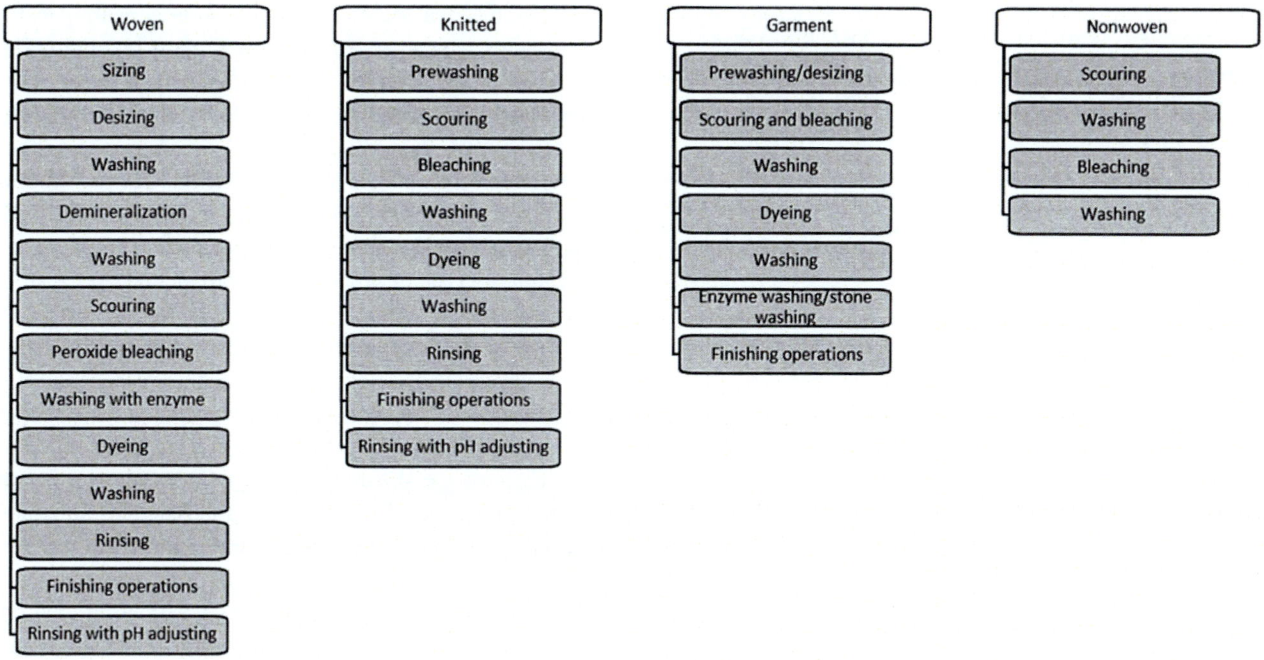

Fig. 2 Unit operations that are involved in different categories of textile processing (Kalliala & Talvenmaa, 2000; Raja et al., 2019)

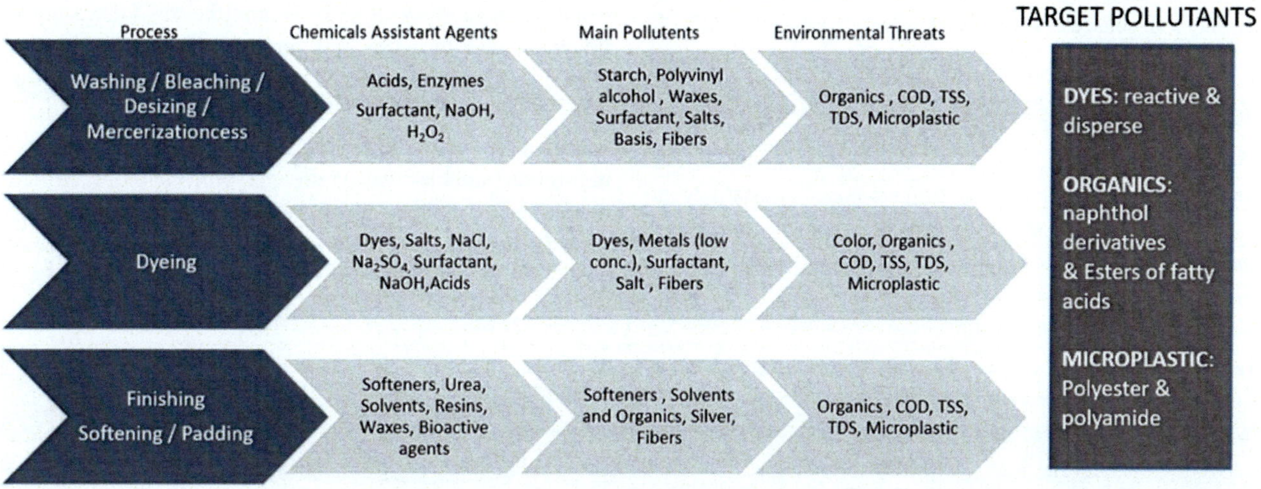

Fig. 3 Textile pollutants matrix (Allègre et al., 2006; Bisschops & Spanjers, 2003; Islam, 2020; Kalliala & Talvenmaa, 2000; Kishor et al., 2021; Ledakowicz & Paździor, 2021; Mirza et al., 2020)

acids, and the by-products of their decomposition. The representatives of this group are naphthol derivatives and esters of fatty acids (Ledakowicz & Paździor, 2021; Mirza et al., 2020). These are the most characteristic because they are ingredients of textile agents and can be residual by-products in the wastewater. The third group of microplastic has been identified quite recently. The impulse for investigation of the microplastic issue was seas deterioration. Since the scientists recognized the sources of microplastic pollution, textiles were found to be one that could not be left out. The origins of textile microplastic are in the primary polyester and polyamide fibers as the standard clothing material (Kishor et al., 2021).

These general groups of impurities recognized in textile wastewater as the primary pollutants are organic-based compounds. But, Table 1 shows total suspended solids (TSS) and total dissolved solids (TDS) as potential environmental threats originating in textile production. The TSS causes a stable suspension in wastewater and can be revealed by turbidity. The source of TSS is fibers mechanically or enzymatic washed out from textiles, next to the inorganic impurities such as dust and sand. Moreover, hydrophobic

Table 1 Global parameters of general textile wastewater (Al-Kdasi et al., 2004; Cay et al., 2009; Holkar et al., 2016; Kishor et al., 2021; Ledakowicz & Paździor, 2021; Mirza et al., 2020; Villegas-Navarro et al., 2001)

Parameter, unit	Value
COD, mg/L	450–3000
BOD_5, mg/L	160–1200
TSS, mg/L	15–330
TDS, mg/L	14–12,000
$N_{Total.}$, mg/L	5–70
P_{Total}, mg/L	1–6
Conductivity, mS/cm	4900–12,500
Chlorides, mg/L	1000–1750
Sulfates, mg/L	250–600
Color by Hazen scale, Pt–Co units	1000–3300
pH	8.5–11

substances such as waxes or resins can form the suspension in textile wastewater. According to Mirza and co-workers (Mirza et al., 2020), the TSS in general textile wastewater can be between 15 and 300 mg/L. However, the TDS range can be as high as 12,000 mg/L (Mirza et al., 2020). As far as, TDS is caused by dissolved, dissociated ions (Na^+, K^+, NH^{4+}, SO_4^{2-}, SO_3^{2-}, ClO^-, Cl^-, CO_3^{2-}, PO_4^{3-}) it makes a high inorganic content (Mirza et al., 2020).

The abovementioned causes of the pollution can conclude that textile wastewater treatment can be demanding. On the one hand, the extensive use of organic-based assistant agents in the production cycle brings many organics, resulting in increased COD value. On the other hand, high values of TSS and TDS mean a lot of inorganics. Consequently, textile wastewater is characterized by a low BOD_5/COD ratio. This ratio is often far below the 0.5 value, which means poor biodegradability (Ledakowicz & Paździor, 2021; Mirza et al., 2020).

The global parameters shown in Table 1 are as informative as the characteristics mentioned earlier, low BOD/COD ratio, intensive color, and high mineral content. However, the feature by type of process in the production cycle can be more informative. When analyzing one of the processes presented in Table 2 in terms of wastewater parameters, it can be observed that some wastewater streams have more mineral content than others. In consequence, it can be more or less susceptible to biological treatment. This observation is the basis of the streaming strategy for textile wastewater treatment proposed by the EU Commission as the most beneficial solution to deal with textile-pollution in wastewater (European Commission, 2003).

Table 3 presents the basis of the UE policy on environmental management in the textile industry. After giving the general tips, this policy includes every field of textile processing. This policy aimed to provide a specific action plan. The tools to deal with environmental aspects were the best available techniques (BATs). Within the tips for wastewater treatment was the streaming strategy. The main idea is to allocate highly contaminated wastewater originating in some specific operations (an example from Table 3 is dyeing wastewater with high chlorides content) and treat non-biodegradable waste streams with appropriate techniques before or instead of the biological treatment (European Parliament, 2002; European Commission, 2003).

Table 2 Parameters of various wastewater streams generated during specific dyeing of cotton fabric with Reactive Black 5 dye. Adapted from Bilińska (2017)

No. of the bath	pH	Conductivity µS/cm	Cl^- mg/L	COD mgO_2/L	BOD_5 mgO_2/L	TC mg/L	TOC mg/L	N_{total} mg/L	P_{total} mg/L
1. Washing with bleaching	9.97	4920	705	2890	–	1197	1065	77.9	8.01
2. Acidification	4.38	2400	79.9	2180	–	749	749	37.6	9.36
3. Rinsing	4.52	1462	65.5	1000	–	370	370	16.8	6.91
4. Dyeing	11.5	61,597	23,275	1200	76	1033	423	24.5	8.52
5. Rinsing	10.6	23,508	7375	490	35	438	170	10.1	3.00
6. Acidification	3.97	11,312	3206	852	280	525	525	6.8	4.01
7. Washing after dyeing	5.5	5449	1456	1175	400	466	466	10.5	0.90
8. Rinsing	6.83	2630	311	536	314	239	217	5.3	0.43
9. Rinsing	7.94	1347	211	188	128	129	79	2.2	0.28
10. Rinsing & neutralization	5.4	1071	104	215	200	119	119	1.2	0.21
11. Mixed	9.85	12,576	3567	956	240	533	407	18.7	4.54

Table 3 Background of UE policy on environmental management in the textile industry prepared based on EU council document on Best Available Techniques in the textile industry (European Parliament, 2002; European Commission, 2003)

IPPC directive 96/61/WE objectives	General BAT for the textile industry	BAT on textile wastewater
Economic growth	Selection of incoming raw material fiber	Wastewater streams strategy
Environment protection	Chemicals management (Selection & use, dosing & dispensing	Allocating highly contaminated wastewater
Social responsibility	Optimization of production processes	Exclusion of wastewater which can cause malfunction of biological treatment
	Water & energy management	Treating non-biodegradable waste streams with appropriate techniques before, or instead of, final biological treatment
	Management of waste streams	
	Wastewater treatment and waste management	

The idea of sustainability for the textile industry was kept in the new EU initiative of the Green Deal. The EU Circular Economy Action Plan encourages industrial operators to keep circularity in textile manufacturing, which should be fulfilled by water recovery (European Commission, 2020). The basis of the streaming strategy for textile wastewater treatment seems to be promising; however, most textile plant operators still use a dilution strategy. The method for diluting contaminants in wastewater means draining all effluents together to lower global pollution at the end of the pipe. After this operation, the wastewater is sent chiefly into the municipal sewer. Choosing the treatment strategy according to the biodegradability factor was discussed in more detail in the following subsection.

Another idea for reducing textile wastewater's detrimental influence is a zero discharge strategy (ZDS). One example of the ZDS is non-water dyeing. The textile materials are dyed in supercritical carbon dioxide instead of water. This chemical processing could be a response to zero discharge and zero-waste production. However, the use of this non-water solvent dyeing technic is minimal. This method uses carbon dioxide as a supercritical fluid, a hydrophobic medium, limiting its use to polyester textiles only. Moreover, special equipment is needed to keep the proper temperature and pressure conditions for this process (Raja et al., 2019). Consequently, the use of water systems is more feasible and common. Therefore, there is no possibility to avoid textile wastewater treatment when eco-friendly production is considered.

3 Textile Wastewater Treatment

Since the water usage in the textile industry is large, and its uptake is in general from aquifers (71.5%), municipal (16%), and surface water (10.4%) sources (Fig. 4a), it is reasonable to apply a circular water economy approach.

Moreover, as can be seen in Fig. 4b the Textiles companies mostly discharge their wastewater to the municipal sewage system (73%). Almost 21% of textile companies discharge wastewater directly to surface water, one company (2.1%) is reusing treated wastewater, and 2 (4.1%) collect wastewater and transport wastewater to the special wastewater treatment plant. For textile wastewater treatment

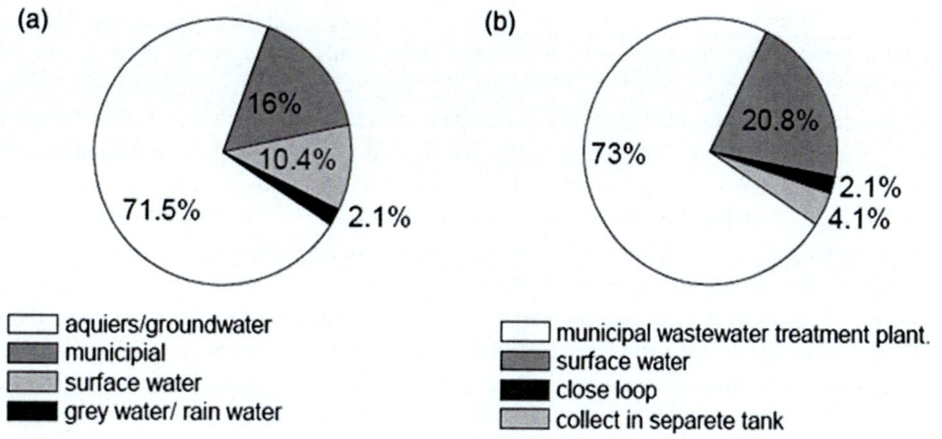

Fig. 4 **a** Water intake in textile companies in Europe, **b** Wastewater discharging (Italy, Belgium, Poland, Romania, Germany, n = 48) (CENTEXBEL, 2017; CENTROCOT, 2017; INCDTP, 2017; Lodz, 2017; Nordwest, 2017; ECWRTI, 2020)

usually, conventional methods (physical, chemical, biological) are applied (Nawaz & Ahsan, 2014; Paździor et al., 2019). However, mostly all of them leads to high efficiency decolorization and low level of organics degradation (Nawaz & Ahsan, 2014; Ogunlaja & Aemere, 2009; Paździor et al., 2017, 2019; Silva et al., 2020). Adsorption, ion exchange, and membrane filtration are used as a preliminary treatment to treat textile wastewater, with a very high decolorization efficiency above 85% (Samsami et al., 2020). Interestingly, membrane filtration, despite its simplicity and high effectivity even toward COD and salts, should not be recommended for high-loaded textile effluents. Moreover, the requirements of the need for periodic membranes replacement ensures that the profits do not exceed costs (Al-Tohamy et al., 2022; Samsami et al., 2020). Coagulation-flocculation is also successfully applied to alter the state of dissolved and suspended solids (TDS and TSS, respectively) in wastewater and separates them through sedimentation. Physical and physico-chemical approaches have numerous advantages (Table 4). As main benefits, a simple design, ease of operation, fewer chemical requirements, and the lack of impact on it by the presence of toxic substances can be considered. However, these methods are frequently not preferred due to many weaknesses like disability to treat reactive dyes, sludge generation (even toxic), problems with safe disposal, and possible secondary pollution associated with the adsorbent regeneration as well as limited applicability. However, it should be mentioned that the specific textile effluents parameters like high temperatures, high chemical oxygen demand, low biological oxygen demand, alkaline environment, dark color, and some heavy metals presence frequently hinder the application.

Biological methods are mostly used as a secondary treatment to treat dissolved matter present in textile wastewater but only if the BOD/COD ratio is suitable for it. A source of successful biological processes toward dye decomposition is attributed to the various groups of microbial cell wall components such as amino, carboxyl, hydroxyl, phosphate, and other charged groups (Jafari et al., 2014; Al-Tohamy et al., 2022). Due to complete dye mineralization, less sludge generates as well as much lower energy consumption those processes are easily and economically applied in developing countries. Moreover, the biological approaches for the treatment of textile wastewater seem to be less expensive, more environmentally safe, and the by-products generated during microbial metabolic activity are non-toxic (Al-Tohamy et al., 2022; Nawaz & Ahsan, 2014; Paździor et al., 2017, 2019). Mostly the activated sludge system and membrane bioreactors are used but algae, yeast, and fungi, as well as enzyme-based systems, are also viable biological candidates for the treatment of textile dye wastewater, converting dye molecules into non-toxic products (Table 4). However, the biggest disadvantages appear when the microbial has been not able to process and when the bacterial community is not able to rebuild. Additionally, since biological technology does not lead to removing non-biodegradable contaminants, secondary effluents still require further treatment to achieve the discharge standards.

The most studied and well-known chemical approach used for textile effluent treatment is the application of advanced oxidation processes (AOPs). This group of techniques that are based on in-situ generation of hydroxyl radicals ($^{\cdot}OH$) is successfully applied for dyes solutions as well as real textile wastewater. The efficiency depends on the method that is applied. The most investigated Photolysis, Photocatalysis, Fenton-based processes, ozone-based processes, and electrochemical processes were considered. These methods are capable of removing several groups of dyes and not only high decolorization but also COD removal is satisfactory. However, like all methods, there are also some limitations and disadvantages: the high cost of energy and reagents, pH dependency, and most seriously toxic by-products generation (Lucyna Al-Tohamy et al., 2022; Bilińska, 2017; Bilińska & Gmurek, 2021; Paździor et al., 2017, 2019; Silva et al., 2020).

The advantages, disadvantages, and conversion pathways for each of the preliminary, secondary, and tertiary treatment processes are summarized in Table 4.

As mentioned in the previous subparagraph, the characteristics of textile effluents have a significant impact on the environment, such effluents without proper treatment cannot be discharged. The high organic matter concentration expressed by COD increases BOD/COD ratio that estimates the biodegradability of effluents. It is well known, that when the ratio is higher than 0.5, a significant content of biologically degradable contaminants is indicated, hence the biological treatment can be implemented (Ceretta et al., 2017; Płuciennik-Koropczuk & Myszograj, 2019). The additional physico-chemical treatments need to be considered when the ratio is between 0.5 and 0.28, which indicates medium biodegradability and a high content of organic substances that are hardly decomposable or biologically indecomposable (Płuciennik-Koropczuk & Myszograj, 2019; Ramos et al., 2021). The high fraction of recalcitrant substance causes that ratio exceeds 3.5, therefore combine processes are required including biological pre-treatment and advanced oxidation or physical–chemical tertiary treatment (Ramos et al., 2021). As can be seen in Fig. 5, the applied treatment process has been correlated with the BOD/COD ratio.

In recent years, the integration of AOPs with membrane processes, electrocoagulation, and biological treatment has received increasing attention for effectively treating industrial wastewater. As textile companies were aware that textile effluents are one of the most contaminated effluents several approaches have been applied. In Italy usually, always physical–chemical treatment processes were applied in

Table 4 Advantages, disadvantages, and conversion pathways for each wastewater treatment method applied to textile wastewater treatment (Al-Tohamy et al., 2022; Behera et al., 2021; Ganiyu et al., 2022; Mirza et al., 2020; Nawaz & Ahsan, 2014; Ogunlaja & Aemere, 2009; Ramos et al., 2021; Samsami et al., 2020)

Treatment	Action pathway	Advantages	Disadvantages
Physical and Physico-chemical—preliminary treatment			
Adsorption	Ions or molecules tend to accumulate and concentrate on the reactive and microporous surface of the coal	Extended surface area, high capacity of adsorption, and a high degree of reactivity.Removal of various dyes	Fast saturation, cost of the adsorbent and difficulty in its regeneration
Filtration	Filtration through membranes with pores in the range of 0.002–0.1 μm (UF), 0.007–2.0 μm (MF), and 0.001–0.07 μm (NF)	MF, UF: Low pressure; NF: Separation of low molecular weight organic compounds and divalent ions from monovalent salts. Treatment of high concentrations	MF, UF: Treated water does not present a good quality. High cost of membranes and possible incrustations; NF: High cost of membranes and possible incrustations
	RO: Membrane filtration to remove dissolved solids alongside ions and larger species from the effluents. Pores smaller than 0.001 μm	Total removal of mineral salts, hydrolysis of reactive dyes, and chemical auxiliary	High pressures, high cost, and possible incrustations
Coagulation-flocculation	The physical–chemical process aimed at removing the particles (suspended solids) that cloud water by precipitating them in the form of sludge	Excellent ability to discolor textile effluents and diminish a total load of suspended solids	Enormous amounts of generated sludge which have a serious disposal problem depend on the characteristics of raw wastewater, especially pH, temperature, the type and the dosage of the used coagulant, as well as the intensity and duration of mixing
Electrocoagulation	Use of current between metallic electrodes generating coagulant species in solution	Dispenses chemicals. Removal of different compounds. Produces less sludge	Periodic replacement of the sacrificial anode. The low content of dissolved solids is a limiting factor. Formation of toxic substances
Biological treatment—secondary treatment			
Activated sludge	The anaerobic processes are responsible for decolorization, while aerobic processes lead to the efficient removal of organic compounds and usually significantly lower color removal	Aerobic bacteria: Discoloration of azo and anthraquinone class dyes; Anaerobic bacteria: Easy operation. Reductive transformation of azo dyes	Aerobic bacteria: Affected by toxic substances. Low rate of discoloration. Need for aeration. Requires additional carbon and energy sources; Anaerobic bacteria: Affected by toxic substances and generates other toxic substances. Low rate of discoloration of azo dyes. Need for post-treatment
Fungi	Enzymatic action, decolorization due to reductive cleavage process	High rates of discoloration of anthraquinone and indigo dyes	Low rate of discoloration of azo dyes. Need for specific bioreactors and external carbon sources. Need a nitrogen-restricted environment. Acidic medium (pH 4.5–5). Inhibition by mixing dyes and chemicals
Microalgae	Through bioconversion, and bioadsorption processes using dead or living biomass	Low-cost wastewater treatment and bioenergy feedstock source	Independent of geographical regions with high temperatures and sunlight fluxes
Biofilm	Due to high diversity of microbial communities, the degradation in a wide range of organic pollutants is possible	No sludge recycling, low head loss, no influence of biomass separation, operational stability, use of whole tank volume for biomass growth, less space, improved mass transfer, and no medium channeling or clogging	Slow biofilm formation, poor permeability, large flow resistance, and easy clogging

(continued)

Table 4 (continued)

Treatment	Action pathway	Advantages	Disadvantages
Advanced oxidation processes—tertiary treatment			
Photolysis	Structural transformation of organics by hydroxylation and electrophilic addition	Limited chemical addition requirement, high treatment capacity, can use renewable energy sources	Inapplicable for dark colorous textile effluents, high energy demand, longer treatment time
H_2O_2/UV	Generation of $^{\cdot}OH$ radicals; dye decomposition by hydroxylation, electrophilic addition and hydrogen abstraction	Easy operation and handling, effective contact between oxidant and pollutant, no sludge generation. UV acts as a disinfectant. Generation of H_2O_2 in-situ	Dependend of turbidity of effluents, partial degradation with limited mineralization, the large amount of H_2O_2 requirements, high-cost operation and risk associated with storage and transportation of H_2O_2, formation of toxic by-products, energy consumption, low efficiency for dark colorous textile effluents
Catalytic ozonation	Generation of Reactive oxygen spices (ROS). Oxidation by oxygen/ozone addition, electrophilic addition unsaturated bonds	Improvement of reaction rate, easily scale-up and implementation	High cost of operation, partial mineralization and limited toxicity reduction, catalyst recovery problems
Photocatalysis	Use of photons to activate the semiconductor catalyst. Superoxide bond cleavage via hydroperoxide intermediate	High efficiency in the degradation and mineralization of organics, no additional chemical requirements, renewable energy sources applicability, no pH barriers, easy to operate and automation if the catalyst is incorporated into the carrier	Inapplicable for dark colorous effluents, cost of artificial radiation, high cost of operation, nanomaterials contaminations, possible toxic metal ions contamination, post-treatment catalyst separation (if suspended), toxic by-products
Fenton-based AOPs	Generation of ROS; dye decomposition by hydroxylation, electrophilic addition and hydrogen abstraction	Efficient discoloration of soluble and insoluble dyes, electro-Fenton is effective for the degradation and mineralization of organics, Fenton process is effective for low-loaded effluents with pH close to optimum, allow maximum contacts between pollutants and generated oxidants, simple and flexible operation with easy implementation on the existing plant, easy-to-handle chemicals and no need for energy input, cheap supplies	Production of iron sludge and post-treatment neutralization. Ineffective for most textile effluents due to acidic condition requirement, high chemical addition requirement, toxic by-products relatively high cost and risk associated with storage and transportation of H_2O_2
Ozone based AOPs	Generation of ROS. Oxidation by oxygen/ozone addition, electrophilic addition unsaturated bonds	No sludge generation, high efficiency in the degradation and mineralization of organics, no additional chemical requirements, no pH barriers, simple and flexible operation with easy implementation on the existing plan	High cost of operation, partial mineralization, and limited toxicity reduction
Electrochemical AOPs	Generation of oxidants on the electrode surface or from chemicals in solution. Dye decomposition by hydroxylation, electrophilic addition, and hydrogen abstraction	Excellent for the degradation and mineralization of organics, easy operation, little or no use of chemicals, stable oxidation potential over a wide pH range except in Fenton-based EAOPs, high versatility, amenability	Formation of toxic by-products, the possibility of chlorinated by-products formation due to high Cl ions, periodic replacement of electrodes due to passivation, inactivation, and polarization problems, high energy consumption

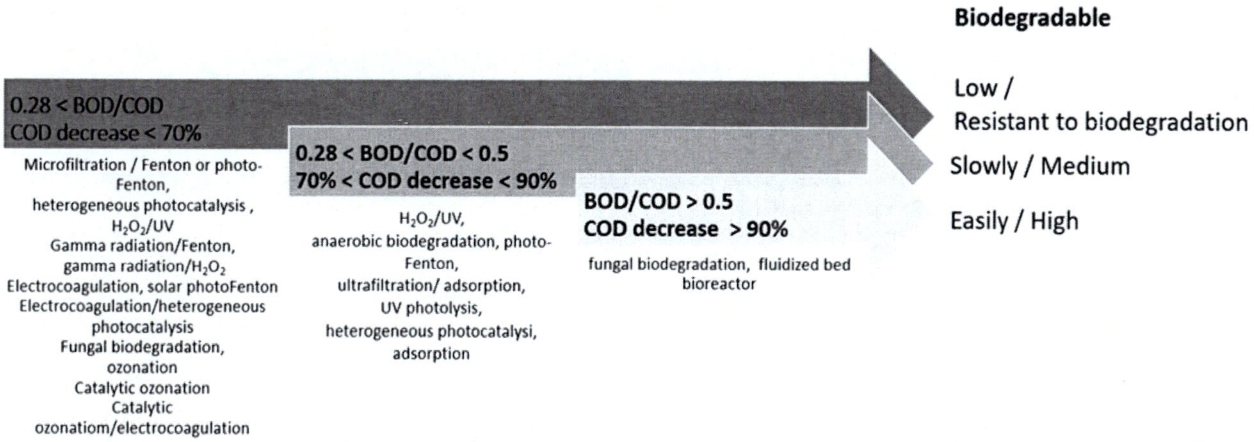

Fig. 5 Assessment of Susceptibility to Biochemical Biodegradation and proposed treatment processes (Ahmed & El-Shishtawy, 2010; Allègre et al., 2006; Al-Tohamy et al., 2022; Bilińska & Gmurek, 2021; dos Santos et al., 2007; Ganiyu et al., 2022; Ledakowicz & Paździor, 2021; Mirza et al., 2020; Nawaz & Ahsan, 2014; Ogunlaja & Aemere, 2009; Paździor et al., 2017, 2019; Raja et al., 2019; Ramos et al., 2021; Samsami et al., 2020; Silva et al., 2020)

textile companies (CENTEXBEL, 2017; CENTROCOT, 2017; INCDTP, 2017; Lodz, 2017; Nordwest, 2017; ECWRTI, 2020). In Poland mostly all textile companies discharge textile effluents to the municipal sewerage to be treated in the municipal wastewater treatment plants, some of them have their treatment plant to polish effluent to achieve legislation level. In Belgium, some textile companies also own on-site wastewater treatment plants and after treatment can direct discharge it to surface waters. While in Romania textiles companies also own on-site pre-treatment plants and but mostly after polishing discharge it to the municipal sewerage for further treatment in municipal facilities. In Germany, the majority of textile companies discharge their wastewater into the sewage system (indirect discharge) for further treatment at the municipal wastewater treatment plants. Wastewater is usually collected in the buffering tanks and treated by physical–chemical (aeration, flocculation, coagulation) and biological processes (activated sludge). These pre-treatment processes are applied mainly to reduce the BOD and COD values, along with other pollutant values.

It should be noted that in Poland, there is one company that has a wastewater treatment plant with a closed water circuit. About 80% of their targeted wastewater is reused. To achieve such a high water recycling level two separate wastewater streams are handled differently. The low-loaded wastewater (mainly from rinsing, acidifying, and washing baths) is treated on-site by biological process (activated sludge chambers), ultrafiltration, and ozonation. After this treatment water can be reused. While the high-loaded wastewater stream is pre-treated by coagulation/flocculation and then discharged to the municipal wastewater treatment plant.

In 2015 United Nations Resolution on the 2030 Agenda for Sustainable Development was signed. One of the 17th Sustainable Development Goals (SDGs)—SDG 6 is dedicated to water aspects, while Subgoal 6.3 is aimed to improve water quality by reducing pollution, eliminating dumping and minimizing the release of hazardous chemicals and materials, halving the proportion of untreated wastewater and substantially increasing recycling and safe reuse globally by 2030 (United Nations, 2015). After the "Blueprint to Safeguard Europe's Water Resources" released by the European Commission in 2012 (European Commission, 2012) where water reuse and recycling were identified as one of the top priorities of the EU, the minimum requirements for water reuse were determined (European Commission, 2018; UNESCO, 2017). Lately, many water conservation and saving programs have been promoted and have been addressed in textile industries. Moreover, if the costs of water uptake, wastewater treatment as well as wastewater discharge are taken into account, the total costs are depended of the discharge approach. In countries where the companies were obligated to polish wastewater before discharging it to the municipal sewage system, the highest cost is due to treatment while discharging is not so much expensive. But in countries where the wastewater is treated in Municipal Wastewater Treatment Plants, the highest costs are linked to discharging (Fig. 6). Therefore, the application of the circular economy approach especially from a close loop of treated wastewater point of view is not only environmental but also economically profitable.

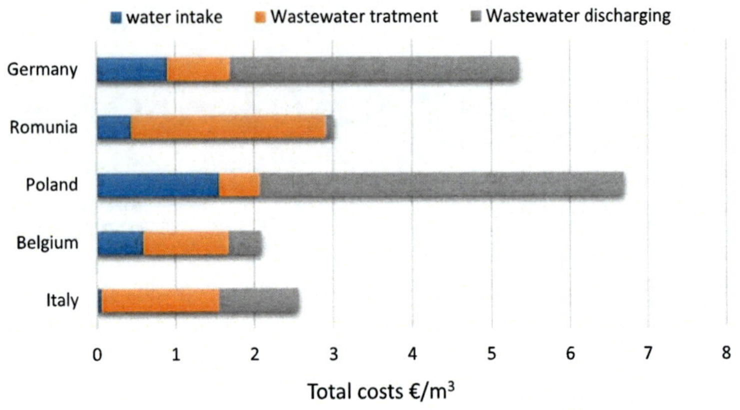

Fig. 6 Total costs for water usage in European textile companies (CENTEXBEL, 2017; CENTROCOT, 2017; INCDTP, 2017; Lodz, 2017; Nordwest, 2017; ECWRTI, 2020)

4 Quality in Water Recycling

As was mentioned in the previous subsection, numerous legal bodies and non-governmental organizations aim to build sustainability in the textile industry by stressing water reuse. However, it should be kept in mind that textiles manufacturers work under the massive pressure of very high-quality requirements for their products. Standards such as Oeko-tex™, Fair Trade™, EU Flower™, Nordic Swan™ RWS™, GRS™ GOTS™, or Bluesign® are only a few commercial quality assurance systems operated by independent certification bodies (Bilińska & Gmurek, 2021). These systems approve textile products by labeling them when they fulfill the strict quality standards. Every standard has its list of hazardous compounds, and they are not allowed to be introduced into textiles during production processes at any stage.

On the other hand, a legal body of the European Commission established the list of substances forbidden for use in textile manufacturing. This legal act contains harmful chemicals that originated in azodyes (European Parliament, 2002). Moreover, all chemical agents for textiles must be registered under REACH legislation when used in the EU. Even though a lot of interest was put into the safety assurance of final textile products, there are still no hints of how should be the quality of recycled water. Since the discussion of possible limits for various textile wastewater pollutants has just been raised, which is also valuable (Mirza et al., 2020), it will take a long time to appoint them for water recycling. However, the step forward has been done by giving minimal quality standards for water reuse in agricultural irrigation by the proposition of the EU Parliament for new regulation in the matter of water recycling (Regulation of the European Parliament and of the Council, 2018).

Nowadays, the most common practice for water reuse in textile processing is the inter-operative rinsing water collection and using it again in afterward operation—the qualification for the reuse is only the lack of color and lack of foaming. The more advanced procedure for water recycling worked out in one of the Polish dyehouses. To keep the standard of recycled water, it must be colorless, non-turbidity, sanitized by ozone, characterized by COD below 100 mg/L, and pH not higher than 8. But, still, this manufacturer uses fresh water in the last stage of textile production to fulfill the high Oeko-tex demands of the products (Paździor et al., 2019).

5 Conclusions

While the twentieth century was the age of massive progress driven by technology, science, and industrialization, the twenty-first century seems to be the age of reflection. Since the heritage of previous decades is the pollution of the environment, we are obligated to act against this damage. As it was discussed, the representatives of numerous legal and non-governmental non-profit organizations try their best to encourage a possibly large group of recipients for pro-environmental behavior. When summarizing this chapter, it can be concluded that water recycling has a crucial role in building a circular economy in the textile industry. Closing the water loop in textile processing was shown as a most beneficial approach that improves sustainability by saving water resources.

The authors' ambition was to give some hints about textile wastewater recycling. Therefore, the short characteristic of textile wastewater was given in this chapter. Global parameters, such as low BOD/COD ratio, intensive color, alkalinity, and high mineral content, were not the only ones considered. The types of pollutants were investigated depending on the type of textile processing. Consequently, particular groups of pollutants were specified. The first group was organics: dyes (reactive and dispersed most often used), naphthol derivatives, and esters of fatty acids, and the last was microplastic (polyester and polyamide fibers mostly). The second specified group was inorganics originating in basis and salts, which cause the extreme TDS in textile wastewater. The authors showed the strategies for wastewater treatment, of which one and the most common is treatment by dilution, and the second is the streaming strategy. The method of diluting wastewater means draining all water together to lower global pollution. The streaming strategy relies on the exclusion of wastewater which can cause malfunction of biological treatment and using a dedicated purification method instead of bio-treatment. Depending on biodegradability and suspended solid fraction, the appropriate treatment methods were discussed in the chapter. The pros and cons of physical, chemical, and biological treatments were discussed. Accordingly, every treatment was assigned as a possible pre-treatment, basic-treatment, and post-treatment step. Consequently, there was a cascade of particular methods shown as the most beneficial in textile wastewater treatment rather than the usage of a single purification method. The most practical methodology for textile wastewater recovery was discussed by giving numerous explanations and examples to give a broader view of this topic.

Acknowledgements The research leading to these results has received funding from the Norway Grants 2014–2021 via the National Centre for Research and Development.

References

Ahmed, N. S. E., & El-Shishtawy, R. M. (2010). The use of new technologies in coloration of textile fibers. *Journal of Materials Science, 45*, 1143–1153. https://doi.org/10.1007/s10853-009-4111-6.

Al-Kdasi, A., Idris, A., Saed, K., & Guan, C. T. (2004). Treatment of textile wastewater by advanced oxidation processes—A review. *Global Nest International Journal, 6*, 222–230.

Al-Tohamy, R., Ali, S. S., Li, F., et al. (2022). A critical review on the treatment of dye-containing wastewater: Ecotoxicological and health concerns of textile dyes and possible remediation approaches for environmental safety. *Ecotoxicology and Environmental Safety, 231*, 113160. https://doi.org/10.1016/j.ecoenv.2021.113160.

Allègre, C., Moulin, P., Maisseu, M., & Charbit, F. (2006). Treatment and reuse of reactive dyeing effluents. *Journal of Membrane Science, 269*, 15–34. https://doi.org/10.1016/j.memsci.2005.06.014.

Behera, M., Nayak, J., Banerjee, S., et al. (2021). A review on the treatment of textile industry waste effluents towards the development of efficient mitigation strategy: An integrated system design approach. *Journal of Environmental Chemical Engineering, 9*, 105277. https://doi.org/10.1016/J.JECE.2021.105277.

Bhatia, D., Sharma, N. R., Kanwar, R., & Singh, J. (2018). Physicochemical assessment of industrial textile effluents of Punjab (India). *Applied Water Science, 8*, 83. https://doi.org/10.1007/s13201-018-0728-4.

Bilińska, L. (2017). Decolorization of textile wastewater by Advanced Oxidation Processes (AOPs) under industrial conditions. Doctoral thesis.

Bilińska, L., & Gmurek, M. (2021). Novel trends in AOPs for textile wastewater treatment. Enhanced dye by-products removal by catalytic and synergistic actions. *Water Resources in India, 26*. https://doi.org/10.1016/j.wri.2021.100160.

Bisschops, I., & Spanjers, H. (2003). Literature review on textile wastewater characterisation. *Environmental Technology (United Kingdom), 24*, 1399–1411. https://doi.org/10.1080/09593330309385684.

Cay, A., Tarakçioğlu, I., & Hepbasli, A. (2009). Assessment of finishing processes by exhaustion principle for textile fabrics: An exergetic approach. *Applied Thermal Engineering, 29*, 2554–2561. https://doi.org/10.1016/j.applthermaleng.2008.12.032.

CENTEXBEL. (2017). Wastewater handling in the Belgian textile industry. A technical and economical assessment. CENTEXBEL.

CENTROCOT. (2017). Wastewater handling in the Italian textile industry. A technical and economical assessment. CENTROCOT.

Ceretta, M. B., Durruty, I., Orozco, A. M. F., et al. (2017). Biodegradation of textile wastewater: Enhancement of biodegradability via the addition of co-substrates followed by phytotoxicity analysis of the effluent. *Water Science and Technology, 2017*, 516–526. https://doi.org/10.2166/wst.2018.179.

dos Santos AB, Cervantes FJ, van Lier JB (2007) Review paper on current technologies for decolourisation of textile wastewaters: Perspectives for anaerobic biotechnology. *Bioresource Technology, 98*, 2369–2385. https://doi.org/10.1016/j.biortech.2006.11.013.0.

European Commission. (2012). A blueprint to safeguard Europe's water resources.

ECWRTI. (2020). Report on wastewater handling in the textile industry in Europe.

European Commission. (2020). Circular economy action plan.

European Commission. (2003). Integrated pollution prevention and control (IPPC). Reference document on best available techniques for the textiles industry.

European Commission. (2018). Annexes to the proposal for regulation of European minimum requirements for water reuse.

European Parliament. (2002). Directive 2002/61/EC of the European Parliament and of the Council of 19 July 2002 amending for the nineteenth time Council Directive 76/769/EEC relating to restrictions on the marketing and use of certain dangerous substances and preparations (Azocoloura). EU Publications.

European Parliament. (2021). The impact of textile production and waste on the environment (infographic)|News|European Parliament. Retrieved April 4, 2022, from https://www.europarl.europa.eu/news/en/headlines/society/20201208STO93327/the-impact-of-textile-production-and-waste-on-the-environment-infographic.

Ganiyu, S. O., Sable, S., & Gamal El-Din, M. (2022). Advanced oxidation processes for the degradation of dissolved organics in produced water: A review of process performance, degradation kinetics and pathway. *Chemical Engineering Journal, 429*, 132492. https://doi.org/10.1016/J.CEJ.2021.132492.

Ghaly, A., Ananthashankar, R., Alhattab, M., & Ramakrishnan, V. (2013). Production, characterization and treatment of textile effluents: A critical review. *Journal of Chemical Engineering & Process Technology, 5*, 1–19. https://doi.org/10.4172/2157-7048.1000182.

Holkar, C. R., Jadhav, A. J., Pinjari, D. V., et al. (2016). A critical review on textile wastewater treatments: Possible approaches. *Journal of Environmental Management, 182*, 351–366.

INCDTP. (2017). *Report on wastewater handling in the Romanian textile industry* (p. 37)

Islam, S. (2020). A study on the solutions of environment pollutions and worker's health problems caused by textile manufacturing operations. *Biomedical Journal of Scientific & Technical Research, 28*. https://doi.org/10.26717/bjstr.2020.28.004692.

Jafari, N., Soudi, M. R., & Kasra-Kermanshahi, R. (2014). Biodegradation perspectives of azo dyes by yeasts. *Microbiology, 83*, 484–497. https://doi.org/10.1134/S0026261714050130.

Kalliala, E., & Talvenmaa, P. (2000). Environmental profile of textile wet processing in Finland. *Journal of Cleaner Production, 8*, 143–154. https://doi.org/10.1016/S0959-6526(99)00313-3.

Kishor, R., Purchase, D., Saratale, G. D., et al. (2021). Ecotoxicological and health concerns of persistent coloring pollutants of textile industry wastewater and treatment approaches for environmental safety. *Journal of Environmental Chemical Engineering, 9*, 105012. https://doi.org/10.1016/j.jece.2020.105012.

Ledakowicz, S., & Paździor, K. (2021). Recent achievements in dyes removal focused on advanced oxidation processes integrated with biological methods. *Molecules, 26*. https://doi.org/10.3390/molecules26040870.

Lodz, T. R. I. (2017). *Report on the wastewater handling in the Polish textile industry. A technical and economical assessment*. Textile Research Institute Lodz.

Mirza, N. R., Huang, R., Du, E., et al. (2020). A review of the textile wastewater treatment technologies with special focus on advanced oxidation processes (AOPs), membrane separation and integrated AOP-membrane processes. *Desalination and Water Treatment, 206*, 83–107. https://doi.org/10.5004/dwt.2020.26363.

Nawaz, M. S., & Ahsan, M. (2014). Comparison of physico-chemical, advanced oxidation and biological techniques for the textile wastewater treatment. *Alexandria Engineering Journal, 53*, 717–722. https://doi.org/10.1016/J.AEJ.2014.06.007.

Ogunlaja, O. O., & Aemere, O. (2009). Evaluating the efficiency of a textile wastewater treatment plant located in Oshodi, Lagos. *African Journal of. Pure and Applied Chemistry, 3*, 189–196.

Paździor, K., Bilińska, L., & Ledakowicz, S. (2019). A review of the existing and emerging technologies in the combination of AOPs and biological processes in industrial textile wastewater treatment. *Chemical Engineering Journal, 376*, 120597. https://doi.org/10.1016/J.CEJ.2018.12.057.

Paździor, K., Wrębiak, J., Klepacz-Smółka, A., et al. (2017). Influence of ozonation and biodegradation on toxicity of industrial textile wastewater. *Journal of Environmental Management, 195*, 166–173. https://doi.org/10.1016/J.JENVMAN.2016.06.055.

Płuciennik-Koropczuk, E., & Myszograj, S. (2019). New approach in COD fractionation methods. *Water (switzerland), 11*, 1–12. https://doi.org/10.3390/w11071484.

Raja, A. S. M., Arputharaj, A., Saxena, S., & Patil, P. G. (2019). *Water requirement and sustainability of textile processing industries*. Elsevier Ltd.

Ramos, M. D. N., Lima, J. P. P., de Aquino, S. F., & Aguiar, A. (2021). A critical analysis of the alternative treatments applied to effluents from Brazilian textile industries. *Journal Water Process Engineering, 43*, 102273. https://doi.org/10.1016/J.JWPE.2021.102273.

Regulation of the European Parliament and of the Council. (2018). Proposal 2018/0169/COD. 0169.

Samsami, S., Mohamadi, M., Sarrafzadeh, M. H., et al. (2020). Recent advances in the treatment of dye-containing wastewater from textile industries: Overview and perspectives. *Process Safety and Environment Protection, 143*, 138–163. https://doi.org/10.1016/J.PSEP.2020.05.034.

Silva, L. G. M., Moreira, F. C., Cechinel, M. A. P., et al. (2020). Integration of Fenton's reaction based processes and cation exchange processes in textile wastewater treatment as a strategy for water reuse. *Journal of Environmental Management, 272*, 111082. https://doi.org/10.1016/J.JENVMAN.2020.111082.

Textil und Bekleidung Nordwest. (2017). Report of wastewater handling in the German textile industry. A technical and economical assessment. Textil und Bekleidung Nordwest.

UNESCO. (2017). *The United Nations World Water Development Report*.

United Nations. (2015). Transforming our world: The 2030 Agenda for Sustainable Development. A/RES/70/1.

Villegas-Navarro, A., Ramírez-M, Y., Salvador-S, M. S., & Gallardo, J. M. (2001). Determination of wastewater LC50 of the different process stages of the textile industry. *Ecotoxicology and Environmental Safety, 48*, 56–61. https://doi.org/10.1006/eesa.2000.1986.

World Trade Organization. (2020). World trade stastical review. Chapter V Statistical tables.

Resource Recovery from Urban Flood, Municipal and Industrial Wastewaters in the Context Remediation Technologies and Circular Economy

Majeti Narasimha Vara Prasad

Abstract

Stormwater is an abundant source of raw water. Stormwater collection, storage treatment and management are important topics particularly in tropical and monsoon dependent countries. In monsoonal countries like India, annual floods during rainy season are a regular feature resulting in urban floods. Example 2005 Bangalore and Mumbai urban flash floods; 2019 and 2021 urban floods of Hyderabad and Chennai. Feasibility of implementing "Sustainable Urban Drainage Systems (SUDS)" is one solution. SUDS is a well-established and operationalized in developed counties. This is a neglected area in the Indian scenario. Urban flood and storm lecwater treatment options and management using macrophytes are described. Bioremediation of industrial waste water, Nutrient removal from urban flood and agricultural waste water, Macrophytes for waste water treatment, resource recovery and value additions, nutrient recovery. Sulphur recovery from distillery waste, Algal biorefinery, ornamentals for treatment of petrochemical and chemicals, food and beverage, wineries and breweries, effluents from aquacultures, distilleries and sugar cane mill effluents, effluents from meat, vegetable and soft drinks processing sectors, agro-industrial wastewater, dairy wastewater treatment, swine wastewaters, acid mine drainage, cork boiling wastewater, tanneries and the textile industry waste waters, pharmaceuticals and personal care products (PPCP) in domestic and industrial effluents are covered. A glossary of terms is also appended.

Keywords

Salt concentrating grasses • Dye decolourization • Sulphur • Distilleries • Biodiesel from algae • Nutrient recovery

1 Introduction

Water is the soul of life. Due to climate change, excessive use, water resources are being polluted. Pollutants that originate mainly from non-point sources, which are difficult to control. Constructed wetlands are designed to intercept and remove a wide range of contaminants from waste water. These wetlands can save time and money by using natural mechanisms to treat non-point source pollution before it reaches lakes, rivers, and oceans. Conventional wastewater treatment plants can effectively remove non-point source pollution, but are expensive to build and operate. Therefore, Local waste water treatment plants are desirable to reuse water. IoT (Internet of Things) technology based urban water management strategies using Indian traditional knowledge systems are gaining importance in contemporary days due to water shortage (Barbhuiya et al., 2022).

2 Sustainable Urban Drainage Systems for Collection of Stormwaters—Water Treatment in Constructed Wetland Using Macrophytes

Constructed and engineered wetlands for water treatment: effective in treating organic matter, nitrogen, phosphorus, decrease the concentrations of heavy metals, organic chemicals, and pathogens.

Natural wetlands: used for wastewater treatment for centuries, uncontrolled discharge irreversible degradation.

M. N. V. Prasad (✉)
School of Life Sciences, University of Hyderabad (An Institution of Eminence), Hyderabad, Telangana 500046, India
e-mail: mnvsl@uohyd.ac.in

M. Smol et al. (eds.), *Water in Circular Economy*, Advances in Science, Technology & Innovation,
https://doi.org/10.1007/978-3-031-18165-8_8

Management of Water and Wastewater Infrastructures in the contemporary period has become a big challenge and is a priority task to all nations (Adepu et al., 2022). Wastewater management has opened up plethora of opportunities in materials and energy sectors (Adepu et al., 2022; Aivazidou, 2022). Currently opportunities for implementing resource recovery from waste waters are emerging as a frontier area in water management and circular economy (Aguilar et al., 2022; Ajala et al., 2022; Capodaglio, 2017; Delgado et al., 2021). Nutrients removal and resource recovery from wastewater is one priority subject (Amin et al., 2022; Tetteh et al. 2022).

3 Potential Role of Aquatic Plants in Phytotechnology for Waste Water Treatment

Phytotechnoloy is the use of plants for environmental cleanup with cogeneration of value additions. Aquatic macrophytes have paramount significance in the monitoring of metals in aquatic ecosystems (eg: *Lemna minor, Eichhornia crassipes, Azolla pinnata*) Aquatic plants are represented by a variety of macrophytic including algal species that occur in various habitats. They are important in nutrient cycling, control of water quality, sediment stabilization and provision of habitat for aquatic organisms. The use of aquatic macrophytes in water quality assessment has been a common practice employing in-situ biomonitors.

The submerged aquatic macrophytes have very thin cuticle and therefore readily take up metals from water through the entire surface. Hence the integrated amounts of bioavailable metals in water and sediment can be indicated to some extent by using macrophytes. Macrophytes with their ability to survive adverse conditions and high colonization rate are excellent tools for phytoremediation. Further they redistribute metals from sediments to water and finally take up in the plant tissues and hence maintain circulation. Benthic rooted macrophytes (both submerged and emergent) play an important role in metal bioavailability from sediments through rhizosphere exchanges and other carrier chelates. This naturally facilitates metal uptake by other floating and emergent forms of macrophytes readily take up metals in their reduced form from sediments, which exist in anaerobic situations due to lack of oxygen and oxidize them in the plant tissues making them immobile and hence bioconcentrate them to a high extent (Pal & Hogland, 2022).

Constructed wetlands are man-made wetlands designed to intercept and remove a wide range of contaminants from water. These wetlands can save you time and money by using natural mechanisms to treat non-point source pollution before it reaches our lakes, rivers, and oceans. oils, nutrients, suspended solids, and other substances. These pollutants originate mainly from non-point sources, which are difficult to control. Conventional wastewater treatment plants can effectively remove non-point source pollution, but are expensive to build and operate. SARS-CoV-2 and other pathogens in municipal wastewater have emerged as an important area of environmental research (Anand et al., 2022; Monteiro et al., 2022; Prasad & Grobelak, 2020). Chapter 1 may please be referred for treatment mechanisms and benefits.

Urban storm water drainage system optimization for sustainability of water is being increasingly considered and practiced (Azari & Tabesh, 2022; Deksissa et al., 2021; Feng et al., 2022; Nepal et al., 2022; Nieuwenhuis et al., 2021). Microbial solutions for the bioremediation-eutrophied ponds and lakes, aquaculture facilities, wastewater facilities, and waste effluents have also been successfully tried at many places (Prasad & Prasad, 2012).

Several approaches and strategies are followed for Water and Wastewater treatment (Alazaiza et al., 2022). In recent times nanotechnology, membrane technology and microbial technology have gained considerable significance (Babu & Vidya, 2021; Chen et al., 2021; Goh et al., 2022; Kushkevych, 2021; Manikandan et al., 2022).

Urban flood, municipal and industrial wastewaters contain many resources. Therefore resource recovery of waste streams is an important strategy of water in circular economy (Fig. 1). In industrial waste waters H_2S is a serious environmental concern. In India several H_2S removal plants linked to power generation plants have been established (Table 1). These installations operate under the supervision of the trained man power. Figure 2 Flow Diagram showing different processes of biological H2S removal plant and Fig. 3 Biogenic sulphur recovery via sulphate reduction (Adams et al., 2008; Namburath et al., 2020).

The above-cited installations established Bioskrubber™ Process which is a biological caustic scrubber to remove H_2S from biogas, in which the spent caustic solution is continuously regenerated in the bioreactor (Henshaw & Zhu, 2001).

$$H_2S_{gas} + OH^- \rightarrow HS^-_{liquid} + H_2O$$
$$HS^- + 1/2O_2 \rightarrow OH^- + S^\circ \downarrow$$

Advantages Bioskrubber™ Process.

Clean technology for H_2S cleaning.

Low operating cost of biogas—upto 92% caustic recycled.

Successful commercially operating plants for last nearly 2 decades.

Projects using Bioskrubber™ process successfully commissioned. Power plants 30 MW.

Very high H_2S removal efficiency—over 99%.

H_2S reduction to <250 ppm guaranteed.

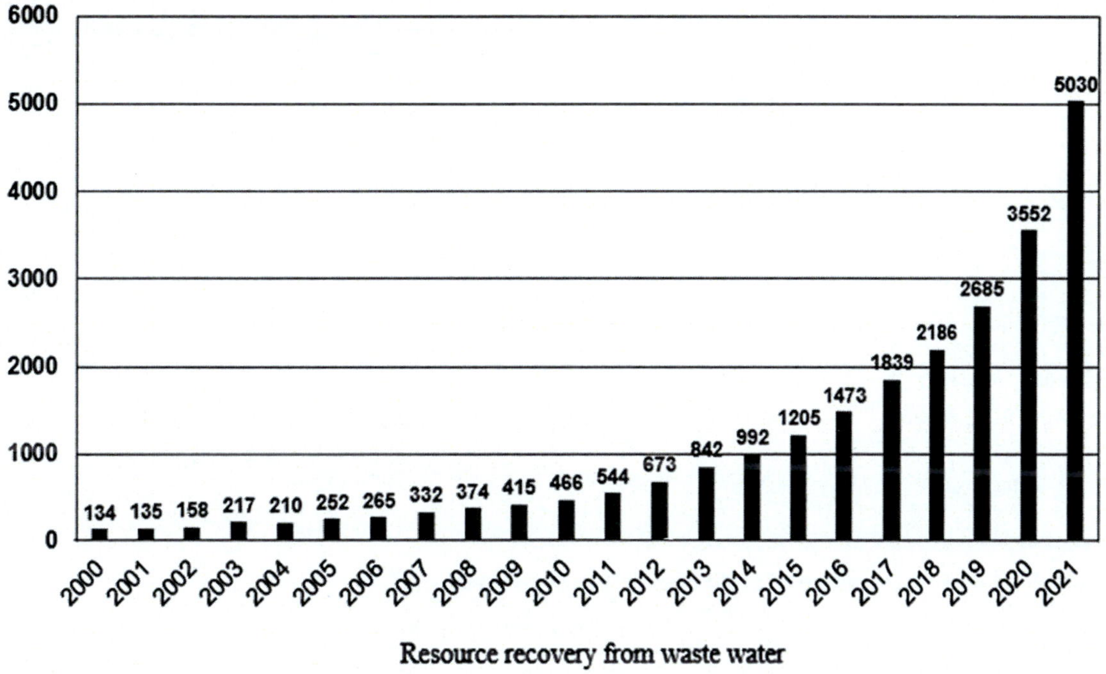

Fig. 1 There has been a steep rise in publications on resource recovery from waste water. *Data source* www.sciencedirect.com. Keywords: resource recovery, waste water

Table 1 News print industries, Water supply system, Steel, Power plants, Chemical Industries and Sugar and distilleries in India have established H_2S removal plants and power generation plants. These installations operate ETP, supervision of the plant with trained man power

Sector	Type of gas	Flow M^3/hr	H_2S In %	H_2S out ppm	Power equivalent MW
Sewage treatment plant	Biogas	381–840	1.0	630–1000	0.60–2.0
Distillery	Biogas	875–1250	2.0–3.0	500–1000	2.0–2.7
Distillery and allied chemicals	Biogas	800	4.6	0.1	1.8
Power plant	Natural gas	1660–3333	0.08	10	5–10
Power plant	Biogas	650	3.5	500	1.30
Starch plant	Biogas	1200	2.0	500	2.0
Distillery	Biogas	750	5.0	500	1.4

No expensive catalyst and chemical required.

Operation at ambient temperature, pressure and pH between 8 and 8.5.

Elemental Sulphur as bi-product with 90% purity.

Additonal Benefits of H_2S Removal.

Impact on Surrounding air.

H_2S is Highly Corrosive—Can damage other equipments and expensive instruments in the vicinity.

Health Hazard and odour nuisance.

Pollution Board Norms—Sulphur emission "Cap" in kgs/day—Today laws are not stringent, Tomorrow?

Commercial Services Turnkey Projects from designing to commissioning.

Supervision on operations of the plant.

Financing the project on a B.O.O.T on a case to case basis.

BOOT = (build, own, operate, transfer).

B.O.O.T Projects—Typical features.

On a case to case basis, Innovative will entirely fund, operate and maintain the Biogas to power project.

Project size—1.4 MW +

Innovative will get Biogas and sell Power to the distillery for captive use or will sell power to a third party.

Gas will be free or at cost.

Transfer period of the project will vary depending on other terms and conditions.

Entire plant will be transferred free of cost on transfer.

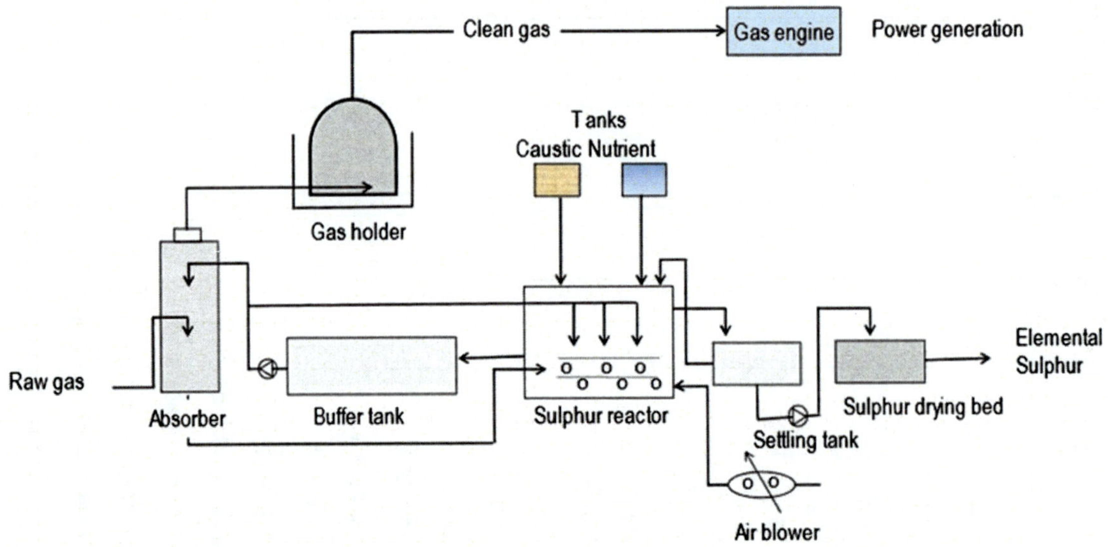

Fig. 2 Flow diagram showing different processes of biological H_2S removal plant

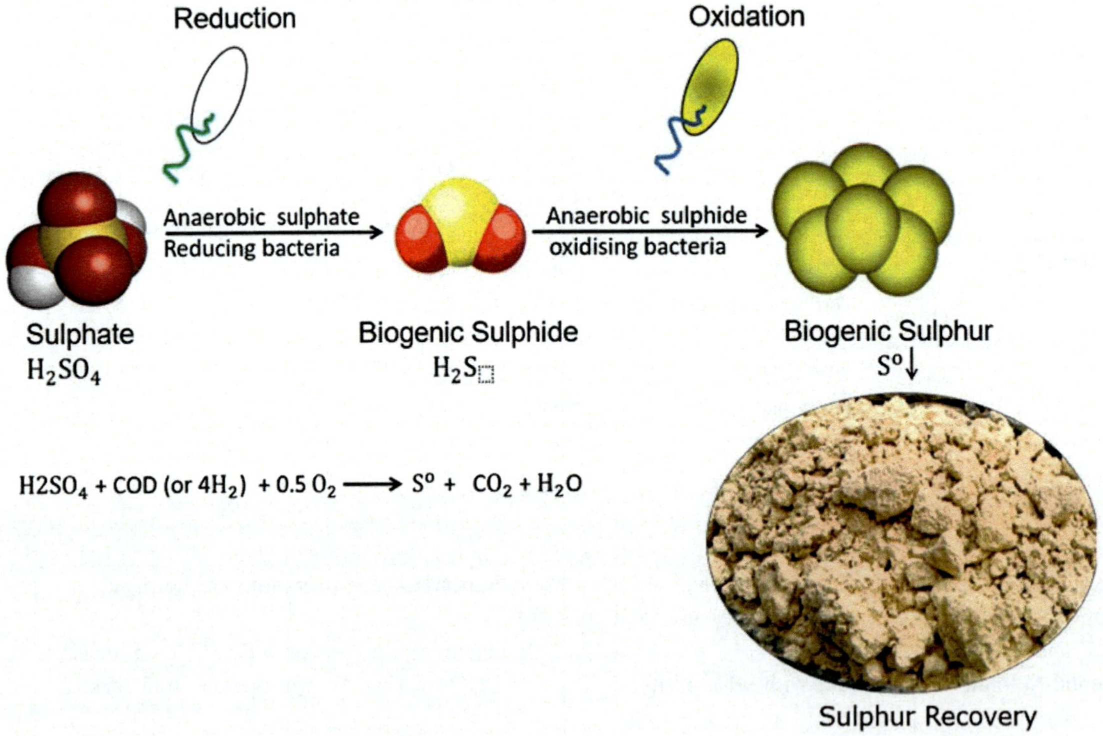

Fig. 3 Biogenic sulphur recovery—sulphate reduction to sulphur

Fig. 4 *Pistia stratiotes* thick cover in fish pond is harvested manually using a small boat. After harvesting the phytomass of *P. stratiotes* involved in the phytofiltration of toxic metals aqua culture is practiced (with permission from Prasad, 2015)

The cleanup of contaminated urban flood water discharged into fishponds by *Pistia stratiotes* (water lettuce/water cabbage), is shown in Fig. 4. For successful spplication of this technology in field, the concept of bioconcentration factor is used, which is the ratio of heavy metals accumulated by plants to that dissolved in the surrounding medium (Fig. 4) (Prasad, 2015).

4 Microalgae Value Chain—Resource Recovery from Urban Flood, Municipal and Industrial Wastewaters

Numerous inhibiting factors were identified for the production and use of microalgae, in particular related to the consumption of resources, acceptance, and competitiveness of algae production. In algae production, there is a considerable need for nutrients, energy, and other inputs. Nitrogen, phosphate, and iron are the main nutrients. If large algae yields are to be achieved, the estimated nutrient requirements are high, but this depends on the type of algae and the cultivation method used (Priefer & Meyer, 2019).

Theoretically, wastewater streams as well as sea and brackish water could be used as nutrient sources. In practice, there are legal barriers because the specifications for nutrient sources used in food production are very high. The use of synthetic fertilizers therefore seems likely. Moreover, further adjuvants and process substances are needed: CO_2, ethanol as an extraction agent, flocculants, disinfectants for the sterilization of reactors, and herbicides. In particular, CO_2 demand is a limiting factor for the cultivation of algae. To ensure the safe use of algae as food, exhaust gases can only be used if no contamination with pollutants can be guaranteed. Therefore, combustion gases cannot be used.Schematic representation of microalgae-mediated remediation of waste water (WW) and valorization of the algal biomass into value-added is shown in Figs. 5 and 6 (Koul et al., 2022; Maurya et al., 2022). Biodiesel from microalgae (Chisti, 2007), algal biopolymers and algal biorefinery models with self-sustainable closed loop approach for blue-bioeconomy (Mohan et al., 2020, 2022) are some of the emerging trends not only for boosting circular economy but also for waste water treatment.

5 Phytoremediation of Brilliant Blue Green Dye Contaminated Water Using *Canna Indica* L

Ornamentals such as *Canna indica* are used for removal of nutrients and colouring material from waste water (Xiao et al., 2010). Disulfonated triphenylmethane disperse dye

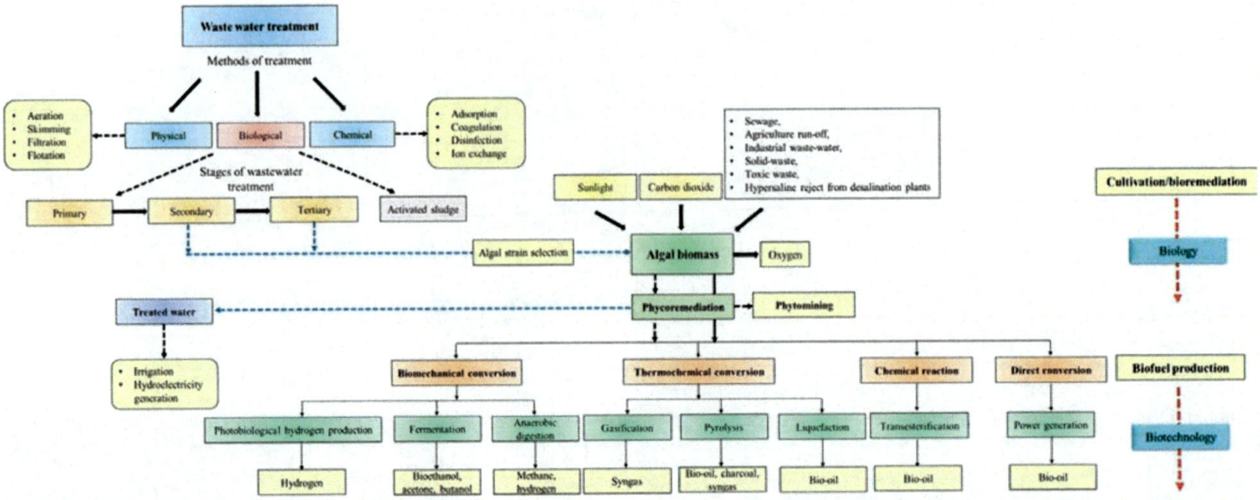

Fig. 5 Schematic representation of microalgae-mediated remediation of waste water (WW) and valorization of the algal biomass into value-added (with permission from Koul et al., 2022)

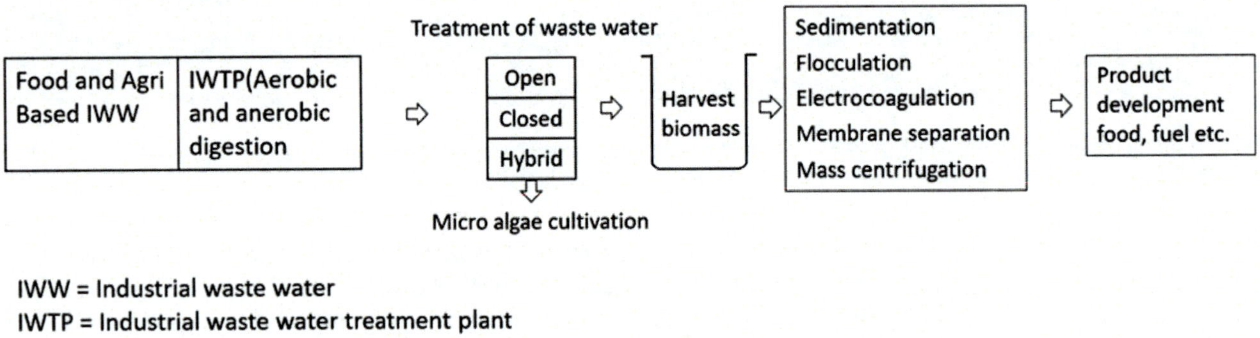

Fig. 6 Recovery of nutrient from food and agriculture waste water using microalgae (Redrawn based on Maurya et al., 2022)

Brilliant Blue G (BBG). The suffix G in the name is an abbreviation for green as the blue color of the dye has a slight greenish tint. Disperse dyes are known to cause allergies and skin irritations. Moreover, it is a potent carcinogen.

Canna indica was used as experimental plants for the decolourization of BBG. Plants (20 g) were treated with Hoagland solutions containing BBG solutions at concentrations of 20 mgL^{-1}. Further standardization experiments were carried out in magenta boxes. Plants of (20 g) were selected and different concentrations of dye solutions were prepared (50, 100, 150, 250 mgL^{-1}). Dye solution without plant was kept as a biotic control, where the biotic control had plant roots immersed in Hoagland solution. Absorbance was measured after an interval of 24 h at 620 nm up to 120 h.

Decolourization % was calculated using the following equation:

$$\% \text{ Decolourization} = (\text{Initial absorbance} - \text{Observed absorbance})/\text{Initial absorbance} \times 100 \text{ Decolourization of BBG using } C. indica$$

From the experiment, it was observed that decolourization efficiency increased up to 97%. The rate of decolourization was affected by the increasing concentration of the dye, ranging from 50 mgL^{-1} up to 250 mgL^{-1}. Decolourization of the dye up to 99% was observed for the concentration of 50 mgL^{-1} within 24 h and 250 mgL^{-1} dye was decolorized up to 97% within 120 h.

Fig. 7 Phytoremediation of Brilliant blue green dye by using *C. indica* (with permission from Prasad 2015)

The phytotransformation of dye was monitored using FTIR. The phytotransformed dye was compared with the untreated sample. Significant disappearance of major peaks and formation of new peaks in the FTIR spectra after dye degradation suggests its potential use in decolourization of Brilliant Blue G (BBG) (Fig. 7).

3D printed bio-based components for water treatment applications (Table 2), Nutrient recovery from industrial wastewater using algae gained significant progress (Figs. 8, 9 and 10). Waste water treatment and biorefinery have been integrated for economic feasibility and are being commercialized (Figs. 11, 12, 13, 14, 15 and 16). Application of microalgae for waste water treatment are shown in Tables 3, 4 and 5).

C. sorokiniana, *M. capsulatus* and a consortium of both of them were grown in triplicates in batch experiments in 600 mL glass bottle at 37 °C. which was connected to an extra compartment of 1 L volume to provide extra gas phase (Fig. 9). Nutrient recovery from industrial wastewater by a co-culture of green microalgae and methanotrophs. Experiments demonstrated that the algae removed 40% of the ammonia and 62% of the phosphorus from wastewater (Rasouli et al., 2018).

The most important role of plants in wetlands is that they increase the residence time of water, which means that they reduce the velocity and thereby increase the sedimentation of particles and associated pollutants. Thus they are indirect involved in water cleaning. Plants also add oxygen providing a physical site of microbial attachment to the roots generating positive conditions for microbes and bioremediation. For efficient removal of pollutants a high biomass per volume of water of the submerged plants is necessary. Uptake of metals in emergent plants only accounts for 5% or less of the total removal capacity in wetlands. Not many studies have been performed on submerged plants, however, higher concentration of metals in submerged than emerged plants has been found and in microcosm wetland the removal by *Elodea canadensis* and *Potamogeton natans* showed up to 69% removal of Zn. (Kadlec 1995; Kadlec and Knight 1996).

In areas where iron is mined, the near by natural water resources are contaminated with leachate. Figure 17 a and b shows river water red in color due to iron in soluble form. Such water can not be used directly for domestic purpose and also for irrigation. Usage of water containing iron and manganese will leave brownish scars on wash room tiles, sanitary ware. Therefore, iron and manganese are removed by oxidation and precipitation and filtration. The scheme shown in Fig. 17 has 3 major steps. Step 1 river water red in color having iron and manganese in solution is in soluble form are aerated excessively and sent to pressure filter.

Manganese is removed by chemical oxidant (chlorine, ozone or potassium permanganate) Most commonly potassium permanganate is used and oxidized water with contaminants which will form precipitates. These water containing iron and manganese precipitates is passed through pressure filter having anthracite coal power and sand filter bed.

Table 2 Summary of the 3D printed bio-based components for water treatment applications (with permission from Fijoł et al. 2022)

Biocomponent	Composition	Water treatment application	Printed component	Water treatment mechanism	Printing method
Chitosan	Chitosan-acrylate-based polymer	Cu (II) removal	Monolithic filter	Adsorption	Stereolitography
	Chitosan—TiO_2	Amoxicillinremova	Scaffold	Photocatalytic degradation	
Alginate	Chitosan—diacrylated Pluronic F-127 Polycaprolactone—sodium alginate	Pb (II), Cu (II), Cd (II), Hg (I) removal	Scaffold	Adsorption	Gel printing
	Sodium alginate-nanoclay-g- poly(acrylic acid)	Pb (II) removal	Scaffold	Adsorption	Gel printing
PLA	PLA—Cu-metal organic frame works	Malachite green removal	Mesh	Adsorption	Gel printing
	PLA—carbon black	Volatile organic compounds removal	Cylindrical sorbent	Adsorption	Fused deposition modelling
	PLA—iron (III) oxide	As(III) removal	Monolithic filter	Adsorption	Fused deposition modelling
	PLA—dopamine/polystyrene nanoparticles	Oil Separation	Mesh	Adsorption	Fused deposition modelling
	PLA—iron powder/hydrogel coating	Oil Separation	Mesh	Filtration	Fused deposition modelling
Cellulose acetate (CA)	CA-ethyl acetate	Oil separation	Mesh	Filtration	Gel printing
	CA-polyvinyl alcohol/silica nanoparticles	Oil separation	Membrane	Filtration	Gel printing

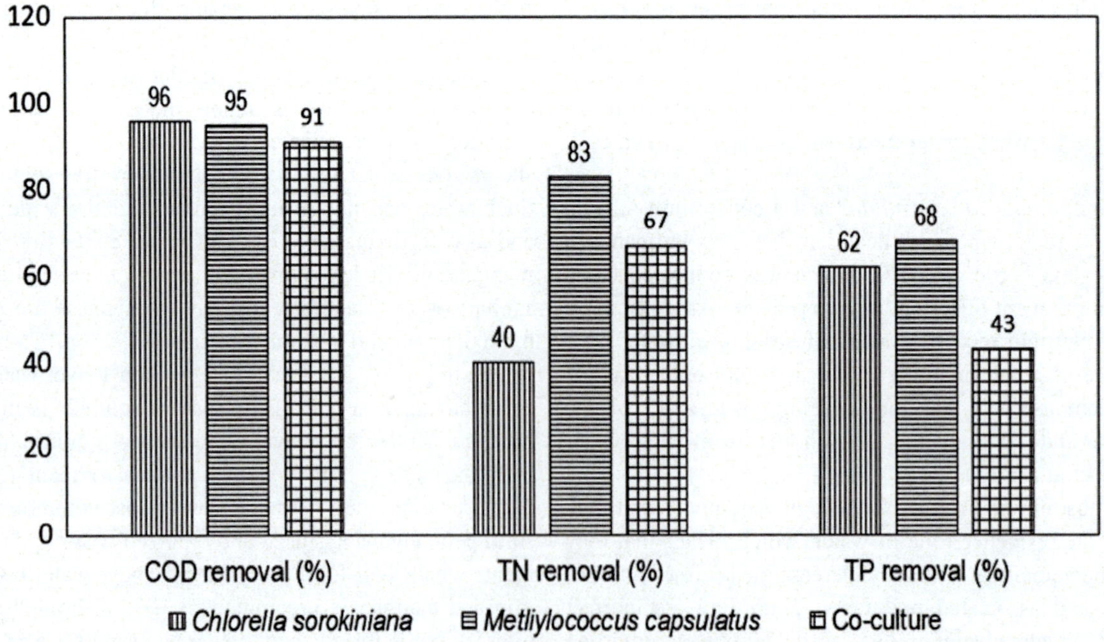

Fig. 8 Percentage removal of nutrients and organic carbon in the different experiments. Note that for nutrients, the percentage is equivalent to the recovery of nutrients from wastewater (*Data source* Rasouli et al., 2018)

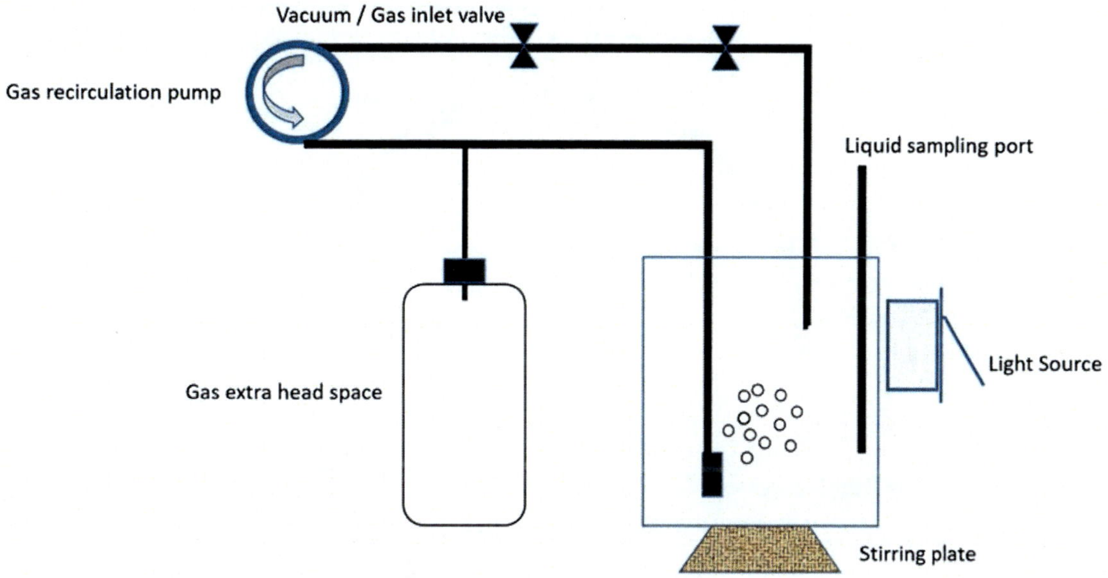

Fig. 9 Schematic representation of the photobioreactor applicable for nutrient recovery from industrial wastewater by a co-culture of green microalgae and methanotrophs (Rasouli et al., 2018)

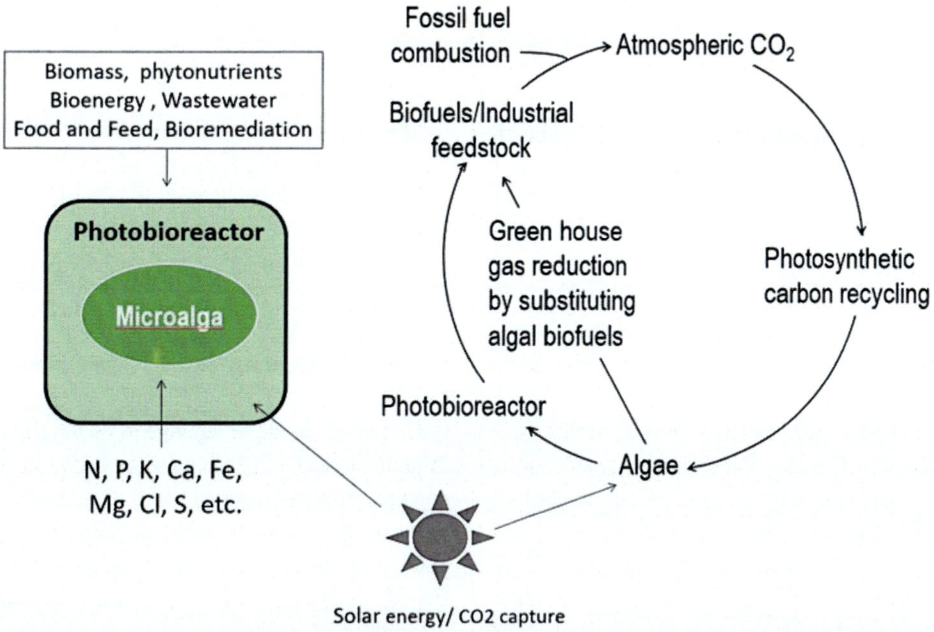

Fig. 10 Schematic view of photobioreactor for cultivation of algae in hazardous waste water (Chisti, 2007) Algae based photobioreactor for cultivation of algae in hazardous waste water for production of bio fuels, industrial feedstock and carbon dioxide sequestration. Leachate from hazardous waste dump sites are being used to establish High Rate Algal Ponds [HRAPs] for production of biodiesel (Park et al., 2011)

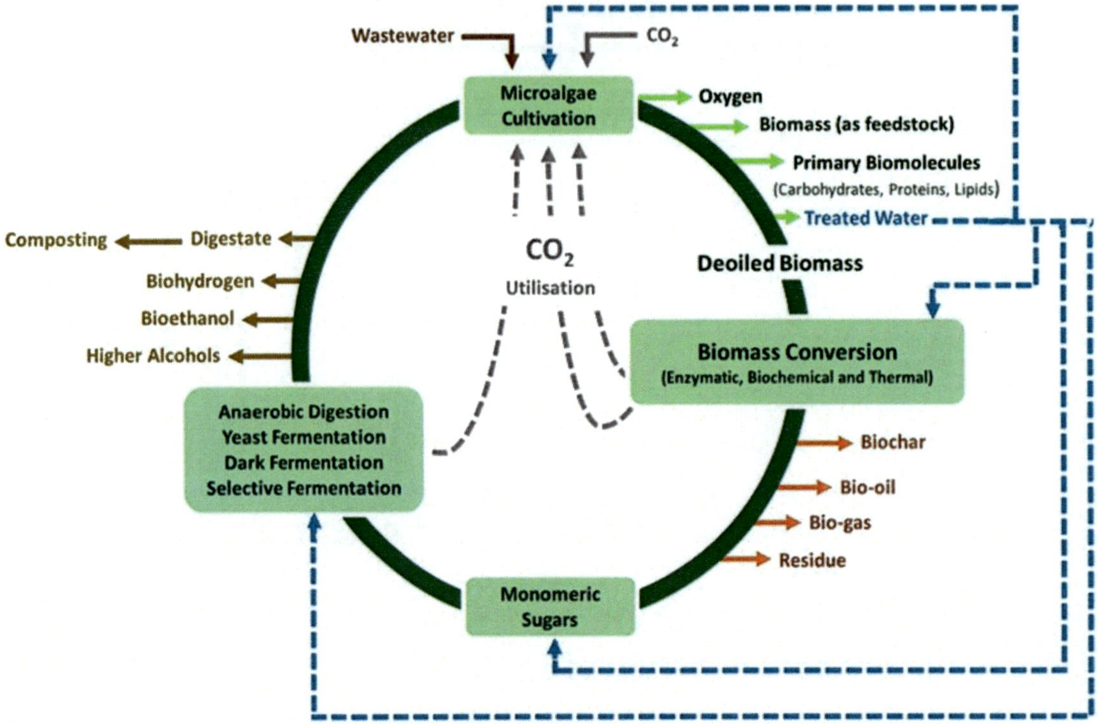

Fig. 11 Self-sustaining biorefinery model in a closed loop approach (with permission from Mohan et al., 2020)

Thus the iron and manganese removed water using the setup/flow scheme in 3 steps 1. Excessive aeration. 2. Chemical oxidation. 3. Pressure filtration with anthracite and sand filter bed. Clean water is sent for public distribution system or for domestic use including irrigation.

6 Conclusions

Nature-Based Solutions for treating waste water from agroecosystems and in Circular Cities have gained considerable importance (Atanasova et al., 2021; Cong & Tomsen, 2021; Freed & Zimmerman, 2022; Kumar et al., 2022; Mitra et al., 2022; Nhubu et al., 2022; Parnian et al., 2022; Shadeed & Stefanakis et al., 2021; Trávníček et al., 2022; Ghafourian et al., 2021; Langergraber et al., 2021; Mancuso et al., 2021; van Hullebusch et al., 2021). Nature's cure for revitalization of water bodies Prasad & Prasad, 2012 is traditionally practiced in many instances. Several potential examples of waste water treatment integrating circular economy to achieve sustainability have been cogently reviewed by Avellán et al. (2022), Bjørnbet et al. (2021) greywater reuse in high-rise buildings for energy recovery and irrigation. Resources, Hadad et al. (2022).

Rethinking the Contribution of Rainwater Harvesting. Huang et al. (2021), Imteaz et al. (2021), Karim et al. (2021), Londra et al. (2021), Oberascher et al. (2021), Ranaee et al. (2021), Shadeed and Alawna, (2021). Technology status and trends of industrial wastewater treatment are reviewed recently (Lavy et al., 2022; Mao et al., 2022; Meilinger & Monstadt, 2021).

Circular Economy in the Water Industry (Mbavarira & Grimm et al., 2021) is gaining considerable momentum in the era of climate change (Water-Energy-Food nexus) (Valencia et al., 2022); water security in the era of climate change (Wallis-Lage & Erdal 2022). In order to achieve green deal and sustainable development, water-wise cities and sustainable water systems are being researched (Wang & Fu 2021). Waste waters contain a wide variety of resources depending upon the type of industry. Resource recovery by applying appropriate technology not only helps in remediation technology, but also recover beneficial resource (Prasad, 2020; Prasad & Grobelak, 2020; Prasad & Shih, 2016; Prasad & Vithanage 2019; Prasad et al., 2019a, 2019b, 2019c). Therefore, resource recovery from urban flood, municipal and industrial wastewaters must continue in the context remediation technologies to boost circular economy.

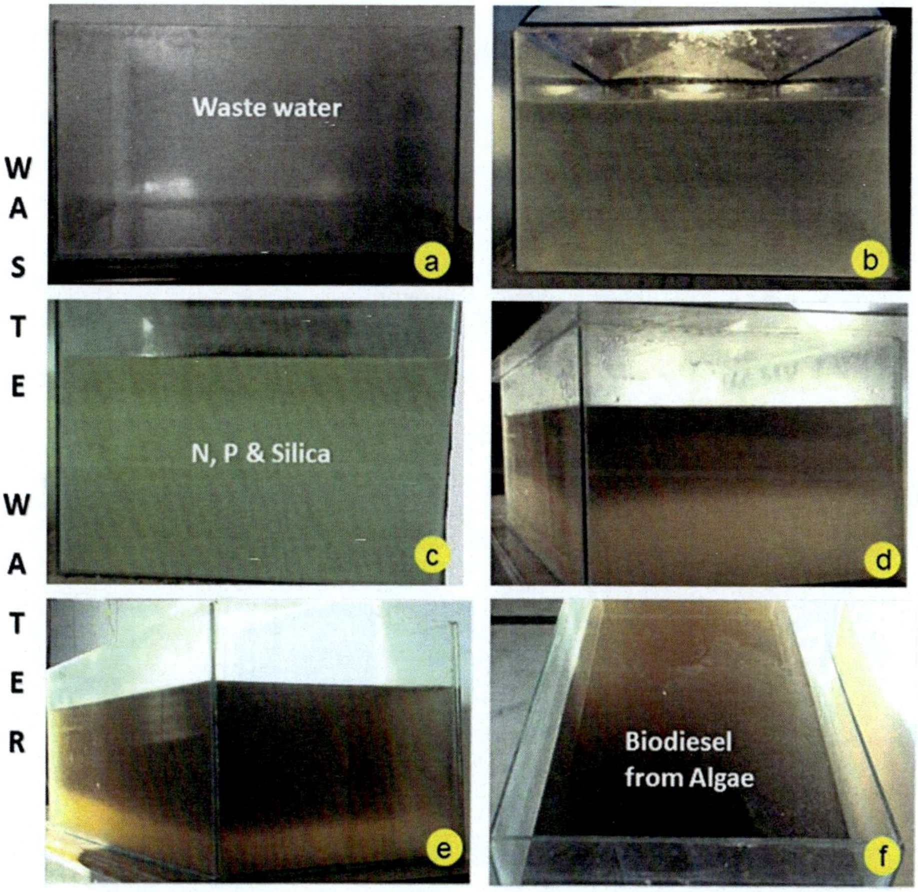

Fig. 12 Biodiesel production using algae and waste water at lab scale (*Source* Mr. M. V. Bhaskar "Kadambari Consultants Pvt Ltd.—Nualgi") (with permission from Prasad, 2015)

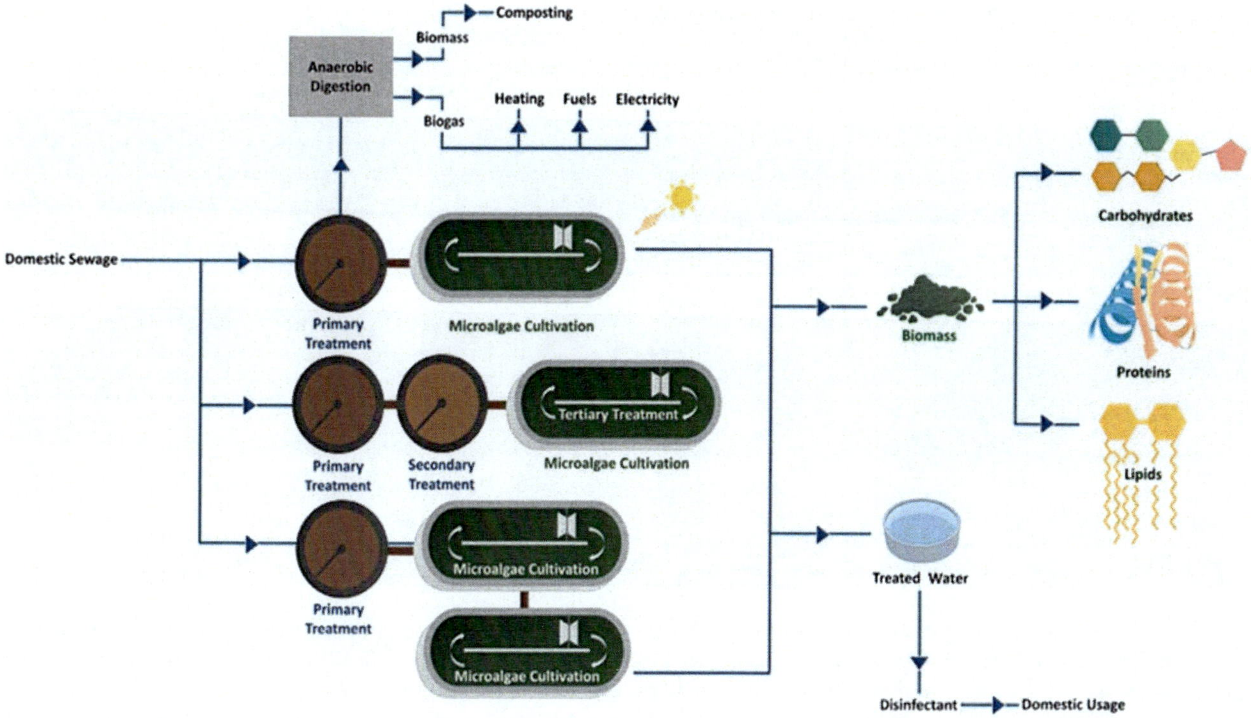

Fig. 13 Integrated model for sewage treatment using microalgae cultivation for the production of biobased products (with permission from Mohan et al., 2020)

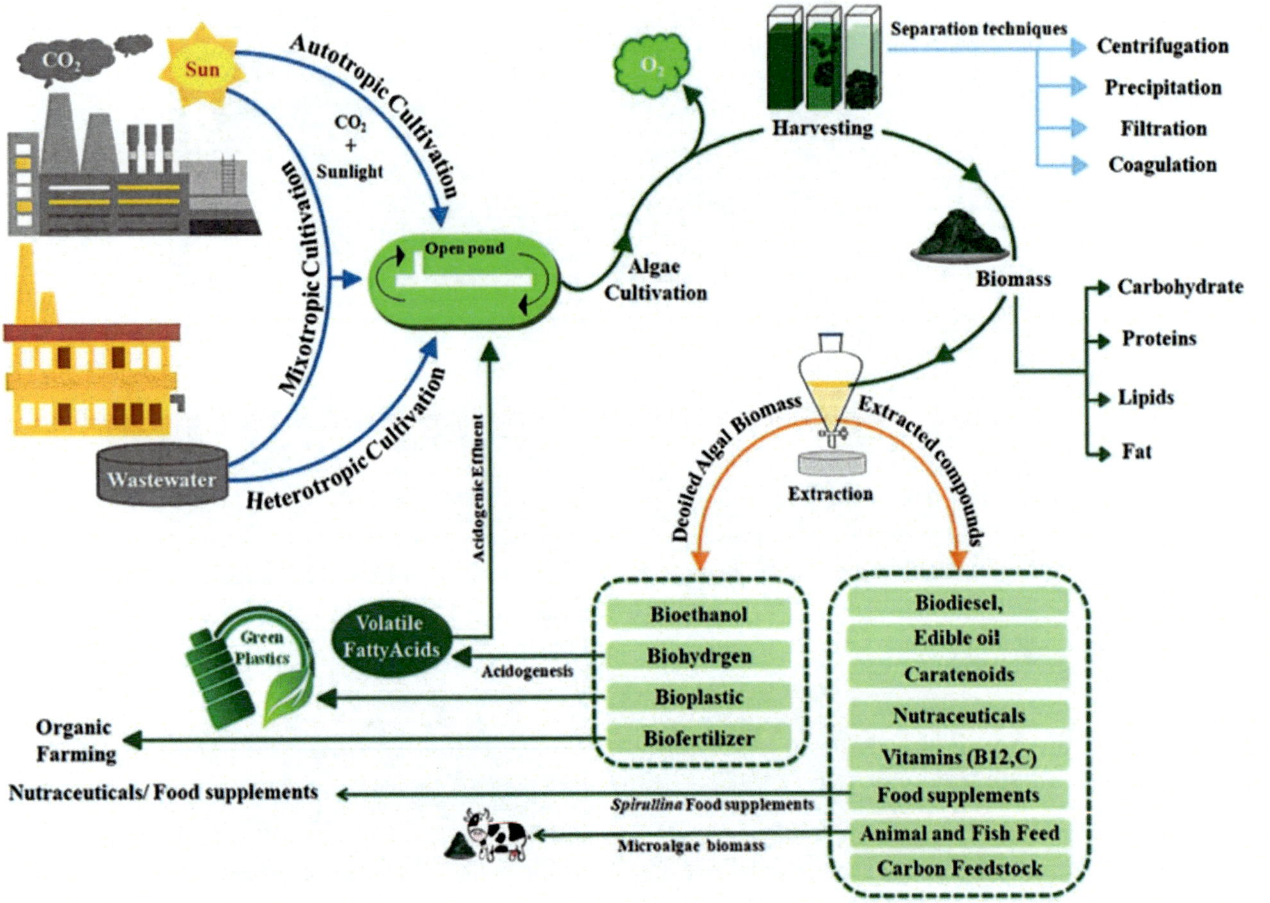

Fig. 14 Integration of wastewater and flue gas for algal biorefinery (with permission from Mohan et al., 2020)

Fig. 15 Nutrient rich and sewage water treatment in constructed engineered wetland using high biomass produced macrophytes viz. *Typha* and *Phragmites*. Production of low cost roofing material and compost from macrophytes used in nutrient rich and sewage water treatment in constructed engineered wetland f. Waste water treatment in High Rate Algal Pond [HRAP] using hazardous waste water (with permission from Prasad, 2015)

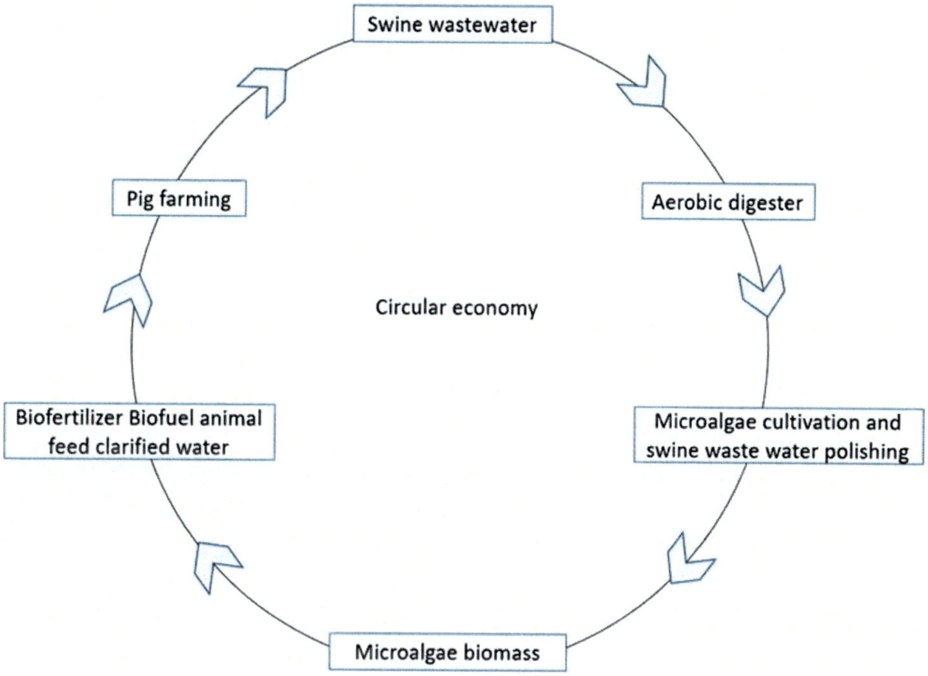

Fig. 16 Valorization of swine wastewater in a circular economy approach (Silveira et al., 2021)

Table 3 Growth of *Chlorella sp* in different wastewaters, and probable polymer products that could be synthesized from the resultant biomass (*Data source* Mohan et al., 2022)

Species	Type of waste water	Composition	Probable polymers and composites
Chlorella sp.	Dairy wastewater	Chloride, sulphide, nitrate, nitrite, phosphate, TDS, TSS, lactose	Algae-polymer composites cellulose Defatted algae-based composites
	Slaughterhouse wastewater	Nitrate, Phosphate, COD, BOD, DS, iron, sulphide, hardness	Algae-polymer composites Cellulose Defatted algae-based composites, Defatted algae-based composites Algae-polymer composites
	Municipal wastewater	COD, TN, TP, Ammonia	Polylactic acid (PLA) Starch
	Swine wastewater	COD, TN, TP, Ammonia, salinity	Algae-polymer composites

Table 4 Published reports on wastewater treatment efficiency of various microalgal species (with permission from Koul et al., 2022)

S. no.	Algal species	Habitat	Type of WW treatment
1	*Prototheca zopfii*	Fresh water	Degrades petroleum hydrocarbons found in Louisiana and utilized 41.4% of crude and 10.7% of motor oils waste
2	*Botryococcus braunii*	Freshwater	Secondary treated sewage wastewater treatment
3	*Oscillatoria* sp.	Freshwater	Municipal wastewater treatment; nitrogen removal 100%; phosphate removal 100%
4	*Scenedesmus dimorphus*	Freshwater	Industrial wastewater treatment; phosphate removal 20–55%
5	*Arthrospira platensis*	Freshwater and brackish water	Industrial wastewater treatment, nitrogen removal 96–100%; phosphate removal 87–99%
6	*Chlorella kessleri*	Freshwater	Artificial medium treatment by algae leads to nitrogen removal 8–19%; phosphate removal 8–20%

(continued)

Table 4 (continued)

S. no.	Algal species	Habitat	Type of WW treatment
7	*Arthrospira (Spirulina)*	Freshwater	Piggery waste treatment; NH_4-N removal up to 84–96% and P removal between 72 and 87%
8	*Acinetobacter haemolyticus, Ralstonia basilensis*	Freshwater and sea water	Treatment of aromatic pollutants present in wastewater
9	*Arthrospira* sp.	Freshwater	Animal wastewater treatment; nitrogen removal 84–96%; phosphate removal 72–87%
10	*Comamonas* sp.	Freshwater	Treatment of acetonitrile present in wastewater
11	*Haematococcus pluvialis*	Freshwater	Primary treated sewage and primary treated piggery wastewater
12	*Chlorella sorokiniana*	Freshwater	Municipal wastewater treatment
13	*Chlorella zofingiensis*	Freshwater	Olive mill waste treatment
14	Consortia algae/diatoms	Freshwater	Dairy wastewater treatment
15	*Oocystis* sp. and *Scenedesmus* sp.	Freshwater	Swine slurry treatment; ammonium is completely consumed; nitrogen removal between 38 and 47%; phosphate removal up to 80%
16	*Chlorella* sp.	Freshwater	Digested manure treatment; nitrogen removal 76–83%; phosphate removal 63–75%
17	*Chlorella reinhardtii*	Freshwater	Artificial medium treatment leads to nitrogen removal 42–83%; phosphate removal 13–14%
18	*Scenedesmus obliquus*	Freshwater	Municipal wastewater treatment; nitrogen removal 79–100%; phosphate removal 47–98%
19	Consortia microalgae	Freshwater	Carpet mill wastewater treatment
20	*Chlamydomonas reinhardtii*	Freshwater and sea water	Municipal wastewater treatment (centrate)
21	*Spirulina platensis*	Marine	Swine wastewater treatment
22	*Ankistrodesmus* and *Scenedesmus quadricauda*	Fresh water	Olive oil mill wastewaters and paper industry wastewater treatment
23	*Botryococcus braunii* and *Chlorella vulgaris*	Freshwater	Domestic sewage treatment
24	*Chlorella pyrenoidosa*	Freshwater	Industrial wastewater treatment
25	*Microalgae consortia*	Freshwater	Poultry waste treatment (diluted 6%)
26	*Chlorella sorokiniana*	Freshwater	Food and municipal waste treatment
27	*Ettlia oleoabundans*	Freshwater	Agricultural anaerobic waste effluent treatment
28	*Nannochloropsis sp.*	Freshwater and sea water	Municipal wastewater treatment
29	*Neochloris oleoabundans*	Freshwater and saline water	Cow manure waste treatment; nitrate utilization up to 90–95%
30	*Monoraphidium braunii*	Freshwater	Organic matter waste treatment
31	*Spirulina platensis*	Marine	Olive-oil mill wastewater treatment
32	*Desmodesmus communis*	Freshwater	Biological wastewater treatment
33	*Scenedesmus acutus*	Freshwater	Municipal wastewater after aerobic treatment
34	*Scenedesmus rubescens, Neochloris vigensis and Chlorococcum sp.*	Freshwater	Synthetic wastewater treatment
35	*Oedogonium sp.*	Freshwater	Piggery wastewater treatment
36	*Chlorella sorokiniana*	Freshwater	Cattle manure effluent treatment
37	*Chlorella sorokiniana*	Freshwater	Wastewater treatment under aerobic dark heterotrophic conditions
38	*Mucidosphaerium pulchellum*	Freshwater	Domestic wastewater treatment

(continued)

Table 4 (continued)

S. no.	Algal species	Habitat	Type of WW treatment
39	*Chlorella pyrenoidosa*	Fresh and brackish	Degrades azo dyes wastewater
40	*Auxenochlorella protothecoides UMN280*	Freshwater	Concentrated municipal wastewater treatment
41	*Gonium sp.*	Freshwater and terrestrial	Textile effluent treatment
42	*Diplosphaera sp. MM1*	Freshwater	Diluted dairy wastewater and winery wastewater treatment
43	*Spirulina platensis*	Freshwater and brackish water	Poultry wastewater treatment
44	*Botryococcus braunii*	Freshwater	Treats secondarily treated sewage in batch and continuous cultures
45	*Consortia (Chlamydomonas subcaudata, Anabaena sp.and Nitzschia sp.)*	Freshwater	Slaughterhouse wastewater treatment
46	*Spirulina platensis*	Marine	Poultry litter wastewater treatment
47	*Arthospiraplantesis*	Freshwater	Poultry litter leachate treatment
48	*Scenedesmus obliquus*	Freshwater	Brewery effluent treatment
49	*Tetraselmis indica*	Marine	Domestic wastewater treatment
50	*Scenedesmus sp.*	Freshwater	Removal of ammonia from anaerobic digestion effluent
51	*Chlorella vulgaris*	Freshwater	Dye wastewater treatment
52	*Chlamydomonas sp. SW15aRL*	Fresh water	Landfill leachate treatment reduced the ammonia nitrogen by 83%
53	*Chlorella sp. A5*	Fresh and marine water	Dairy wastewater treatment
54	*Chlorella variabilis*	Freshwater	Dairy wastewater treatment
55	*Nannochloropsis sp.*	Freshwater and sea water	Tannery effluent treatment
56	*Chlorella sorokiniana UKM2, Chlorella pyrenoidosa UKM7 and Coelastrella sp. UKM4*	Freshwater	Reduces pollutants in palm oil mill effluent (POME) and also fixes CO_2
57	*Chlorella vulgaris MBFJNU-1*	Freshwater	Swine slurry treatment
58	*Chlorella variabilis, Scenedesmus obliquus*	Freshwater	Dairy wastewater treatment
59	*Chlorella pyrenoidosa*	Freshwater	Synthetic wastewater and sewage treatment

Table 5 Macromolecular composition of *Chlorella sorokiniana*, *Methylococcus capsulatus*, algal–bacterial consortium. methanotrophs (*Data source* Rasouli et al., 2018)

	Proteins	Lipids	Carbohydrates	Chlorophylls	Carotenoids
Chlorella sorokiniana	45.03	29.88	22.65	5.84	0.27
Methylococcus capsulatus	32.52	21.82	16.36	–	0.1
Algal consortium	27.62	34.5	31.96	2.19	0.22

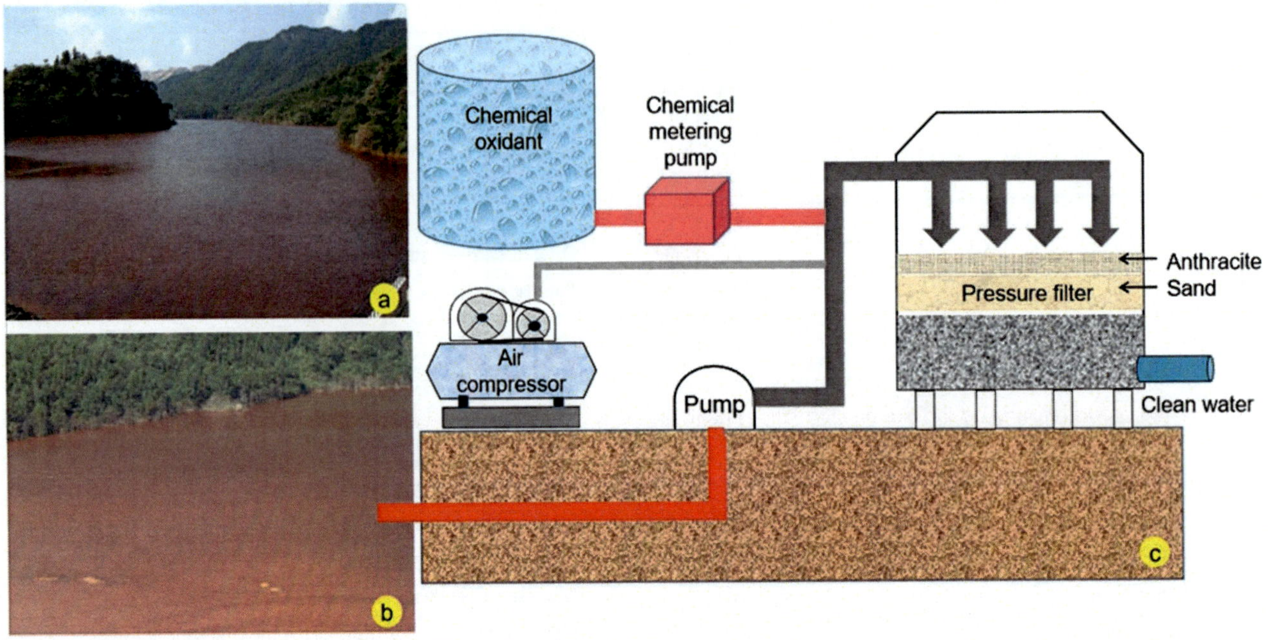

Fig. 17 Open cast mining of iron and manganese leachate pollute near by rivers. Please note the red color of the water which is abundantly available but can not be used for domestic and agriculture. Such water can be treated in the following steps to remove iron and manganese. **a** Pumping the water with aeration to cause precipitation of iron and manganese, **b** Potassium permanganate solution is pumped into the water circulation system to oxidise manganese and to precipitate, **c** Pumping air using air compressor and circulationg the water with iron and manganese precipitates through pressure filter (anthracite coal poder and sand filter bed) and **d** Chemically and physically precipitated iron and manganese salts are sent through pressure filter

Acknowledgements Thanks are due to Elsevier for according permission to reproduce Figs. 6, 11, 13, 14 and Tables 2 and 4; Springer Nature for Figs. 4, 7, 10, 12 and 15.

References

Adams, M., Lawrence, R., & Bratty, M. (2008). Biogenic sulphide for cyanide recycle and copper recovery in gold–copper ore processing. *Minerals Engineering, 21*(6), 509–517.

Adepu, N., Kermanshachi, S., Safapour, E., & Pamidimukkala, A. (2022, March). Challenges and risks in resilience management of water and wastewater infrastructures. In *Proceedings of the ASCE Construction Research Congress (CRC)*, Arlington, VA, USA (pp. 9–12).

Aguilar, M. G., Jaramillo, J. F., Ddiba, D., Páez, D. C., Rueda, H., Andersson, K., & Dickin, S. (2022). Governance challenges and opportunities for implementing resource recovery from organic waste streams in urban areas of Latin America: Insights from Chía, Colombia. *Sustainable Production and Consumption, 30*, 53–63.

Aivazidou, E. (2022). Water management in agriculture and industry: Challenges, trends, and opportunities. *Sustainability, 14*(1), 66.

Ajala, O. J., Tijani, J. O., Bankole, M. T., & Abdulkareem, A. S. (2022). Wastewater treatment technologies. In *Inorganic-organic composites for water and wastewater treatment* (pp. 1–28). Springer.

Alazaiza, M. Y., Albahnasawi, A., Ali, G. A., Bashir, M. J., Nassani, D. E., Al Maskari, T., Amr, S. S. A., & Abujazar, M. S. S. (2022). Application of natural coagulants for pharmaceutical removal from water and wastewater: A review. *Water, 14*(2), 140.

Amin, H. M., Gad, A. A. R., Abdel-Ghany, U. A. S., & Saddek, R. A. (2022). Nutrients removal from partially treated wastewater using soil aquifer treatment system. *Journal of Advanced Engineering Trends, 42*(1), 1–12.

Anand, U., Li, X., Sunita, K., Lokhandwala, S., Gautam, P., Suresh, S., Sarma, H., Vellingiri, B., Dey, A., Bontempi, E., & Jiang, G. (2022). SARS-CoV-2 and other pathogens in municipal wastewater, landfill leachate, and solid waste: A review about virus surveillance, infectivity, and inactivation. *Environmental Research, 203*, 111839.

Atanasova, N., Castellar, J. A., Pineda-Martos, R., Nika, C. E., Katsou, E., Istenič, D., Pucher, B., Andreucci, M. B., & Langergraber, G. (2021). Nature-Based Solutions and Circularity in Cities. *Circular Economy and Sustainability*, 1–14.

Avellán, T., Nagabhatla, N., Jalan, I., & Liao, D. (2022). Integrating circularity to achieve sustainability: Examples of various wastewater treatment systems. In *Circular economy and sustainability* (pp. 21–35). Elsevier.

Azari, B., & Tabesh, M. (2022). Urban storm water drainage system optimization using a sustainability index and LID/BMPs. *Sustainable Cities and Society, 76*, 103500.

Babu, L. R., & Vidya, K. V. (2021). Nanotechnology for wastewater treatment. *Advanced Research in Material and Chemical Science, 1*, 41.

Barbhuiya, M. R., Bhardwaj, M., Shukla, S., Kibret, A. T., & Panda, G. (2022). IoT technology-based urban water management strategies using Indian traditional knowledge system. In *IoT and IoE driven smart cities* (pp. 275–291). Springer.

Bjørnbet, M. M., Skaar, C., Fet, A. M., & Schulte, K. Ø. (2021). Circular economy in manufacturing companies: A review of case study literature. *Journal of Cleaner Production*, 126268.

Capodaglio, A. G. (2017). Integrated, decentralized wastewater management for resource recovery in rural and peri-urban areas. *Resources, 6*(2), 22.

Chen, L., Zhai, Y., van der Mark, E., Liu, G., van der Meer, W., & Medema, G. (2021). Microbial community assembly and metabolic function in top layers of slow sand filters for drinking water production. *Journal of Cleaner Production, 294*, 126342.

Chisti, Y. (2007). Biodiesel from microalgae. *Biotechnology Advances, 25*(3), 294–306.

Cong, R. G., & Thomsen, M. (2021). Review of ecosystem services in a bio-based circular economy and governance mechanisms. *Ecosystem Services, 50*, 101298.

Deksissa, T., Trobman, H., Zendehdel, K., & Azam, H. (2021). Integrating urban agriculture and stormwater management in a circular economy to enhance ecosystem services: Connecting the dots. *Sustainability, 13*, 8293. https://doi.org/10.3390/su13158293.

Delgado, A., Rodriguez, D. J., Amadei, C. A., & Makino, M. (2021). *Water in circular economy and resilience (WICER)*. World Bank.

Feng, W., Liu, Y., & Gao, L. (2022). Stormwater treatment for reuse: Current practice and future development–A review. *Journal of Environmental Management, 301*, 113830.

Fijoł, N., Aguilar-Sánchez, A., & Mathew, A. P. (2022). 3D-printable biopolymer-based materials for watertreatment: A review. *Chemical Engineering Journal,* 430, 132964.

Freed, A., & Zimmerman, R. (2022). Urban systems and services: Vulnerabilities and impact. In Climate change and US cities: Urban systems, sectors, and prospects for action (p. 79).

Ghafourian, M., Stanchev, P., Mousavi, A., & Katsou, E. (2021). Economic assessment of nature-based solutions as enablers of circularity in water systems. *Science of The Total Environment*, 148267.

Goh, P. S., Wong, K. C., & Ismail, A. F. (2022). Membrane technology: A versatile tool for saline wastewater treatment and resource recovery. *Desalination, 521*, 115377.

Hadad, E., Fershtman, E., Gal, Z., Silberman, I., & Oron, G. (2022). Simulation of dual systems of greywater reuse in high-rise buildings for energy recovery and potential use in irrigation. *Resources, Conservation and Recycling, 180*, 106134.

Henshaw, P. F., & Zhu, W. (2001). Biological conversion of hydrogen sulphide to elemental sulphur in a fixed-film continuous flow photo-reactor. *Water Research, 35*(15), 3605–3610.

Huang, Z., Nya, E. L., Rahman, M. A., Mwamila, T. B., Cao, V., Gwenzi, W., & Noubactep, C. (2021). Integrated water resource management: Rethinking the contribution of rainwater harvesting. *Sustainability, 13*, 8338. https://doi.org/10.3390/su13158338.

Imteaz, M. A., Bayatvarkeshi, M., & Karim, M. R. (2021). Developing generalised equation for the calculation of payback period for rainwater harvesting systems. *Sustainability, 13*, 4266. https://doi.org/10.3390/su13084266.

Kadlec, R. H. (1995). Overview: Surface flow constructed wetlands. *Water Science and Technology, 32*(3), 1–12.

Kadlec, R. H., & Knight, R. L. (1996). *Treatment Wetlands,* 893. Boca Raton, FL: CRC Press.

Karim, M. R., Sakib, B. M. S., Sakib, S. S., & Imteaz, M. A. (2021). Rainwater harvesting potentials in commercial buildings in Dhaka: Reliability and economic analysis. *Hydrology, 8*, 9. https://doi.org/10.3390/hydrology8010.

Koul, B., Sharma, K., & Shah, M. P. (2022). Phycoremediation: A sustainable alternative in wastewater treatment (WWT) regime. *Environmental Technology and Innovation, 25*, 102040.

Kumar, S., Doshi, S., Mishra, G., & Iyer, M. (2022). Making Indian cities water-sensitive: A critical review of frameworks. *Climate Change and Water Security*, 277–290.

Kushkevych, I. (2021). The application of microorganisms in wastewater treatment. *Processes, 9*, 1914. https://doi.org/10.3390/pr9111914.Specialissue.

Langergraber, G., Castellar, J. A. C., Pucher, B., Baganz, G. F. M., Milosevic, D., Andreucci, M.-B., Kearney, K., Pineda-Martos, R., & Atanasova, N. (2021). A framework for addressing circularity challenges in cities with nature-based solutions. *Water*, 13, 2355. https://doi.org/10.3390/w131723.

Lavy, B. L., Weaver, R. C., & Hagelman, R. R. (2022). Using the Change Point Model (CPM) framework to identify windows for water resource management action in the lower Colorado River Basin of Texas, USA. *Water, 14*(1), 18.

Londra, P. A., Kotsatos, I.-E., Theotokatos, N., Theocharis, A. T., & Dercas, N. (2021). Reliability analysis of rainwater harvesting tanks for irrigation use in greenhouse agriculture. *Hydrology, 8*, 132. https://doi.org/10.3390/hydrology8030132.

Mancuso, G., Bencresciuto, G. F., Lavrnić, S., & Toscano, A. (2021). Diffuse water pollution from agriculture: A review of nature-based solutions fornitrogen re and recovery. *Water*, 13, 1893. https://doi.org/10.3390/w13141893.

Manikandan, S., Subbaiya, R., Saravanan, M., Ponraj, M., Selvam, M., & Pugazhendhi, A. (2022). A critical review of advanced nanotechnology and hybrid membrane based water recycling, reuse, and wastewater treatment processes. *Chemosphere, 289*, 132867.

Mao, G., Han, Y., Liu, X., Crittenden, J., Huang, N., & Ahmad, U. M. (2022). Technology status and trends of industrial wastewater treatment: A patent analysis. *Chemosphere, 288*, 132483.

Maurya, R., Zhu, X., Valverde-Pérez, B., Kiran, B. R., General, T., Sharma, S., Sharma, A. K., Thomsen, M., Mohan,S. V., Mohanty, K. and Angelidaki, I. (2022). Advances in microalgal research for valorization of industrialwastewater. *Bioresource Technology, 343*, p.126–128.

Mbavarira, T. M., & Grimm, C. A. (2021). Systemic view on circular economy in the water industry: Learnings from a Belgian and Dutch case. *Sustainability, 13*, 3313. https://doi.org/10.3390/su13063313.

Meilinger, V., & Monstadt, J. (2021). From the sanitary city to the circular city? Technopolitics of wastewater restructuring in Los Angeles, California. *International Journal of Urban and Regional Research.*

Mitra, P., Haldar, A., & Banerjee, P. (2022). Public participation in restoration and sustainable use of wetland ecosystem services in India. In *Handbook of research on monitoring and evaluating the ecological health of Wetlands* (pp. 280–303).

Mohan, A. A., Antony, A. R., Greeshma, K., Yun, J. H., Ramanan, R., & Kim, H. S. (2022). Algal biopolymers as sustainable resources for a net-zero carbon bioeconomy. *Bioresource Technology, 344*, 126397.

Mohan, S. V., Hemalatha, M., Chakraborty, D., Chatterjee, S., Ranadheer, P., & Kona, R. (2020). Algal biorefinery models with self-sustainable closed loop approach: Trends and prospective for blue-bioeconomy. *Bioresource Technology, 295*, 122128.

Monteiro, S., Rente, D., Cunha, M. V., Gomes, M. C., Marques, T. A., Lourenço, A. B., Cardoso, E., Álvaro, P., Silva, M., Coelho, N., & Vilaça, J. (2022). A wastewater-based epidemiology tool for COVID-19 surveillance in Portugal. *Science of the Total Environment, 804*, 150264.

Namburath, M., Papirio, S., Moscariello, C., Di Costanzo, N., Pirozzi, F., Alappat, B. J., & Sreekrishnan, T. R. (2020). Effect of nickel on the comparative performance of inverse fluidized bed and continuously stirred tank reactors for biogenic sulphur-driven autotrophic denitrification. *Journal of Environmental Management, 275*, 111301.

Nepal, M., Bharadwaj, B., Karki Nepal, A., Khadayat, M. S., Pervin, I. A., Rai, R. K., & Somanathan, E. (2022). Making urban waste management and drainage sustainable in Nepal. In *Climate change and community resilience* (pp. 325–338). Springer.

Nhubu, T., Muzenda, E., & Belaid, M. (2022). The water-waste nexus: Sustainability implications for integrated water resources and waste management in Harare. In *Handbook of research on resource management and the struggle for water sustainability in Africa* (pp. 368–387). IGI Global.

Nieuwenhuis, E., Cuppen, E., Langeveld, J., & de Bruijn, H. (2021). Towards the integrated management of urban water systems: Conceptualizing integration and its uncertainties. *Journal of Cleaner Production, 280*, 124977.

Oberascher, M., Dastgir, A., Li, J., Hesarkazzazi, S., Hajibabaei, M., Rauch, W., & Sitzenfrei, R. (2021). Revealing the challenges of smart rainwater harvesting for integrated and digital resilience of urban water infrastructure. *Water, 13*, 1902. https://doi.org/10.3390/w13141902.

Pal, D., & Hogland, W. (2022). An overview and assessment of the existing technological options for management and resource recovery from beach wrack and dredged sediments: An environmental and economic perspective. *Journal of Environmental Management, 302*, 113971.

Park, J. B. K., Craggs, R. J., & Shilton, A. N. (2011). Wastewater treatment high rate algal ponds for biofuel production. *Bioresource Technology, 102*, 35–42.

Parnian, A., Furze, J. N., & Parnian, A. (2020). Water out of waste–solar-desalination-humidification-dehumidification auxiliaries processing extractive industrial operations. *Water Productivity Journal, 1*(1), 31–38.

Prasad, M. N. V. (2015). Phytoremediation crops and biofuels. In E. Lichtfouse (Ed.), *Sustainable agriculture reviews 17* (pp. 159–261). © Springer International Publishing. https://doi.org/10.1007/978-3-319-16742-8_7.

Prasad, M. N. V. (ed.). (2020). *Disinfection byproducts in drinking water: Detection and treatment Butterworth-Heinemann.*

Prasad, M. N. V., & Grobelak, A. (eds.). (2020). *Waterborne pathogens: Detection and treatment.* Butterworth-Heinemann

Prasad, M. N. V., & Prasad, R. (2012). Nature's cure for cleanup of contaminated environment—A review of bioremediation strategies. *Reviews on Environmental Health, 27*(4), 181–189.

Prasad, M. N. V., & Shih, K. (eds.). (2016). *Environmental materials and waste: Resource recovery and pollution prevention.* Academic Press.

Prasad, M. N. V., & Vithanage, M. (eds.). (2019). *Electronic waste management and treatment technology.* Butterworth-Heinemann.

Prasad, M. N. V., de Campos Favas, P. J., Vithanage, M., & Mohan, S. V. (eds.). (2019a). *Industrial and municipal sludge: Emerging concerns and scope for resource recovery.* Butterworth-Heinemann.

Prasad, M. N. V., Vithanage, M., & Borthakur, A. (eds.). (2019b). *Handbook of electronic waste management: International best practices and case studies.* Butterworth-Heinemann.

Prasad, M. N. V., Vithanage, M., & Kapley, A. (eds.). (2019c). *Pharmaceuticals and personal care products: Waste management and treatment technology: Emerging contaminants and micro pollutants.* Butterworth-Heinemann.

Priefer, C., & Meyer, R. (2019). One concept, many opinions: How scientists in Germany think about the concept of bioeconomy. *Sustainability, 11*(15), 4253.

Ranaee, E., Abbasi, A. A., Tabatabaee Yazdi, J., & Ziyaee, M. (2021). Feasibility of rainwater harvesting and consumption in a middle eastern semiarid urban area. *Water, 13*, 2130. https://doi.org/10.3390/w13152130.

Rasouli, Z., Valverde-Pérez, B., D'Este, M., De Francisci, D., & Angelidaki, I. (2018). Nutrient recovery from industrial wastewater as single cell protein by a co-culture of green microalgae and methanotrophs. *Biochemical Engineering Journal, 134*, 129–135.

Shadeed, S., & Alawna, S. (2021). Optimal sizing of rooftop rainwater harvesting tanks for sustainable domestic water use in the West Bank, Palestine. *Water, 13*, 573. https://doi.org/10.3390/w13040573.

Silveira, C. F., de Assis, L. R., de Sousa Oliveira, A. P., & Calijuri, M. L. (2021). Valorization of swine wastewater in a circular economy approach: Effects of hydraulic retention time on microalgae cultivation. *Science of The Total Environment*, 147861.

Stefanakis, A. I., Calheiros, C. S., & Nikolaou, I. (2021). Nature-based solutions as a tool in the new circular economic model for climate change adaptation. *Circular Economy and Sustainability, 1*, 303–318. https://doi.org/10.1007/s43615-021-00022-3.

Tetteh, E. K., Asante-Sackey, D., Armah, E. K., & Rathilal, S. (2022). Tapping wastewater resource: Why and how?. In *Handbook of biofuels* (pp. 125–146). Academic Press.

Trávníček, P., Junga, P., Kotek, L., & Vítěz, T. (2022). Analysis of accidents at municipal wastewater treatment plants in Europe. *Journal of Loss Prevention in the Process Industries, 74*, 104634.

Valencia, A., Zhang, W., & Chang, N. B. (2022). Sustainability transitions of urban food-energy-water-waste infrastructure: A living laboratory approach for circular economy. *Resources, Conservation and Recycling, 177*, 105991.

van Hullebusch, E. D., Bani, A., Carvalho, M., Cetecioglu, Z., De Gusseme, B., Di Lonardo, S., Djolic, M., van Eekert, M., Griessler Bulc, T., Haznedaroglu, B. Z., et al. (2021). Nature-based units as building blocks for resource recovery systems in cities. *Water, 13*, 3153.

Wallis-Lage, C., & Erdal, Z. K. (2022). Resilience through systems thinking for water infrastructure. In *Water security under climate change* (pp. 41–67). Springer.

Wang, X. C., & Fu, G. (2021). *Water-Wise Cities and Sustainable Water Systems: Concepts, Technologies, andApplications*, 451. IWA Publishing.

Xiao, H., Cheng, S., & Wu, Z. (2010). Microbial community variation in phytoremediation of triazophos by Canna indica Linn. in a hydroponic system. *Journal of Environmental Science*, 22, 1225–1231.

Preliminary Microbiological Risk Assessment for Local Water Reuse in a Small MBR Wastewater Treatment Plant

Klara Ramm

Abstract

The article presents an analysis of the efficiency of a small wastewater treatment plant (WWTP) located in a ski resort. Due to the lack of a treated wastewater receiver, a water recovery system has been designed. The WWTP technology is based on a membrane bioreactor with microfiltration and UV disinfection. The reclaimed water is directed to a retention tank where it is mixed with the water taken from a stream. The tank supplies snow cannons in winter, and in summer, it is used to water the slope. The main objective of the research was to assess the microbial risk for the developed water reuse solution and to check the suitability of the guidelines in the EU Regulation on the quality of reclaimed water (European Union, Regulation 2020/741 of the European Parliament and of the Council of 25 May 2020 on minimum requirements for water reuse. OJ L 177, 2020) when water is not used in agriculture. The study also tested the risk analysis method used in the drinking water supply sector. The results show that the solution to use reclaimed water locally for snow production (winter) and irrigation (summer) has a controllable microbial risk.

Keywords

Water reuse • Reclaimed water • MBR • Microbiological risk

K. Ramm (✉)
Chamber of Commerce Polish Waterworks, Warsaw University of Technology, Warsaw, Poland
e-mail: k.ramm@igwp.org.pl

Warsaw University of Technology, The Faculty of Building Services, Hydro and Environmental Engineering, Warsaw, Poland

1 Introduction

Sustainable management of water resources requires the implementation of solutions rationalizing and optimizing the use of available resources with significant attention to the environment. The use of alternative water sources (other than surface or groundwater), such as recycled water, brackish water, seawater, and stormwater, has been practiced in many regions of the world and for many types of irrigation such as agricultural crops, golf courses, forests, and open-space landscapes (Hardy et al., 2015; Maniam et al., 2022; United Nations, 2021). Recycling of water from municipal wastewater is nothing new, but it is developing mainly in countries with a shortage of fresh water. It can be very useful all over Europe if the water is properly prepared for the needs.

The reuse of water for irrigation in agriculture and green spaces is part of the circular economy. Nitrogen and phosphorus, as well as micronutrients and organic matter in reclaimed water, can increase soil nutrition and fertility during irrigation and have beneficial effects on plant growth. However, there is a risk of nutrient imbalance and nitrate contamination of groundwater due to irrational management of the reclaimed water.

Appropriate water preparation and risk assessment are essential. In EU countries where water reuse has been used for many years, there are already certain guidelines and legal acts strengthening the safe use of reclaimed water (Paranychianakis et al., 2015).

In Spain reuse of water is of great importance, especially in areas where the shortage of conventional resources is a structural problem. According to Rodríguez-Villanueva and Sauri (2021) Spain leads Europe in water reuse. Agricultural irrigation is the destination of 41% of treated wastewater, followed by irrigation of golf courses and public parks (31%). These activities are concentrated in the Mediterranean, especially in the regions of Murcia and Valencia.

M. Smol et al. (eds.), *Water in Circular Economy*, Advances in Science, Technology & Innovation,
https://doi.org/10.1007/978-3-031-18165-8_9

The use of reclaimed water for irrigation is not the exclusive domain of Southern Europe. From 1978 Vernon BC (Canada) stores the recovered water in the reservoir and uses it for watering green areas, mostly golf courses (Water Canada, 2021).

In Singapore reclaimed water named NEWater is recycled from treated wastewater basing on a 3-step treatment process: ultrafiltration/microfiltration, reverse osmosis, and ultraviolet disinfection. According to the National Water Agency of Singapore, compared to desalination, NEWater is more energy efficient and profitable (Singapore's National Water Agency, 2022).

California in the US has focused on water reclamation as it can improve the resources balance. Melbourne in Australia has been using recycled water from both wastewater and gray water since the 1970s to improve its water safety, and is now one of the most active cities using treated wastewater for irrigation (Radcliffe & Page, 2020). In Australia the first recycled water recommendations were elaborated in 1977.

Water recovery is therefore becoming popular, but it is important to ensure its safety.

Maiolo and Pantusa (2018) highlight the necessity to reduce the impact on water bodies. The possibility to reuse wastewater also implies, however, a series of problems concerning technical, economic, and sanitary aspects that need to be suitably considered. Maiolo and Pantusa describe an interesting case study from Italy where reclaimed water during vegetation season can be used to irrigate crops, and during winter it can be used to produce artificial snow. Thinking of melting snow, which effectively produces irrigation of the slopes, wastewater reuse for snowmaking can be reasonably considered as a type of reuse for irrigation of green areas, or for recreational and sporting activities. The production of artificial snow is one of those applications that can have a significant impact on saving resources, provided that the risks are mitigated. In addition, the maintenance of ski runs requires a significant amount of water. According to Hahn (2004) with 1 m^3 of water, on average 2–2.5 m^3 of snow can be produced. For the snow base, about 30 cm of snow, a ski run of 1 ha requires at least 1000 m^3 of water, while the subsequent snowmaking requires, depending on the situation, a considerably higher water consumption. However, the Mount Buller operator in Australia needs 1300 m^3 of water to cover 1 ha with 30 cm of snow. Every hour, the system can convert 720 m^3 of water to snow (Mtbuller, 2021). According to Paccard (2011) in the snowmaking of 1 ha of ski slopes in France, approximately 3000–4000 m^3 of water were used.

Mount Buller continues to invest in water reuse infrastructure. The system enables to recycle up to 2000 m^3 of wastewater per day, by treating wastewater with a stringent disinfection process that results in the production of Class A recycled water. Class A is the highest achievable Australian standard in recycled water (Environment Protection Authority Victoria, 2021). Mt Buller's recycled water is collected in the Sun Valley Reservoir, significantly increasing the level of water in this dam.

In 2012, Snowbowl in Northern Arizona (USA), after 10 years of legal disputes, became the first ski resort in the world to produce 100% of artificial snow from treated wastewater (Rulseh, 2013).

Any action must be based on a risk assessment to mitigate and minimalize the risk for humans and the environment. Exall (2004) highlights that the nutrient value of the reclaimed water may be beneficial to plants and grasses, and significant savings in fertilizer costs may be achieved. Van Rossum (2020) emphasizes the importance of a small, onsite system. Velasquez and Yanful point out public acceptance and trust of consumers in the quality of reclaimed water are considered by many to be the most important factor determining the outcomes of water reclamation projects.

Many countries have developed different standards and guidelines. Water recovery is especially popular in southern Europe. Therefore, there are guidelines in these countries. For example, in France, agricultural and green area (lawns, parks, golf courses) irrigation with treated wastewater is authorized and regulated by the 2 August 2010 national regulatory decision (France, 2010).

In turn, Canadian standard, CAN/-CSA-B128.1–06/B128.2–06 (R2016) can be used for non-potable water systems for different applications including irrigating lawns and gardens. Canadian regions have developed guidelines to help water operators find the weaknesses of the system so that a thorough risk analysis can be performed. In 2020 the Australian Environment Protection Authority of Victoria reviewed, updated, and consolidated guidance for water recycling. It requires risk management based on the HACCP (Hazard Analysis Critical Control Points) elements. The outcome of using the framework is a risk management plan that describes the nature of a recycled water system and how it should be operated and managed. The plan is necessary for operators—it provides a 'living' document that can be reviewed and audited, both internally and externally.

Because the monitoring of all the pathogens occurring in recycled water is not feasible "indicator organisms" have been employed to evaluate its microbiological quality. Total coliforms, fecal coliforms, or *Escherichia coli* are the most used indicators, with the latter to be considered as the most representative of fecal contamination (Paranychianakis et al., 2015). Moreover, new contaminants occur (e.g., viruses of COVID-19).

Most of the functioning guidelines require monitoring of basic parameters such as total suspended solids (TSS), turbidity, biochemical oxygen demand (BOD), and microbial quality including fecal pathogens and indicators. Chemical requirements are also important, particular attention should be paid to salinity. Harivandi (2004) drew attention to the salinity of the soil when reclaimed water is used for irrigation of golf courses. Qian and Mecham (2005) raise attention to the problems with salinity and relatively high Na and B accumulation in the soil.

Lyu and Chen (2016) conducted a survey in seven parks in Beijing. Field samples were collected from seven parks in Beijing that irrigated reclaimed water with different irrigation histories in 2011 and 2014 to evaluate the long-term impacts of reclaimed water irrigation on soil quality. Results showed an increase of nutrients in the soil, such as nitrogen and phosphorus, with an increase of the organic substance of 6–17%, without significant pH changes.

Adequate technology is needed to prepare water of the right quality for irrigation of green spaces or to produce snow. Experience from the operation of membrane bioreactor (MBR)—based wastewater treatment plants shows that it is a very effective technology.

It combines the activated sludge process with microfiltration (ultra- and nanofiltration are also used) and is widely recognized as an effective tool for wastewater treatment and water reuse due to the high reclaimed water quality. According to Diao (2020), MBR is more suitable for small site with stable raw wastewater quality. For example, for university campuses that are built far from the city center or urban wastewater treatment plant, costs of sewage transportation can be saved thanks to the on-campus wastewater treatment. MBR technology can guarantee the quality of wastewater that can be directly recycled locally. Taesopapong and Ratanatamskul (2020) reviewed Eco Biofilter/Membrane Bioreactor (MBR) as a compact wastewater treatment and recycling plant. They also confirm that this technology works well in small centers.

The aim of Purnell et al. (2015, 2016) was to assess the microbiology removal efficiency of the MBR treatment plant. The mean levels of fecal coliform bacteria dropped to 0.3 CFU/100 ml in treated wastewater and were not detected in the samples taken after the chlorination step. A relatively large reduction (5.3 log) in the number of somatic coliphages was also observed. MBR proves to be a suitable basis for many water recovery applications, especially when combined with other technologies (Zhang and Farahbakhsh, 2007). Comerton et al. (2005), Bunani et al. (2018), Joss et al. (2011) combined MBR with additional treatment to investigate system-wide water quality, focusing on waterborne pathogens, disinfection by-products, nitrates, and other contaminants. The results gave good water quality for all analyzed parameters. Membrane bioreactors also perform well during peak flux (Hirani et al., 2010).

The use of MBR as a primary technology in water recovery is not a new idea. Membrane biological reactor is recognized as an efficient and steady wastewater treatment process. However, treated wastewater is usually not suitable for direct use in nature. Hence, the "water reclamation plant" appears in the Regulation 2020/741highlighting the importance of disinfection. There are still no requirements in EU legislation regarding the quality of reclaimed water for natural use. Legislation of some Member States exists, and the risk management guidelines are still being developed by the Joint Research Center, although preliminary studies were already prepared several years ago (Alcalde Sanz & Gawlik, 2014).

The purpose of this pilot is to check the usefulness of the guidelines from Regulation 2020/741 and to use the existing knowledge in the field of risk analysis in the water supply. As the solution at Kasina Ski is a precedent in Poland, it is an interesting case study for further research.

2 Materials and Methods

The analysis is based on a wastewater treatment plant serving the Kasina Ski Resort. A local wastewater treatment plant treats wastewater from toilets, restaurants, and break rooms for station service. Therefore, it was selected to implement the risk management plan resulting from the EU Regulation 2020/741, and ISO standards.

The wastewater treatment plant in MBR technology was built by Schwander Polska on the basis of Alfa Laval microfiltration membranes type MFP_2, with a pore size of 0.2 μm, transmembrane pressure during operation of 0.01–0.04 bar and a typical net flux range of 10–30 L/m^2/h (Schwander, 2021). The plant has a capacity of Q_{max} = 78 m^3/d.

The scope of water use includes the discharge of treated domestic (urban) wastewater to the ground in the amount of Q_{maxs} = 0.002528 m^3/s, Q_{avg} = 78 m^3/d, Q_{maxy} = 28470 m^3/y.

The quality balance of the wastewater inflow into the WWTP is presented in Table 1.

The operator of the ski resort and the WWTP applied for a water permit according to an obligatory Polish procedure. The treatment plant is not subject to the urban wastewater treatment directive's requirements (European Union, 1991),

Table 1 The load of the wastewater inflow in Kasina Ski

Number of people (p. e)	Daily average amount of wastewater (m^3/d)	BOD_5 (kg/d)	COD (kg/d)	Total suspended solids (kg/d)	Total nitrogen (kg/d)	Total phosphorus (kg/d)
390	78.0	23.4	48.8	25.4	4.3	0.7

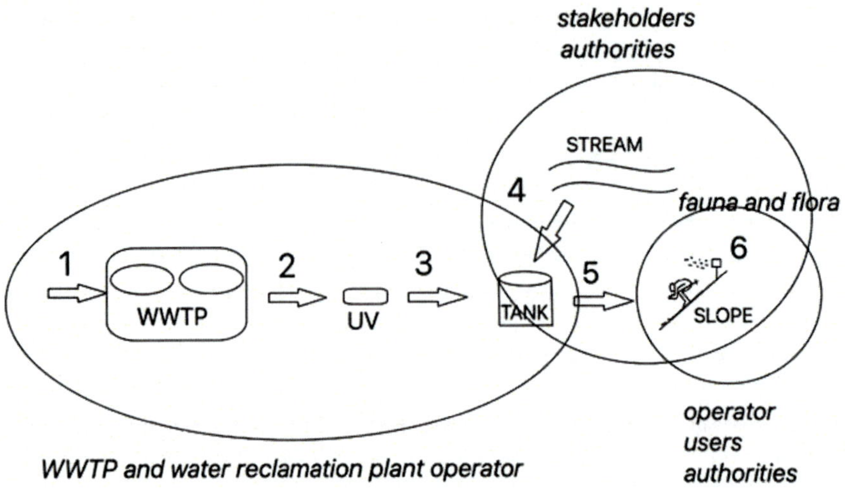

Checkpoints: 1: raw waste water, 2: treated waste water, 3: treated waste water after UV, 4: water in the Kasinianka stream and stress factors, 5 water quality in the reservoir, 6 slope condition and stress factors.

Fig. 1 Risk analysis checkpoints

as it does not serve an agglomeration but a ski resort. The operator has developed an operating manual for the whole wastewater treatment and reuse system and is responsible for its efficient functioning. The permit includes the consent to discharge treated wastewater into the ground.

For the needs of snowmaking, a water intake from the stream Kasinianka, situated below the ski resort was made. Water is pumped from the stream into a retention reservoir with a volume of about 2.5 thousand m^3, which also receives wastewater treated from the treatment plant. Before, treated wastewater is disinfected by UV lamp. From the reservoir, water is distributed on the surface of the ski slope. In the period from April 1 to October 30, it is pumped and soaked through the ground and then disappears into the ground. In the period from November 1 to March 30, the water is distributed by snow cannons and then introduced into the ground. The area of the ski slope is 4.96 ha.

According to the local Spatial Development Plan (SDP), irrigated area is intended for sports and tourist services (Mszana Dolna, 2013). Field conditions related to the use of reclaimed water were assessed. Consequently, the treatment plant was built in accordance with the SDP recommendation to eliminate sources of water pollution.

For risk assessment, measurement campaigns were launched to assess the quality of treated wastewater. By law, samples should be taken 4 times a year. However, for risk assessment purposes, an extended bi-weekly measurement cycle was initiated.

Initial analyzes are based on the guidelines in the Regulation 2020/741. Reclaimed water must be safely used and managed.

The risk assessment elaborated for Kasina Ski WWTP describes the entire system and responsibilities as shown in the schematic diagram in Fig. 1.

A very important element is the identification of potential threats related to pollution and infrastructure failures. In 2020, the European Union adopted the Regulation 2020/741 specifying the basic principles of using water for crop irrigation (European Union, 2020). It is based on risk management. The Regulation obviously presents some general assumptions and stresses the need to consider the requirements of many EU directives. The European Commission is developing detailed guidelines. Such guidelines have been in use in the world for years, where water recovery is more commonly used. Guidelines for risk management base on logic already known by water utilities, as it is described by the World Health Organization (2017), Zhang and Farahbakhsh (2007) and the European standard EN 15975-2:2013-12 “Security of drinking water supply—Guidelines for risk and crisis management—Part 2: Risk management”.

According to it, the classification tables for consequences and probability are used for hazard assessment and risk assessment. When analyzing the water reuse system in accordance with Regulation 2020/741, a 5 × 5 risk matrix can be adopted as proposed in Table 2. The risk classification includes 5 levels:

- 1–5 low risk
- 6–10 medium risk
- 12–15 high risk
- 16–20 very high risk
- 25 unacceptable risk.

Table 2 Typical risk matrix used for analyses in Kasina

Probability	Consequences					
		Weight of the consequences				
		Negligible	Slight	Moderate	Major	Catastrophic
		1	2	3	4	5
Improbable	1	1	2	3	4	5
Unlikely	2	2	4	6	8	10
Moderately likely	3	3	6	9	12	15
Likely	4	4	8	12	16	20
Almost certain	5	5	10	15	20	25

Risk analysis and estimation are performed based on a qualitative method. Assessment of the probability of certain hazardous events and their consequences is performed using an expert method, based on the analysis of laboratory results, documents, and stakeholder consultations.

Moreover, the International Organization for Standardization has issued several standards for water reuse, such as ISO 16075 series: "Guidelines for treated wastewater use for irrigation projects" or ISO 20419 "Treated wastewater reuse for irrigation". All these documents were used in the risk analysis, although initially only in microbiological hazards.

Regulation 2020/741 also emphasizes the importance of evaluating EU directives. A crucial element of the risk assessment is the microbiological hazard analysis. In this article, the author focuses on this very problem.

The analytical process includes testing the quality of raw wastewater and treated wastewater after the UV lamp. The analyzed basic and microbiological parameters result from the requirements of Regulation 2020/841 for class A reclaimed water, which concerns the irrigation of plants for consumption. This, of course, is not the case with Kasina. However, the results are compared to the strictest requirements.

Analyzed parameters according to the Regulation are *E.coli*, turbidity, Biological Oxygen Demand (BOD_5), Total Suspended Solids (TSS). *Legionella* should be verified only if there is a risk of aerosolization. Intestinal nematodes (helminth eggs) should be analyzed if the reclaimed water is used for irrigation of pastures or forage.

For the Kasina Ski case study *Legionella* is measured, but intestinal nematodes are not.

For microbiological risk analysis, which is a part of the risk assessment, following parameters were measured: colony of microorganisms in 36 °C, and in 22 °C, *Pseudomonas aeruginosa*, Fecal enterococci, Fecal coliforms, Coliforms, group of *Salmonella* spp., *Ascaris* sp., *Trichuris* sp. and *Toxocara* sp., *Clostridium perfringens* with spores.

The water permit issued to the operator of the Kasina Ski treatment plant includes the following parameters:

BOD_5—25 mgO_2/L,
COD_{Cr}—125 mgO_2/L,
TSS—35 mg/L.

3 Results

The Kasina Ski WWTP has variability depending on the ski and bike seasons. In January 2022 it treated approximately 20 m^3/day of wastewater. It was a high season because of the Polish winter holidays. The treated wastewater after UV disinfection is considered reclaimed water. Results of reclaimed water analysis are presented in Table 3.

This table compares the results with the parameters required by Annex I of Regulation 2020/741 for water intended for plant irrigation. The results were compared with the highest class of recovered water A. None of the quality indicators exceeds the parametric value Intestinal nematodes have not been measured as this parameter applies to plants eaten raw.

Results of additional microbiology analysis of raw wastewater and reclaimed water are presented in Table 4.

The table presents the results of microbiological analyzes carried out in the winter season (December, January, and February) for raw wastewater and reclaimed water. The content of microorganisms in the wastewater at 36 and 22 °C is in the range of 10^5–10^6. MBR + UV reduce them to a different extent. The remaining parameters are properly reduced.

4 Discussion

The ski resort is located in the basin of the Kasinianka stream (tributary of Raba, tributary of Vistula). The Local Development Plan forbids changes to water conditions. That is why water intake from Kasinianka stream is located below the ski station and the ski resort remains in the stream's catchment area.

Table 3 Results of reclaimed water (MBR + UV) analysis

Parameter	E.coli	BOD5	TSS	Turbidity	Legionella if risk of aerosolization	Intestinal nematodes
	pcs/100 mL	mg/L	mg/L	NTU	CFU/L	egg/L
Requirement of 2020/741 class A	≤ 10	≤ 10	≤ 10	≤ 5	<1000	≤ 1
17 January 2021	0	<3	<2	1	0	n/d
2 December 2021	<1	0.9	<2	1.06	0	n/d
15 December 2021	<1	1	5	2.29	0	n/d
5 January 2022	<1	0.9	<2	<0.1	0	n/d
17 January 2022	<1	0.8	<2	<0.1	22	n/d
14 February 2022	<1	1	<2	0.12	0	n/d

Table 4 Microbiology of wastewater inflow to the WWTP and reclaimed water outflow to the retention tank

	Parameter								
Date	Microorg. at 36 °C	Microorg. at 22 °C	Fecal enterococci	Fecal coliforms	Coliforms	*Salmonella* spp.	*Pseudo*monas aeruginosa	*Ascaris* sp., *Trichuris* sp., *Toxocara* sp.	*Clostridium perfringens* with spores
	CFU/1 mL	CFU/1 mL	MPN/100 mL	MPN/100 mL	MPN/100 mL		CFU/1 mL	Numer/L	CFU/1 mL
2 December 2021									
Raw ww	33 * 10^4	71 * 10^4	9.8 * 10^3	16 * 10^4	24 * 10^5	Not det	460	0	2800
Recl. Water	122	345	0	0	0	Not det	0	0	2
5.December 2021									
Raw ww	46 * 10^4	64 * 10^4	17 * 10^3	5.2 * 10^4	10 * 10^5	Not det	140	0	600
Recl. Water	13 000	15,000	0	0	33	Not det	0	0	250
5 January 2022									
Raw ww	40 * 10^4	120 * 10^4	20 * 10^3	17 * 10^4	2.4 * 10^5	Not det	940	0	13,000
Recl. water	14000	4200	0	0	30	Not det	0	0	2
14 February 2022									
Raw ww	360 * 10^4	260 * 10^4	49 * 10^3	330 * 10^4	170 * 10^5	Not det	9000	0	220
Rcl water	191	445	0	0	0	Not det	0	0	0

According to hydrogeological data, the study area is situated on sedimentary rocks of the Cretaceous and Paleogene ages, consisting of alternating sandstones and shales. Clay rubble and clay appear. The whole area is covered with stony soil 0.2 m thick. No surface watercourses flow through the area. There are Tertiary and Quaternary waters under the surface. According to the Regulation on discharging sewage into waters or into the ground (Poland, 2019), domestic sewage may be discharged into the ground if the place where it is discharged, or the bottom of the water system is separated by a layer of soil with a thickness of at least 3 m from the highest usable groundwater aquifer. Underground water in the vicinity of the WWTP is present in the lower part of the station at a depth of 2.6 m below ground level, in the form of seepage. However, the highest aquifer is located below 3 m below ground level. Due to the fact that the treated wastewater will be discharged gradually into the ground, the condition was met. Therefore, there is no microbiological risk to groundwater.

The treatment plant is underloaded, even on the peak days of the ski season. In January 2022 when the ski lift was very charged the whole day the inflow to the wastewater treatment was 20 m^3/d. Therefore, it should be possible to connect to it at least a part of a nearby village, which does not have any wastewater network and own wastewater treatment plant. Even though it is not fully loaded, the Kasina WWTP works efficiently. The applied MBR technology combined with UV disinfection turns out to be effective in the context of the requirements of the Regulation 2020/741 (Table 2). In the analyzed winter season the reclaimed water didn't contain E.coli. BOD was under 3 mg/L, Total Suspended Solids didn't exceed 5 mg/L, and turbidity 3 NTU. Legionella appeared once but in safe amounts. Besides, Legionella is not a threat in this case. The technologies used will not pose a risk of creating an aerosol. According to the Regulation Intestinal nematodes (helminth eggs) should be analyzed if the reclaimed water is used for irrigation of pastures or forage. This is not the case of Kasina Ski.

A very important challenge is to maintain the microbiological safety of the reclaimed water. Therefore, a campaign to measure several microbiological parameters was launched, which should last the entire year. In the winter season, the technology turns out to be effective as it is indicated in Table 5.

The research carried out in the winter season shows that the MBR + UV reduces well the most of analyzed indicators. Moreover, reusing water for snowmaking would provide more nutrients for the vegetation on the ski slope (Sidan & Weiping, 2016). However, Cernusca and Tappeiner (1990) drew attention to lower soil water absorption and greater runoff as a result of the modification of ski slopes. Rixen et al. (2004) observed that with natural soil snow, frost lasts longer than with artificial snow. However, the growing season is shorter.

It should be noted that the research should be carried out longer to capture the seasonality.

The pore size can be a key factor to assess the effectiveness in microbiology reduction. According to Arevalo et al. (2012) the capacity of the MBR systems to eliminate organic matter depends on the biological activity in the active sludge, and it is not affected by the type of membrane. However according to their study microbiology is better reduced during nanofiltration which seems obvious because the pore size is 0.034 µm.

The ISO 16075 standard indicates *E. coli* and fecal coliforms as an important indicator. In the analyzed period, the reduction of these parameters turned out to be 100%.

Reduction of coliforms in the WWTP is at 5.108 log. Coliforms are a group of bacteria including *E.coli* but also *Klebsiella*, *Enterobacter*, *Citrobacter*, as well as *Serratia* and *Hafnia* (Leclerc et al., 2001). They are used as an indicator of water quality. However, their presence in water can be of both fecal and environmental origin. Some of them multiply in water (especially warm water), soil, and plant material. Therefore, this group cannot directly serve as a specific indicator of fecal water pollution, but it can, like the total number of microorganisms, be a criterion for assessing the cleanliness and integrity of water (mainly drinking) distribution systems. A more important indicator is therefore *E. coli*, which signals the fecal contamination of the water. Both parameters are eliminated in Kasina Ski WWTP.

Because *Clostridium perfringens* spores are both specific to sewage contamination and environmentally stable, they are considered as possible conservative indicators of human fecal contamination and possible surrogates for environmentally stable pathogens (Stelma, 2018). *C. perfringens* spores could

Table 5 Microbiology reduction values in Kasina wastewater treatment plant (MBR + UV)

	Average in wastewater	Average in reclaimed water	Log reduction	% reduction
Microorganisms in 36 °C CFU/1 mL	1,197,500	682.825	2.244	99.43
Microorganisms in 22 °C CFU/1 mL	1,287,500	4997.5	2.411	99.612
Fecal enterococci MPN/100 mL	23,950	0	∞	100
Fecal coliforms MPN/100 mL	920,500	O	∞	100
Coliforms MPN/100 mL	5,160,000	40.25	5.108	100
Salmonella spp.	Not detected	Not detected	–	–
Pseudomonas aeruginosa CFU/1 mL	2635	o	∞	100
Ascaris sp., *Trichuris* sp., *Toxocara* sp. Number/L	0	0	–	–
Clostridium perfringens with spores CFU/1 mL	4155	63.5	1.8158	98.472
E.coli pcs/100 mL	839,500	0	∞	100

Table 6 Basic risk factors in the context of microbiological safety of reclaimed water in Kasina Ski

Sources of threats and hazardous events	Characteristics of a probable hazardous event. Local conditions	Probability	Consequence	Risk
Deteriorating quality of raw wastewater	High microbiological contamination related to an epidemic or infection of users. Such an event is possible, but the technology used shows the effectiveness of microbiology reduction. In addition, reclaimed water is disinfected and mixed with stream water	4	2	8
Inadequate capacity of the treatment plant	Underload or overload may result in an insufficient reduction of micropollutants. Currently, despite the underload, the system shows a high microbiology reduction	2	3	6
Plant contamination	In the summer season, the natural vegetation growing on the slope will be watered. The water will seep into the soil. The presence of bystanders during irrigation is reduced to a minimum	2	2	4
Diffusion of responsibility for the functioning of the system	The same entity is responsible for the efficient functioning of the treatment plant, as well as for reclaimed water storage and irrigation	2	2	4
Failure of the wastewater treatment plant	Failure may cause local soil contamination. However, a reservoir where reclaimed water is mixed with the stream water is a safety buffer. Then, however, it is impossible to produce snow or water. The WWTP has an operating and emergency response manual	4	3	12
Exposure of the population to contact with microbiologically contaminated water	The exposure of the population is limited to the users of the slope, who can be easily evacuated in the event of a system failure	2	3	6
Exposure of animals to contact with microbiologically contaminated water	There are no farm animals in the area. The risk of wild animal contact is low	2	2	4

be used as a microbiological indicator parameter and should be measured in drinking water. The reference value should be 0 CFU/100 mL in drinking water. No reference value should be mentioned in raw water as, based on studies in fresh and marine waters or brackish water, the presence of *C. perfringens* spores is not correlated with the presence of *Cryptosporidium* and *Giardia* (oo)cysts (Lamy et al., 2020).

The risk assessment was based on the matrix presented in Table 2, based on qualitative assessment. The main risk factors are summarized in Table 6.

The table summarizes an analysis of the main hazardous events related to microbial contamination. The value of probability and consequences are determined by an expert method. The probability is the product of these two values. Similar events related to incidents, dilution of responsibility, and microbial contamination are indicated in the risk analysis of drinking water supply (World Health Organization, 2017) and water reuse systems described, inter alia, by the National Research Council (2012) in the USA or Water Directors in the EU (European Union, 2016).

It is necessary to check the efficiency of the system in the summer season, when the reclaimed water temperature is higher, and the vegetation season is in progress. It seems appropriate to match the quality of reclaimed water to the needs of the system, but the physicochemical risks need to be assessed. Therefore, further research is necessary (emerging pollutants, nitrogen, phosphorus, salination), as highlighted by Zhang & Wang, (2014), Gu et al. (2019).

The risk assessment should also include elements of the cost–benefit analysis that is sometimes very difficult as mentioned by Declercq (2017), Qin and Horvath (2020).

5 Conclusions

More and more local problems and challenges mobilize to act toward water reuse. A small MBR treatment plant with a microfiltration membrane, equipped with a UV lamp, effectively removes microbiological contamination in the winter season. In Poland, it is necessary to build hundreds of small sewage treatment plants serving agglomerations below 2000 p.e., hotels and other resorts; MBR + UV technology may become one of the solutions.

The greatest risk may be related to the failure of the wastewater treatment plant including the UV lamp. Therefore, there should be a double disinfection system.

The retention tank consists of a significant barrier to mitigate the risk of contamination.

Other microbiological risks are also associated with technology; therefore, it is important to constantly monitor the quality of raw and treated wastewater.

Acknowledgements The author would like to thank Schwander Polska for the provided data.

References

Alcalde Sanz, L., & Gawlik, B. (2014). *Water reuse in Europe—Relevant guidelines, needs for and barriers to innovation*. Publications Office of the European Union, JRC9258.

Arévalo, J., Ruiz, L. M., Parada-Albarracín, J. A., González-Pérez, D. M., Pérez, J., Moreno, B., & Gómez, M. A. (2012). Wastewater reuse after treatment by MBR. Microfiltration or ultrafiltration? *Desalination, 299*, 22–27.

Bunani, S., Yörükoğlu, E., Sert, G., Kabay, N., Yüksel, Ü., Yüksel, M., Egemen, Ö., & Taylan, P. T. (2018). Utilization of reverse osmosis (RO) for reuse of MBR-treated wastewater in irrigation-preliminary tests and quality analysis of product water. *Environmental Science and Pollution Research, 25*, 3030–3037. https://doi.org/10.1007/s11356-015-4199-y.

Cernusca, A., & Tappeiner, U. (1990). Implanti di risalita e piste da sci. Provincia Autonoma di Trento Assessorato al Territorio, Ambiente e Foreste.

Comerton, A., Andrews, R. C., & Bagley, D. M. (2005). Evaluation of an MBR–RO system to produce high quality reuse water: Microbial control, DBP formation and nitrate. *Water Research, 39/16*. https://doi.org/10.1016/j.watres.2005.07.014.

Declercq, R., Loubier, S., Condom, N., & Molle, B. (2017). Socio-economic interest of treated wastewater reuse in agricultural irrigation and indirect potable water reuse: Clermont-Ferrand and Cannes case studies' cost-benefit analysis. *Irrigation and Drainage, 69*, 194–208. https://doi.org/10.1002/ird.2205.

Diao, L. (2020). Application of mbr technology in schoolyard domestic sewage treatment. In *IOP Conference Series: Earth Environmental Science* (Vol. 526, pp. 012002).

Environment Protection Authority of Victoria. (2021). Victorian Guideline for water recycling. State of Victoria.

European Union. (1991). Council Directive 91/271/EEC of 21 May 1991 concerning urban waste-water treatment. OJ L 135, Retrieved May 30, 1991.

European Union. (2016). Guidelines on integrating water reuse into water planning and management in the context of the WFD. Retrieved March 1, 2022, from https://ec.europa.eu/environment/water/pdf/Guidelines_on_water_reuse.pdf.

European Union. (2020). Regulation 2020/741 of the European Parliament and of the Council of 25 May 2020 on minimum requirements for water reuse. OJ L 177. Retrieved June 5, 2020.

Exall, K. (2004, February 1). A review of water reuse and recycling, with reference to Canadian practice and potential: 2. Applications. *Water Quality Research Journal, 39*(1), 13–28. https://doi.org/10.2166/wqrj.2004.004.

France. (2010). Arrêté du 2 août 2010 relatif à l'utilisation d'eaux issues du traitement d'épuration des eaux résiduaires urbaines pour l'irrigation de cultures ou d'espaces verts. JORF no 0201 du 31 août 2010.

Gu, X., Xiao, Y., Yin, S., Liu, H., Men, B., Hao, Z., Qian, P., Yan, H., Hao, Q., Niu, Y., Huang, H., & Pei, Q. (2019). Impact of long-term reclaimed water irrigation on the distribution of potentially toxic elements in soil: An in-situ experiment study in the North China plain. *International Journal of Environmental Research and Public Health, 16*, 649. https://doi.org/10.3390/ijerph16040649.

Hahn, F. (2004). Innevamento artificiale Nelle Alpi (Snowmaking in the Alps), Report CIPRA-International. CIPRA International, Schaan, Liechtenstein. Retrieved February 5, 2022, from https://www.cipra.org/it/pubblicazioni/2709/dateien/454_it/@@download/file/Dossier_Kunstschnee_I.pdf?inline=true.

Hardy, D., Cubillo, F., Han, M., & Li, H. (2015). *Alternative water resources: A review of concepts, solutions and experiences*. International Water Association.

Harivandi, M. A. (2004). Evaluating recycled waters for golf course irrigation. *USGA Green Section Record, 42*(6), 25–29.

Hirani, Z. M., DeCarolis, J. F., Adham, S. S., & Jacangelo, J. G. (2010). Peak flux performance and microbial removal by selected membrane bioreactor systems, *Water Research, 44/8*. https://doi.org/10.1016/j.watres.2010.01.003.

Joss, A., Baenninger, C., Foa, P., Koepke, S., Krauss, M., McArdell, C. S., Rottermann, K., Wei, Y., Zapata, A., & Siegrist, H. (2011). Water reuse: >90% water yield in MBR/RO through concentrate recycling and CO_2 addition as scaling control. *Water Research, 45/18*. https://doi.org/10.1016/j.watres.2011.09.011.

Lamy, M.-C., Sanseverino, I., Niegowska, M., & Lettieri, T. (2020). *Microbiological parameters under the drinking water directive. Current state of art on somatic coliphages and Clostridium perfringens and spores*. EUR 29932 EN. Publications Office of the European Union. https://doi.org/10.2760/005492.

Leclerc, H., Mossel, D. A., Edberg, S. C., & Struijk, C. B. (2001). Advances in the bacteriology of the coliform group: Their suitability as markers of microbial water safety. *Annual Review of Microbiology, 55*, 201–234. https://doi.org/10.1146/annurev.micro.55.1.201.

Lyu, S., & Chen, W. (2016). Soil quality assessment of urban green space under long-term reclaimed water irrigation. *Environmental Science and Pollution Research, 23*, 4639–4649. https://doi.org/10.1007/s11356-015-5693-y.

Maiolo, M., & Pantusa, D. (2018, December 1). A proposal for multiple reuse of urban wastewater. *Journal of Water Reuse and Desalination, 8*(4), 468–478. https://doi.org/10.2166/wrd.2017.144.

Maniam, G., Zakaria, N. A., Leo, C. P., Vassilev, V., Blay, K. B., Behzadian, K., & Poh, P. E. (2022). An assessment of technological development and applications of decentralized water reuse: A critical review and conceptual framework. *Wires Water, 9*(3), e1588. https://doi.org/10.1002/wat2.1588.

Mszana Dolna. (2013, August 26). Resolution No. XXXVI/403/2013 of the Mszana Dolna Commune Council of August 14, 2013. on the amendment of the local spatial development plan of the Mszana Dolna commune of the village of Kasina Wielka. Official *Journal of the Lesser Poland Voivodeship*, 5269.

MtBuller. (2021). Retrieved December 15, 2021, from https://www.mtbuller.com.au.

National Research Council. (2012). *Water reuse: Potential for expanding the nation's water supply through reuse of municipal wastewater*. National Academies Press. https://doi.org/10.17226/13303.

Paccard, P. (2011). *Gestion durable de l'eau en montagne: Le cas de la production de neige en stations de sports d'hiver*. Université de Savoie.

Paranychianakis, N. V., Salgot, M., Snyder, S. A., & Angelakis, A. N. (2015). Water reuse in EU States: Necessity for uniform criteria to mitigate human and environmental risks. *Critical Reviews in Environmental Science and Technology, 45*(13), 1409–1468. https://doi.org/10.1080/10643389.2014.955629.

Poland. (2019). Regulation of the Minister of Maritime Economy and Inland Navigation of 12 July 2019 on substances particularly harmful to the aquatic environment and the conditions to be met when discharging sewage into waters or ground, as well as when discharging rainwater or meltwater into waters or into aquatic devices (oryginal title: Rozporządzenie Ministra Gospodarki Morskiej i Żeglugi Śródlądowej z dnia 12 lipca 2019 r. w sprawie substancji szczególnie szkodliwych dla środowiska wodnego oraz warunków, jakie należy spełnić przy wprowadzaniu do wód lub do ziemi ścieków, a także przy odprowadzaniu wód opadowych lub roztopowych do wód lub do urządzeń wodnych). *Polish Official Journal*, 1311.

Purnell, S., Ebdon, J., Buck, A., Tupper, M., & Taylor, H. (2015). Bacteriophage removal in a full-scale membrane bioreactor (MBR)—Implications for wastewater reuse. *Water Research*. https://doi.org/10.1016/j.watres.2015.01.019.

Purnell, S., Ebdon, J., Buck, A., Tupper, M., & Taylor, H. (2016). Removal of phages and viral pathogens in a full-scale MBR: Implications for wastewater reuse and potable water. *Water Research*. https://doi.org/10.1016/j.watres.2016.05.013.

Qian, Y., & Mecham, B. (2005). Long-term effects of recycled wastewater irrigation on soil chemical properties on golf course fairways. *Agronomy Journal, 97*. https://doi.org/10.2134/agronj2004.0140.

Qin, Y., & Horvath, A. (2020). Use of alternative water sources in irrigation: Potential scales, costs, and environmental impacts in California. *Environmental Research Communications, 2*, 055003.

Radcliffe, J. C., & Page, D. (2020). Water reuse and recycling in Australia—History, current situation and future perspectives. *Water Cycle*. https://doi.org/10.1016/j.watcyc.2020.05.005.

Rixen, C., Haeberli, W., & Stoeckli, V. (2004). Ground temperatures under Ski Pistes with artificial and natural snow. *Arctic, Antarctic, and Alpine Research, 36*(4), 419–427. https://doi.org/10.1657/1523-0430(2004)036[0419:GTUSPW]2.0.CO;2.

Rodríguez-Villanueva, P., & Sauri, D. (2021). Wastewater treatment plants in mediterranean Spain: An exploration of relations between water treatments, water reuse, and governance. *Water, 13*(12), 1710. https://doi.org/10.3390/w13121710.

Rulseh, T. (2013). The fire chief project: Pure as the driven snow? *Treatment Plant Operator*. Retrieved March 10, 2022, from https://www.tpomag.com/blog/2013/11/the_fire_chief_project_pure_as_the_driven_snow.

Schwander. (2021). Retrieved March 13, 2022, from https://www.schwander.pl/pl/technologia.

Sidan, L., & Weiping, C. (2016) *Environmental science and pollution research international*. Heidelberg Tom 23, Nr/wydanie 5. https://doi.org/10.1007/s11356-015-5693-y.

Singapore's National Water Agency. (2022). Retrieved March 13, 2022, from https://www.pub.gov.sg/watersupply/fournationaltaps/newater.

Stelma, G. N., Jr. (2018). Use of bacterial spores in monitoring water quality and treatment. *Journal of Water and Health, 16*(4), 491–500. https://doi.org/10.2166/wh.2018.013.

Taesopapong, S., & Ratanatamskul, C. (2020). Innovative eco biofilter/membrane bioreactor (MBR) technology for community wastewater recycling. In *IOP Conference Series: Earth and Environmental Science* (Vol. 427, p. 012011).

United Nations. (2021). *Handbook on water allocation in a transboundary context*. United Nations Publications UNECE. Retrived July 7, 2022, from https://unece.org/sites/default/files/2021-12/ECE_MP.WAT_64_Handbook%20on%20water%20allocation%20in%20a%20the%20transboundary%20context.pdf.

Van Rossum, T. (2020). Water reuse and recycling in Canada—history, current situation and future perspectives. *Water Cycle, 1*. https://doi.org/10.1016/j.watcyc.2020.07.001.

Velasquez, D., & Yanful, E. K. (2015). Water reuse perceptions of students, faculty and staff at Western University, Canada. *Journal of Water Reuse and Desalination, 5*(3), 344–359. https://doi.org/10.2166/wrd.2015.126.

Water Canada. (2021). Retrieved March 13, 2022, from https://www.watercanada.net/change-expected-at-vernon-water-reclamation-centre/.

World Health Organization. (2017). Guidelines for drinking-water quality, 4th edition, incorporating the 1st addendum.

Zhang, K., & Farahbakhsh, K. (2007). Removal of native coliphages and coliform bacteria from municipal wastewater by various wastewater treatment processes: Implications to water reuse. *Water Research 41/12*. https://doi.org/10.1016/j.watres.2007.03.010.

Zhang, C. M., & Wang, X. C. (2014). Distribution of enteric pathogens in wastewater secondary effluent and safety analysis for urban water reuse. *Human and Ecological Risk Assessment: An International Journal, 20*(3), 797–806. https://doi.org/10.1080/10807039.2013.810957.

Can Remote Sensing Be a Useful Tool to Increase Water Reuse in Agriculture?

Anas Tallou, Ayoub Haouas, Hassan Erraji, Anas Driouich, Francisco Pedrero Salcedo, Gaetano Alessandro Vivaldi, and Faissal Aziz

Abstract

The reuse of treated wastewater has been increasingly recommended. As accessible water supplies remain scarcer, treated wastewater becomes more significant for irrigation in agriculture. The quality of water is a leading determinant of the environmental system. For various uses, especially irrigated agriculture, high accuracy information on water quality is crucial. Autonomous unmanned aerial vehicle (UAV) data on water quality in the field might assist in reducing geographic information shortages by covering a big space in a short period, High-Resolution Remote Sensing Data, spatially distributed necessary data for water management. Additionally, (UAV) devices have a set of constraints and obstacles. Autonomous vehicles must be provided with extra navigating equipment or acoustic localization systems. This chapter provides a multispectral water quality surveillance approach that relies on UAV technologies that can efficiently examine water and perform multi-temporal surveys regarding important water quality indicators that can also be applied to wastewater treatment and reuse in agriculture. The results can lead to a powerful compromise base for the treatment of agricultural water reuse.

Keywords

Wastewater treatment • Remote sensing • Water reuse • Reclaimed water • Multispectral images • Precision Agriculture • Unmanned aerial vehicle

1 Introduction

Water scarcity is one of the top global threats, affecting people's livelihoods and well-being much beyond socio-economic and environmental issues (World Economic Forum (WEF), 2020). Water scarcity and deterioration of water quality are projected to exacerbate the worldwide water problem in dry and densely populated regions (UN-Water, 2020).

Despite increasing water shortages and deterioration of water quality, irrigation has played a significant part in feeding the world's growing population and is anticipated to continue to do so in the future (World Bank, 2017). Arid and

A. Tallou (✉)
Bioprocess and Environmental Chemistry, Sultan Moulay Slimane University of Beni Mellal (FST), Beni-Mellal, Morocco
e-mail: anas.tallou@usms.com

A. Haouas
Centre d'Agrobiotechnologie et Bioingénierie, Unité de Recherche Labellisée CNRST (Centre AgroBiotech-URL-CNRST-05), Cadi Ayyad University, Marrakech, Morocco

H. Erraji
Laboratory of Bioresources, Biotechnology, Ethnopharmacology and Health, Faculty of Sciences, Mohamed Premier University, Oujda, Morocco

A. Driouich
Laboratory of Process Engineering and Environment, Faculty of Sciences and Technology, University Hassan II, 28806 Mohammedia, Morocco

F. P. Salcedo
Department of Irrigation, CEBAS-CSIC, Campus Universitario de Espinardo, 30100 Murcia, Spain

G. A. Vivaldi
Dipartimento di Scienze Agro-Ambientali e Territoriali, Università degli Studi di Bari Aldo Moro, Via Amendola 165/A, 70126 Bari, Italy

F. Aziz
Laboratory of Water, Biodiversity and Climate Change, Semlalia Faculty of Sciences, University Cadi Ayyad, Marrakech, Morocco

National Centre for Research and Studies on Water and Energy (CNEREE), University Cadi Ayyad, Marrakech, Morocco

M. Smol et al. (eds.), *Water in Circular Economy*, Advances in Science, Technology & Innovation,
https://doi.org/10.1007/978-3-031-18165-8_10

semi-arid areas, which have limited freshwater supplies, depend on irrigation to support the productivity and sustainability of agricultural production systems. Water reuse is not a new method; wastewater treatment and reuse insights have been accumulated throughout human history. Human wasteland utilization is really an old practice that has gone through several levels of development from ancient to modern times (Angelakis & Rose, 2014).

There are several techniques for monitoring wastewater treatment plants (WWTPs) depending on visual observations, monitoring solids or removal efficiency of different pollutants and toxic compounds (e.g., Biological Oxygen Demand (BOD), heavy metals, etc.) (US Environmental Protection Agency, 2004). To reduce the cost of maintenance and inspection processes, operators are required to visit sites with cameras and video surveying equipment. Some WWTPs, for example, used a remote camera to evaluate network piping status and irregular connections (US Environmental Protection Agency, 2004; Suzana et al., 2012). Unmanned aircraft systems (UASs) have been increasingly popular for asset surveying and monitoring in recent decades. Wind turbines, solar panels, dams, and pipe leaks are some of the facilities that UASs have been used to investigate in the energy and water sectors. However, the potential for such technologies to assist in the inspection of WWTPs is still undetermined (Sancho Martínez et al., 2020).

Likewise, image processing methods have been investigated over the last few years to detect failures and problems in WWTPs. Machine learning algorithms, for example, are employed to model urban runoff in Switzerland (Tokarczyk et al., 2015). UASs have been used to monitor the temperature of wastewater collection containers in Sweden, with image processing utilized to identify any runoff water (Panasiuk et al., 2015). Nevertheless, the use of image processing and unmanned aerial systems (UAS) for monitoring and evaluating WWTP installations is still to be reported in the research literature (Sancho Martínez et al., 2020).

The present chapter discussed the worldwide problem of water shortage in agriculture as it is the big consumer of water resources. As a consequence of water scarcity and availability for use in agriculture, reclaimed water reuse in agriculture becomes one of the most important strategies to reduce pressure on freshwater. In addition, the chapter presents the application of remote sensing in different aspects of agriculture starting from crop growth and production, control of irrigation and identification of crop disease and problems. The last part of the present chapter is an overview of the employment of unmanned aircraft vehicles (UAVs) as a tool for better reuse of reclaimed water in agriculture.

2 Water: Biggest Agriculture Challenge

Extreme weather events as a result of climate change have resulted in water scarcity and environmental degradation and deterioration, but also as a consequence of human activities which will be among the most likely threats in the near future (The Global Risks Report, 2021). Water is crucial for all life aspects, from socio-economic development to the conservation of ecosystems. Water exists in several forms on planet Earth, including the atmosphere, the ocean, rivers, lakes, rocks, etc. On the Earth's surface, 96% of the water is saline water, which forms oceans, while the rest represents freshwater. The majority of the world's freshwater is preserved in the ground (USGS, 1984), while the rest bulk is trapped in glaciers, ice caps, and permafrost. Only about 1% of the world's freshwater is readily available for human consumption. Indeed, according to global climate projections, the Mediterranean region will face severe environmental challenges such as droughts, irregular rainfall events, ocean acidification, and sea-level rise, all of which will have environmental consequences on resource depletion, food security, and human health (Cramer et al., 2018). As a result, agriculture's biggest challenge nowadays is to produce more with less water. In all aspects of life, water is a basic requirement. It is the most precious gift to all creatures for the continuity of life on this planet. Because their bodies are mostly made of water, all organisms require more water to survive. As a result, water management would be a critical component of sustainable development, allowing current needs to be met without negatively impacting future generations' ability to meet their own (World Commission on Environment and Development, 1987).

Food demand is likely to rise dramatically as a result of the projected growth in the global population. As a consequence, increased pressure on the use of water resources in electricity, agriculture, industrial processes, and daily demand is estimated. Agriculture is known to be the field that consumes more water, accounting for over 70% of worldwide freshwater (IPCC, 2019). Water shortage is thought to be a limiting factor for long-term development, particularly in the Mediterranean region, where agriculture plays a major role (United Nations World Water Assessment Programme, 2017).

The agricultural sector is critical to increasing food availability and consequently ensuring food security. Food production has increased by more than 100% over the last 30 years (United Nations, 2017). FAO estimates that by 2050, 60% of more food will be required to meet the rising demand of a growing worldwide population. Agriculture activities should be carried out in accordance with FAO's sustainable development goals in order to tackle future difficulties. Our world has a limited number of arable farmlands, the majority of which is either cultivated or has been degraded by erosion or overexploitation, and so the lookout for new lands may result in deforestation and the displacement of communities. Agriculture must focus on creative approaches that allow for the efficient use of both land and natural resources in order to meet future difficulties. The use of Decision Support Systems or rational land management has become critical in this regard. Artificial intelligence technologies (AIT) with prospective Precision Agriculture (PA) innovations are a crucial element in this regard. AIT is becoming increasingly popular in today's agriculture due to its benefits of producing healthier crops, controlling pests, monitoring soil and growing conditions, managing information for farmers, reducing workload, and optimizing a large range of agriculture-related activities throughout the food supply chain (Pedrero el al., 2021). Continued investment in agricultural knowledge, research, and farming techniques will allow producers to respond to environmental issues while ensuring revenue, and developing alternatives to increasing stresses on natural resources and ecosystem degradation and deterioration.

3 Reclaimed Water Reuse in Precision Agriculture

The 2030 Agenda for Sustainable development Goals (SDGs) aims to ensure long-term development, notably zero hunger (Goal 2) and clean water and sanitation (Goal 6). The paper "The World in 2050" (IIASA, 2018) proposed six model transitions for attaining long-term sustainability by 2050, integrating resource distribution to address food and water challenges (Water, Energy, Food Nexus). According to the World Economic Forum's "Global Risks Report 2020" (WEF, 2019), water and food crises are among the top ten long-term risks in the impact category during the next ten years. Due to limited freshwater resources, several semi-arid and arid countries, particularly in the Middle East and Africa (Hussain et al., 2019), are now experiencing severe water stress. As a result, alternative water sources must be established to solve water scarcity issues in agriculture.

Generally, wastewaters are any waters that have had their quality degraded due to human activity. Agricultural, municipal, and industrial processes can all produce wastewater. In scarce regions, using reclaimed municipal water (RMW) in agriculture is becoming more common (Pedrero et al., 2010; Tallou et al., 2020). Several research on the reuse of different wastewater for irrigation has been done in recent years (Asano et al., 2007; Jiménez & Asano, 2008; Pescod, 1992; Rhoades et al., 1992; Levy et al., 2011; Tallou et al., 2020, 2021). They realized that wastewater reuse for agriculture has much promise because it offers many advantages. It satisfies rising water demands, conserves drinking water supplies, minimizes polluted effluent disposal into surface water bodies, lowers treatment costs, and enhances farmers' economic benefits by reducing fertilizer application rates through fertigation (Jiménez-Cisneros, 1995; Paranychianakis et al., 2006; Romero-Trigueros et al., 2019), In the context of a circular economy, this is very significant. Despite these advantages, treated wastewater can be utilized for irrigation in regulated environments to reduce the risk of pathogenic and hazardous pollutants in agricultural chemical products, soils, and groundwater (Al-Nakshabandi et al., 1997). However, certain health dangers are associated with using treated wastewater for irrigation, including the possibility of wastewater contaminating crops or groundwaters since its chemical composition differs from that of most natural waters utilized in irrigation (Pereira, & Oliva-Teles 2002).

Municipal wastewater typically contains significant levels of suspended and dissolved particles, including micro and macronutrients, with little additional salt being removed during traditional (secondary and tertiary) treatment. However, it is crucial to note that some pathogenic species, such as Helminths, Enteric bacteria, and viruses, present a dangerous health risk when wastewater is used for irrigation. Furthermore soil properties, particularly Electric Conductivity (EC), Nitrogen content, and Specific Absorption Rate (SAR) values are particularly susceptible to wastewater chemicals, necessitating long-term monitoring of these metrics (Salcedo et al., 2022). Several researchers (Pescod, 1992; Tarchitzky et al., 1984; Vinten et al., 1991) have shown that increased biomass and suspended particles cause a partial biological blockage of soil pores, resulting in reduced hydraulic conductivity in wastewater-irrigated soil. Finally, to develop an effective strategy for promoting and expanding the use of wastewater as an alternative water source, monitoring systems for nutrients, microorganisms, and toxic compounds are required, which would significantly reduce the risk of nutritional imbalances and, as a result, production deterioration. Access to long-term water services is critical for climate change adaptation in the agricultural sector. Instead of typical wastewater treatment, reclaiming non-conventional wastewater for irrigation is especially beneficial to building a circular water economy. Systematic economic evaluation, including sensitivity and reliability evaluations, should be carried out thoroughly, taking into account a variety of elements such as varied economic and water prices (Amos et al., 2018), as well as the payback period (Kuntz Maykot & Ghisi, 2020).

The diversity of production parameters in space and time within cultivated plots receives little attention in traditional agricultural field management. Precision agriculture is defined as "using the appropriate amount in the right place at the right time,' according to a commonly accepted definition

(Gebbers & Adamchuk, 2010). To increase the efficiency of the inputs in the management process dynamics, it considers the variability in time and space of the influencing elements on the agricultural production process.

Precision agriculture is a systematic strategy for managing soils and crops that aims to reduce choice uncertainty by better understanding and managing the most relevant spatial and temporal variability for decision-making. It's being utilized for site-specific management, and it's thought to make the best use of technology possible by taking a multidisciplinary approach. Thus, from crop production to the construction of water resources and their quality levels to climate change mitigation, rational and multidisciplinary agriculture would provide tremendous and irrefutable benefits to humanity (Mastrorilli et al., 2015). Conversely, these advantages are wonderful illustrations of how economic and environmental concerns may coexist.

Since the early 1990s, precision agricultural technology has been commercially available, but adoption has been slow (McBride & Daberkow, 2003). The three most popular PA information technologies are global positioning systems (GPS), guidance systems, GPS yield and soil monitors/maps, and variable-rate input application technologies. PA uses data sensors, connected devices, remote control tools, and other advanced technologies to provide farmers greater control over their fields and teams. As a result, the use of proximal and remote sensors is becoming more common in today's PA. Proximal sensors are employed on the soil (for example, soil moisture sensors or electromagnetic induction sensors for detecting soil electrical conductivity), on the plant (for example, dendrometers and sap flow sensors), and the canopy (for example, dendrometers and sap flow sensors). In reality, remote sensing allows you to record a signal from the surface being examined or from any object. This allows for the detection of spatial and temporal variability and a significant reduction in work expenses and time spent on extensive field surveys. The sensor's distance from the target might range from a few meters (UAV) to hundreds of kilometers (satellite sensors in orbit on satellite platforms). The use of sensors enables the PA cycle to begin, which starts with data acquisition, continues with the extraction of useful information (geostatistics programs, GIS) and decision-making, and concludes with field operations, which are still required for the validation of newly collected data.

4 Application of Remote Sensing in Precision Agriculture

With the Explorer, TIROS (Television Infrared Observation Satellite) series, Corona, and later Landsat missions, satellite remote sensing was a successor to airborne remote sensing in the 1960s. Remote sensing begins with electromagnetic radiation from the Sun (passive remote sensing) or from the satellite (active remote sensing). When incident radiation interacts with the Earth's surface, it is reflected, absorbed, and transmitted. The satellite sensor measures these reflected radiations and contains information about the terrestrial activities occurring during the satellite's overhead at a given location. The hydrological cycle, vegetation processes, interactions with water bodies, geomorphology, and topography all fall under terrestrial processes. Measurements in only specific wavelengths/frequencies may be sensitive to each of the terrestrial processes. As a result, it's critical to find the right satellite sensors for the job. The spatial, temporal, spectral, and radiometric resolutions associated with sensor setup also influence satellite observations (Karthikeyan et al., 2020).

In this chapter, the three main aspects of using remote sensing in PA are (1) crop growth and yield evaluation, (2) qualitative and quantitative irrigation assessment, and (3) identification of crop damage due to pests, infections, and crop lodging, and weeds. Figure 1 represent the different application of remote sensing in food security.

To estimate crop yield and leaf area index (LAI), for example, regression relationships with indices such as normalized difference vegetation index (NDVI), ratio vegetation index (RVI), and vegetation condition index (VCI) are used. For this reason, both multispectral and hyperspectral imageries have been proven to be beneficial. Physical models like the PROSAIL model (which mimics canopy reflectance as a function of LAI and other variables or parameters) are commonly employed to calculate LAI. Crop simulation models also consider climate and agricultural factors while modeling crop variables. Data assimilation has proven to synergistically be beneficial for combining model simulations with remote sensing data to improve crop state variables. With growing efforts to gather soil moisture readings at global scales (using microwave sensors), attempts are being made to assimilate soil moisture into crop simulation

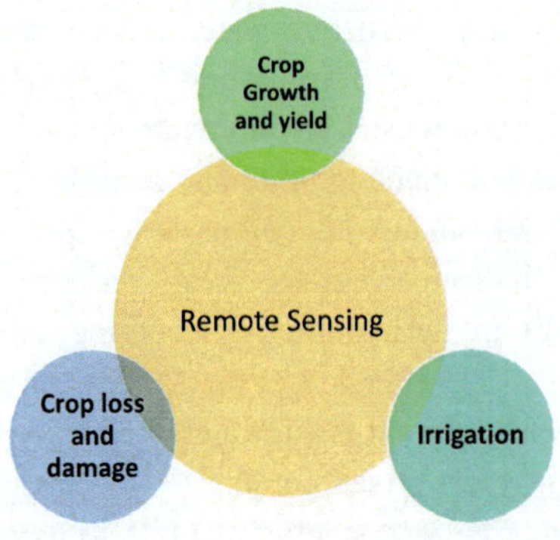

Fig. 1 Most important aspect of remote sensing in PA (Karthikeyan et al., 2020)

models (in a data assimilation framework) to enhance crop yield estimation. Solar-induced fluorescence (SIF) is one of the newest applications to estimate crop yield. It can detect and determine the crop yield through regression analysis (photosynthesis activity is known to influence biomass production, and therefore it can be detected using SIF). Radar microwave waves are commonly employed for agricultural growth assessment due to their ability to penetrate the atmosphere. Crop height estimation is also used for crop monitoring, and it can be detected and monitored using InSAR, tandem interferometry, PolSAR, and PolInSAR techniques (Karthikeyan et al., 2020).

Irrigation research benefits from remote sensing data by identifying irrigated regions and estimating the amount of water required/supplied. Optical/thermal reflectance data gathered over multiple years are treated with either supervised or unsupervised classification methods to identify irrigated regions. For this purpose, derived products such as NDVI, normalized difference moisture index (NDMI), enhanced vegetation index (EVI), and land surface water index (LSWI) are also utilized in categorization schemes (Romero-Trigueros et al., 2017). Similar categorization algorithms identify irrigation using backscatter time series from radar sensors. The soil moisture measured by passive microwave sensors monitors moisture variations due to rainfall and irrigation events. As a result, efforts are conducted to determine the geographical and temporal patterns of irrigated regions using soil moisture products. For calculating the irrigation water needs, evapotranspiration (ET) is an important parameter. Using optical/thermal reflectance from satellites, energy balance approaches are common in estimating ET.

On the other hand, soil moisture is a significant factor in determining how much irrigation water is delivered to a farm. Recent attempts have concentrated on extracting irrigation information using satellite soil moisture and precipitation data sets. The majority of crop loss experiments are now using Unmanned Aerial Vehicles (UAVs), which have extremely high resolution (Romero-Trigueros et al., 2017). There is a need to establish a framework for combining information from UAVs on a local scale with global-reachivariouste sensors for real-time monitoring of agricultural losses on a big scale. A crop may be subjected to various pressures as it progresses through its life cycle. To properly design crop protection, efforts must be taken to separate the pressures utilizing satellite sensors (Karthikeyan et al., 2020; Pedrero et al., 2021).

Crop losses and damage due to pests and diseases are detected at a large scale using optical/thermal reflectance data. Besides vegetation indices, disease detection indices such as the aphid index (AI) and disease water stress index (DWSI), among others, are used. The spread of pests and diseases is identified by applying classification algorithms to multi-temporal data. However, such detection is possible only when the reflectance from pest/disease-affected plants exhibits a significantly different spectral pattern than that of a healthy leaf (Romero-Trigueros et al., 2017). Although efforts are being made to use microwave or SIF data for pests and disease detection, the experiments are confined to field scale (using field/airborne sensors). Crop lodging is another factor that contributes to crop losses. Crop lodging is now identified via backscatter measurements. Because crop height is connected to lodging, the detecting algorithms derive crop height and associated factors (such as inclination). Finally, vegetation indicators and classification algorithms are used to detect weeds that cause crop losses and help detect damage early for better intervention and protection of crop yield (Karthikeyan et al., 2020; Romeros et al., 2016).

5 UAVs Remote Sensing for Reclaimed Water

Due to water shortage, irrigation water is not always accessible (especially in the summer) in semi-arid Mediterranean locations (Fonseca-Pedrero et al., 2015). As a result, irrigation timing must be exact, necessitating irrigation water productivity measures (Tapsuwan et al., 2014). The controlled deficit irrigation method, in which water deficits are applied only during the crop developmental stages that are least vulnerable to water stress, is one technique used nowadays (Chalmers et al., 1981). Furthermore, current climate change projections imply that drought spells will become more frequent and intense (García-Galiano et al., 2015; Stocker et al., 2013). Non-conventional water sources, such as reclaimed water (RW), might be used to help farmers solve this problem. On the one hand, because of its high concentration of macronutrients (N, P, K), RW can be advantageous for crops (Pedrero et al., 2013) and yet, an excess of these nutrients may be lost due to leaching and other processes (Romero-Trigueros et al., 2014; Haouas et al., 2021). However, due to its high salt concentration, RW may pose a concern to agriculture. As a result, poor irrigation management with RW can worsen secondary salinization and soil degradation over time, resulting in detrimental effects on crop physiology, growth, and quality in the long run, etc. (Romero-Trigueros et al., 2014).

Remote sensing techniques and enhanced agricultural management require a reliable assessment of plant water status in order to be effective. This is accomplished by taking measurements at the leaf scale and then upscaling the data to the field level. Measuring the spectral response of canopies is a quick and non-destructive way to detect stress in orchards (Jones & Vaughan, 2010). When compared to more time-consuming and arduous field procedures based on leaf sampling, remote sensing (RS) techniques have shown to be more beneficial and cost-effective (González-Dugo et al., 2013a).

Traditional RS methods have a number of flaws: satellite photography is sometimes hampered by cloud cover, and remote sensors mounted on towers inside agricultural fields are extremely expensive when data from several plots are required (Anderson & Gaston, 2013). However, due to technology advancements, cost reductions, and the miniaturization of sensors, the employment of unmanned aircraft vehicles (UAVs) has expanded in recent years. Farmers might use these UAVs to assess crop characteristics like water stress and then change their water management strategies as needed. As a result, by enhancing both the spatial and temporal resolution of data on the vegetative condition, UAVs technology can bridge the knowledge gap between the leaf and the canopy (Gago et al., 2015). However, the accuracy of aerial RS methods must be tested in the field by plant-truth data, such as measurements of plant water status (leaf water potential), gas exchange (net photosynthesis and stomatal conductance), chlorophyll concentration, and leaf shape (Tallou et al., 2021; Berni et al., 2009; Luque-Contreras et al., 2014; Gago et al., 2013; González-Dugo et al., 2013a; 2013b; Zarco-Tejada et al., 2012).

Imagery RS technologies are mostly based on the wavelength reflectance of canopies in the visible, such as red, green, and blue, and non-visible, such as near-infrared ranges of the spectrum (NIR). Visible, multispectral, and hyperspectral cameras are routinely used for remote monitoring of these particular reflectances (Baluja et al., 2012a, 2012b; Zarco-Tejada et al., 2012; Zarco-Tejada et al., 2013). Because of its link with leaf pigment composition, plant biophysical or structural factors, and physiological health, this reflectance can be utilized as a plant status indicator (Jones & Vaughan, 2010). Leaf chlorophyll absorbs red wavelengths (from 660 to 680 nm) in particular (Ollinger, 2011). Because saline environments damage the functioning and concentration of chlorophyll in the leaves, the reflectance of the leaves may be lowered correspondingly. The spectral response in the near-infrared (750–1400 nm) domain is mostly regulated by the internal structure of the leaf, such as mesophyll thickness and water content (Bonilla et al., 2015).

The Normalized Difference Vegetation Index (NDVI), which combines data from both domains, has demonstrated favorable relationships with water stress indicators (water potential and stomatal conductance) in a number of crops (Gago et al., 2015; Glenn et al., 2009). The majority of indicators employed for this purpose are connected to canopy structural changes over the course of a year or growth season, although techniques based on diurnal physiological changes over a single day are uncommon (González-Dugo et al., 2014).

In recent years, research has centered on comparing several vegetation indices obtained from UAVs equipped with multi-spectral cameras to field-collected measurements of plant-physiological and structural parameters (Berni et al., 2009; Luque-Contreras et al., 2014; Zarco-Tejada et al., 2013). Drought is a well-studied stressor (Baluja et al., 2012a, 2012b; Gago et al., 2015; Rodríguez-Pérez et al., 2007; Stagakis et al., 2012; Zarco-Tejada et al., 2012); however, research on saline stress from RW employing UAV technology still limited (Luque-Contreras et al., 2014). Furthermore, because of the expense and time necessary for lengthy durations, investigations that examine saline and/or water stress tolerances are rare (Romero-Trigueros et al., 2017).

6 Conclusion

This chapter discussed the importance of water for all creatures, especially for humankind. As it is obvious, the water demand is still rising nowadays due to the water, energy, and food challenges (WEF Nexus). For this reason, reusing reclaimed water in agriculture is considered one of the best alternatives to deal with water scarcity. In addition, using remote sensing technology for water management and precision agriculture is very important. It will help reduce water consumption, improve irrigation with reclaimed water, avoid contamination of crops due to reclaimed water pollutants, and detect water loss, crop damage, and diseases. Despite this, there are still some uncertainties in the irrigation evaluation to measure both needs and consumption using satellite remote sensing. In situ data on irrigation water delivered to fields is required to address this issue. Most crop loss research is now done with Unmanned Aerial Vehicles (UAVs), which have extremely high resolution. A framework for combining UAV information on a local scale with satellite sensors with a worldwide reach for large scale, real-time monitoring of agricultural losses is needed.

References

2019—IPCC. (2019). Retrieved March 31, 2022, from https://www.ipcc.ch/2019/

Al-Nakshabandi, G. A., Saqqar, M. M., Shatanawi, M. R., Fayyad, M., & Al-Horani, H. (1997). Some environmental problems associated with the use of treated wastewater for irrigation in Jordan. *Agricultural Water Management, 34*(1), 81–94. https://doi.org/10.1016/S0378-3774(96)01287-5

Amos, C. C., Rahman, A., & Gathenya, J. M. (2018). Economic analysis of rainwater harvesting systems comparing developing and developed countries: A case study of Australia and Kenya. *Journal of Cleaner Production, 172,* 196–207. https://doi.org/10.1016/j.jclepro.2017.10

Anderson, K., & Gaston, K. J. (2013). Lightweight unmanned aerial vehicles will revolutionize spatial ecology. *Frontiers in Ecology and the Environment, 11*(3), 138–146. https://doi.org/10.1890/120150

Angelakis, A. N., & Rose, J. B. (2014). *Evolution of Sanitation and Wastewater Technologies through the Centuries*. https://doi.org/10.2166/9781780404851.

Baluja, J., Diago, M. P., Balda, P., Zorer, R., Meggio, F., Morales, F., & Tardaguila, J. (2012b). Assessment of vineyard water status variability by thermal and multispectral imagery using an unmanned aerial vehicle (UAV). *Irrigation Science, 30*(6), 511–522. https://doi.org/10.1007/s00271-012-0382-9

Baluja, J., Diago, M. P., Goovaerts, P., & Tardaguila, J. (2012a). Assessment of the spatial variability of anthocyanins in grapes using a fluorescence sensor: Relationships with vine vigour and yield. *Precision Agriculture, 13*(4), 457–472. https://doi.org/10.1007/s11119-012-9261-x

Berni, J. A. J., Zarco-Tejada, P. J., Suarez, L., & Fereres, E. (2009). Thermal and narrowband multispectral remote sensing for vegetation monitoring from an unmanned aerial vehicle. *IEEE Transactions on Geoscience and Remote Sensing, 47*(3), 722–738. https://doi.org/10.1109/TGRS.2008.2010457

Bonilla, F. A., Khan, D. A., Ballas, Z. K., Chinen, J., Frank, M. M., Hsu, J. T., Keller, M., et al. (2015). Practice parameter for the diagnosis and management of primary immunodeficiency. *The Journal of Allergy and Clinical Immunology, 136*(5), 1186–1205. e1–78. https://doi.org/10.1016/j.jaci.2015.04.049.

Chalmers, T. C., Smith, H., Blackburn, B., Silverman, B., Schroeder, B., Reitman, D., & Ambroz, A. (1981). A method for assessing the quality of a randomized control trial. *Controlled Clinical Trials, 2* (1), 31–49. https://doi.org/10.1016/0197-2456(81)90056-8

Cramer, W., Guiot, J., Fader, M., Garrabou, J., Gattuso, J.-P., Iglesias, A., Lange, M. A., et al. (2018). Climate change and interconnected risks to sustainable development in the Mediterranean. *Nature Climate Change, 8*(11), 972–980. https://doi.org/10.1038/s41558-018-0299-2

Fonseca-Pedrero, E., Ortuño-Sierra, J., Sierro, G., Daniel, C., Cella, M., Preti, A., Mohr, C., & Mason, O. J. (2015). The Measurement invariance of schizotypy in Europe. *European Psychiatry: The Journal of the Association of European Psychiatrists, 30*(7), 837–844. https://doi.org/10.1016/j.eurpsy.2015.07.005

Gago, J., Douthe, C., Coopman, R. E., Gallego, P. P., Ribas-Carbo, M., Flexas, J., Escalona, J., & Medrano, H. (2015). UAVs challenge to assess water stress for sustainable agriculture. *Agricultural Water Management, 153*(May), 9–19. https://doi.org/10.1016/j.agwat.2015.01.020

Gago, E. J., Roldan, J., Pacheco-Torres, R., & Ordóñez, J. (2013). The city and urban heat islands: A review of strategies to mitigate adverse effects. *Renewable and Sustainable Energy Reviews, 25* (September), 749–758. https://doi.org/10.1016/j.rser.2013.05.057

García-Galiano, S., Giménez, P., & Giraldo-Osorio, J. D. (2015). Assessing nonstationary spatial patterns of extreme droughts from long-term high-resolution observational dataset on a semiarid basin (Spain). *Water, 7*https://doi.org/10.3390/w7105458.

Gebbers, R., & Adamchuk, V. I. (2010). Precision agriculture and food security. *Science (New York, N.Y.), 327*(5967), 828–31. https://doi.org/10.1126/science.1183899.

Glenn, A. L., Raine, A., & Schug, R. A. (2009). The neural correlates of moral decision-making in psychopathy. *Molecular Psychiatry, 14* (1), 5–6. https://doi.org/10.1038/mp.2008.104

Gonzalez-Dugo, V., Zarco-Tejada, P. J., & Fereres, E. (2014). Applicability and limitations of using the crop water stress index as an indicator of water deficits in citrus orchards. *Agricultural and Forest Meteorology, 198–199*(November), 94–104. https://doi.org/10.1016/j.agrformet.2014.08.003

Gonzalez-Dugo, V., Zarco-Tejada, P., Nicolás, E., Nortes, P. A., Alarcón, J. J., Intrigliolo, D. S., & Fereres, E. (2013). Using high resolution UAV thermal imagery to assess the variability in the water status of five fruit tree species within a commercial orchard. *Precision Agriculture, 14*(6), 660–678https://doi.org/10.1007/s11119-013-9322-9.

González-Dugo, M., Escuin, S., Cano, F., Cifuentes, V., Padilla, F. L. M., Tirado, J. L., Nicolás Oyonarte, P. F., & Mateos, L. (2013a). Monitoring evapotranspiration of irrigated crops using crop coefficients derived from time series of satellite images. II. Application on basin scale. *Agricultural Water Management, 125*(July), 92–104. https://doi.org/10.1016/j.agwat.2013.03.024.

Haouas, A., El Modafar, C., Douira, A., Ibnsouda-Koraichi, S., Filali-Maltouf, A., Moukhli, A., & Amir, S. (2021, September). Phosphate sludge: Opportunities for use as a fertilizer in deficient soils. *Detritus*. https://doi.org/10.31025/2611-4135/2021.15112.

Hussain, H. A., Men, S., Hussain, S., Chen, Y., Ali, S., Zhang, S., Zhang, K., et al. (2019). Interactive effects of drought and heat stresses on morpho-physiological attributes, yield, nutrient uptake and oxidative status in maize hybrids. *Scientific Reports, 9*(1), 3890. https://doi.org/10.1038/s41598-019-40362-7

Jiménez, B., & Asano, T. (2008). Water reuse: An international survey of current practice. *Issues and Needs*. https://doi.org/10.2166/9781780401881

Jiménez-Cisneros, B. (1995). Wastewater reuse to increase soil productivity. *Water Science and Technology, 32*, 173–180. https://doi.org/10.1016/0273-1223(96)00152-7

Jones, H. G., & Vaughan, R. A. (2010). *Remote sensing of vegetation*. Oxford University Press.

Karthikeyan, L., Chawla, I., & Mishra, A. K. (2020). A review of remote sensing applications in agriculture for food security: Crop growth and yield, irrigation, and crop losses. *Journal of Hydrology, 586*(July), 124905. https://doi.org/10.1016/j.jhydrol.2020.124905

Levy, S., et al. (2011). The competitive advantage of a dual-transporter system. *Science, 334*(6061), 1408–1412.

Luque-Contreras, D., Carvajal, K., Toral-Rios, D., Franco-Bocanegra, D., & Campos-Peña, V. (2014). Oxidative stress and metabolic syndrome: Cause or consequence of Alzheimer's disease? *Oxidative Medicine and Cellular Longevity, 2014*, 497802. https://doi.org/10.1155/2014/497802

Mastrorilli, P., Dell'Anna, M. M., Rizzuti, A., Mali, M., Zapparoli, M., & Leonelli, C. (2015). Resin-immobilized palladium nanoparticle catalysts for organic reactions in aqueous media: Morphological aspects. *Molecules (Basel, Switzerland), 20*(10), 18661–18684. https://doi.org/10.3390/molecules201018661

Maykot, J. K., & Ghisi, E. (2020). Assessment of a rainwater harvesting system in a multi-storey residential building in Brazil. *Water, 12*(5), 1482. https://doi.org/10.3390/w12051482

McBride, W., & Daberkow, S. G. (2003). Information and the adoption of precision farming technologies. *Journal of Agribusiness, 21*(1). https://econpapers.repec.org/article/agsjloagb/14671.htm.

Metcalf & Eddy, Inc an AECOM Company, Asano, T., Burton, F., & Leverenz, H. (2007). *Water reuse: Issues, technologies, and applications*. McGraw-Hill Education. https://www.accessengineeringlibrary.com/content/book/9780071459273.

Ollinger, S. V. (2011). Sources of variability in canopy reflectance and the convergent properties of plants. *New Phytologist, 189*(2), 375–394. https://doi.org/10.1111/j.1469-8137.2010.03536.x

Panasiuk, O., Hedström, A., Marsalek, J., Ashley, R. M., Viklander, M. (2015). Contamination of stormwater by wastewater: A review of detection methods. *Journal of Environmental Management, 152*, 241–250. [CrossRef] [PubMed].

Paranychianakis, N. V., Nikolantonakis, M., Spanakis, Y., & Angelakis, A. N. (2006). The effect of recycled water on the nutrient status of Soultanina grapevines grafted on different rootstocks.

Agricultural Water Management, 81(1), 185–198. https://doi.org/10.1016/j.agwat.2005.04.013

Pedrero, F., Kalavrouziotis, I., Alarcón, J. J., Koukoulakis, P., & Asano, T. (2010). Use of treated municipal wastewater in irrigated agriculture—review of some practices in Spain and Greece. *Agricultural Water Management, 97*(9), 1233–1241. https://doi.org/10.1016/j.agwat.2010.03.003

Pedrero, F., Alarcón, J. J., Nicolás, E., & Mounzer, O. (2013). Influence of irrigation with saline reclaimed water on young grapefruits. *Desalination and Water Treatment 51*(10–12), 2488–2496. https://doi.org/10.1080/19443994.2012.747654

Pedrero Salcedo, F., Alarcón Cabañero, J. J., & Pérez Cutillas, P. (2021). *Assessing soil salinity using remote sensing in Campo de Cartagena.* EGU General Assembly 2021, online, 19–30 Apr 2021, EGU21-14478. https://doi.org/10.5194/egusphere-egu21-14478

Pereira, T. G., & Oliva-Teles, A. (2002). Preliminary evaluation of pea seed meal in diets for gilthead sea bream (Sparus aurata) juveniles. *Aquaculture Research, 33*(14), 1183–1189.

Pescod, M. B. (Warren). (1992). The urban water cycle, including wastewater use in agriculture. *Outlook on Agriculture, 21*(4), 263–70.https://doi.org/10.1177/003072709202100404.

Publications of the U.S. Geological Survey, 1984. (1984). USGS Unnumbered Series. *Publications of the U.S. Geological Survey, 1984*. Publications of the US Geological Survey. U.S. Government Printing Office. https://doi.org/10.3133/70043714.

Rhoades, J. D., Kandiah, A., & Mashali, A. M. (1992). The use of saline waters for crop production. FAO Irrigation and Drainage, FAO, Rome (p. 48).

Rodríguez-Pérez, J. R., Riaño, D., Carlisle, E., Ustin, S., & Smart, D. R. (2007). Evaluation of hyperspectral reflectance indexes to detect grapevine water status in vineyards. *American Journal of Enology and Viticulture, 58*(3), 302–317.

Romero-Trigueros, C., Vivaldi, G. A., Nicolás, E. N., Paduano, A., Salcedo, F. P., & Camposeo, S. (2019). Ripening indices, olive yield and oil quality in response to irrigation with saline reclaimed water and deficit strategies. *Frontiers in Plant Science, 10*, 1243. https://doi.org/10.3389/fpls.2019.01243

Romero-Trigueros, C., Nortes, P. A., Alarcón, J. J., Hunink, J. E., Parra, M., Contreras, S., Droogers, P., & Nicolás, E. (2017). Effects of saline reclaimed waters and deficit irrigation on citrus physiology assessed by UAV remote sensing. *Agricultural Water Management, Special Issue: Advances on ICTs for Water Management in Agriculture, 183* (March), 60–69https://doi.org/10.1016/j.agwat.2016.09.014.

Romero Trigueros, C., Nortes Tortosa, P. A., Alarcón Cabañero, J. J., Nicolás Nicolás, E. (2014, October). *Determination of 15N stable isotope natural abundances for assessing the use of saline reclaimed water in grapefruit.* 10/Full/12_668_Romero-Trigueros_14.pdf.

Salcedo, F. P., Cutillas, P. P., Cabañero, J. J. A., & Vivaldi, A. G. (2022). Use of remote sensing to evaluate the effects of environmental factors on soil salinity in a semi-arid area. *Science of the Total Environment, 815*(April), 152524. https://doi.org/10.1016/j.scitotenv.2021.152524

Sancho Martínez, J., Fernández, Y. B., Leinster, P., & Casado, M. R. (2020). Combining unmanned aircraft systems and image processing for wastewater treatment plant asset inspection. *Remote Sensing, 12*(9), 1461https://doi.org/10.3390/rs12091461.

Secretary-General, U., & World Commission on Environment and Development. (1987, August). *Report of the world commission on environment and development*. https://digitallibrary.un.org/record/139811.

Stagakis, S., Gonzalez-Dugo, V., Cid, P., GuillenCliment, M. L., & Zarco-Tejada, P. J. (2012). Monitoring water stress and fruit quality in an orange orchard under regulated deficit irrigation using narrow-band structural and physiological remote sensing. *Journal of Photogrammetry and Remote Sensing, 71*, 47–61. https://doi.org/10.1016/j.isprsjprs.2012.05.003

Stocker, T. F., Qin, D., Plattner, G.-K., Alexander, L. V., Allen, S. K., Bindoff, N. L., Bréon, F.-M., Church, J. A., Cubasch, U., Emori, S., Forster, P., Friedlingstein, P., Gillett, N., Gregory, J. M., Hartmann, D. L., Jansen, E., Kirtman, B., Knutti, R., Krishna Kumar, K., Lemke, P., Marotzke, J., Masson-Delmotte, V., Meehl, G. A., Mokhov, I. I., Piao, S., Ramaswamy, V., Randall, D., Rhein, M., Rojas, M., Sabine, C., Shindell, D., Talley, L. D., Vaughan, D. G., & Xie, S.-P. (2013). Technical summary. In T. F. Stocker, D. Qin, G.-K. Plattner, M. Tignor, S. K. Allen, J. Doschung, A. Nauels, Y. Xia, V. Bex, & P. M. Midgley (Eds.), *Climate change 2013: The physical science basis. Contribution of working group I to the fifth assessment report of the intergovernmental panel on climate change* (pp. 33–115). Cambridge University Press. https://doi.org/10.1017/CBO9781107415324.005.

Suzana, D., Iulia, N. (2012). Efficient management of small and medium sized enterprises for a sustainable development of rural entrepreneurship. *Lucrări Ştiinţifice Management Agricol, 15*, 205–211.

TWI2050 Report: Transformations to Achieve the Sustainable Development Goals-Report 2018-IIASA. (2018). Retrieved March 31, 2022, from https://previous.iiasa.ac.at/web/home/research/twi/Report2018.html.

Tallou, A., Salcedo, F. P., Haouas, A., Jamali, M. Y., Atif, K., Aziz, F., & Amir, S. (2020). Assessment of biogas and biofertilizer produced from anaerobic co-digestion of olive mill wastewater with municipal wastewater and cow dung. *Environmental Technology & Innovation, 20*(November), 101152. https://doi.org/10.1016/j.eti.2020.101152

Tallou, A., Aziz, F., Garcia, A. J., Salcedo, F. P., El Minaoui, F. E., Amir, S. (2021, April). Bio-fertilizers issued from anaerobic digestion for growing tomatoes under irrigation by treated wastewater: Targeting circular economy concept. *International Journal of Environmental Science and Technology*.https://doi.org/10.1007/s13762-021-03265-7.

Tapsuwan, S., et al. (2014). Adapting to less water: Household willingness to pay for decentralised water systems in urban Australia. *Water Resources Management, 28*, 1111–1125https://doi.org/10.1007/s11269-014-0543-0.

Tarchitzky, J., Banin, A., Morin, J., & Chen, Y. (1984). Nature, formation and effects of soil crusts formed by water drop impact. *Geoderma, 33*, 135–155. https://doi.org/10.1016/0016-7061(84)90025-9

The Global Risks Report 2021. (2021). *Global Risks Report 2021* (blog). Retrieved March 31, 2022, from https://wef.ch/3c51TbB.

Tokarczyk, P., Leitao, J. P., Rieckermann, J., Schindler, K., Blumensaat, F. (2015). High-quality observation of surface imperviousness for urban runoff modelling using UAV imagery. *Hydrology and Earth System Sciences, 19*, 4215–4228.

UN World Water Development Report 2020. (2020). *UN-Water* (blog). Retrieved March 31, 2022, from https://www.unwater.org/publications/world-water-development-report-2020/.

United States Environmental Protection Agency. (2004). NPDES compliance inspection manual. IEEE.

Vinten, A. J. A., Frenkel, H., Shalhevet, J., & Elston, D. A. (1991). Calibration and validation of a modified steady-state model of crop response to saline water irrigation under conditions of transient root zone salinity. *Journal of Contaminant Hydrology, Validation of Flow and Transport Models for the Unsaturated Zone, 7*(1), 123–144. https://doi.org/10.1016/0169-7722(91)90041-X

World Bank. (2017). *The World Bank annual report 2017*. Annual report. World Bank. https://doi.org/10.1596/978-1-4648-1119-7.

World Economic Forum Annual Meeting Davos 2020. (2019). World Economic Forum. Retrieved March 31, 2022, from https://www.weforum.org/events/world-economic-forum-annual-meeting-2020/.

WWAP (United Nations World Water Assessment Programme). (2017). *The United Nations World Water Development Report 2017*. Wastewater: The Untapped Resource. Paris, UNESCO.

Zarco-Tejada, P. J., González-Dugo, V., & Berni, J. A. J. (2012). Fluorescence, temperature and narrow-band indices acquired from a UAV platform for water stress detection using a micro-hyperspectral imager and a thermal camera. *Remote Sensing of Environment, Remote Sensing of Urban Environments, 117* (February), 322–337. https://doi.org/10.1016/j.rse.2011.10.007

Zarco-Tejada, P. J., González-Dugo, V., Williams, L. E., Suárez, L., Berni, J. A. J., Goldhamer, D., & Fereres, E. (2013). A PRI-based water stress index combining structural and chlorophyll effects: Assessment using diurnal narrow-band airborne imagery and the CWSI thermal index. *Remote Sensing of Environment, 138*(November), 38–50. https://doi.org/10.1016/j.rse.2013.07.024

Nutrients Recovery in the Water and Wastewater Sector

Revanth Mamidala, Kiruthika Eswari Velmaiel, Santhoshi Chitthaluri, Vineeth Manthapuri, Kasadi Naveen, and P. RajaSekhar

Abstract

The concept of circular economy applied to wastewater treatment sectors is a step toward sustainability. Considering the limitation on the availability of raw materials and resources for fertilizer manufacturing, nutrient recovery from wastewater is self-reliant from an economic and environmental perspective. In addition to that, eutrophication of water bodies is avoided by reducing the pollutant load on water bodies. This review described the methods and mechanisms of physicochemical, biological, and hybrid technologies for nutrient recovery from wastewater. The contemporary technologies have advantages over conventional methods due to the selective separation of nutrients in complex wastewater matrices containing other organic and metal ions. Bio-electrochemical systems (BES) and osmotic membrane bioreactors (OMBR) are very efficient and economic due to their low energy consumption. The economic dissection emphasizes the need for commercialization and quality checks on the recovered products. The efficiency of the nutrient recovery system and the quality of recovered products varies with the feedstock. The addition of a pre-treatment unit in the treatment or recovery train may enhance the efficiency of the nutrient recovery system.

Keywords

Nutrient recovery • Membrane-based technologies • Wastewater • Circular economy

1 Introduction

The rapid population growth and intensive agricultural practices have increased the demand for fertilizer. The increase in demand may increase the import of fertilizers which further burdens the economy within the country. It is reported that 30 percent of nitrogen and 16 percent of phosphorous in fertilizers after their application end up in wastewater streams (Xie et al., 2016). Wastewater could be a valuable resource for nutrient recovery.

The uncontrolled wastewater discharges and the incomplete removal of nutrients (ammonium, nitrogen, and phosphate) have elevated levels of nutrients in the waterbodies. This results in the eutrophication of aquatic systems which exerts a detrimental effect on the natural environment. The removal of nutrients from wastewater by a conventional process like nitrification–denitrification, and chemical precipitation requires additional energy and cost. The removal of nitrogen from wastewater and its release into the atmosphere as gaseous nitrogen requires 45 MJ energy per kg of nitrogen in wastewater. This process also contributes to greenhouse gas emissions. The conventional method of fertilizer production emits 0.1–25 kg CO_{2eq} per Kg of fertilizers (Gowd et al., 2022). A sustainable solution to address the pollutant load on waterbodies and fertilizers demand is to recover and recycle nutrients from wastewater. The nutrients recovered from wastewater could be alternate fertilizers and reduces the demand for fossil-based fertilizers.

R. Mamidala (✉) · P. RajaSekhar (✉)
Water Resources Engineering, Department of Civil Engineering, Osmania University, Hyderabad, India
e-mail: revanth.12@outlook.com

P. RajaSekhar
e-mail: prscivilou@osmania.ac.in

K. E. Velmaiel · V. Manthapuri (✉)
Environmental and Water Resource Engineering Division, Department of Civil Engineering, IIT Madras, Chennai, India
e-mail: ce18s006@smail.iitm.ac.in

S. Chitthaluri
Environmental Engineering, Department of Civil Engineering, NIT Warangal, Hyderabad, India

K. Naveen
Water Resources Engineering, Department of Civil Engineering, KU Leuven, Leuven, Belgium

M. Smol et al. (eds.), *Water in Circular Economy*, Advances in Science, Technology & Innovation,
https://doi.org/10.1007/978-3-031-18165-8_11

Circular economy in wastewater treatment promotes the concept of reduction in waste generation and employment of wastewater as a secondary source for energy and value-added products recovery without the involvement of toxic chemicals and excessive energy. The main driver for the circular economy concept is the limited availability of resources. The linear economic model considers the treatment of waste only when it poses a potential threat to humans and the environment (Robles et al., 2020). The conventional treatment technologies aim to reduce the chemical and biological oxygen demand to permissible limits. The circular economy model makes the nutrient recovery from wastewater, a closed-loop as it involves the manufacture of fertilizers, usage in agriculture, nutrients end up in wastewater, wastewater collection, treatment and recovery, and use of recovered nutrients in agriculture. The circular economy model employed in wastewater treatment aims to maximize the benefits of resource recycling which paved the way for the development of new nutrient recovery technologies. Thus, this study aims to review the existing conventional technologies and recently developed physico-chemical, biological, and hybrid technologies for nutrient recovery from wastewater and their economic feasibility.

The objective of the study is to focus on the advanced and effective methodology adopted in contemporary technologies in extracting nutrients from wastewater and also promote the widespread usage of recovered nutrients in fertilizers etc.

2 Methods and Mechanism

Nutrient Extraction techniques are broadly classified into basic conventional technologies and Contemporary technologies as shown in Fig. 1.

a. **Conventional Technologies**

The conventional technologies basically work on the process involving chemical and biological actions.

i. Chemical Action

Nutrient recovery from wastewater by chemical action mainly includes reactions such as chemical precipitation and chemical adsorption (Safie & Zahrim, 2021). In the chemical precipitation process, the compounds of calcium and magnesium are commonly used to react with nutrients present in wastewater to form precipitates such as Hydroxyapatite ($Ca_5(OH)(PO_4)_3$) and Struvite ($MgNH_4PO_4.6H_2O$). The reactions involved are shown in Eqs. 1 and 2 (Liu & Wang, 2019).

$$Mg^{2+} + PO_4{}^{3-} + PH_4{}^{-} + 6H_2O_2 \rightarrow MgNH_4PO_4.6H_2O \downarrow \quad (1)$$

$$5Ca^{2+} + 3PO_4{}^{2-} + OH^{-} \rightarrow Ca_5(OH)(PO_4)_3 \downarrow \quad (2)$$

Due to the presence of rich amounts of nitrogen and phosphorous in struvite, it could be used as an agricultural fertilizer; whereas, hydroxyapatite is a primary raw material for phosphorous manufacturing.

(a) Chemical Precipitation

The chemical precipitation process is influenced by various factors such as temperature, the concentration of compounds, and foreign substances. The ideal pH range for the chemical precipitation is reported to be 7–11 (Kabdasli et al., 2022). The variation in the concentration affects the profiles

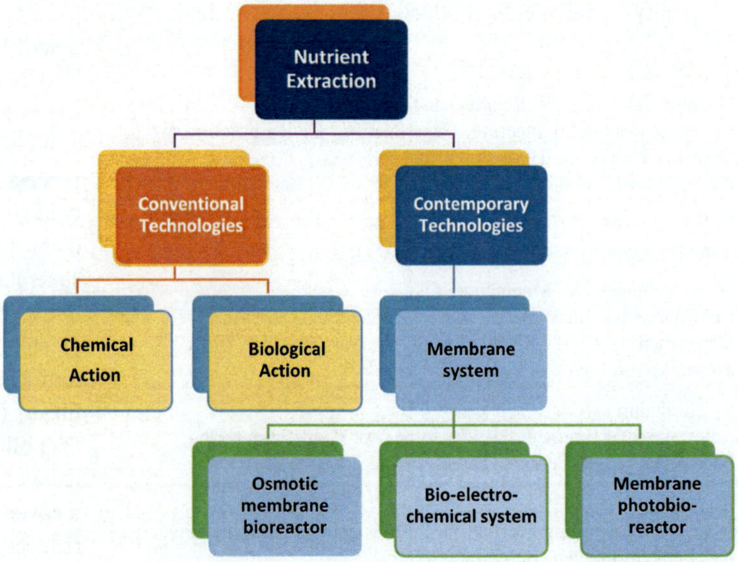

Fig. 1 Flowchart of Nutrient Extraction Technologies (Kamilya et al., 2022)

of ammonium and phosphate ions, with the formation of unexpected metal-based precipitates at higher pH and the precipitation rate is dampened at low pH values. At the same time, even with lower pH values, the increase in temperature improved the amount of formation of precipitates. However, for the formation of struvite precipitate, low temperatures (less than 15 °C) are found to be beneficial (Krishnamoorthy et al., 2021). The presence of Calcium ions notably has detrimental effects on struvite precipitation. Therefore, the optimum pH and the dosages of Mg or Ca in the chemical precipitation process depend mostly on the composition of the wastewater. Some organics like pharmaceuticals and hormones were also detected in the recovered precipitates, which could affect the environment and health of the consumer if utilized as a fertilizer (Enyemadze et al., 2021).

(b) Chemical Adsorption

Chemical adsorption, is another reliable process for the extraction of nutrients, due to the simplicity of its design and operation procedure, which is cost-effective and highly stable. In the recovery of phosphate, metal-based adsorbents are employed due to their ease of access and efficiency. Phosphate adsorption can be achieved by adopting the mechanisms such as electrostatic attraction, ion exchange, and surface precipitation, which can also be used parallel to the chemical precipitation process by employing Mg^{-} and ca^{-} based adsorbents. The metal- and biochar-based adsorbents are dominantly used in the adsorption process. The chemical reactions of Fe^{2+}/Fe^{3+} in the extraction of phosphorous from wastewater are as shown in Eqs. 3 and 4 (Zeng et al., 2021).

$$Fe^{3+} + PO_4^{3-} \rightarrow FePO_4 \downarrow \quad (3)$$

$$3Fe^{2+} + 2PO_4^{3-} \rightarrow Fe_3(PO_4)_2 \downarrow \quad (4)$$

Similar to chemical precipitation, adsorption also depends on various factors such as pH value, temperature, the process of desorption, and other ions in the wastewater. The pH value heavily affects the phosphate adsorption, higher values of which enhance the adsorption process, in contrast, a lower pH would inhibit the repulsion of phosphate from negatively charged adsorbents (Shyam et al., 2022).

Phosphate adsorption being an endothermic process, the rise of temperature facilitates the adsorption mechanism. The presence of anions in domestic and industrial wastewater is very common. The anions such as CO_3^{2-}, NO_3^-, F^-, OH^-, and SO_4^{2-} would compete for the adsorption sites along with phosphate ions. Their effect majorly depends on the mechanism of adsorption adopted and the properties of the adsorbent (Shyam et al., 2022).

The chemical adsorption process is followed by the desorption of the nutrients, which results in a product, which is rich in nutrients. The main desorption methods used are solvent washing and calcination. The desorption mechanism is determined by the property of the adsorbent and also the adsorption mechanism. Unlike phosphate ions, ammonium ions are adsorbed by acid solutions like sulphuric acid. Before adsorption, the ammonium ions are converted into volatile ammonia at high temperatures and pH. After adsorption, the volatile ammonia is converted into ammonium salts such as ammonium sulphate ($(NH_4)_2.2SO_4$) (Glover et al., 2012) which is commonly used as a soil fertilizer since it contains nitrogen at 21% (w/w) and sulphur at 24% (w/w) (Speight, 2017).

Overall, a comprehensive study and thorough understanding of the behavior of compounds in both the processes of chemical precipitation and adsorption along with desorption is necessary to develop efficient dosages of feed provided to the system.

ii. Biological Action

Typical biomass in the activated sludge process will contain a nitrogen content of 3–10% and phosphorous content of 1.5–2.5% based on the dry weight. The biological Action of nutrient recovery can be done by various methods as shown in Fig. 2.

Phosphorous recovery in the AO process can be divided into two phases; the anaerobic phase followed by the aerobic phase. Phosphorous Accumulating Organisms (PAO) have a crucial role in the anaerobic phase where cells release the phosphate ions into the solution which results in the accumulation of phosphate in wastewater. During this process, other metal ions such as Mg^{2+} and K^+ could also be released into the wastewater along with some energy. The energy thus released in this process is used by carbon compounds (volatile fatty acids) and then it is stored as PHAs (Poly-β-hydroxy alkanoates), which are utilized by the phosphate when transferred to the aerobic phase. Acid and alkaline solutions are mixed with the sludge to transport the phosphate ions from the sludge phase to the liquid phase. The type of acid and alkaline solution depends on the properties and components of sludge. To increase the quality of the phosphate, chloride additives such as $MgCl_2$ and $CaCl_2$ are added at high temperatures. Thus, the phosphate in the solution is easily available for the plants as a fertilizer (Donatello & Cheeseman, 2013).

Recovery of Nitrogen from activated sludge typically includes oxidation and nitrification in favourable conditions. Two such methods involving this mechanism are suspended growth (bacteria kept in suspension) and the Fixed film

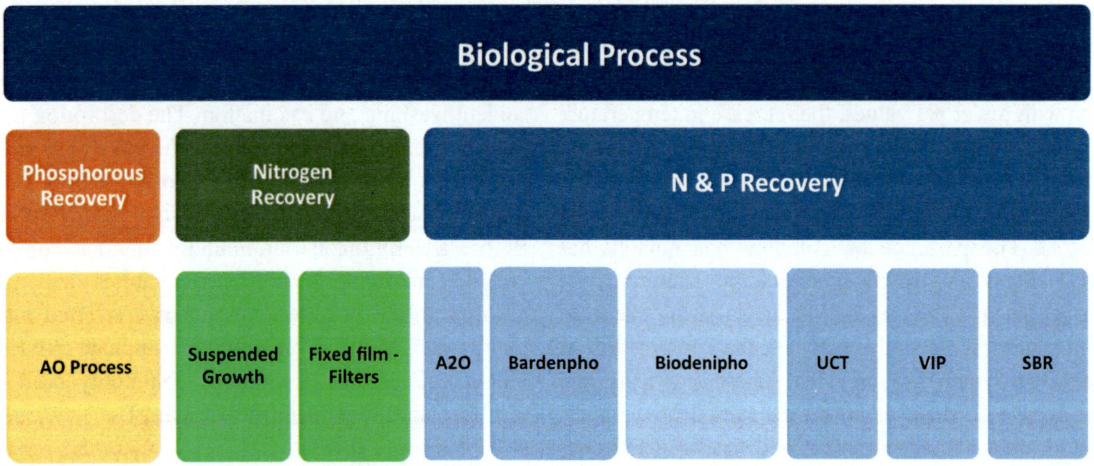

Fig. 2 Overview of nutrient removal by biological processes. (*Source* Department of the Army, 2001, *Public Works Technical Bulletin 420–49-39)*

process (nitrification in trickling filters). To recover both nitrogen and phosphorous from wastewater, nitrification and denitrification are followed by phosphorous removal by PAO. Because, in the anaerobic phase of phosphate removal, it is crucial to avoid dissolved oxygen and nitrogen to obtain higher efficiency.

b. **Contemporary Technologies**

The contemporary nutrient recovery methods are affected by many factors, one of the main factors indeed is the presence of foreign materials and other ions in the wastewater that compete with the nutrients that are to be extracted. These coexisting substances and other toxic materials have detrimental effects on the quality of the recovered products. The membrane system is one of the effective processes in the separation of nutrients from wastewater, despite the presence of foreign substances and other metal ions.

(a) **Membrane system**

Based on the type of membrane and the methodology adopted, forward osmosis (FO), membrane distillation (MD) and Electrodialysis (ED) are the three major technologies employed for the extraction of nutrients. Figure 3 illustrates the schematic diagrams of nutrient recovery by a membrane system. In the forward osmosis technique, a semi-permeable membrane is placed between the feed solution and a draw solution. Between both the solutions, an osmotic pressure gradient is applied that forces the transfer of water from the feed solution to the draw solution. The basic principle in the FO method is that the membrane adopted in the FO process blocks the flow of nutrients from the feed side to the draw side, thus increasing the concentration of nutrients on the feed side (Nguyen et al., 2018). The nutrient-enriched feed solution is then used to extract nutrients. It was found that, in comparison to ammonium, higher amounts of phosphate can be trapped in the feed solution, due to the large radius of hydrated ammonium ions (Ray et al., 2020). Moreover, the phosphate ions are repelled from the membrane surface due to its negatively charged surface, thus facilitating the accumulation of phosphate in the feed solution. In contrast, this harms the concentration of ammonium ions on the feed side. For enrichment of both phosphate and ammonium, a neutral pH value is favourable since ammonium converts into volatile ammonia at higher pH values.

In the membrane distillation process, volatile substances are generated by gently heating the feed solution which is then transferred to the draw solution through a membrane. Thus, the ammonium in the feed solution is converted into volatile ammonia, which is then received on the draw side by the acid solutions. The ammonia gas reacts with acids like sulphuric acid (H_2SO_4) and forms into ammonium salts (Gao et al., 2020). This process is majorly affected by the temperature and pH levels of the solution. In the electrodialysis process, two membranes namely cation exchange membrane (CEM) and anion exchange membrane (AEM) are adopted to separate phosphate and ammonium ions from the feed solution by applying a current field. This increases the concentration of ammonium ions in the anode chamber and the phosphate ions are accumulated in the cathode chamber, which forms the solutions nutrient-rich and sent for further recovery (Kedwell et al., 2021).

(b) **Osmotic Membrane Bio-Reactor**

Osmotic Membrane Bio-Reactor (OMBR) is based on the principle of Forward Osmosis (FO) technology. In OMBR, the FO process and aerobic biological action are integrated,

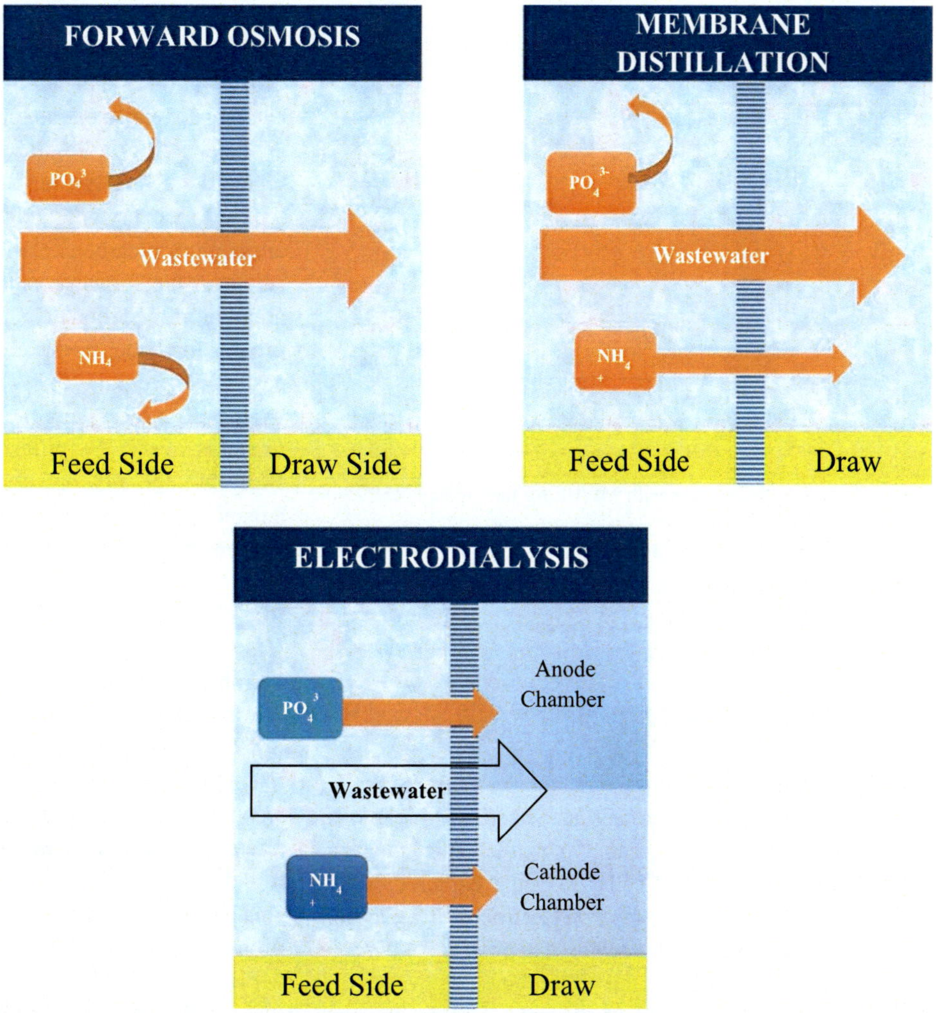

Fig. 3 Schematic diagram of various membrane technologies to enhance nutrient concentration (Nguyen et al., 2018; Reig et al., 2021; Gao et al., 2020; Kedwell et al., 2021)

which enables the direct recovery of nutrients from wastewater. OMBR process can achieve the phosphate and ammonium concentration of about >95% making the feed solution is enriched with nutrients and they are extracted in the form of struvite ($MgNH_4PO_4.6H_2O$) or calcium phosphate ($Ca_3(PO_4)_2$). The most important fact is that the process does not involve the addition of any chemicals for nutrient recovery, except for the addition of alkaline chemicals to raise the pH levels. The use of $MgCl_2$ as the solute on the draw side supplements the Mg^{2+} ions for nutrient recovery by the process of chemical precipitation. This condition is known to be reverse draw flux which is a property of the FO process. Further, NH_4CO_3 can also be used as a draw solute as it can be recovered by applying moderate heat (Nguyen et al., 2018).

The purity of the extracted nutrients can be enhanced by adopting the membranes of microfiltration (MF) or ultrafiltration (UF) which are to be installed on the feed side in addition to the FO membrane as shown in Fig. 4. The additional membrane (MF/UF) extract the nutrients from the concentrated solution from the FO membrane resulting in fewer foreign substance passing through them. This ensures the potential of nutrient extraction which is economically feasible. To make the OMBR system more cost-effective, an MD or RO membrane can be employed to extract the water from the draw solution thus making it recyclable (Yang et al., 2021). The setback of this method can be the fouling of the FO membrane. This can lead to increased internal concentration polarization which is directly proportional to the nutrients permeating through the membrane into the draw solution (Nguyen et al., 2018). Installing a fixed bed biofilm on the feed side of OMBR can decrease the suspended growth and also decrease the fouling of the FO membrane (Li et al., 2018).

The efficiency of the OMBR hybrid system depends on the properties of the membrane, composition of draw solution, hydraulic retention time, and the concentration of the wastewater. Higher pH values form unexpected precipitates. Flat sheets of cellulose tri-acetate membranes are found to have a higher rate of repulsion of ammonia ions when compared to the membranes made of thin-film composites (Tange et al., 2017). The lower retention time of wastewater

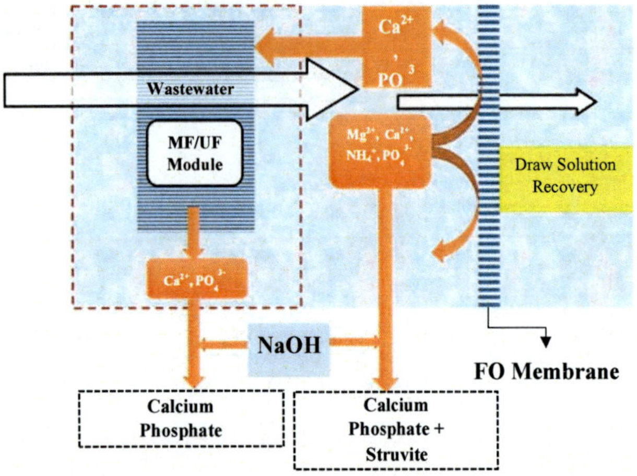

Fig. 4 OMBR hybrid system with MF/UF for the nutrients extraction (Ye et al., 2021)

increases the risk of accumulation of salts and fouling of the membrane, thus resulting in higher operating costs (Pourabdollah et al., 2014).

(c) **Bio-electro-chemical System**

Bio-electrochemical systems work on electricity generated by the metabolism of microbes and electrochemical reactions. Over the past few years, to extract nutrients from wastewater, Microbial fuel cells are being explored, which constate an original type of Bio-electrochemical system (BES). Further, the studies were extended to Microbial Electrolysis cells (MEC) and Microbial Recovery Cell (MRC) (Zhang et al., 2019). A typical microbial fuel cell consists of an anode chamber and a cathode chamber separated by a cation exchange membrane. The electrons and protons are generated in the anode chamber, in which, the electrons move towards electron acceptors to complete the electrical loop.

The chemical reactions of a conventional microbial fuel cell (carbon source is taken as glucose) are shown in Eqs. 5 and 6 (Logan et al., 2006).

$$C_6H_{12}O_6 + 6H_2O \rightarrow CO_2\uparrow + 24H^+ + 24e^- \quad (5)$$

$$2H_2O + O_2 + 4e^- \rightarrow 4OH^- \quad (6)$$

As shown in Fig. 5, the ammonium ions can be transferred to the cathode chamber from the anode chamber through the cation exchange membrane by diffusion induced by the concentration gradient and electric current driven migration. This results in the increased concentration of ammonium ions in the cathode chamber. This ammonium thus accumulated is transformed into volatile ammonia due to the cathode reaction, which in turn, increases the pH around the cathode region (Zhang et al., 2022). The volatile ammonia is recovered as ammonium salts by air stripping the cathode chamber and introducing the ammonia onto the acid solution. Whereas, the chemical precipitation method can be adopted to extract phosphate as the pH in the cathode chamber is high (Goglio et al., 2022). The phosphate precipitates are mostly found to be on the cathode electrode.

The recovery of volatile ammonia in the form of ammonium bicarbonate will facilitate its usage as the draw solute in the forward osmosis process. On the other hand, ammonium bicarbonate is a promising draw solute because, on the moderate application of heat, it can be recovered easily (Feng et al., 2018). Thus, BES can also be supported by the FO process for more efficient recovery of nutrients (Yuan et al., 2015) by feeding the treated wastewater from BES into the FO process. Their integration can help in enhancing the economic and technical feasibility of the nutrient extraction system.

(d) **Membrane Photobioreactor**

Membrane Photobioreactor (MPBR) is a combination of Photobioreactor (PBR) integrated with membrane technology, to decouple the solid retention time and hydraulic retention time as illustrated in Fig. 6. In the last decade, PBR technology is being explored with the addition of the microalgae process (Keyvan Hosseini et al., 2020). MPBR is found to have higher nutrient recovery efficiency and a smaller footprint than a PBR, by enhancing the accumulation of biomass in the PBR. Further, the operational costs of MPBRs (US \$0.113 per m^3) are very less when compared to that of PBRs (US \$0.65–0.96 per m^3) (Sheng et al., 2017). In the MPBR, organics are converted into hydrogen by the process of photo-digestion which utilizes energy from light sources (e.g., Solar Energy) (Peter et al., 2021). The light energy is stored by the microalgae into which the carbon dioxide and nutrients are incorporated. The resultant nutrient-rich microalgae can be converted into products such as fertilizer, feed for animals, and even biogas plants.

Therefore, the membrane system consists of only physical separation and does not require any biological action. Hence, it is more suitable to use these membrane systems for the extraction of nutrients in countries having insufficient sources of energy and decentralized wastewater treatment systems or in places without a centralized wastewater treatment system. They can also be adopted in regions where the biological action cannot be conducted due to extreme climatic conditions (Peter et al., 2021). For the places where the wastewater treatment plants are located near the sea, the forward osmosis process is highly recommended as it is economical to use, the readily available seawater as the draw solution thus, reducing the overall costs of providing Mg^{2+} ions. On the other hand, the membrane distillation filtration

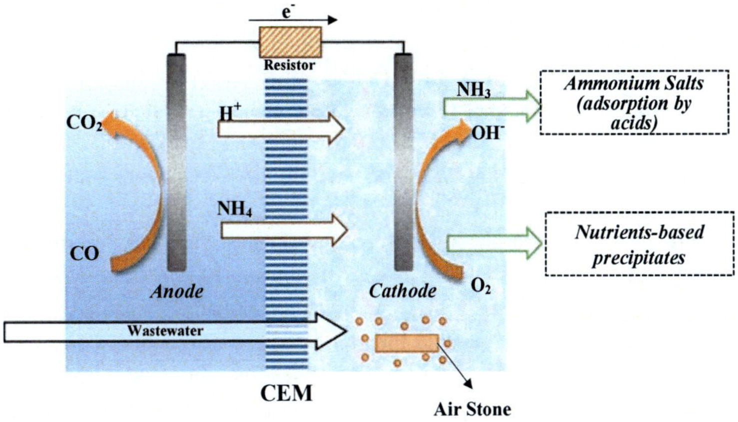

Fig. 5 Schematic representation of the MFC for the nutrient extraction (Ye et al., 2021)

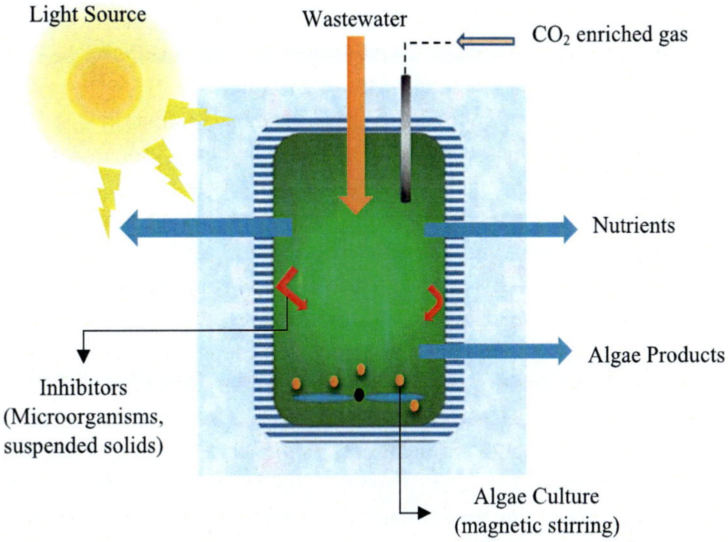

Fig. 6 Schematic diagram of membrane photobioreactor (Kumar et al., 2020)

process can be adopted to extract the ammonium from complex industrial wastewater, as the FO method is found to be ineffective in this case (Alkhatib et al., 2021). In addition, the MD process is economic in the extraction of ammonium from wastewater having higher temperatures, as it is not necessary to provide external thermal energy from an artificial source and solar energy (a Low-grade energy source) is sufficient (Peter et al., 2021).

Membrane systems are often integrated with biological action to improve their range of application in the extraction of nutrients, which can also be adopted in centralized wastewater treatment facilities. The OMBR system, in specific, is employed to extract nutrients with a reduced potential of membrane fouling. In the OMBR system, improvements can be made in the recovery system for economic feasibility. Besides, the complex wastewater from industries can be treated by BES containing an anaerobic process, to extract nitrogen in the form of ammonium and phosphorous in the form of phosphate. This facilitates the recovery through the chemical precipitation process. BES is an energy-neutral wastewater treatment process as it produces electrical energy and the important point is that it requires biodegradable organic matter for energy recovery. At the same time, despite the lower environmental footprint, MBPR may take longer durations to extract nutrients with additional light sources.

3 Analysis and Discussion

a. Economic Expediency of Nutrient Recovery Systems

The struvite precipitates formed inside the pipelines and machinery are removed by spending around US $72,284.55 by some of the treatment plants in Europe. Nutrient recovery can drastically reduce these costs and in addition, it provides supplementary fertilizers for agricultural uses. In addition, it could also facilitate decreasing the surplus dry sludge that is produced during the treatment process by 2–8%. The subsequent decrease in the costs involved in sludge management is about US $130.57/ton.P and US $0.78/kg. struvite (Jeanmaire & Evans, 2001 and Shu et al., 2006). The study conducted by Dockhorn, (2009) elucidates that the cost of phosphate recovery is less (US $2.2–3.3/kg.P) than that of the removal of the same from wastewater. Further, a reduction of 27% in energy demand was observed by saving an amount of US $4092.16 per day when shifted from phosphate removal to phosphate recovery (Daneshgar et al., 2019 and Levlin & Hultman, 2003). Furthermore, the reduction of nutrients may also facilitate the prevention of eutrophication in the lakes and the estimated environmental benefits, otherwise termed as shadow price by phosphate recovery are found to be US $47.23 per kg of Phosphate (Molinos-Senante et al., 2011). The environmental benefits of nutrients extraction from wastewater include: (i) Decrement in the formation of unexpected precipitates, (ii) Reduction in the surplus sludge generated, and (iii) Fall in the eutrophication potential of the wastewater.

The extraction of nutrients by the chemical action is found to cost less when the wastewater is in the liquid phase rather than that in the sludge phase with an average cost of US $8.9 for the liquid phase and US $14 for the sludge phase per kg recovery of Phosphate. This might be due to the additional treatment of sludge to make the phosphate ions in the solution more accessible before recovery (Latif et al., 2015). Also, a decrease in the phosphate content in the backflow in the liquid phase reduces the costs of operation, whereas the phosphate ions in the sludge phase can lead to unexpected compounds of heavy metals that require an additional system for their disposal. The cost range of phosphate extraction by chemical action estimated by processes and technologies is reported to be US $2.45–9.79/kg.P (Desmidt et al., 2015).

The general pH of wastewater is acidic and the costs of alkaline chemicals added to bring up the pH values, the additional chemicals also improve the ionic strength of the solution, thus enhancing the dissolution potential of struvite, yet they account for more than 90% of the operational costs along with the energy utilized in mixing process (Sakthivel et al., 2012). Aeration is proposed to be an economical alternative to improve the pH by removing the CO_2, thus decreasing the requirement for additional alkaline chemicals by more than 50% (Corona et al., 2021). The aeration method is affected by the composition of the influent wastewater and the rate of airflow. The investigation results of Huang et al., (2017) also agree with the statement that aeration costs less than additional chemicals (NaOH) in improving the pH values of the wastewater (Table 1). When compared to $CaCl_2$ as a precipitator, $MgCl_2$ resulted in a lower cost of nutrient extraction, at the same time, Fe and Al-based chemical precipitators are more expensive as shown in Table 1 and the resultant Phosphate is not suitable for application on the ground directly. Another economical solution, especially for wastewater from swine farms was proposed by Wang et al., (2019) by using plant ash for increasing the pH levels and Mg metal pellets as the precipitators in recovering the phosphate with an evaluated cost of US $0.62 per kg.P., excluding the cost of manpower and maintenance.

Besides, the usage of hydroxides of Mg/Ca than the other compounds as the precipitators can enhance the chemical precipitation process by supplementing additional ions and increasing the pH of the solution (Daneshgar et al., 2019). The lower solubility and poor alkalinity of $Mg(OH)_2$ require the addition of acid solution to dissolve it, which lowers the pH of the solution. Moreover, it is not possible to maintain the pH and ratios of Ca:P or Mg:P:N simultaneously.

To decrease the cost of chemicals, various sources of magnesium were used in the chemical precipitation of struvite by investigating (Sakthivel et al., 2012) on source-separated urine and the results suggest the addition of $MgSO_4$ may elevate the concentration of SO_4 in the solution which has a detrimental effect on the quality of precipitates of struvite. As it can be seen in Table 2, the addition of MgO incurs a low cost, at the same time, it could be advantageous in increasing the pH and also as a source of magnesium. As far as Bittern is concerned, it is available free of cost as it is a waste product but the cost of transportation is high, thus limiting its usage. $MgCl_2$ as a precipitator overcomes the drawbacks of other chemicals except for its cost. Although the cost of the process using wood ash is very low, it was found to accumulate heavy metals in the struvite precipitate,

Table 1 Variation of cost of nutrient extraction with various precipitators (Huang et al., 2017)

S.no.	Precipitator	Cost of pH increase by Aeration (US $/kg·P)	Cost of pH increase by Additional NaOH (US $/kg·P)
1	$CaCl_2$	1.56	1.85
2	$MgCl_2$	0.82	0.95
3	MgO	0.38	–
4	$FeSO_4$	–	1.99
5	$FeCl_3$	–	3.13
6	$AlCl_3$	–	3.54

Table 2 Estimated struvite production costs using different magnesium materials (*Data Source* Etter et al., 2011 and Sakthivel et al., 2012)

Magnesium Source	Mg:P Ratio	Costs Incurred (US \$/kg·struvite)
Bittern (Mg^{2+})	1.1	0.19
Magnesite ($MgCO_3$)	1	0.11
Magnesium sulphate ($MgSO_4{\cdot}7H_2O$)	1.1	0.32
Wood ash (Solid MgO)	1.9	0.024
Magnesium chloride ($MgCl_2{\cdot}6H_2O$)	1.1	0.13
Magnesium sulfate ($MgSO_4{\cdot}7H_2O$)	1.1	0.16
Magnesium oxide (MgO)	1.5	0.10
Bittern (Mg^{2+})	1.1	0.0046–0.19

thus decreasing the product quality and the prospect of its application in agriculture as a fertilizer. Hence, additional processing is necessary to eliminate these foreign chemicals in the compounds before utilizing them as a source of magnesium.

The use of seawater as a source of magnesium is very economical but the quality of the recovered struvite is low due to high salinity. The estimated cost of extracted struvite was US \$0.55 per kg. struvite (Kumashiro et al., 2001). Hence, the chemical opted as the source of magnesium should be able to balance the overall input costs and the quality of the product. Phosphoric acid with a greater market price can be extracted by the wet process of chemical treatment for nutrient extraction from the sludge. On one hand, the usage of microorganisms in the biological action to extract phosphates from the sludge has been found to lower the costs, while the dependability of bacteria is being challenged (Keyvan Hosseini et al., 2020). On the other hand, heat exchanger facilities are being adopted in the thermochemical treatment, to reduce the consumption of energy, at the same time methane gas can be used as an alternative source of energy.

An economically feasible and widely explored technology, for the recovery of ammonia from wastewater, is a combination of ammonia stripping along with the process of acid absorption. The same methodology was adopted by Pradhan et al. (2017) and succeeded in the extraction of above 85% of ammonia from the pure urine sample with a profit of about US \$2.5 per m^3. The product of this process (ammonium sulphate) is estimated to have a market value of about US \$1.11 per kg. of nitrogen.

It can be noted that the increased concentration of phosphate ions from 50 to 800 mg. phosphate per litre can lower the overall costs from US \$3113.6 to US \$578.4 (Dockhorn, 2009). In the same way, the costs of ammonium recovery can be brought down from US \$11.9 to US \$2.92 per kg. nitrogen by rising the ammonium concentration from 539 to 2470 mg. of ammonium per litre (De Vrieze et al., 2016). Hence, it is essential to improve the concentration of nutrients in the recovery systems for the economic viability of the process. In the membrane-based system of nutrient extraction, the struvite precipitation hybrid is found to consume low input energy (260 kWh/kg.P), which could bring down the input costs very effectively (Bradford-Hartke et al., 2012). But the cost of membranes and membrane fouling control poses a challenge for their practical application.

The electrodialysis process was adopted for the extraction of ammonium by Ward et al. (2018) and observed a power consumption in the range of 3.4 kWh–6.4 kWh per kg. ammonium. It was concluded that the process is more competitive for ammonium concentrations above 1500 mg. N per litre. The nutrient extraction by use of calcium-activated synthetic zeolites is economically feasible only in combination with an ultrafiltration unit and in a large urban treatment plat (Shaban et al., 2017). In the OMBR method, as it is not necessary to add additional salts for the precipitates, it reduces the costs. whereas, methane fermentation could generate biogas and energy during the anaerobic process in the OMBR, thus reducing some of the energy costs (Nie et al., 2017). But the biogas thus generated, need further processing to remove the toxic substances before it is used to generate electricity.

Although bio-electrochemical systems are more economically feasible due to the generation of electricity and increase of pH by cathode reactions, the formation of precipitates on the surface of the cathode may have detrimental effects on the performance of the BES in extracting nutrients (Zhang et al., 2022). Renewable energy such as solar energy (Zhang et al., 2013) and waste heat energy (Qin & He, 2014) can supplement the energy source for the extraction of nutrients, which can also reduce the energy costs.

b. **Economic Dissection of Extracted Nutrients**

The most common precipitates of nutrients that are recovered from wastewater are calcium phosphate and struvite, of which struvite has been proven to be a promising fertilizer due to its composition containing phosphate and ammonium in the same compound. Further, the plants' phosphate absorbing capacity is enhanced by the magnesium content present in the struvite (González-Ponce et al., 2009). But due to the stoichiometric ratio of the precipitation process of struvite, it is not an easy product when compared to calcium phosphate. Although the presence of high P content and lower solubility also makes calcium phosphate a good fertilizer (Römer & Steingrobe, 2018), its precipitates are found to be more effective in lower pH ranges (acid soils) in which the disintegration of P is stronger (Bach et al., 2022). In addition, impurities such as heavy metals are observed in extracted nutrients, yet their content is less than the impurities in a commercial fertilizer (Latifian et al., 2012).

Due to less commercialization of extracted nutrients, the current market values of these products cannot be considered the exact values. Nevertheless, few investigators have reported the cost of struvite as shown in Table 3. While the cost of phosphate-based commercial fertilizers is US $2.11–3.67/kg. P stands on top against the extracted nutrient fertilizers. At the same time, the market value of triple superphosphate (TSP) which can be formed from calcium phosphate was reported as US $553.78/ton (Desmidt et al., 2015). Hence, the phosphate rock can be replaced by calcium phosphate precipitates as raw material as it costs US $922.96–1245.44)/ton·P. Besides, struvite production is suggested over the production of P-Fertilizers such as TSP due to the lower consumption of energy (Daneshgar et al., 2019).

The plant absorption capacity, also known as bioavailability of the extracted nutrients can be tested in two methods, of which struvite's uptake by the plant is usually tested by pot and field trials. The results show a high bioavailability of struvite in a wide range of pH and types of soil (Corona et al., 2021) which is similar performance to TSP, whereas the calcium precipitates are effective only in acidic soils (Cabeza et al., 2011). The other test is to use chemical analogues as extractants for the acquisition of plants (Melia et al., 2017). Two contradictory statements were made over the performance of extracted struvite, one reporting the increased growth of plants (Ryu et al., 2012), while the other claimed to have a lower yield, probably due to the poor quality of the struvite obtained (Ackerman et al., 2013). Another study by Bauer et al. (2007) on the phosphate extracted from swine wastewater, reported the product obtained could be soluble in citrate and also contains high levels of P available to plants.

Direct land application of phosphate extracted by thermochemical treatment has a detrimental effect on the land due to the presence of more than 10% of input chloride additives observed in the product (Adam et al., 2007). Although the thermochemical treatment method using $MgCl_2$ has a higher bioavailability than adopting $CaCl_2$, the extracted product is highly soluble in acidic soils with 88% efficiency when compared to a P-fertilizer soluble in water, whereas the same is about 71.2% in soils with neutral pH and 4% in alkaline soils (Nanzer et al., 2014).

Arif et al. (2017) observed that, though the phosphates extracted using biochar adsorbents help in the restoration of soil, they are effective in alkaline soils only. At the same time, MgO-coupled magnetic biochar (MMSB) for the extraction of phosphate has been proven to be of higher efficiency in the pot experiment (Li et al., 2016). The factors affecting the effectiveness of biochar adsorbents are the concentration of adsorbent, availability of phosphate, the quantity of phosphate adsorbed, and the synergy between the phosphate and other available nutrients. Calcination of phosphate-based adsorbents could improve the quality of the products extracted by eliminating the organic impurities from the extracted phosphate (Xie et al., 2015). CaO-MgO-based hybrid carbon composite adsorbents were also developed, which were found to perform better and the phosphate thus recovered could be utilized as a fertilizer in substitution for chemical fertilizers (Li et al., 2018).

c. **Prospective Outlook**

Minerals and nutrient wealth on earth are precious resources. The rise in demand for nutrients due to the rapid growth in population and costly production process of the nutrients along with the scarce resources have forced the necessity for additional technology to supplement the conventional nutrient supply. This also facilitates in reducing the heavy metal contamination caused by intense mining activity (Sartorius et al., 2012) and reducing the global warming caused due to the greenhouse gases produced by large-scale ammonium production (Ye et al., 2018). Hence, the nutrient extraction at the treatment stage of wastewater is advantageous in supplementing the chemical fertilizers, but the cost of recovery of the same may be surpassed by chemical fertilizers and they may not cover all the operational costs with the current technology.

Table 3 Market Values of Extracted Struvite reported by various sources

S.no.	Reported price in (US $)	References
1	344.72/ton	Etter et al. (2011)
2	2.89/kg.P	Desmidt et al. (2015)
3	244.64/ton in Japan	Münch and Barr (2001)
4	207.04–345.07/ton in Australia	

Although the extraction of nutrients from wastewater is not feasible in terms of economy, there are indirect benefits such as reduction of sludge and unwanted precipitates, thus reducing the cost of disposal and also the maintenance cost is reduced as the scaling is also reduced. In addition, the extraction of nutrients also decreases the possibility of eutrophication in the downstream aquatic environment. Thus, nutrient extraction can benefit the environment, but the government must modify certain policies and provide incentives for full-scale production to reach the market with a fair price and quality.

It has been established that the membrane-based technologies are feasible in terms of both economic and technical aspects, within which the BES and OMBR-based nutrient recovery systems are more advantageous due to low energy consumption and the decreased potential of membrane fouling. Despite the advantages of the AnOMBR system over the Aerobic OMBR system, few investigations were carried out on the AnOMBR hybrid system. Therefore, further research on the application of the AnOMBR system towards the extraction of nutrients is necessary. Furthermore, the studies on the combination of FO, ED, and MD processes with MFC are to be extended to analyse the efficiency of the hybrid system in enriching the nutrients from diluted wastewater, which is a drawback of the MFC system. Besides quality, the focus should also be on increasing the release of phosphate during the anaerobic process. Feed solution plays an important role in determining the performance of a nutrient extraction system using membrane methods. Hence, it is crucial to research the variation of efficiency with various combinations of wastewater as feed solution to improve the nutrient removal by performing a pre-treatment of the same. Thus, the technical performance of the system can be improved.

4 Conclusions

Nutrient extraction from wastewater has been found to be effective in reducing the effluent nutrient concentration and inturns helps in the reduction of eutrophication in the downstream water bodies. Besides, nutrient extraction enables in establishing a sustainable framework where the extracted nutrients can replace the conventional fertilizers and thus generating a revenue from wastewater treatment. Both conventional and contemporary technologies were found to be effective in nutrient extraction by integrating them with membrane technology. Further studies to cut down the operation costs of nutrient extraction technologies are essential. The products from nutrient extracted are to be supported and encouraged by government agencies by introducing relevant policies and providing incentive/subsidy for nutrient extraction units.

Acknowledgements The authors would like to acknowledge the efforts and publications of all the researchers who have contributed in enhancing the nutrient extraction technologies.

References

Ackerman, J. N., Zvomuya, F., Cicek, N., & Flaten, D. (2013). Evaluation of manure-derived struvite as a phosphorus source for canola. *Canadian Journal of Plant Science, 93*(3), 419–424. https://doi.org/10.4141/cjps2012-207

Adam, C., Kley, G., & Simon, F. G. (2007). Thermal treatment of municipal sewage sludge aiming at marketable P-fertilisers. *Materials Transactions, 48*(12), 3056–3061. https://doi.org/10.2320/matertrans.MK200707

Alkhatib, A., Ayari, M. A., & Hawari, A. H. (2021). Fouling mitigation strategies for different foulants in membrane distillation. *Chemical Engineering and Processing-Process Intensification, 167*, 108517. https://doi.org/10.1016/j.cep.2021.108517

Arif, M., Ilyas, M., Riaz, M., Ali, K., Shah, K., Haq, I. U., & Fahad, S. (2017). Biochar improves phosphorus use efficiency of organic-inorganic fertilizers, maize-wheat productivity and soil quality in a low fertility alkaline soil. *Field Crops Research, 214*, 25–37. https://doi.org/10.1016/j.fcr.2017.08.018

Bach, I. M., Essich, L., Bauerle, A., & Müller, T. (2022). Efficiency of phosphorus fertilizers derived from recycled biogas digestate as applied to maize and ryegrass in soils with different pH. *Agriculture, 12*(3), 325. https://doi.org/10.3390/agriculture12030325

Bauer, P. J., Szogi, A. A., & Vanotti, M. B. (2007). Agronomic effectiveness of calcium phosphate recovered from liquid swine manure.https://doi.org/10.2134/agronj2006.0354

Bradford-Hartke, Z., Lant, P., & Leslie, G. (2012). Phosphorus recovery from centralised municipal water recycling plants. *Chemical Engineering Research and Design, 90*(1), 78–85. https://doi.org/10.1016/j.cherd.2011.08.006

Cabeza, R., Steingrobe, B., Römer, W., & Claassen, N. (2011). Effectiveness of recycled P products as P fertilizers, as evaluated in pot experiments. *Nutrient Cycling in Agroecosystems, 91*(2), 173–184. https://doi.org/10.1007/s10705-011-9454-0

Corona, F., Hidalgo, D., Martín-Marroquín, J. M., & Antolín, G. (2021). Study of the influence of the reaction parameters on nutrients recovering from digestate by struvite crystallisation. *Environmental Science and Pollution Research, 28*(19), 24362–24374. https://doi.org/10.1007/s11356-020-08400-4

Daneshgar, S., Buttafava, A., Callegari, A., & Capodaglio, A. G. (2019). Economic and energetic assessment of different phosphorus recovery options from aerobic sludge. *Journal of Cleaner Production, 223*, 729–738. https://doi.org/10.1016/j.jclepro.2019.03.195

De Vrieze, J., Smet, D., Klok, J., Colsen, J., Angenent, L. T., & Vlaeminck, S. E. (2016). Thermophilic sludge digestion improves energy balance and nutrient recovery potential in full-scale municipal wastewater treatment plants. *Bioresource Technology, 218*, 1237–1245. https://doi.org/10.1016/j.biortech.2016.06.119

Department of the Army. (2001, September). Biological Nutrient Removal, Public Works Technical Bulletin 420-49-39. https://www.wbdg.org/FFC/ARMYCOE/PWTB/pwtb_420_49_39.pdf.

Desmidt, E., Ghyselbrecht, K., Zhang, Y., Pinoy, L., Van der Bruggen, B., Verstraete, W., ... & Meesschaert, B. (2015). Global phosphorus scarcity and full-scale P-recovery techniques: A review. *Critical Reviews in Environmental Science and Technology, 45*(4), 336–384.https://doi.org/10.1080/10643389.2013.866531

Dockhorn, T. (2009, April). About the economy of phosphorus recovery. In *International conference on nutrient recovery from wastewater streams* (pp. 145–158). IWA Publishing.

Donatello, S., & Cheeseman, C. R. (2013). Recycling and recovery routes for incinerated sewage sludge ash (ISSA): A review. *Waste Management, 33*(11), 2328–2340. https://doi.org/10.1016/j.wasman.2013.05.024

Etter, B., Tilley, E., Khadka, R., & Udert, K. M. (2011). Low-cost struvite production using source-separated urine in Nepal. *Water Research, 45*(2), 852–862. https://doi.org/10.1016/j.watres.2010.10.007

Feng, L., Xie, L., Suo, G., Shao, X., & Dong, T. (2018). Influence of temperature on the performance of forward osmosis using ammonium bicarbonate as draw solute. *Transactions of Tianjin University, 24*(6), 571–579. https://doi.org/10.1007/s12209-018-0159-1

Gao, L., Yang, G., Zhang, J., & Xie, Z. (2020). De-ammonification using direct contact membrane distillation–an experimental and simulation study. *Separation and Purification Technology, 250*, 117158. https://doi.org/10.1016/j.seppur.2020.117158

Glover, T., & Peterson, G., DeCoste, J., & Browe, M. (2012). Adsorption of ammonia by sulfuric acid treated zirconium hydroxide. *Langmuir: The ACS Journal of Surfaces and Colloids, 28*, 10478–87. https://doi.org/10.1021/la302118h

Goglio, A., Marzorati, S., Zecchin, S., Quarto, S., Falletta, E., Bombelli, P., ... & Schievano, A. (2022). Plant nutrients recovery from agro-food wastewaters using microbial electrochemical technologies based on porous biocompatible materials. *Journal of Environmental Chemical Engineering, 10*(3), 107453. https://doi.org/10.1016/j.jece.2022.107453

González-Ponce, R., López-de-Sá, E. G., & Plaza, C. (2009). Lettuce response to phosphorus fertilization with struvite recovered from municipal wastewater. *HortScience, 44*(2), 426–430. https://doi.org/10.21273/HORTSCI.44.2.426

Gowd, S. C., Ramakrishna, S., & Rajendran, K. (2022). Wastewater in India: An untapped and under-tapped resource for nutrient recovery towards attaining a sustainable circular economy. *Chemosphere, 291*, 132753. https://doi.org/10.1016/j.chemosphere.2021.132753

Huang, H., Zhang, D., Zhao, Z., Zhang, P., & Gao, F. (2017). Comparison investigation on phosphate recovery from sludge anaerobic supernatant using the electrocoagulation process and chemical precipitation. *Journal of Cleaner Production, 141*, 429–438. https://doi.org/10.1016/j.jclepro.2016.09.127

Enyemadze, I., Momade, F. W., Oduro-Kwarteng, S., & Essandoh, H. (2021). Phosphorus recovery by struvite precipitation: A review of the impact of calcium on struvite quality. *Journal of Water, Sanitation and Hygiene for Development, 11*(5), 706–718. https://doi.org/10.2166/washdev.2021.078

Jeanmaire, N., & Evans, T. (2001). Technico-economic feasibility of P-recovery from municipal wastewaters. *Environmental Technology, 22*(11), 1355–1361. https://doi.org/10.1080/09593332208618194

Kabdasli, I., Kuşçuoğlu, S., Tünay, O., & Siciliano, A. (2022). Assessment of K-struvite precipitation as a means of nutrient recovery from source separated human urine. *Sustainability, 14*(3), 1082. https://doi.org/10.3390/su14031082

Kamilya, T., Gautam, R. K., Muthukumaran, S., Navaratna, D., & Mondal, S. (2022). Technical advances on current research trends and explore the future scope on nutrient recovery from waste-streams: A review and bibliometric analysis from 2000 to 2020. *Environmental Science and Pollution Research*, 1–19.https://doi.org/10.1007/s11356-022-20895-7

Kedwell, K. C., Jørgensen, M. K., Quist-Jensen, C. A., Pham, T. D., Van der Bruggen, B., & Christensen, M. L. (2021). Selective electrodialysis for simultaneous but separate phosphate and ammonium recovery. *Environmental Technology, 42*(14), 2177–2186. https://doi.org/10.1080/09593330.2019.1696410

Keyvan Hosseini, M., Pajoum Shariati, F., Bonakdarpour, B., & Heydarinasab, A. (2020). The effect of mechanical cleaning technology (MCT) on membrane fouling in a novel hybrid membrane photobioreactor (HMPBR) containing Arthrospira platensis (Spirulina). *Journal of Applied Phycology, 32*(1), 83–91. https://doi.org/10.1007/s10811-019-01944-0

Krishnamoorthy, N., Dey, B., Unpaprom, Y., Ramaraj, R., Maniam, G. P., Govindan, N., ... & Paramasivan, B. (2021). Engineering principles and process designs for phosphorus recovery as struvite: A comprehensive review. *Journal of Environmental Chemical Engineering, 9*(5), 105579. https://doi.org/10.1016/j.jece.2021.105579

Kumar, R., Ghosh, A. K., & Pal, P. (2020). Synergy of biofuel production with waste remediation along with value-added co-products recovery through microalgae cultivation: A review of membrane-integrated green approach. *Science of the Total Environment, 698*, 134169. https://doi.org/10.1016/j.scitotenv.2019.134169

Kumashiro, K., Ishiwatari, H., & Nawamura, Y. (2001, March). A pilot plant study on using seawater as a magnesium source for struvite precipitation. In: *Second international conference on recovery of phosphates from sewage and animal wastes, Noordwijkerhout, Holland.*

Latif, M. A., Mehta, C. M., & Batstone, D. J. (2015). Low pH anaerobic digestion of waste activated sludge for enhanced phosphorous release. *Water Research, 81*, 288–293. https://doi.org/10.1016/j.watres.2015.05.062

Latifian, M., Liu, J., & Mattiasson, B. (2012). Struvite-based fertilizer and its physical and chemical properties. *Environmental Technology, 33*(24), 2691–2697. https://doi.org/10.1080/09593330.2012.676073

Levlin, E., & Hultman, B. (2003, March). Phosphorus recovery from phosphate rich side-streams in wastewater treatment plants. In *Polish Swedish seminar, Gdansk March* (pp. 23–25).

Li, R., Wang, J. J., Zhou, B., Awasthi, M. K., Ali, A., Zhang, Z., ... & Mahar, A. (2016). Recovery of phosphate from aqueous solution by magnesium oxide decorated magnetic biochar and its potential as phosphate-based fertilizer substitute. *Bioresource Technology, 215*, 209–214.https://doi.org/10.1016/j.biortech.2016.02.125

Li, Y., Zhang, B., Li, G., & Luo, W. (2018). Osmotic membrane bioreactor and its hybrid systems for wastewater reuse and resource recovery: Advances, challenges, and future directions. *Current Pollution Reports, 4*(1), 23–34. https://doi.org/10.1007/s40726-018-0080-1

Liu, X., & Wang, J. (2019). Impact of calcium on struvite crystallization in the wastewater and its competition with magnesium. *Chemical Engineering Journal, 378*, 122121. https://doi.org/10.1016/j.cej.2019.122121

Logan, B. E., Hamelers, B., Rozendal, R., Schröder, U., Keller, J., Freguia, S., ... & Rabaey, K. (2006). Microbial fuel cells: methodology and technology. *Environmental Science & Technology, 40*(17), 5181–5192.https://doi.org/10.1021/es0605016

Melia, P. M., Cundy, A. B., Sohi, S. P., Hooda, P. S., & Busquets, R. (2017). Trends in the recovery of phosphorus in bioavailable forms from wastewater. *Chemosphere, 186*, 381–395. https://doi.org/10.1016/j.chemosphere.2017.07.089

Molinos-Senante, M., Hernández-Sancho, F., Sala-Garrido, R., & Garrido-Baserba, M. (2011). Economic feasibility study for phosphorus recovery processes. *Ambio, 40*(4), 408–416. https://doi.org/10.1007/s13280-010-0101-9

Münch, E. V., & Barr, K. (2001). Controlled struvite crystallisation for removing phosphorus from anaerobic digester sidestreams. *Water Research, 35*(1), 151–159. https://doi.org/10.1016/S0043-1354(00)00236-0

Nanzer, S., Oberson, A., Berger, L., Berset, E., Hermann, L., & Frossard, E. (2014). The plant availability of phosphorus from thermo-chemically treated sewage sludge ashes as studied by 33P labeling techniques. *Plant and Soil, 377*(1), 439–456. https://doi.org/10.1007/s11104-013-1968-6

Nie, Y., Kato, H., Sugo, T., Hojo, T., Tian, X., & Li, Y. Y. (2017). Effect of anionic surfactant inhibition on sewage treatment by a submerged anaerobic membrane bioreactor: Efficiency, sludge activity and methane recovery. *Chemical Engineering Journal, 315*, 83–91. https://doi.org/10.1016/j.cej.2017.01.022

Nguyen, N. C., Chen, S. S., Jain, S., Nguyen, H. T., Ray, S. S., Ngo, H. H., ... & Duong, H. C. (2018). Exploration of an innovative draw solution for a forward osmosis-membrane distillation desalination process. *Environmental Science and Pollution Research, 25*(6), 5203–5211.https://doi.org/10.1007/s11356-017-9192-1

Peter, A. P., Khoo, K. S., Chew, K. W., Ling, T. C., Ho, S. H., Chang, J. S., & Show, P. L. (2021). Microalgae for biofuels, wastewater treatment and environmental monitoring. *Environmental Chemistry Letters, 19*(4), 2891–2904. https://doi.org/10.1007/s10311-021-01219-6

Pourabdollah, M., Torkian, A., Hashemian, S. J., & Bakhshi, B. (2014). A triple fouling layers perspective on evaluation of membrane fouling under different scenarios of membrane bioreactor operation. *Journal of Environmental Health Science and Engineering, 12*(1), 1–11. https://doi.org/10.1186/2052-336X-12-91

Pradhan, S. K., Mikola, A., & Vahala, R. (2017). Nitrogen and phosphorus harvesting from human urine using a stripping, absorption, and precipitation process. *Environmental Science & Technology, 51*(9), 5165–5171. https://doi.org/10.1021/acs.est.6b05402

Qin, M., & He, Z. (2014). Self-supplied ammonium bicarbonate draw solute for achieving wastewater treatment and recovery in a microbial electrolysis cell-forward osmosis-coupled system. *Environmental Science & Technology Letters, 1*(10), 437–441. https://doi.org/10.1021/ez500280c

Ray, H., Perreault, F., & Boyer, T. H. (2020). Ammonia recovery from hydrolyzed human urine by forward osmosis with acidified draw solution. *Environmental Science & Technology, 54*(18), 11556–11565. https://doi.org/10.1021/acs.est.0c02751

Reig, M., Vecino, X., Gibert, O., Valderrama, C., & Cortina, J. L. (2021). Study of the operational parameters in the hollow fibre liquid-liquid membrane contactors process for ammonia valorisation as liquid fertiliser. *Separation and Purification Technology, 255*, 117768. https://doi.org/10.1016/j.seppur.2020.117768

Römer, W., & Steingrobe, B. (2018). Fertilizer effect of phosphorus recycling products. *Sustainability, 10*(4), 1166. https://doi.org/10.3390/su10041166

Robles, Á., Aguado, D., Barat, R., Borrás, L., Bouzas, A., Giménez, J. B., & Seco, A. (2020). New frontiers from removal to recycling of nitrogen and phosphorus from wastewater in the Circular Economy. *Bioresource Technology, 300*, 122673. https://doi.org/10.1016/j.biortech.2019.122673

Ryu, H. D., Lim, C. S., Kang, M. K., & Lee, S. I. (2012). Evaluation of struvite obtained from semiconductor wastewater as a fertilizer in cultivating Chinese cabbage. *Journal of Hazardous Materials, 221*, 248–255. https://doi.org/10.1016/j.jhazmat.2012.04.038

Safie, N. N., & Zahrim, A. Y. (2021). Recovery of nutrients from sewage using zeolite-chitosan-biochar adsorbent: Current practices and perspectives. *Journal of Water Process Engineering, 40*, 101845. https://doi.org/10.1016/j.jwpe.2020.101845

Sakthivel, S. R., Tilley, E., & Udert, K. M. (2012). Wood ash as a magnesium source for phosphorus recovery from source-separated urine. *Science of the Total Environment, 419*, 68–75. https://doi.org/10.1016/j.scitotenv.2011.12.065

Sartorius, C., von Horn, J., & Tettenborn, F. (2012). Phosphorus recovery from wastewater—Expert survey on present use and future potential. *Water Environment Research, 84*(4), 313–322. https://doi.org/10.2175/106143012X13347678384440

Shaban, M., AbuKhadra, M. R., Nasief, F. M., & El-Salam, A. (2017). Removal of ammonia from aqueous solutions, ground water, and wastewater using mechanically activated clinoptilolite and synthetic zeolite-a: Kinetic and equilibrium studies. *Water, Air, & Soil Pollution, 228*(11), 1–16. https://doi.org/10.1007/s11270-017-3643-7

Sheng, A. L. K., Bilad, M. R., Osman, N. B., & Arahman, N. (2017). Sequencing batch membrane photobioreactor for real secondary effluent polishing using native microalgae: Process performance and full-scale projection. *Journal of Cleaner Production, 168*, 708–715. https://doi.org/10.1016/j.jclepro.2017.09.083

Shu, L., Schneider, P., Jegatheesan, V., & Johnson, J. (2006). An economic evaluation of phosphorus recovery as struvite from digester supernatant. *Bioresource Technology, 97*(17), 2211–2216. https://doi.org/10.1016/j.biortech.2005.11.005

Shyam, S., Arun, J., Gopinath, K. P., Ribhu, G., Ashish, M., & Ajay, S. (2022). Biomass as source for hydrochar and biochar production to recover phosphates from wastewater: A review on challenges, commercialization, and future perspectives. *Chemosphere, 286*, 131490. https://doi.org/10.1016/j.chemosphere.2021.131490

Speight, J. G. (2017). Industrial inorganic chemistry. *Environmental Inorganic Chemistry for Engineers*, 1.https://doi.org/10.1016/B978-0-12-849891-0.00003-5

Tange, Y., Takesawa, S., & Yoshitake, S. (2017). Asymmetric triacetate membrane keeps high water flux during ultrafiltration: In vitro study. *Journal of Artificial Organs, 20*(4), 399–402. https://doi.org/10.1007/s10047-017-0971-8

Wang, F., Wei, J., Zou, X., Fu, R., Li, J., Wu, D., ... & Chen, H. (2019). Enhanced electrochemical phosphate recovery from livestock wastewater by adjusting pH with plant ash. *Journal of Environmental Management, 250*, 109473. https://doi.org/10.1016/j.jenvman.2019.109473

Ward, A. J., Arola, K., Brewster, E. T., Mehta, C. M., & Batstone, D. J. (2018). Nutrient recovery from wastewater through pilot scale electrodialysis. *Water Research, 135*, 57–65. https://doi.org/10.1016/j.watres.2018.02.021

Xie, T., Reddy, K. R., Wang, C., Yargicoglu, E., & Spokas, K. (2015). Characteristics and applications of biochar for environmental remediation: A review. *Critical Reviews in Environmental Science and Technology, 45*(9), 939–969. https://doi.org/10.1080/10643389.2014.924180

Xie, M., Shon, H. K., Gray, S. R., & Elimelech, M. (2016). Membrane-based processes for wastewater nutrient recovery: Technology, challenges, and future direction. *Water Research, 89*, 210–221. https://doi.org/10.1016/j.watres.2015.11.045

Yang, Y. L., Wu, Y., Lu, Y. X., Cai, Y., He, Z., Yang, X. L., & Song, H. L. (2021). A comprehensive review of nutrient-energy-water-solute recovery by hybrid osmotic membrane bioreactors. *Bioresource Technology, 320*, 124300. https://doi.org/10.1016/j.biortech.2020.124300

Ye, Y., Ngo, H. H., Guo, W., Chang, S. W., Nguyen, D. D., Varjani, S., ... & Nguyen, D. P. (2021). Bio-membrane based integrated systems for nitrogen recovery in wastewater treatment: Current applications and future perspectives. *Chemosphere, 265*, 129076. https://doi.org/10.1016/j.chemosphere.2020.129076

Ye, Y., Ngo, H. H., Guo, W., Liu, Y., Chang, S. W., Nguyen, D. D., ... & Wang, J. (2018). A critical review on ammonium recovery from wastewater for sustainable wastewater management. *Bioresource Technology, 268*, 749–758.https://doi.org/10.1016/j.biortech.2018.07.111

Yuan, H., Lu, Y., Abu-Reesh, I. M., & He, Z. (2015). Bioelectrochemical production of hydrogen in an innovative pressure-retarded osmosis/microbial electrolysis cell system: Experiments and modeling. *Biotechnology for Biofuels, 8*(1), 1–12. https://doi.org/10.1186/s13068-015-0305-0

Zeng, W., Li, B., Lin, X., Lv, S., Yin, W., Li, P., ... & Wu, J. (2021). Enhanced phosphate removal by zero valent iron activated through oxidants from water: batch and breakthrough experiments. *RSC Advances, 11*(63), 39879–39887.https://doi.org/10.1039/D1RA05664F

Zhang, J., Yuan, H., Deng, Y., Abu-Reesh, I. M., He, Z., & Yuan, C. (2019). Life cycle assessment of osmotic microbial fuel cells for simultaneous wastewater treatment and resource recovery. *The International Journal of Life Cycle Assessment, 24*(11), 1962–1975. https://doi.org/10.1007/s11367-019-01626-6

Zhang, Y., Pinoy, L., Meesschaert, B., & Van der Bruggen, B. (2013). A natural driven membrane process for brackish and wastewater treatment: Photovoltaic powered ED and FO hybrid system. *Environmental Science & Technology, 47*(18), 10548–10555. https://doi.org/10.1021/es402534m

Zhang, Z., Sarkar, D., Li, L., & Datta, R. (2022). Contaminant removal and resource recovery in bioelectrochemical wastewater treatment. *Current Pollution Reports*, 1–18https://doi.org/10.1007/s40726-022-00218-7

Technologies for Nutrient Recovery from Municipal Wastewater

Dominika Sobotka, Bogna Śniatała, and Jacek Mąkinia

Abstract

The increasing cost of phosphate fertilizer, the scarcity of high-quality phosphate rock, as well as the natural gas (CH_4) used in the energy-intensive Haber-Bosh process for liquid ammonia (NH_3) production, together with increasing water body eutrophication are driving a need to accelerate nutrient recovery—one of the twenty-first century's greatest global sustainability challenges. The adoption of changing paradigms, which involve the transition of wastewater treatment plants (WWTPs) to water resource recovery facilities (WRRFs), requires understanding the available recovery and recycling technologies. Nutrients are present in both liquid (e.g., secondary effluent, anaerobic digestion side stream, and sludge liquor) and solid residues (e.g., chemically/biologically bound in primary, secondary, or digested sludge, or in sludge ash after incineration). Since waste stream characteristics are varied, they must be processed with different methods. Currently, many techniques for macronutrient recovery are available. In this chapter, the most widely applied and promising strategies for nutrient recovery from municipal wastewater (both mainstream and sidestream) are characterized in terms of recovery efficiencies, energy requirements, and additional criteria (e.g., necessary reagents, conditions, simplicity of operation). This chapter also highlights the possible uses of recovered by-products as fertilizer and the requirement to transform current regulations to the CE concept.

Keywords

Nutrient recovery • Nitrogen • Phosphorus • Circular economy • Wastewater • WRRF

D. Sobotka (✉) · B. Śniatała · J. Mąkinia
Faculty of Civil and Environmental Engineering, Gdansk University of Technology, Gdańsk, Poland
e-mail: domsobot@pg.edu.pl

1 Introduction

The increasing pressure on crops production influencing rapidly growing fertilizers demand, the essential macronutrients, i.e., nitrogen (N) and phosphorus (P) supply uncertainty due to P mineral deposits depleting and energetically intensive, dependent on the price of natural gas, Haber–Bosch process used for the industrial synthesis of ammonia (NH_3), accompanied with observable and more acute effects of unsustainable use of the environment advocate searching for alternative nutrients sources (Sengupta et al., 2015). Since humans and animals consume nutrients from crops and produce nutrient-rich waste streams from food processing, wastewater treatment plants (WWTPs) can become mines of nitrogen and phosphorus compounds (Mehta et al., 2015), and the sludge digester liquors are recognized as the greatest recovery potential source (Munasinghe-Arachchige & Nirmalakhandan, 2020).

Wastewater remains a hardly untapped source of water, energy, nutrients, organic matter, and other useful by-products. Among the major nutrients, 16.6 mln tons of nitrogen are embedded in wastewater produced worldwide annually, together with 3 mln tons of phosphorus and 6.3 mln tons of potassium (K) (Qadir et al., 2020). Therefore, following the Circular Economy (CE) model framework in the European water and wastewater sector (European Commission, 2020), traditional wastewater treatment plants are increasingly regarded as "water resource recovery facilities" to reflect the value of water, nutrients, energy, and other resources, besides ensuring the required effluent quality.

Nitrogen and phosphorus forms are the major nutrients in wastewater. Both nutrients can be recovered from the wastewater in the solid and liquid forms. The most well-known solid forms are struvite (magnesium ammonium phosphate, MAP), apatite (calcium phosphate), and vivianite (ferrous phosphate). While phosphoric acid and ammonium sulfate ($(NH_4)_2SO_4$) are the primary liquid forms of recovered nutrients (Brennan et al., 2021; Liu et al., 2021a, 2021b).

M. Smol et al. (eds.), *Water in Circular Economy*, Advances in Science, Technology & Innovation,
https://doi.org/10.1007/978-3-031-18165-8_12

Nutrient recovery technologies are rapidly expanding due to the need for appropriate recycling of the key elements from waste resources in accordance with the CE concept. Recovery of nutrients from municipal wastewater would not only prevent the increasing risk of anoxia and eutrophication in the receiving water bodies, but also encourage the transition towards a CE-based development model (Rout et al., 2021). New technologies allow for recovering nutrients from wastewater as slow-release fertilizers with a negligible heavy metal content (i.e., crystallization, electrodialysis) or even as pure substances (ammonia stripping, bioelectrochemical systems) (Liu et al., 2022; Ye et al., 2020). However, nutrient recovery technologies are not yet fully mature, as further research is needed to optimize process efficiency, improve economical feasibility and enhance their commercial applicability.

In this chapter, possibilities of nutrient recovery from municipal wastewater mainstreams and sidestreams are described in terms of recovery efficiencies and energy requirements. Moreover, the possibility of using recovered by-products as a fertilizer will be discussed.

2 Nutrient Content of Wastewater (Nutrient Concentrations in Wastewater)

Wastewater contains inorganic and organic nutrients and suspended solids. Nitrogen, phosphorus, and potassium—there are valuable nutrients contained in wastewater. In general, there are three types of municipal wastewater in WWTPs: primary effluent, secondary effluent, and sludge liquors generated from the dewatering of anaerobic-digested activated sludge (Constantine, 2006). The concentrations of N and P in wastewater samples may vary depending on the wastewater characteristics and the type of applied treatment (AlMomani & Örmeci, 2016).

The major forms of nitrogen present in the primary effluent are organic nitrogen and ammonium. According to EPA (2000) primary effluents has total N concentrations ranging from 20 to 85 mg/L with approximately 60 percent ammonia nitrogen, 40 percent organic nitrogen, and small quantities of nitrates. During secondary treatment, these two major forms of nitrogen are rapidly converted to nitrate by nitrifying bacteria (ammonium oxidizing bacteria and nitrite oxidizing bacteria) or complete ammonia oxidizers (comammox bacteria) (Rahimi et al., 2020). The concentration of ammonium nitrogen (NH_4^+–N) and phosphate phosphorous (PO_4^{3-}–P) in the secondary effluent range from 40 to 48 mg NH_4^+–N/L and 9–29.6 mg PO_4^-–P/L, respectively (Table 1). The sludge digester liquors typically contain high concentrations of dissolved ammonium nitrogen, phosphorus, and suspended and colloidal solids. The total N (which includes organic nitrogen and ammonia nitrogen) and P concentrations can reach values up to 1500 mg N/L and 200 mg P/L, respectively (Table 1).

3 Possibilities of Nutrients Recovery from Wastewater

In wastewater treatment plants (WWTPs), nutrients can be recovered in different locations, including raw wastewater, sludge liquor, fermented/nonfermented sludge, or ash. In general, recovery from a solution that is more concentrated than municipal wastewater is preferred, e.g., anaerobic zones of EBPR systems or anaerobic digester effluents. The

Table 1 N and P concentration in different types of municipal wastewater in WWTPs

Type of wastewater	N-concentration (mg/L)	P-concentration (mg/L)	References
Primary effluent	20–85 mg N/L 66.9–113 mg N/L 40.8 ± 0.4 mg N/L 42.22 mg N/L 28–38 mg N/L	– – 10.0 mg P/L 3.25 mg P/L 3–5.3 mg P/L	EPA (2000) Czerwionka et al. (2012) AlMomani and Örmeci (2016) Perera and Englehardt (2020) Zaman et al. (2021)
Secondary effluent	40–48 mg NH_4^+–N/L 44.0 ± 0.2 mg N/L 42.1–43.4 mg NH_4^+–N/L	9–12 mg PO_4^-–P/L 26.0 mg P/L 27–29.6 mg PO_4^-–P/L	Martinez et al. (2000) AlMomani and Örmeci (2016) Bali and Gueddari (2019)
Sludge digester liquors	355–1170 mg NH_4^+–N/L 130 ± 2 mg N/L 1460–1750 mg NH_4^+–N/L 287–832 mg NH_4^+–N/L	79–169 mg P/L 200.0 mg P/L 15.1–23.3 mg PO_4–P/L 0.3–112.5 mg PO_4–P/L	Pastor et al. (2010) AlMomani and Örmeci (2016) Xu et al. (2018) Siomoes et al. (2020)

recovery rate of phosphorus from the liquid phase is lower (10–60% from wastewater treatment plant influent), than from sludge (35–70%) and from sludge ashes (70–98%) (Chrispim et al., 2019). P-recovery from the liquid phase is a relatively simple process, whereas P-recovery from sewage sludge ash requires more energy and costs more. In medium and large WWTPs, sludge digester liquors are the primary source of nitrogen for recovery. It is estimated that the digestate has a potential of recovering 27% of the incoming nitrogen to the WWTP (Van der Hoek et al., 2018). A variety of recovery methods are used for this purpose, including chemical precipitation, chemical extraction, gas-permeable membrane separation and adsorption, electrodialysis or microalgae (Fig. 1).

a. **Sewage sludge**

Sewage sludge (SS) is the solid, semisolid, or slurry residual material generated during the wastewater treatment process. This residue is frequently divided into two types: primary sludge and waste activated sludge. Primary sludge is generated from chemical precipitation, sedimentation, and other

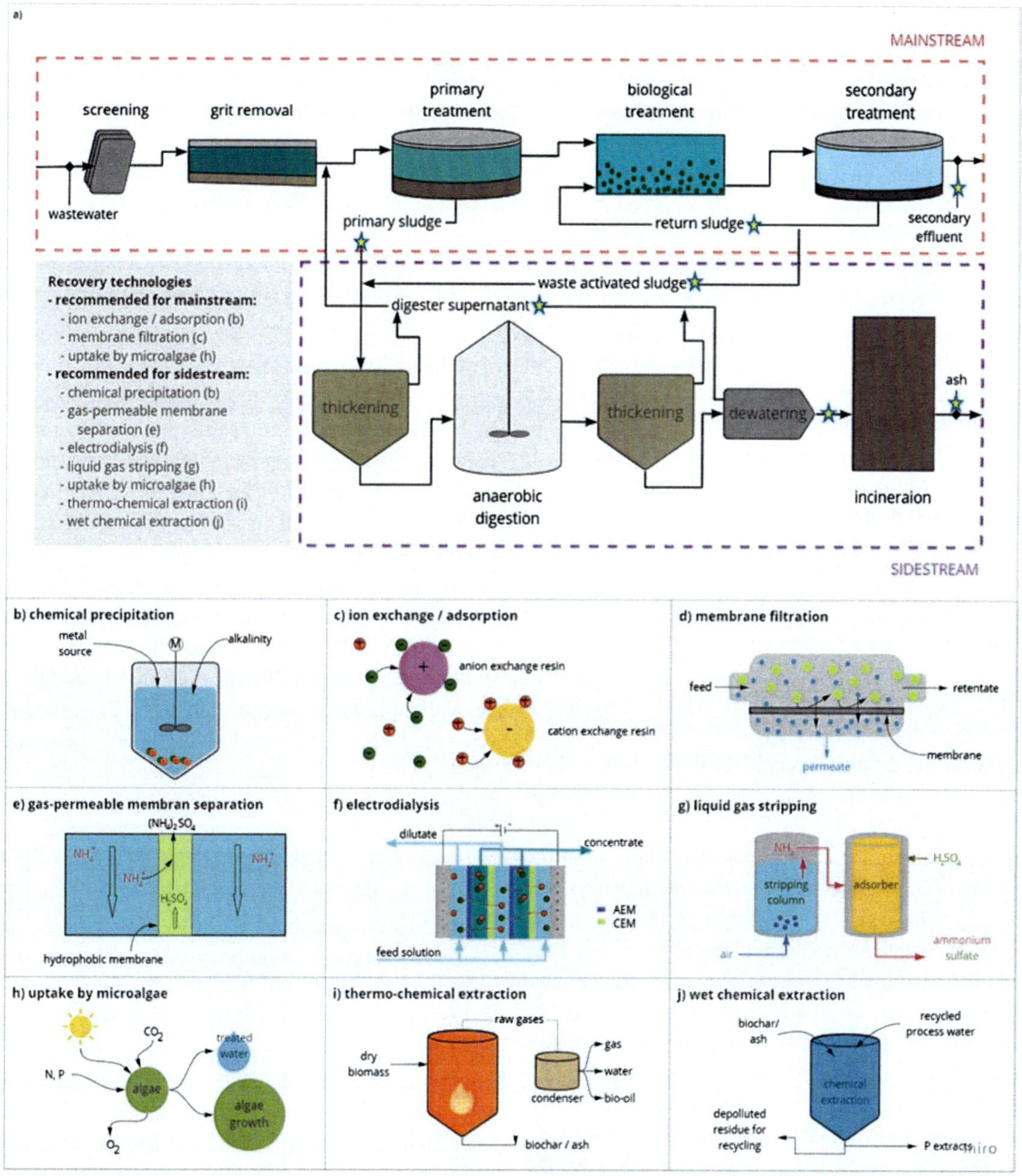

Fig. 1 Potential location (marked with an asterisk) of nutrient recovery during the wastewater treatment process (**a**), based on Egle et al. (2016) and the mechanism of the highly efficient nutrient recovery processes (**b–j**)

primary processes, whereas waste activated sludge is the activated waste biomass resulting from biological treatments. Currently, some nutrients in SS are recycled to agricultural soils via direct land application, typically following composting, liming, and/or anaerobic digestion. This strategy, which represents a traditional nutrient recovery approach, also has limitations, such as a lack of complete confidence in the uniformity, nutrient content, and availability of land-applied sludge, as well as human health hazards associated with the presence of viruses, organic pollutants, and heavy metals (Chojnacka et al., 2020; Awasthi et al., 2020; Sikosana et al., 2016).

b. **Main-stream wastewater**

Due to the relatively low concentrations of nutrients in the mainstream effluent, this effluent is not considered as the main medium for nutrient recovery in a WWTP. However, in combination with methods that increase phosphorus concentration in wastewater, e.g., EBPR process, mainstream wastewater can also be a valuable source of recovered nutrients (Zhang et al., 2022). At low concentrations of ammonia (40–60 mg N/L), the only ammonium recovery techniques that have shown feasibility are assimilative uptake in microbial cells, which recovers biomass N but not ammonium; and reverse osmosis, which is non-selective toward ammonium and requires substantial pretreatment (Cruz et al., 2019).

c. **Sludge digester liquors**

In municipal WWTPs employing anaerobic sludge digestion, sludge digester liquors are concentrated streams which contain high N concentrations (mainly in the form of ammonium) and phosphorus, suspended solids, and alkalinity, while the concentrations of biodegradable organic carbon are low. Typically, 15–25% of the influent nitrogen and phosphorus loads are recirculated back to the mainstream with the return liquors from sludge dewatering processes (Jardin et al., 2006).

The growing attention receives additional sludge digester liquors treatment in the sidestream, before recirculating it to the mainstream. Due to significantly lower costs (compared to physical and chemical methods), biological processes have most often been used for this purpose. However, physical and chemical methods can be a viable alternative due to the potential of recovery of nitrogen and phosphorus as a fertilizer. It is estimated that the phosphorus recovery potential from the sludge digester liquors (after fermentation and dehydration) is 50–60% (Urdalen, 2013). In Europe, an average of 230,000 tons of phosphate per year can be recovered in this way. The most desirable and commonly recovered product from digester supernatant is struvite (Zangarini et al., 2020; Magrí et al., 2020; Chrispim et al., 2019).

d. **Sewage sludge ash**

The phosphorus recovery from sewage sludge ash (SSA) is generally divided into two approaches: the wet- and thermo-chemical processes. Because phosphorus is more integrated and bound in SSA, extraction, and reprecipitation are more expensive. Although there are various reasons to extract phosphorus from ash. First, with a 90% recovery share, this technique has more potential than sludge recovery. Second, the high temperatures used in combustion destroy toxic organic matter in sludge. Third, most heavy metals may be extracted from sludge ash using a thermal-metallurgical method. The treated ash can be used directly as a fertilizer and no waste is formed. This explains why phosphorus recovery from sludge ash is a more promising approach than phosphorus extraction from sludge (Smol et al., 2020; Herzel et al., 2016).

4 Nutrient's Recovery Techniques

a. **Chemical precipitation**

Chemical precipitation converts dissolved components into particulate inorganic compounds for separation from the liquid bulk (Fig. 1b). Changing the temperature, pH, or adding metal ions to the waste streams creates supersaturated conditions, which are a thermodynamic driving force for precipitation. With these conditions, certain products can be precipitated (Rugaika et al., 2019; Günther et al., 2018). This technique is used to simultaneously recover N and P from nutrient-rich streams. Struvite contains about 12% P and 6% N, with almost no heavy metals or biological contaminants (Gell et al., 2011). Magnesium is typically deficient in waste streams and must be added as magnesium chloride ($MgCl_2$), magnesium hydroxide ($Mg(OH)_2$), or magnesium oxide (MgO) to achieve supersaturation. The struvite formation reaction requires high reactant (NH_4^+, PO_4^{3+}, and K^+) concentrations and a high pH (pH > 8.0) (Bradford-Hartke et al., 2021; Sengupta et al., 2015; Zhang et al., 2019).

Chemical precipitation is capable of removing 80–90% of soluble phosphates and 20–30% of soluble ammonia from waste streams. Due to the specific gravity of struvite, the crystals can easily be separated from the liquid bulk using gravity settling, mechanical separation (filter press), or an integrated crystallization and separation process. Struvite crystallization is mostly used for phosphorous recovery in waste streams with a low solids concentration (2000 mg/L), but a relatively high phosphorous concentration ($P\text{–}PO_4$ >)

50 mg/L). Typically, the process is run at a short hydraulic residence time of approximately 60 min, a moderately alkaline pH of 8.0–9.0, and an unregulated temperature of 25–35 °C. Due to struvite's slow crystal growth rate, long SRTs (>10 days) are required, resulting in the development of larger aggregated crystals or granules (Mehta et al., 2015).

Phosphate is most commonly recovered as magnesium, calcium, and ferric salts (Zangarini et al., 2020). By using different magnesium-based reagents, it is possible to obtainmagnesium ammonium phosphate (MAP, struvite), magnesium potassium phosphate (MPP, K-struvite), magnesium hydrogen phosphate trihydrate (newberyite), and/or trimagnesium phosphate octahydrate (bobierrite) (Magrí et al., 2020; Robles et al., 2020). They may be used as a single dosage of slow-release fertilizer in several years. Similarly, various calcium-based reagents involve the formation of different compounds: hydroxyapatite (HAP or acalcium phosphate, CaP), tricalcium phosphate (TCP), amorphous calcium phosphate (ACP), octacalcium phosphate (OCP), dicalcium phosphate (DCPA), dicalcium phosphate dihydrate (DCPD, brushite) (Magrí et al., 2020; Robles et al., 2020). Moreover, the calcium phosphate recovered as hydroxyapatite is very similar in composition to the mined phosphate rock (Magrí et al., 2020; Sengeputa et al., 2015). The application of iron reagents is less common since it is strongly bound to phosphorus, making it less bioavailable (Chrispim et al., 2019). However, ferrous iron phosphate (vivianite, FeP) recovery is promising due to the process natural ubiquity, easy accessibility, and foreseeable economic value (Jupp et al., 2021).

b. **Uptake by microalgae**

Microalgae are photoautotrophic microorganisms that use solar energy and carbon dioxide as energy and carbon source to reduce inorganic nutrients such as nitrogen and phosphorus to organic matter thus producing biomass (Fig. 1h) (Ali et al., 2021; Muhammad et al., 2020). The nutrient-rich algae can be processed with nutrient release techniques, such as anaerobic digestion or thermo-chemical methods or may be used directly as animal feed or fertilizer (Mehta et al., 2015). Chemical fertilizers pollute and contaminate soils, whereas biofertilizers do not. This makes biofertilizers one of the most common commercial products made from recovered value-added products. The efficiency and quality of the microalgae downstream processing are important determinants of the final product quality. There are several reasons why fertilizers have become so popular, including their ease of transportation, ability to dissolve in soil, and ability to accurately predict the needed dose. Soil depletion is caused by the absence of micronutrients, such as magnesium and iron in synthetic fertilizers, which can be replenished by using microalgae.

Nutrient recovery by microalgae uptake has been studied with several wastewater types including manure, in batch and continuous cultivations (Acién Fernández et al., 2018). Microalgae's commercial potential has long been recognized. However, the commercial-scale production of microalgae biomass requires a substantial amount of water and nutrients. It was estimated that 5–10% replacement of fossil crude requirement with algal biofuels would require ~952–1903 billion m^3 of water, 10–21 billion tons of N, and 2–4 billion tons of P fertilizers (Renuka et al., 2021). Hence, coupling wastewater treatment and microalgae production is suggested as a synergetic alternative for wastewater utilization, minimizing freshwater consumption, reducing GHG production, nutrient costs, and biomass cultivation for value-added products (Acién Fernández et al., 2018; Renuka et al., 2021). However, contamination of algal biomass growth in wastewater is still recognized as a challenge.

c. **Chemical extraction**

Wet-chemical and thermo-chemical extractions are used to extract phosphate from sludge, making it more available to plants and crops. Sludge-to-liquid phosphate ion transfer is frequently carried out using acid or alkaline solutions. A type of solution should be selected based on the properties of the sludge, including its content and treatment technology. Thermo-chemical extraction is dedicated for phosphate-containing sludge, and performed at high temperatures (800–1000 °C) in the presence of chloride additives (e.g., $MgCl_2$ and calcium chloride $CaCl_2$) (Fig. 1i). Because heavy metals may be removed in this process, the recovered phosphate could be of higher quality and more accessible to plants (Ye et al., 2020). During the wet-chemical extraction processes (also known as leaching), phosphorus should first be dissolved using strong acids, heat, and/or pressure (Fig. 1j). After that, separation of metal ions and phosphorus is necessary (Sartorius et al., 2011). The wet-chemical extraction requires fermentation of the sludge, which is not economical on a small scale. If the sludge is mono-incinerated (i.e., with no other fueling substances present), wet-chemical phosphorus recovery from the ash is also possible. Wet-chemical treatment starts with the extraction of phosphorus and all metals contained in the sludge. After being dissolved, phosphate and metal ions have to be precipitated separately in different steps to obtain an uncontaminated phosphate product. This requires large amounts of chemicals and acid resistant equipment. The process also produces many different residuals (metals, sludge with low pH, leachates) that have to be neutralized and disposed at high cost.

Both technologies can produce fertilizer products for direct use in agriculture.

d. **Gas-permeable membrane separation**

The use of hydrophobic gas permeable membranes (GPMs) for ammonia recovery is a relatively new approach. Studies have demonstrated that on the order 95% of free ammonia can be recovered from both gaseous phases and liquid phases (Perera et al., 2019). Ammonia is transferred across a membrane by convection and diffusion. Volatilized ammonia is either condensed or absorbed into an acidic solution (Mehta et al., 2015). Typically, sulfuric acid (H_2SO_4) is recirculating in tubular membranes submerged in ammonia-rich wastewater or vice versa, which allows for ammonia recovery in the form of ammonium sulfate ($(NH_4)_2SO_4$) (Fig. 1e).

A higher ammonia concentration in wastewater improves performance since the process is driven by partial pressure differences. This method uses hydrophobic membranes made of silica, ceramic, polyvinylidene difluoride (PVDF), polypropylene (PP), polytetrafluoroethylene (PTFE), or polymer composites. In comparison to symmetric membranes, asymmetric membranes, such as PVDF, have a lower mass transfer resistance. Membranes can be made of hollow fiber, tubular flat sheet, or spiral coiled cylinders, and the membranes are employed submerged or externally (Mehta et al., 2015).

The choice of membrane materials and combinations is based on fouling resistance, flexibility, texture, cost, availability, and supply. The poor absorption rate per a unit surface area and high capital and operating expenditures per unit volume of wastewater streams are the key barriers for implementation. Due to difficulties with membrane pore clogging and fouling, this approach may not be suited for recovering ammonia from complex matrices containing hydrophobic substances, such as fats, oils, and grease.

e. **Liquid gas stripping**

Gas stripping is the mass transfer of ammonia from the liquid to the gas phase (Fig. 1g). Unlike gas-permeable membrane processing, all waste stream elements (not only ammonia) are free to exert their own partial pressure, while reducing the mass transfer efficiency. An extractant gas (typically air) is used to extract dissolved ammonia from the wastewater. This transfer is only relevant in instances when the effluent ammonia concentration is high (NH_4–N > 2000 mg/L). Air-stripping followed by acid adsorption is a well-established method that has been investigated extensively over the last decade and has shown recovery efficiency of up to 94% (Perera et al., 2019). However, requires the addition of extra base (to raise the pH to 9.5–12) and/or heating to 25–85 °C to covert ammonium (NH_4^+) to its volatile form (NH_3), resulting in high energy consumption (3.9–28.2 kWh/kg N) (Zhang et al., 2018).

The stripped ammonia is recovered by condensation, absorption, or oxidation to make a concentrated fertilizer. Ammonium sulfate ($(NH_4)_2SO_4$), various ammonia salts, or a concentrated ammonia solution are all products of gas stripping. These include the high specific operating costs, need for a concentrated and pretreated feed, and creation of a spent waste stream (low in ammonia and with high pH) unsuitable for lagoon storage or land application. A few facilities worldwide have implemented ammonia–nitrogen stripping and recovery from the sidestreams, but this technology has been proven to be not economically feasible (Beckinghausen et al., 2020).

f. **Electrodialysis**

Electrodialysis (ED) is an extraction technology enabling simultaneous but selective recovery of specific cations (e.g., ammonium NH_4^+, potassium K^+, sodium Na^+, calcium Ca^+) and anions (e.g., phosphate PO_4^{3-}, sulfate SO_4^{2-}, chlorine Cl^-) from one solution (the diluate) to another solution (the concentrate) in the presence of electric field (Kedwell et al., 2021). An ED unit typically has a series of anion (AEMs) and cation (CEMs) exchange membranes located between electrodes (Fig. 1f). In effect, concentrated solutions of the process-oriented anions and cations are obtained separately.

Even though ED has the potential to recover selective nutrients from the different types of wastewaters and produce either liquid or precipitated to solid form fertilizers, it is most applicable for N and K (Perera et al., 2019). When applied for P-recovery, it requires using a sacrificing metal anode release metal ions (e.g., magnesium Mg^{2+}) under the influence of the current and therefore consumes additional electric energy (Lin et al., 2015). ED is also used to concentrate P such that it can be subsequently recovered, typically as struvite (Perera et al., 2019).

g. **Membrane filtration (forward osmosis—FO)**

Forward osmosis (FO) is a unique membrane method that uses osmosis to transfer water over a semipermeable membrane (Fig. 1d). The FO process itself requires little external energy to run, which is a major benefit. Further treatment of the draw solution is necessary to extract fresh water and can be performed using pressure or thermally driven membrane technologies. The forwarded osmotic membrane reactor (OMBR) is an example of FO-based nutrient recovery technology, achieving a nutrient recovery efficiency of over 95%. In the OMBR, nutrients are recovered in the form of struvite/calcium phosphate precipitates without additional chemical reagents (excluding alkaline chemicals for pH correction). Purification of recovered solution can be performed using micro- or ultra-filtration membranes installed on the feed side of OMBR (Ye et al., 2020).

5 Discussion

A variety of physical, chemical, and biological strategies for nitrogen and phosphorus recovery from wastewater have been identified, with approximately 50 nitrogen recovery technologies and more than 50 phosphorus recovery technologies (Beckinghausen et al., 2020; Egle et al., 2016). The most promising recovery techniques are presented in Table 2. The selection of a suitable and efficient technology for nutrient recovery from wastewater depends on many factors, e.g., the type of recovered ion and the waste stream characteristics, including the particular ion concentration. Therefore, the potential location of the process in the wastewater treatment line is crucial. Possible effect of nutrient recovery technologies on wastewater treatment is often neglected (Egle et al., 2016).

From the process sustainability perspective, essential are the reduction of nutrient concentrations in effluent (at least to a level acceptable by law), GHG emissions, and chemicals used. Therefore, biological processes are the most eco-friendly. The involvement of microorganisms which feed on inorganic carbon, N, and P turning them into organic compounds that build their biomass with the simultaneous release of oxygen significantly reduces the number of pollutants released in effluents. What is more, since nutrient-rich waters are their natural environment, the process requires negligible amounts of chemicals (except of small doses to maintain appropriate environmental conditions) (Günther et al., 2018). In contrast, physicochemical technologies necessitate additional chemical input (i.e., precipitant agent in chemical precipitation), regeneration associated with pollutant release (i.e., adsorptive media or electrodialysis electrodes), and may have unsatisfactory achievable effluent concentrations (i.e., electrodialysis) or, like membrane processes, all of these drawbacks (Perera et al., 2019). However, some unsustainable factors may also appear if specifically consider the end product orientation of biological recovery technologies.

Except for membrane procedures, which are associated with highly costly and easily fouled and clogged membrane modules, all of the proposed recovery technologies are economically efficient. The necessity to handle produced wastes and by-products alters the operational costs of particular technologies. The most practical methods are chemical precipitation and adsorption. Pumps, mixers, and thermostats are major energy consumers in processes. Furthermore, although simple filtering happens naturally, membrane distillation requires sufficient absorbent flux and consistent process temperature (Ezugbe and Rathilal, 2020; Huang et al., 2020; Molinos-Senante et al., 2011).

Hydrothermal liquefaction (HTL) of the algal biomass allows for the reuse and recycling of nutrients accumulated in the algal. In this method wet algae (around 10–30 wt.%) are converted into a bio-crude oil with a high heating value of 35–40 MJ/Kg in hot liquid water between 250–374 °C and 4–22 MPa. The HTL aqueous phase can include up to 50% nitrogen and 85–100% phosphorus, depending on the procedure and circumstances (Munasinghe-Arachchige & Nirmalakhandan, 2020). Obtained HTL water can be reused as a source of nutrients for algal cultivation. Although algal techniques appear to be more profitable (nutrients elimination, less chemicals, energy efficiency, product variety), their application may be constrained by space availability. Furthermore, the financial, energy, and chemical requirements for biomass harvesting (up to 30% of total biomass production cost) and valorization must be considered (Acién Fernández et al., 2018). Since phosphorus accumulating organisms (PAOs) biomass requires less time to release nutrients than algae, it has recently been identified as a favored biological strategy (Perera et al., 2019).

6 Perspectives and Future Outlook

A wide spectrum of circular nature-based technologies, aiming at recovering macronutrients in new bio-based fertilizers is proposed by the scientists all over the world. Recently, the awareness of the need to recover micronutrients and organic matter is also increasing. However, most of the promising technologies still require deeper pilot and full-scale application tests. Nevertheless, it is already clear that in order to close nutrient cycles, no single technique will be the ultimate solution. As a specific product supply should match its demand in time and space, a technology that meets the local available resources, the desired product purity, and intended purpose will have to be chosen (Günther et al., 2018).

Still, the production of high-efficiency products from waste sources does not guarantee their commercial success. Thus, a "concept-to-market" strategy is required. The specific technology end-products need to be fully characterized, evaluated, and adequately tested (Theregowda et al., 2019). It is important to understand, among others, the associated risks, side effects and long-term cost (Reid et al., 2018). While planning a suitable market entry, raising awareness and replicating successfully implemented examples may aid to overcome the skepticism of waste-derived products usage (Chojnacka et al., 2019). New business models and infrastructure for supplying a chain for

Table 2 Characteristic of the highly efficient nutrient recovery techniques (Egle et al., 2016; Kogler et al., 2021; Sartorius et al., 2011; Xie et al., 2016)

Recovery technique	Energy efficiency	Costs reduction	Chemicals reduction	Reduced GHG emissions	Low N and P effluent	Easier sludge management	Simple operation	Waste reduction	Recovered nutrients	
									P	N
Chemical precipitation	–	–	–	+	+	+	+	+	+	+
Wet-chemical extraction	+	–	–	+	–	+	+	–	+	–
Thermo-chemical extraction	+	–	+	–	–	+	+	–	+	–
Ion exchange/adsorption	+	+	+	–	+	+	+	+	+	–
Gas-Permeable membrane separation	–	–	–	–	–	–	+	–	–	+
Gas stripping	–	+	+	–	–	–	+	–	–	+
Membrane filtration	+	–	–	–	–	–	+	–	+	+
Uptake by microalgae	+	+	+	+	+	+	+	–	+	+
Electrodialysis	+	+	+	–	–	–	+	–	+	+

(continued)

Table 2 (continued)

Recovery technique	Commercial technology	Conditions and reagents	Efficiency %	Final product
Chemical precipitation	OSTARA (Pearl®)	$MgCl_2$, NaOH	95 (P) 15 (N)	MAP
	PHOSPAQ™	O_2, MgO	75–81 (P)	MAP
	ANPHOS®	O_2, $Mg(OH)_2$	80–90 (P)	MAP
	NuReSYS	O_2, $Mg(Cl)_2$	96 (P)	MAP
	AirPrex®		80–90	MAP
	MULTIFORM™	$MgCl_2$, NaOH	80–90	MAP
Wet-chemical extraction	BioCon	H_2SO_4	–	Phosphoric acid
	SEABORNE	H_2SO_4, H_2S-rich biogas, $Mg(OH)_2$, NaOH	>90 (P)	MAP, ammonium sulfate
	SEPHOS	H_2SO_4, NaOH	–	Calcium phosphate (12% P)
	PASH	HCl, NaOH, H_3PO_4, H_2SO_4	70–80	Phosphorus potential 90%
Thermo-chemical extraction	ASH DEC	Up 1050° C	98	20% P_2O_5
	RecoPhos	1900° C and H_2PO_4	–	P_2O_5–22%
	Bama/AshDac	850–1000° C and Chlorine donor	–	P_2O^5 content of 13–18%
	Merphec	2000° C	60–80	4.6–12% P_2O_5 with over 90% citric acid solubility
Ion exchange/adsorption	RECYPHOS	–	–	FeP (no fertilizer)
	P-ROC	CSH as absorbent	85–95	CaP, CaP on CSH
	PHOSIEDI	–	–	CaP
Gas-Permeable membrane separation	–	H_2SO_4, NaOH	95	$(NH_4)_2SO_4$
Gas stripping	–	Air, adsorber+H_2SO_4,	90	$(NH_4)_2SO_4$
Membrane filtration	–	OMBR[a]	90–95 (P)	Amorphous calcium phosphate
Uptake by microalgae	–	HTL[b]	74 (P) 31 (N)	MAP, ammonium sulfate
Electrodialysis	–	IEM[c]	72–75 (P) 72.9 (N)	Phosphoric acid

[a] OMBR—forwarded osmotic membrane bioreactor
[b] HTL—Hydrothermal liquefaction
[c] IEM—ion exchange membrane

recovered materials need to be provided (Liu et al., 2021a, 2021b). To overcome these socio-economic, environmental, technical, and legal aspects a transdisciplinary approach is needed (Jama-Rodzeńska et al., 2021).

7 Conclusions

Due to the CE concept, recovering nutrients from wastewater becomes more profitable than their removal. One potential reason is that this recovery may serve as a backup supply of fertilizer, relieving some of the pressures associated with rising demand for food production. Moreover, nitrogen recovery has the potential to mitigate the negative effects of wastewater treatment on the environment and save cost. Therefore, the nutrient recovery method should be chosen carefully to provide good nutrients removal from wastewater while maximizing its recovery rate, destroying hazardous substances, and producing a material that could be applied with low environmental risks, good fertilizing effects, and high economic efficiency.

Acknowledgements The research leading to these results has received funding from the Norway Grants 2014–2021 via the National Centre for Research and Development under the grant number NOR/SGS/INPORR/0074/2020-00.

References

Acién Fernández, F. G., Gómez-Serrano, C., & Fernández-Sevilla, J. M. (2018). Recovery of nutrients from wastewaters using microalgae. *Frontiers in Sustainable Food Systems, 2*, 59.

Ali, S., Peter, A. P., Chew, K. W., Munawaroh, H. S. H., & Show, P. L. (2021). Resource recovery from industrial effluents through the cultivation of microalgae: A review. *Bioresource Technology, 337*, 125461.

AlMomani, F. A., & Örmeci, B. (2016). Performance of Chlorella Vulgaris, Neochloris Oleoabundans, and mixed indigenous microalgae for treatment of primary effluent, secondary effluent and centrate. *Ecological Engineering, 95*, 280–289.

Awasthi, M. K., Sarsaiya, S., Patel, A., Juneja, A., Singh, R. P., Yan, B., Awasthi, S. K., Jain, A., Liu, T., Duan, Y., Pandey, A., Zhang, Z., & Taherzadeh, M. J. (2020). Refining biomass residues for sustainable energy and bio-products: An assessment of technology, its importance, and strategic applications in circular bio-economy. *Renewable and Sustainable Energy Reviews*, 127.

Bali, M., & Gueddari, M. (2019). Removal of phosphorus from secondary effluents using infiltration–percolation process. *Applied Water Science, 9*(3), 1–8.

Beckinghausen, A., Odlare, M., Thorin, E., & Schwede, S. (2020). From removal to recovery: An evaluation of nitrogen recovery techniques from wastewater. *Applied Energy, 263*, 114616.

Bradford-Hartke, Z., Razmjou, A., & Gregory, L. (2021). Factors affecting phosphorus recovery as struvite: Effects of alternative magnesium sources. *Desalination*, 504.

Brennan, B., Lawler, J., & Regan, F. (2021). Recovery of viable ammonia–nitrogen products from agricultural slaughterhouse wastewater by membrane contactors: A review. *Environmental Science: Water Research & Technology, 7*(2), 259–273.

Chojnacka, K., Gorazda, K., Witek-Krowiak, A., & Moustakas, K. (2019). Recovery of fertilizer nutrients from materials-contradictions, mistakes and future trends. *Renewable and Sustainable Energy Reviews, 110*, 485–498. https://doi.org/10.1016/j.rser.2019.04.063

Chrispim, M. C., Scholz, M., & Nolasco, M. A. (2019). Phosphorus recovery from municipal wastewater treatment: Critical review of challenges and opportunities for developing countries. *Journal of Environmental Management, 248*.

Constantine, T. A. (2006). North American experience with centrate treatment technologies for ammonia and nitrogen removal. *Proceedings of the Water Environment Federation, 2006*(7), 5271–5281.

Cruz, H., Law, Y. Y., Guest, J. S., Rabaey, K., Batstone, D., Laycock, B., ... & Pikaar, I. (2019). Mainstream ammonium recovery to advance sustainable urban wastewater management. *Environmental Science & Technology, 53*(19), 11066–11079.

Czerwionka, K., Makinia, J., Pagilla, K. R., & Stensel, H. D. (2012). Characteristics and fate of organic nitrogen in municipal biological nutrient removal wastewater treatment plants. *Water Research, 46* (7), 2057–2066.

Egle, L., Rechberger, H., Krampe, J., & Zessner, M. (2016). Phosphorus recovery from municipal wastewater: An integrated comparative technological, environmental and economic assessment of P recovery technologies. *Science of the Total Environment, 571*, 522–542.

EPA, U. (2000). *Wastewater technology fact sheet: Trickling filter*. Environmental Protection Agency.

European Commission, Directorate-General for Environment (2020) *Circular Economy Action Plan: International aspects*. Publications Office.

Ezugbe, E. O., & Rathilal, S. (2020). Membrane technologies in wastewater treatment: A review. *Membranes, 10*(5).

Gell, K., Ruijter, F. J. de, Kuntke, P., Graaff, M. de, & Smit, A. L. (2011). Safety and effectiveness of struvite from black water and urine as a phosphorus fertilizer. *Journal of Agricultural Science, 3* (3).

Günther, S., Grunert, M., & Müller, S. (2018). Overview of recent advances in phosphorus recovery for fertilizer production. In *Engineering in Life Sciences* (Vol. 18, Issue 7, pp. 434–439). Wiley-VCH Verlag.

Herzel, H., Krüger, O., Hermann, L., & Adam, C. (2016). Sewage sludge ash–A promising secondary phosphorus source for fertilizer production. *The Science of the Total Environment, 542*(Pt B), 1136–1143.

Huang, X., Guida, S., Jefferson, B., & Soares, A. (2020). Economic evaluation of ion-exchange processes for nutrient removal and recovery from municipal wastewater. *Npj Clean Water, 3*(1), 1–10.

Jama-Rodzeńska, A., Białowiec, A., Koziel, J. A., & Sowiński, J. (2021). Waste to phosphorus: A transdisciplinary solution to P recovery from wastewater based on the TRIZ approach. *Journal of Environmental Management, 287*.

Jardin, N., Thöle, D., & Wett, B. (2006). Treatment of sludge return liquors: Experiences from the operation of full-scale plants. *Proceedings of the Water Environment Federation, 2006*(7), 5237–5255.

Jupp, A. R., Beijer, S., Narain, G. C., Schipper, W., & Slootweg, J. C. (2021). Phosphorus recovery and recycling-closing the loop. In

Chemical Society Reviews (Vol. 50, Issue 1, pp. 87–101). Royal Society of Chemistry.

Kedwell, K. C., et al. (2021). Selective electrodialysis for simultaneous but separate phosphate and ammonium recovery. *Environmental Technology (United Kingdom), 42*(14), 2177–2186.

Kogler, A., Farmer, M., Simon, J. A., Tilmans, S., Wells, G. F., & Tarpeh, W. A. (2021). Systematic evaluation of emerging wastewater nutrient removal and recovery technologies to inform practice and advance resource efficiency. *ACS ES&T Engineering, 1*(4), 662–684.

Lin, H., Gan, J., Rajendran, A., Reis, C. E. R., & Hu, B. (2015). *Phosphorus removal and recovery from digestate after biogas production*. IntechOpen.

Liu, H., Hu, G., Basar, I. A., Li, J., Lyczko, N., Nzihou, A., & Eskicioglu, C. (2021a). Phosphorus recovery from municipal sludge-derived ash and hydrochar through wet-chemical technology: A review towards sustainable waste management. *Chemical Engineering Journal, 417*, 129300.

Liu, Q., Yang, L., & Yang, M. (2021b). Digitalisation for water sustainability: Barriers to implementing circular economy in smart water management. *Sustainability (Switzerland), 13*(21).

Liu, Y., He, L. F., Deng, Y. Y., Zhang, Q., Jiang, G. M., Liu, H. (2022). Recent progress on the recovery of valuable resources from source-separated urine on-site using electrochemical technologies: A review. *Chemical Engineering Journal*, 136200.

Magrí, A., Carreras-Sempere, M., Biel, C., & Colprim, J. (2020). Recovery of phosphorus from waste water profiting from biological nitrogen treatment: Upstream, concomitant or downstream precipitation alternatives. *Agronomy, 10*(7), 1039.

Martınez, M. E., Sánchez, S., Jimenez, J. M., El Yousfi, F., & Munoz, L. (2000). Nitrogen and phosphorus removal from urban wastewater by the microalga Scenedesmus obliquus. *Bioresource Technology, 73*(3), 263–272.

Molinos-Senante, M., Hernández-Sancho, F., Sala-Garrido, R., & Garrido-Baserba, M. (2011). Economic feasibility study for phosphorus recovery processes. *Ambio, 40*(4), 408.

Muhammad, G., Alam, M., Xiong, W., Lv, Y., & Xu, J. L. (2020). Microalgae biomass production: An overview of dynamic operational methods. *Microalgae Biotechnology for Food, Health and High Value Products*, 415–432.

Munasinghe-Arachchige, S. P., & Nirmalakhandan, N. (2020). Nitrogen-fertilizer recovery from the centrate of anaerobically digested sludge. *Environmental Science & Technology Letters, 7*(7), 450–459.

Pastor, L., Mangin, D., Ferrer, J., & Seco, A. (2010). Struvite formation from the supernatants of an anaerobic digestion pilot plant. *Bioresource Technology, 101*(1), 118–125.

Perera, M. K., & Englehardt, J. D. (2020). Simultaneous nitrogen and phosphorus recovery from municipal wastewater by electrochemical pH modulation. *Separation and Purification Technology*, 250.

Perera, M. K., Englehardt, J. D., & Dvorak, A. C. (2019). *Technologies for recovering nutrients from wastewater: A critical review. Environmental Engineering Science* (pp. 511–529). Mary Ann Liebert Inc.

Qadir, M., Drechsel, P., Jiménez Cisneros, B., Kim, Y., Pramanik, A., Mehta, P., Olaniyan, O (2020) Global and regional potential of wastewater as a water nutrient and energy source. *Natural Resources Forum, 44*(1), 40–51. https://doi.org/10.1111/1477-8947.12187

Rahimi, S., Modin, O., & Mijakovic, I. (2020). Technologies for biological removal and recovery of nitrogen from wastewater. *Biotechnology Advances, 43*, 107570.

Reid, K., Canada, A., Zhang, T., & Goel, P. (2018). Nutrient recovery and reuse in Canada: Foundations for a national framework the international institute for sustainable development (IISD) would like to acknowledge review and inputs from a steering committee comprising Brad Bass, Environment and Climate Change Canada. https://iisd.org/event/national-nutrient-reuse-and-recovery-forum

Renuka, N., Ratha, S. K., Kader, F., Rawat, I., & Bux, F. (2021). Insights into the potential impact of algae-mediated wastewater beneficiation for the circular bioeconomy: A global perspective. *Journal of Environmental Management, 297*, 113257.

Robles, Á., Aguado, D., Barat, R., Borrás, L., Bouzas, A., Giménez, J. B., ... & Seco, A. (2020). New frontiers from removal to recycling of nitrogen and phosphorus from wastewater in the Circular Economy. *Bioresource Technology, 300*, 122673.

Rout, P. R., Shahid, M. K., Dash, R. R., Bhunia, P., Liu, D., Varjani, S., Zhang, T. C., Surampalli, R. Y. (2021). Nutrient removal from domestic wastewater: A comprehensive review on conventional and advanced technologies. *Journal of Environmental Management*, 296.

Rugaika, A. M., van Deun, R., Njau, K. N., & van der Bruggen, B. (2019). Phosphorus recovery as calcium phosphate by a pellet reactor pre-treating domestic wastewater before entering a constructed wetland. *International Journal of Environmental Science and Technology, 16*(7), 3851–3860.

Sartorius, C., von Horn, J., & Tettenborn, F. (2011). Phosphorus recovery from wastewater-state-of-the-art and future potential. *Proceedings of the Water Environment Federation, 2011*(1), 299–316.

Sengupta, S., Nawaz, T., & Beaudry, J. (2015). Nitrogen and phosphorus recovery from wastewater. *Current Pollution Reports, 1*(3), 155–166.

Sikosana, M., Randall, D. G., Petrie, D. J., South Africa. Water Research Commission, University of Cape Town. Department of Chemical Engineering. Environmental & Process Systems Engineering, Eidgenössische Anstalt für Wasserversorgung, A. und G. P. E., Aurecon, & Københavns universitet. Department of Plant and Environmental Sciences. (2016). Nutrient and energy recovery from sewage: Towards an integrated approach: Report to the Water Research Commission.

Smol, M., Adam, C., & Anton Kugler, S. (2020). Inventory of Polish municipal sewage sludge ash (SSA)-Mass flows, chemical composition, and phosphorus recovery potential. *Waste Management (New York, N.Y.), 116*, 31–39.

Theregowda, R. B., González-Mejía, A. M., Ma, X., & Garland, J. (2019). Nutrient recovery from municipal wastewater for sustainable food production systems: An alternative to traditional fertilizers. *Environmental Engineering Science, 36*(7), 833–842.

Urdalen, I. (2013). *Phosphorus recovery from municipal wastewater*. Norwegian University of Science and Technology.

Van der Hoek, J. P., Duijff, R., & Reinstra, O. (2018). Nitrogen recovery from wastewater: Possibilities, competition with other resources, and adaptation pathways. *Sustainability, 10*(12), 4605.

Xie, M., Shon, H. K., Gray, S. R., & Elimelech, M. (2016). Membrane-based processes for wastewater nutrient recovery: Technology, challenges, and future direction. *Water Research, 89*, 210–221.

Xu, X., Wang, G., Zhou, L., Yu, H., & Yang, F. (2018). Start-up of a full-scale SNAD-MBBR process for treating sludge digester liquor. *Chemical Engineering Journal, 343*, 477–483.

Ye, Y., Ngo, H. H., Guo, W., Chang, S. W., Nguyen, D. D., Zhang, X., ... & Liang, S. (2020). Nutrient recovery from wastewater: From technology to economy. *Bioresource Technology Reports, 11*, 100425.

Zaman, M., Kim, M., & Nakhla, G. (2021). Simultaneous nitrification-denitrifying phosphorus removal (SNDPR) at low DO for treating carbon-limited municipal wastewater. *Science of the Total Environment, 760*, 143387.

Zangarini, S., Sciarria, T. P., Tambone, F., & Adani, F. (2020). Phosphorus removal from livestock effluents: Recent technologies and new perspectives on low-cost strategies.

Zhang, C., Guisasola, A., & Baeza, J. A. (2022). A review on the integration of mainstream P-recovery strategies with enhanced biological phosphorus removal. *Water Research*, 118102.

Zhang, C., Ma, J., Song, J., He, C., & Waite, T. D. (2018). Continuous ammonia recovery from wastewaters using an integrated capacitive flow electrode membrane stripping system. *Environmental Science & Technology, 52*(24), 14275–14285.

Zhang, H., Gong, W., Luo, X., Xie, B., Li, G., & Liang, H. (2019). Obtaining high-purity struvite from anaerobically digested wastewater: Effects of pH, Mg/P, and Ca 2+ interactions. *Environmental Engineering Science, 36*(1), 102–113.

Circular Economy in Domestic and Industrial Wastewaters: Challenges and Opportunities

Majeti Narasimha Vara Prasad

Abstract

Population explosion, climate change and lack of adequate infrastructure for drainage and water distribution systems imposed a shortage of water for domestic and industrial usage. This has forced water resource managers to shift from traditional linear, reuse patterns to a circular economy (CE). British Standards Institution (BSI) developed and launched the first circular-economy standard BS 8001:2017 (6Rs) Framework for implementing the principles of the circular economy in the management of the water sector. Due to global challenges like climate change, shift in rainfall, ever-increasing pollution and emissions, biodiversity loss, wide variety of waste generation, the challenges and opportunities have become very critical for deriving anticipated results in the successful implementation of CE in the water sector. Circular Economy attained a new and sustainable dimension which works on multiple "R"s Multiple "R"s have an overriding impact over 6Rs and/or 3Rs. Therefore, the importance of multiple "R"s, viz., Reduce, Reclaim, Recovery, Recycle, Refuse, Remediation, Renovation, Replenish, Resilience, Restore, Reuse, Reverence to nature, Reclamation, Repurposing and Rainwater harvest is being increasingly considered in various parts of the world. In this context, Resource Recovery and energy recovery from wastewater, removal/recovery of beneficial minerals from waste streams, renewables from wastewater using algae, etc., are covered in this paper with future perspectives. In this chapter, bioremediation of domestic and industrial wastewater is chosen as examples for discussing the opportunities and challenges.

M. N. V. Prasad (✉)
School of Life Sciences, University of Hyderabad (an Institution of Eminence), Hyderabad, Telangana 500 046, India
e-mail: mnvsl@uohyd.ac.in

Keywords

Multiple "R"s • Domestic • Industrial • Wastewaters • Effluents

1 Introduction

Water is the soul of life

Water bodies (rural and urban) rejuvenation and management must be a global priority (Achillas & Bochtis, 2020). Earth's total global stock of fresh water for human use is 2.5% while 97.5% is saline. From the 2.5%, the largest part (~70%) is frozen in polar regions and 30% are confined to aquifers of difficult access. Thus, only 0.007% of the total global water is directly accessible for human use. Unfortunately, part of this water is polluted by industries and domestic use (Azma et al., 2021). Hence, water reclamation and circular economy are of paramount importance in water security. Figure 1 shows the exponential increase in publications on resource recovery from wastewater from 2000 to 2020.

Macrophytes have attracted the attention of global scientists for their potential application in the remediation of wastewater (Fig. 2). Constructed wetlands have been designed and engineered using native diversity of macrophytes for treating industrial and domestic wastewaters (Fig. 3).

Integrating circularity to achieve sustainability citing examples of various wastewater treatment systems (Avellan et al., 2022). Bioremediation of domestic and industrial wastewaters integrated with enhanced biodiesel production using novel oleaginous microalgae is progressing globally (Arora et al., 2016). Industrial and domestic wastewater remediation is being successfully deployed by various industries and water management agencies (Adelodun et al.,

M. Smol et al. (eds.), *Water in Circular Economy*, Advances in Science, Technology & Innovation,
https://doi.org/10.1007/978-3-031-18165-8_13

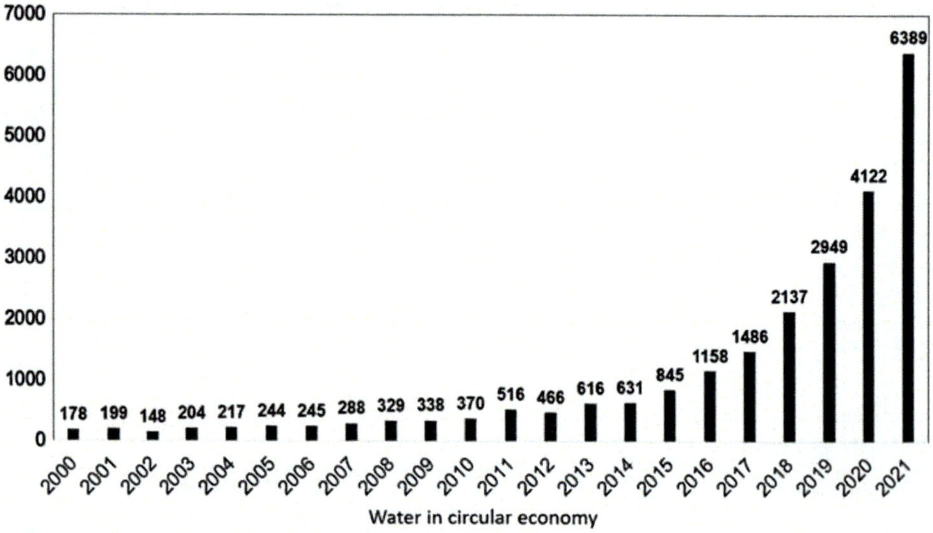

Fig. 1 There has been a gradual rise in publications on water in circular-economy wastewater. *Data source* www.sciencedirect.com. Keywords: circular economy, water

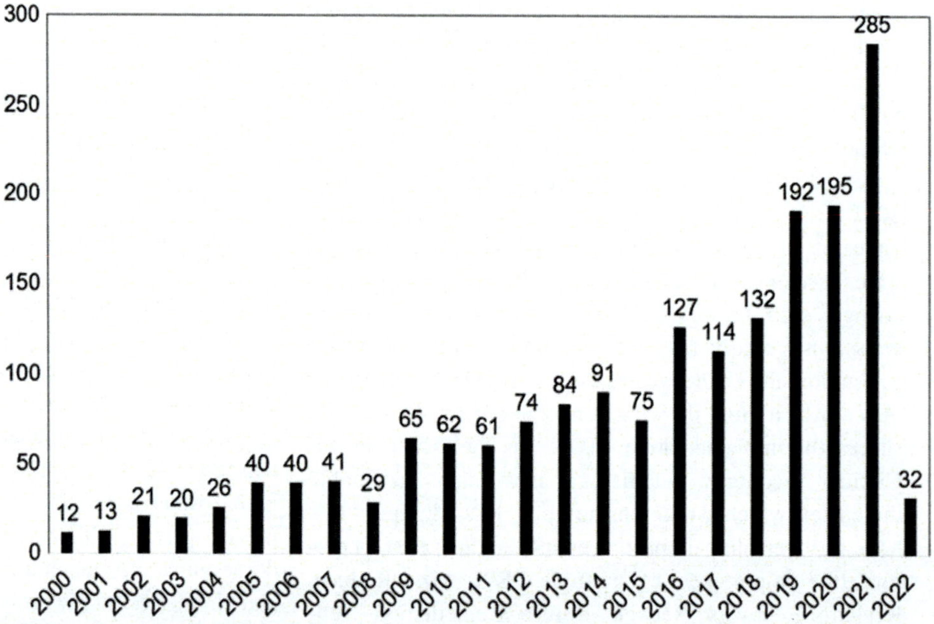

Fig. 2 There has been a gradual rise in publications using macrophytes for the remediation of wastewater. *Data source* www.sciencedirect.com. Keywords: Macrophytes, remediation

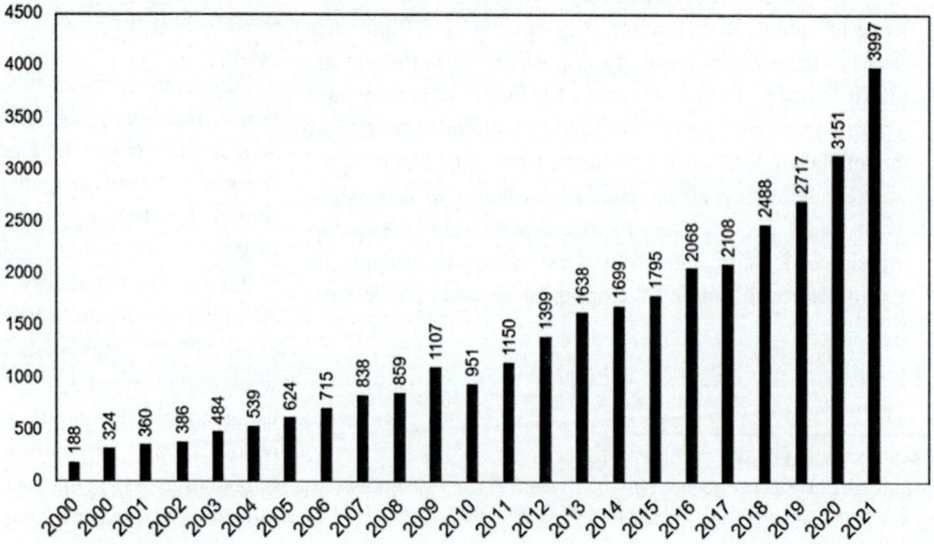

Fig. 3 There has been a gradual rise in publications on various aspects of constructed wetlands and their ecosystem service. *Data source* www.sciencedirect.com. Keywords: Constructed wetlands

Fig. 4 **a** Wettest place on planet earth **b** Rain clouds in the state of Meghalaya

2021; Ahmad et al., 2021; Ali et al., 2021; Ammeri et al., 2021).

The climate of India is broadly described as tropical monsoonal type with four major seasons: Winter, Hot summer, Rainy (South Western monsoon) and Post monsoon (North East monsoon). The highest annual rainfall recorded was 24,555 mm in 1974 and is labelled as the "Wettest place on earth" in the State of Meghalaya, North East India (Fig. 4a, b).

Circular economy is a practical solution for promoting entrepreneurial sustainability, economic growth, environmental resilience and a better quality of life for all. Common weaknesses and gaps in current bioeconomy strategies are seen in terms of the sound use of land and water resources, waste management along the whole value chain, possible competition between the different biomass end-use sectors and mechanisms to benefit farmers.

In high rainfall area, rainwater harvesting (RWH) is advantageous for water security. Rainwater harvesting (RWH) is the process of accumulating incident raindrops on ground surfaces and roofs, with the help of cisterns, tanks and underground check dams (Fig. 5). RWH has been applied for centuries by humankind to meet water supply needs and nowadays can still represent an important practice to improve the efficiency of the use of water in the urban future by reducing household expenditures on water consumption and offering many opportunities for domestic and agricultural needs (Helmreich & Horn, 2009; Krishna et al., 2020; Pari et al., 2021).

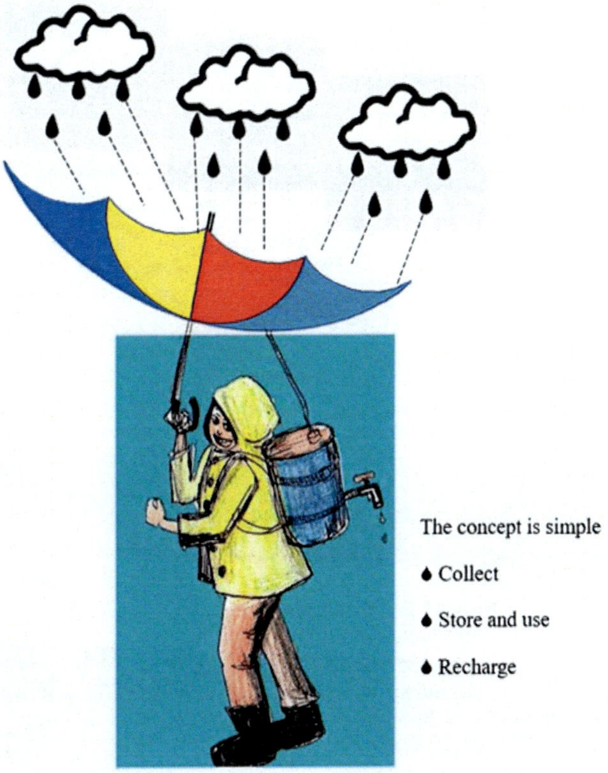

Fig. 5 Cartoon showing simplistic water harvesting for domestic use

Broadly rainwater can be harvested for two purposes: (a) Storing rainwater for ready use in containers above or below ground and (b) Charging into the soil for withdrawal later (groundwater recharging). The storage of rainwater on the surface is a traditional technique and the structures used were underground tanks, ponds, check dams, etc.

2 Circular Economy in the Urban Water Sector

The water sector involves the exchange of water within various components such as surface water, groundwater, and urban water. These exchanges induce a change in the quantity and quality of water in these components. The water is used by humans, and hence the behavioural and management aspects also become dominant in the urban water sector. A conventional flow of water and its exchange between various components is illustrated with the help of Fig. 6. Circular economy is game-changing municipal wastewater treatment technology towards energy and carbon neutrality (Zhang & Liu, 2022) and novel perspectives for urban water resource management have gained paramount significance (Smol, 2022; Zhou et al., 2021).

Multiple "R"s approach for the sustainability of water including natural resources is explained in Table 1 The Confluence of 6Rs and BS8001:2017 principles for circular economy in water are detailed in Table 2.

Schematic presentation of urban water cycle considering 6Rs strategies of Circular Economy in the urban water sector is shown in Fig. 7.

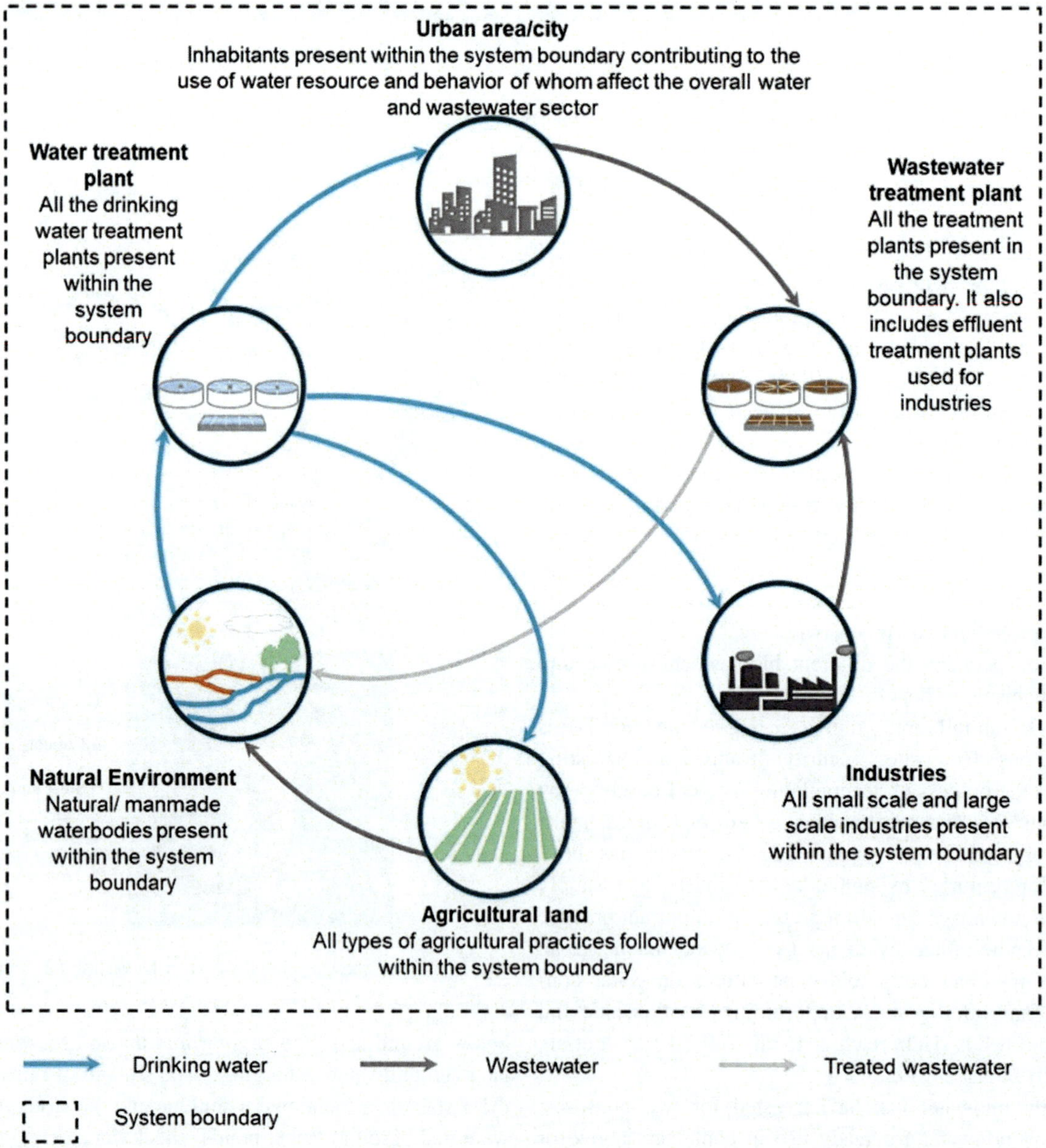

Fig. 6 Conventional urban water cycle with the exchange of water and wastewater within the system boundary. Reproduced with permission from Kakwani and Kalbar (2020)

Table 1 Multiple "R"s for the sustainability of water including natural resources. Reproduced with permission from Kakwani and Kalbar (2020)

Reduce	Decrease in the consumption of freshwater by creating awareness in consumers, applying efficient fixtures or appliances and reducing leakages, thereby reducing wastewater generation
Reclaim	Treatment of wastewater and its use outside the loop/process
Recovery	Extraction of valuable resources (e.g. energy, material) present in wastewater
Recycle	Use of treated wastewater within the same loop or in the same process
Refuse	Nonhazardous waste that requires collection and transport to a processing site
Remediation	Action in particular of reversing or stopping environmental deterioration or reducing the toxicity of a substrate
Renovation	Modification of any existing structure, or portion of a structure to a new look
Replenish	to fill with
Resilience	The ability to withstand adversity and bounce back from the stress
Restore	Replenish the water resources through artificial interventions (e.g. Managed aquifer recharge, rainwater harvesting, afforestation, rejuvenation of waterbodies) Rainwater harvest
Reuse	Reuse of used water in its crude form (no treatment/processing involved) for multiple purposes inside or outside the loop
Reverence to nature	To make full or complete again, as by supplying what is lacking, used up,
Reclamation	Restoration of land to its original condition
Repurposing	Repurposing is the modification or revision of a product to function for a new application

Repurposing is the modification or revision of a product to function for a new application or in other domains that helps to reduce the end-of-pipe waste, redeems cost/materials and increase renewability and durability. The repurposing of the products through chemical/thermal/biochemical methods reduces the requirement of primary feedstock, optimizes the resource yields and increases the multifunctionality of products.

Multiple "R"s for sustainability of water is documented in Tables 3, 4, 5, 6, 7, 8 and 9. Tables 3, 4 and 5 enumerate examples of wastewater remediation, Table 6 Resilience of water, Table 7 Rainwater harvesting, Table 8 Water reclamation and reuse and Table 9 Water reuse and recycle.

3 Wastewater Treatment

3.1 Innovative Circular-Economy Solutions Co-developed by Water Industry Sector

Wastewater treatment plants (WWTPs) either do not exist in many developing countries, or the capacity of the installed WWTP is not adequate for the growing population and wastewater generated. Therefore, transition from a linear to a circular-economy operation/design concept is an immediate necessity. Such an approach would help in resource recovery and more sustainable management of waste and natural resources. However, the transition towards a circular economy is a complex and long process due to the existence of technical, economic, social and regulatory barriers. These existing barriers are critical challenges for a modern and sustainable WWTP concept.

In this context, the European Union's Horizon 2020 project "Achieving wider uptake of water-smart solutions - WIDER UPTAKE" aims to overcome the existing barriers (technological, regulatory, organizational, social and economic) towards the transition froma linear to a circular-economy model for WWTPs. The flow scheme shown in Fig. 8 is aimed at increasing the awareness of the existing barriers to a circular economy and summarizes the key contributions of the WIDERUPTAKE project in terms of water reuse, sludge reuse and nutrient recovery.

In climate constrained world, the importance of wetlands is gaining considerably. In the last five years, the publications on various aspects of wetlands have been steeply growing. Aquatic macrophytes and microalgae have paramount significance in the bioremediation of aquatic ecosystems (e.g., *Typha latifolia, Eichhornia crassipes, Ipomea* sp. *Lemna minor, Polygonum sp, Alternanthera philoxeroides, Phragmites* sp.) Aquatic macrophytes and microalgae are represented by a wide taxonomic variety of life forms. They are important in nutrient cycling, controlling water quality, sediment stabilization and provision of habitat for a host of aquatic organisms.

Table 2 Confluence of 6Rs and BS8001:2017 principles for CE in water. Reproduced with permission from Kakwani and Kalbar (2020)

<table>
<tr><th></th><th>Systems thinking</th><th>Value optimization</th><th>Innovation</th><th>Collaboration</th><th>Stewardship</th><th>Transparency</th></tr>
<tr><td>Reduce</td><td colspan="2">Optimal utilization of freshwater by using water-efficient fixtures at household level enhances the overall availability of freshwater, reducing wastewater generation</td><td colspan="2">Collaboration between consumers, suppliers and the government to provide innovative devices (e.g., smart meters) can reduce the water demand and enhance the value of water, which was otherwise ignored</td><td colspan="2">Monitoring reduction at source level, thus assessing direct and indirect impacts of reducing consumption with an authentic and transparent exchange of information internally and externally</td></tr>
<tr><td>Reuse</td><td colspan="2" rowspan="3">Cascading/reuse, recycling, and/or treatment/reclamation at the household or industry level enhances the utility and value of used water, thus significantly reducing the overall fresh water demand</td><td colspan="2">Collaboration between water consumer and water utilities to design the reuse activity considering sustainable management of resources. For example, reusing water within the industry for cooling purposes enhances the value of used water</td><td colspan="2" rowspan="3">Assessing positive and negative impacts of wastewater reuse in its crude form and hence maintaining transparency among authorities about the characteristics of wastewater and % reuse adopted internally as well as externally</td></tr>
<tr><td>Recycle</td><td colspan="2" rowspan="2">Collaboration between WWTP government and consumer to design efficient and innovative recycling facilities to enhance value creation</td></tr>
<tr><td>Reclaim</td></tr>
<tr><td>Recover</td><td colspan="2">Wastewater is a potential resource for water, energy, and material, thus recovery is an opportunity to keep the waste (resources) present in wastewater at its highest utility and value, eventually reducing the extraction of Fresh resources</td><td colspan="2">Collaboration between waste management authorities, market, and consumer to design reclamation and/or recovery facilities to generate/recover products as per the market value and consumer requirement considering the sustainability aspect</td><td>Monitoring recycle, reclamation and recovery activities including direct and indirect impacts from economic, environmental and social aspects</td><td>Transparency to exchange information about the methodology and extent of recycle, reclamation and recovery activities practiced</td></tr>
<tr><td>Restore</td><td colspan="2">Utilizing rainwater or treated wastewater for groundwater restoration can enhance the groundwater level reducing water scarcity in the dry season. Another example is afforestation which helps in restoring the water cycle, thus increasing the value of land as well as reducing environmental pollution</td><td colspan="2">Collaboration between government authorities and customers to devise innovative groundwater restoration models to enhance the availability of fresh water in the future, e.g., rainwater harvesting</td><td>Responsible action towards the restoration activities at small scale as well as large scale for promoting water restoration in urban water cycle</td><td>Transparency to share information about the activities involved in the restoration and the extent of restoration achieved on a timely basis</td></tr>
<tr><td>Repurposing</td><td colspan="2">Grey water use for irrigation</td><td colspan="2"></td><td></td><td></td></tr>
</table>

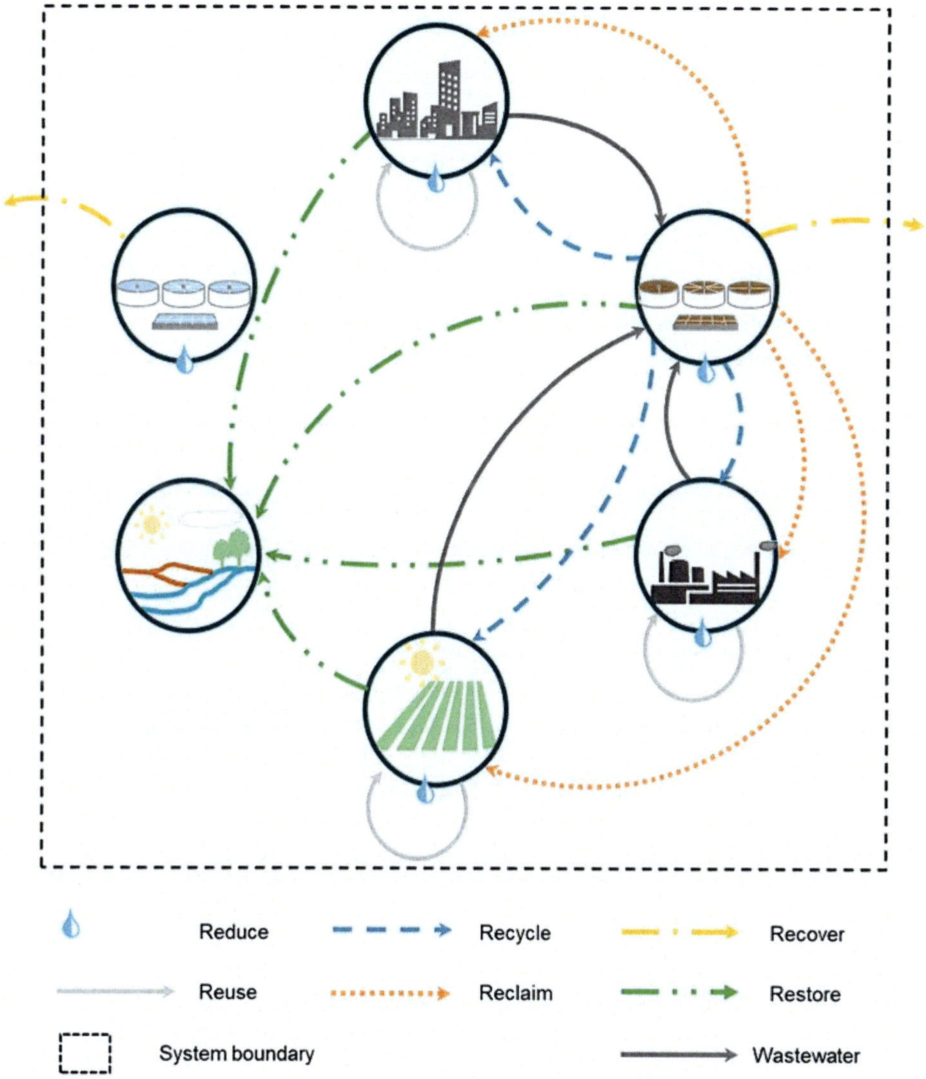

Fig. 7 Urban water cycle considering 6Rs strategies of Circular Economy in the urban water sector (Reproduced with permission from Kakwani & Kalbar, 2020)

3.2 Wetlands for Wastewater Treatment

Two kinds of wetlands are in the service of mankind (a) Natural wetlands—used for wastewater treatment for centuries; (b) Constructed wetlands—effective in treating organic matter, nitrogen and phosphorus, thereby decreasing the concentrations of trace metals and organic chemicals (Kadlec & Knight, 1996). The submerged aquatic macrophytes have very thin cuticles, and therefore, readily take up metals from water through the entire surface. Macrophytes possess an extraordinary ability to survive the adverse conditions of pollution and possess a high colonization rate which are virtual tools of excellence for phytoremediation. Further, they redistribute metals from sediments to water, finally accumulate in the plant tissues, accelerate biogeochemical processes and hence maintain homoeostasis. Both submerged and emergent macrophytes play an important role in metal bioavailability from sediments through rhizosphere exchanges and other carrier chelates. These phenomena facilitate metal uptake by other floating and emergent forms of macrophytes. Macrophytes readily take up metals in their reduced form sediments, where anaerobic situations prevail due to lack of oxygen and oxidize them in the plant tissues making them immobile and thus get bio-concentrated in their tissues).

3.3 Weathering of Sulphidic Mine Tailings

A special group of wetland plants reduces element leakage from mine tailings by phytostabilisation. Plants that are tolerant to elements of high concentrations have been found useful for the reclamation of dry mine tailings containing elevated levels of metals and other elements. Mine tailings rich in sulphides, e.g., pyrite, can form **Acid Mine Drainage** (AMD) if it reacts with atmospheric oxygen and water, which may also promote the release of metals and As. To prevent AMD formation, mine tailings rich in sulphides may

Table 3 Macrophytes for wastewater treatment

	Author(s)	Year
Phytoremediation potentials of *Eichhornia crassipes* for nutrients and organic pollutants from textile wastewater	Adelodun et al.	(2021)
Macrophyte and indigenous bacterial co-remediation process for pentachlorophenol removal from wastewater	Ammeri et al.	(2021)
Lead phytoremediation potentials of four aquatic macrophytes under hydroponic cultivation	Das et al.	(2021)
Biomass of the macrophyte remedies and detoxifies Cd (II) and Pb (II) in aqueous solution	de Freitas et al.	(2021)
Chloride accumulation in above ground biomass of three macrophytes (*Phragmites Australis, Juncus Maritimus, and Typha Latifolia*) Depending on their growth stages and salinity exposure application for Cl-removal and phytodesalinization	Delattre et al.	(2021)
Arsenic phytoremediation in contaminated and flooded soil: accumulation and translocation in two macrophytes	Ferreira et al.	(2021)
Potentially toxic elements contamination in indigenous aquatic macrophytes of the Bahmanshir River, Iran	Haghnazar et al.	(2021)
Removal potential of multiple perfluoroalkyl acids (PFAAs) by submerged macrophytes (Vallisneria natans) in aquatic environments	Hua et al.	(2021)
Constructed wetland, an eco-technology for wastewater treatment (macrophyte, biolfilm and substrate)	Kataki et al.	(2021)
Phytoremediation of partially treated MSW leachate by selected free floating and emergent macrophytes in subsurface vertical flow constructed wetlands	Perera and Yatawara	(2021)
Azolla farming for sustainable environmental remediation	Sebastian et al.	(2021)
Composting: a low-cost biotechnological approach to ameliorating macrophyte nuisance in fresh waters	Udume et al.	(2021)
Exploring the phytoremediation potential of macrophytes for treating sewage effluent through constructed wetland technology (CWT) for sustainable agriculture	Suganya et al.	(2022)
Textile wastewater bioremediation using immobilized *Chlorella* sp	Wu et al.	(2021a, 2021b)
Economic potential of bioremediation using immobilized microalgae-based microbial fuel cells	Wu et al.	(2021a, 2021b)

Table 4 Bioremediation of wastewater and value additions

	Author(s)	Year
Potential of microalgae in bioremediation of wastewater	Ahmad et al.	(2021)
Valorizing lignin-like dyes and textile dyeing wastewater by a newly constructed lipid-producing and lignin modifying oleaginous yeast consortium valued for biodiesel and bioremediation	Ali et al.	(2021)
Sarcocornia neei: a novel halophyte species for bioremediation of marine aquaculture wastewater and production diversification in integrated systems	Beyer et al.	(2021)
Spirulina sp. as a bioremediation agent for aquaculture wastewater: production of high added value compounds and estimation of theoretical biodiesel	Cardoso et al.	(2021)
Promotion of bioremediation performance in constructed wetland microcosms for acid mine drainage treatment by using organic substrates and supplementing domestic wastewater and plant litter broth	Chen et al.	(2021)
Enhanced *Arthrospira platen*sis biomass production combined with anaerobic cattle wastewater bioremediation	de Souza et al.	(2021)
Improving microbial bioremediation efficiency of intensive aquacultural wastewater based on bacterial pollutant metabolism kinetics analysis	Dong et al.	(2021)
Use of brown algae for green synthesis, characterization applications of iron oxide nanoparticles for antialgal and wastewater bioremediation	El-Sheekh et al.	(2021)
Green technology for bioremediation of the eutrophication in aquatic ecosystems	El-Sheekh et al.	(2021)
Resilience of *Dreissena polymorpha* in wastewater effluent: use as a bioremediation tool	Géba et al.	(2021)
Bioremediation of heavy metals from wastewater using nanomaterials	Kaur Nagi et al.	(2021)
Bioremediation of coke plant wastewater from steel industry with mixed activated sludge–microalgal consortium in lab-scale semi-continuous mode	Kaur Nagi et al.	(2021)

(continued)

Table 4 (continued)

	Author(s)	Year
Microalgal-bacterial consortia as future prospect in wastewater bioremediation, environmental management and bioenergy production	Khoo et al.	(2021)
Bioremediation mechanism and potential of copper by actively growing fungus *Trichoderma lixii* CR700 isolated from electroplating wastewater	Kumar and Dwivedi	(2021)
Microbial communities of the olive mill wastewater sludge stored in evaporation ponds as resource for sustainable bioremediation	Martínez-Gallardo et al.	(2021)
Enhanced bioremediation of heavy metals from phosphate processing wastewater–using the indigenous *Bacterium Serratia rubidaea* NCTC12971	Moula et al.	(2021)
Indigenous bacteria as bioremediation agent in the wastewater of sugar factory	Pardamean et al.	(2021)
An evolved native microalgal consortium for the bioremediation of biogas and centrate wastewater: start-up, optimization and stabilization	Qiu et al.	(2021)
Simultaneous bioremediation of Cyanide, Phenol, and Ammoniacal-N from synthetic coke-oven wastewater using *Bacillus* sp. NITD 19	Rai et al.	(2021)
Olive mill wastewater-evaporation ponds long term stored: integrated assessment of in situ bioremediation strategies based on composting and vermicomposting	Sáez et al.	(2021)
Bioremediation of hexavalent chromium from wastewater using bacteria	Seragadam et al.	(2021)
Critical review on microbial community during in situ bioremediation of heavy metals from industrial wastewater	Sharma et al.	(2021)
Bioremediation of lignin derivatives and phenolics in wastewater with lignin modifying enzymes	Singh et al.	(2021)
Importance of algae and bacteria in the bioremediation of heavy metals from wastewater treatment plants	Ummalyma et al.	(2021)
Resource recovery through bioremediation of wastewaters and waste carbon by microalgae: a circular bioeconomy approach	Ummalyma et al.	(2021)
Textile wastewater bioremediation using immobilized *Chlorella* sp. Wu-G23 with continuous culture	Wu et al.	(2021a, 2021b)
Economic potential of bioremediation using immobilized microalgae-based microbial fuel cells	Wu et al.	(2021a, 2021b)
Bioremediation of wastewater containing mercury using three newly isolated bacterial strains	Zhao et al.	(2021)
Bioremediation of copper polluted wastewater-like nutrient media and simultaneous synthesis of stable copper nanoparticles by a viable green alga	Žvab et al.	(2021)
Importance of algae and bacteria in the bioremediation of heavy metals from wastewater treatment plants	Ummalyma, and Singh	(2021)
Resource recovery through bioremediation of wastewaters and waste carbon by microalgae: a circular bioeconomy approach	Ummalyma et al.	(2021)
Recent advancements in microbial bioremediation of industrial effluents	Choudhary et al.	(2022)

be saturated with water to reduce the penetration of atmospheric oxygen. An organic layer with plants on top of the mine tailings would consume oxygen, as would plant roots through respiration. Thus, phytostabilisation on water-covered mine tailings may further reduce the oxygen penetration into the mine tailings and prevent the release of elevated levels of elements into the surroundings. Metal tolerance can be evolutionarily developed while some plant species seem to have an inherent tolerance to trace metals. Since some wetland plant species have been found with the latter property, for example, *Thypha latifolia*, *Glyceriafluitans* and *Phragmitesaustralis*, wetland communities may easily establish on submerged mine tailings, without prior development of metal tolerance. Some plant species have mechanisms that make it possible to cope with high external levels of elements. Low accumulators are plants that can reduce the uptake when the substrate has high element concentrations, or have a high net efflux of the element in question, thus the plant tissue concentration of the element is low even though the concentration in the substrate is high (Prasad 2015).

Mine tailings weather by penetration of oxygen thus forming free metal ions and sulfuric acid in acidic mine drainage (AMD) water. Wetland plants have the ability to either take up oxygen from the air or use photosynthetic oxygen and translocate the oxygen to the roots and into the rhizosphere. Thereby, they will increase the redox potential and thus also decrease the pH and increase the release of

Table 5 Waste water remediation

	Authors	Year
Layered double hydroxides/biochar composites as adsorbents for water remediation applications: recent trends and perspectives	dos Santos et al.	(2021)
Electric-field-driven nanoparticles produce dual-functional bipolar electrodes and nanoelectrolytic cells for water remediation	Guo et al.	(2021)
Recent advances in cobalt-activated sulfate radical-based advanced oxidation processes for water remediation: a review	Hou et al.	(2021)
MoS2-based nanocomposites: synthesis, structure, and applications in water remediation and energy storage: a review	Maksoud et al.	(2021)
Adsorbents for real-scale water remediation: gaps and the road forward	Mudhoo et al.	(2021)
The enhancement of treatment capacity and the performance of phytoremediation system by fed batch and periodic harvesting	Ng and Chan	(2021)
MOF based engineered materials in water remediation	Rego et al.	(2021)
The journey of water remediation through biomimetic strategies: a mechanistic insight	Sen Gupta et al.	(2021)

Table 6 Resilience of water

	Authors	Year
Evaluating the resilience of water resources management using the evidential reasoning approach–the case of Zarrinehrud river basin	Behboudian and Kerachian	(2021)
Economic factors affecting water supply resilience to disasters	Balaei et al.	(2021)
Water for Bongo: creative adaptation, resilience and Dar es Salaam's water supply	Bender et al.	(2021)
Resilience of cold water aquaculture: a review of likely scenarios as climate changes in the Gulf of Maine	Bricknell et al.	(2021)
Water quality resilience prediction A novel machine learning application	Imani et al.	(2021)
Resilience of cold water aquaculture	Bricknell et al.	(2021)

Table 7 Rainwater harvesting

	Authors	Year
Design of a passive rainwater harvesting system with green building approach	Kucukkaya et al.	(2021)
Rainwater harvesting as social capital for urban water supply: mitigation of floods and droughts	Maryono et al.	(2021)
Impact of roof rain water harvesting of runoff capture and household consumption	Meshram et al.	(2021)
Automated irrigation and rainwater harvest for sustainable community farming: a review	Mortaza et al.	(2021)
Assessing the viability of rainwater harvesting system for sustainable manufacturing in water stressed regions	Ogwu et al.	(2021)

metals. Aquatic plants can tolerate a very low pH, which can be necessary when treating AMD. *Carex rostrata, Phragmites australis, Typha angustifolia, T. latifolia,* have been found growing under field conditions in pH as low as 2–4.4. Thus, several of the wetland plants serve as fast track botanical cleanup crews.

Natural wetlands are being used for wastewater treatment for centuries. Constructed wetlandsare effective in treating organic matter, nitrogen and phosphorus, and decrease the concentrations of heavy metals and pathogens. Babatunde (2022) detailed the design concept and performance of a constructed wetland system engineered for the circular

Table 8 Water reclamation and reuse

	Authors	Year
Wastewater reclamation and reuse potentials in agriculture: towards environmental sustainability	Fito and Van Hulle	(2021)
Development of a community system for water reclamation from grey water in Gujba: a conceptual method	Lawan and Surendran	(2021)
Waste-water reclamation–urban and agricultural	McGauhey	(2021)
Solar processes and ozonation for fresh-cut wastewater reclamation and reuse: assessment of chemical, microbiological and chlorosis risks of raw-eaten crops	Nahim-Granados et al.	(2021)
Upcoming challenges of water reclamation from unconventional sources	Rögener	(2021)
Water reclamation and reuse for environmental conservation	Shammas et al.	(2021)

Table 9 Water reuse and recycle

	Authors	Year
Development on sustainable water recycle, reuse and management	Azma et al.	(2021)
Advanced control system for reverse osmosis optimization in water reuse systems	Galizia et al.	(2021)
An integrated biorefinery approach via material recycle/reuse networks for the extraction of value-added components from shrimp: computer-aided simulation and environmental assessment	Moreno-Sader et al.	(2021)
Identification of surrogates for rapid monitoring of microbial inactivation by ozone for water reuse: a pilot-scale study	Shi et al.	(2022)
Extending the life of water reuse reverse osmosis membranes using chlorination	Souza-Chaves et al.	(2022)
Monitoring of human enteric virus and coliphages throughout water reuse system of wastewater treatment plants to irrigation endpoint of leafy greens	Truchado et al.	(2021)
Wastewater reclamation and reuse potentials in agriculture	Fito and Hulle	(2021)

economy. In Circular Economy and Sustainability, the treatment mechanisms operating in these are: Filtration and uptake of contaminants, Settling of suspended solids due to decreased, Water velocity and trapping action of plants, leaves, and stems, Precipitation, adsorption, and sequestration of metals, Microbial decomposition of petroleum hydrocarbons and other organics.

Surface flow constructed wetlands are being designed for the treatment of municipal wastewaters in developed nations. The use of constructed wetlands is spreading rapidly in developed nations due to their cost-effectiveness in the treatment of non-point source pollution. However, in tropical nations due to water scarcity and high surface evapo-transpitration, the constructed wetlands for the treatment of wastewaters are not gaining importance. Further, due to water scarcity, wetland technology in many tropical countries is facing hurdles from a practical point of view (Fig. 9a, b). However, Karnal technology developed in Indian town (Karnal) attained national importance (Fig. 10).

Arsenic contamination of ground and surface waters—cleanup with naturally occurring and synthetic abiotic resources:

Arsenic (As) contamination of ground and surface waters is due to the weathering and internal dissolution processes in the source materials. Sorption/precipitation, ion exchange, membrane separation, etc. are some of the methods adopted

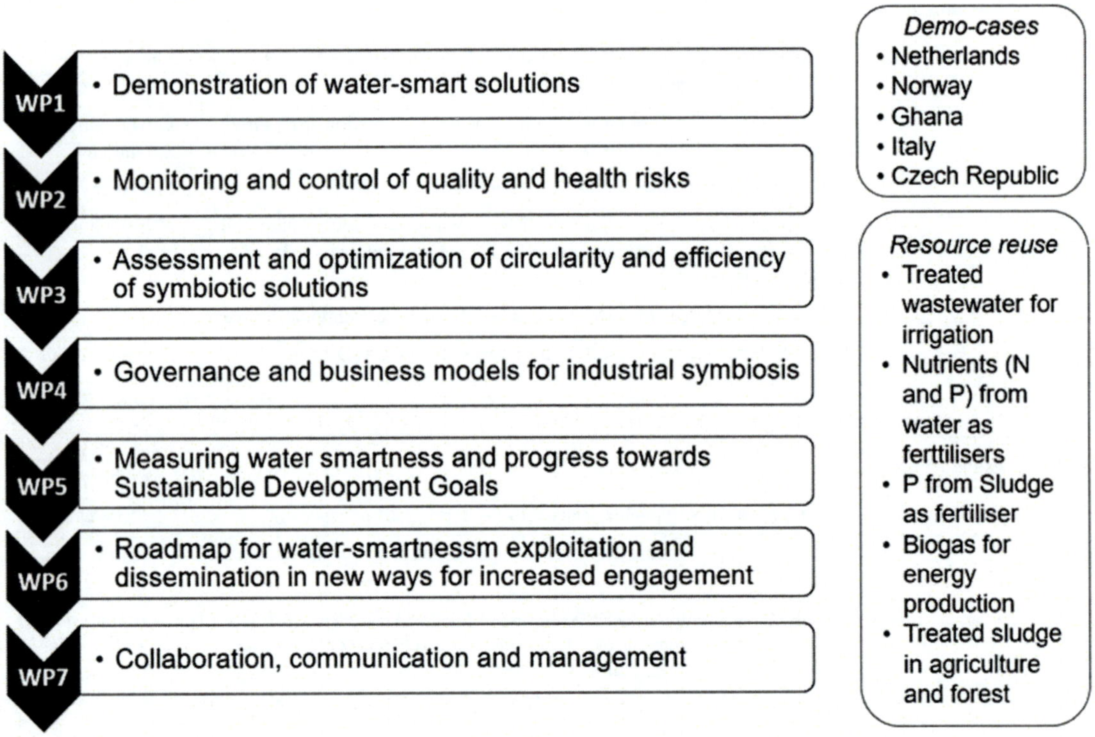

Fig. 8 Summary of WIDER UPTAKE (an international project financed by the European Union's Horizon 2020 Research and Innovation Programme focusing on innovative circular-economy solutions co-developed by water utilities and private businesses from industry sectors (Redrawn from Mannina et al., 2021)

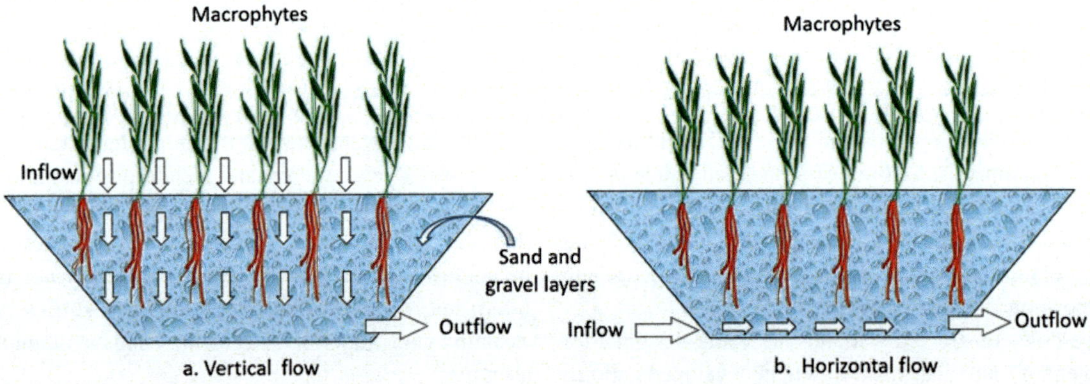

Fig. 9 **a** Schematic view of a generalized constructed wetland of vertical flow **b.** Schematic view of a generalized constructed wetland of horizontal flow

for the removal of As from aqueous solutions (Seidel et al., 2005). Sorption technique is generally considered to be promising, and a wide range of adsorbents have been developed and tested which include mesoporous silica, activated-alumina and carbon and modified zeolites. Several ferric (hydr)oxides were ideally used as adsorbents for removing both As(V) and As(III) from aqueous solutions. Table 10 details the naturally occurring resources that have been tested in laboratory conditions for removal of Arsenic from aqueous solution.

4 Plants and Plant Products for the Treatment of as Contaminated Water

Recently, Kumari et al. (2005) examined the property of *Moringa oleifera* Lamarck (Drumsticks) seed powder for removal of arsenic [arsenite, As(III); and arsenate, As(V)] from water. The efficiency of the desorption of the metals from loaded biosorbent for the reusability of the biosorbent

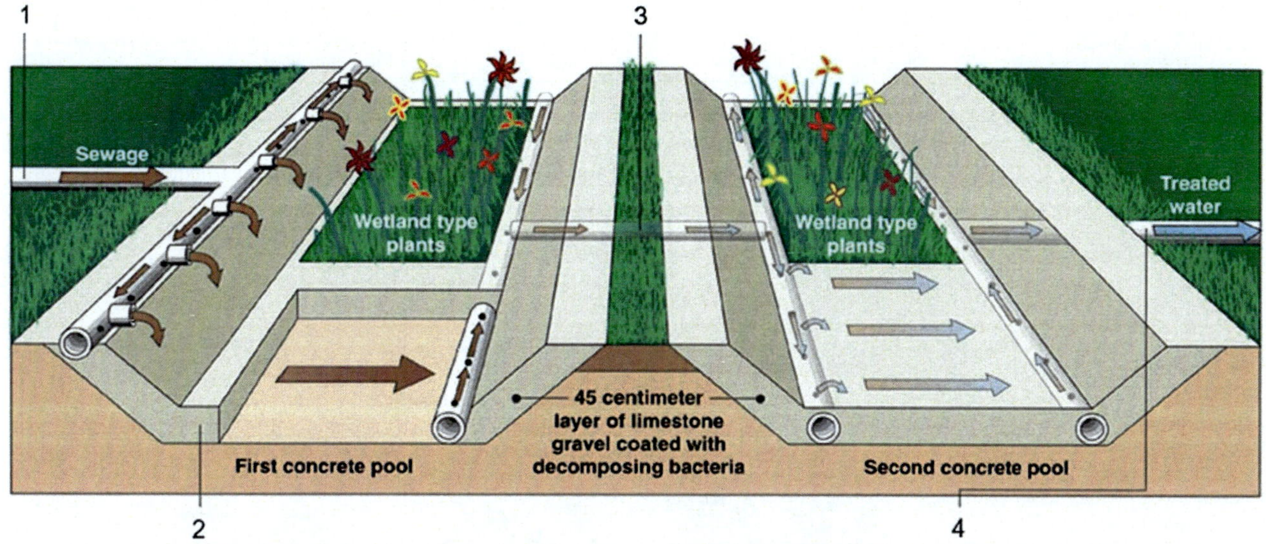

Fig. 10 Karnal technology (Karnal is the name of a town in the northern part of India, where this technology was developed for sewage treatment). 1. Raw sewage drains by Gravity into the first pool and flows through a long perforated PVC pipe into a bed of limestone gravel, 2. Microbes in the limestone gravel, 3. Wastewater flows through anology was developed) nother perforated PVC pipe into a second pool, where the same process is repeated, Break down the sewage into chemicals that can be adsorbed by the plant roots and the gravel absorbs phosphorus and 4. Treated water flowing from the second pool is nearly free of bacteria and nutrients. Treated water can be recycled for irrigation and flushing toilets

Table 10 Selected potential plants for the treatment of Arsenic contaminated water

Plant name	Author(s)	Year
Moringa oleifera	Kumari et al.	(2006)
Ferns for phytofiltration	Huang et al.	(2004)
A. thaliana	Sauge-Merle et al.	(2003)
Arabidopsis thaliana (transgenic)	Dhankher et al.	(2002)
Chinese brake fern (*Pteris vittata*)	Ma et al.	(2001)
Tamarisk (*Tamarix parviflora*)	Tossell et al.	(2000)
Eucalyptus (*Eucalyptus camaldulensis*)	Tossell et al.	(1998)

has also been examined. *Moringa oleifera* carbon was found to be comparable to commercial powdered activated water treatment carbons. *Moringa oleifera* can easily be cultivated on marginal lands in India and other countries with limited sources of irrigation and fertilizers. Hence, in developing countries like India, it would be an economically feasible option for the purification of As contaminated water (Tables 10 and 11, Figs. 11 and 12).

WWTPs have become an important nexus of SMART city. Figure 13 shows the intelligent wastewater system concept in which WWTPs not only treat wastewater with an efficiency that allows effluent reuse, but also generate energy and produce fertilizers (Figs. 8 and 9) (Prasad et al., 2019).

Notable natural coagulants used in treating wastewater from industries and merits and demerits of notable purification techniques are shown in Tables 12 and 13, respectively.

5 Desalination and Water Reclamation

In many arid areas of the world, desalination and water reclamation are of paramount importance for water security. Among various technologies (Babuand & Vidya, 2021, Manikandan et al., 2022), membrane technologies are being followed which play a significant role in water and energy sustainability. Desalination removes salts from seawater. Conventional methods have high energy requirements and hence solarenergy-based desalination plants are preferred which efficiently remove salt from the water. Humidification dehumidification (HDH) systems provide an optimal solution that couples solar desalination with air humidification and air dehumidification (Parnian et al., 2020; Xie et al., 2022).

The seawater greenhouse is considered an HDH system, that is partially or entirely powered by solar energy. It uses

Table 11 Potential biomass for biosorption of Arsenic

Sorbent	As specieis	Author(s)	Year
Waste tea fungal biomass	Arsenic	Murugesan et al.	(2006)
Lessonia nigrescens	Arsenic(V)	Hansen et al.	(2005)
Methylated yeast biomass	Arsenic(V)	Seki et al.	(2005)
Chlorella vulgaris	Arsenic(III)	Kumaresan and Riyazuddin	(2005)
Garcinia cambogia	Arsenic(III)	Kamala et al.	(2005)
Sorghum biomass	Arsenic	Cano-Aguiler et al.	(2005)

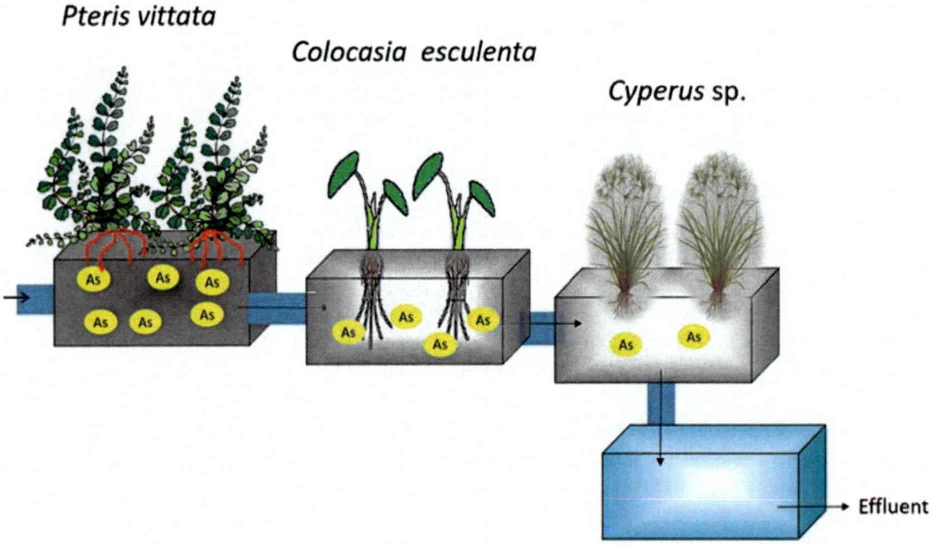

Fig. 11 Feasible model with arsenic accumulators for ex situ phytofiltration, e.g., Taro (*Colaocasia esculenta*) and *Pteris vitata* in a series of tanks. Certain plants act as a biofilter to remove Arsenic (phytofiltation) (Reproduced with permission from Prasad, 2015)

seawater to humidify and cool the greenhouse atmosphere, consequently, freshwater is produced and 'extracted' by plants, in warm arid seashores, with great benefits to productivity (Figs. 14 and 15). Various types of HDH have been conceived and advanced; additionally, different forms of condensers have been used in seawater greenhouses. The seawater greenhouse system uses a water heater-air humidifier constructed on a shallow water reservoir filled with saline water (Figs. 14 and 15), which achieves a higher gradient of salt removal compared to seawater-greenhouse systems. In Figs. 14 and 15, saline water flows into a lined base pool covered by double sheets of clear plastic curtains to form a greenhouse; water is pumped and sprayed into the air by vortex nuzzles. In the humidifier of the system, water and the atmosphere of the semi-greenhouse are partially warmed by the sun's radiation, producing moist atmospheric conditions. Fans evacuate the humid air to the dehumidifier section (ground dehumidifier) and freshwater is produced, which flows to the reservoir. The approach of Figs. 14 and 15 develops the HDH and would provide a low cost and efficient system with benefits that specialize it in water evaporator-production systems (Parnian et al., 2020).

Currently, 3D printing techniques occupy 3.5% of the deployment of technologies in Environmental Projects 2021 (Data source Environmental Business International Inc. San Diego, CA, United States). Some research groups are working on the usage of 3D printing for water remediation, wastewater treatment and sorption (Fijoł et al., 2022). 3D printing techniques such as Fused deposition modelling (FDM), Direct ink wetting (DIW) and Stereolithography (SLA) with nature-derived biopolymers and biopolymer-based materials including chitosan, Polylactic acid (PLA), alginate, and Cellulose acetate (CA) for their potential application within the water treatment industry (Figs. 15 and 16) (Fijoł et al., 2022).

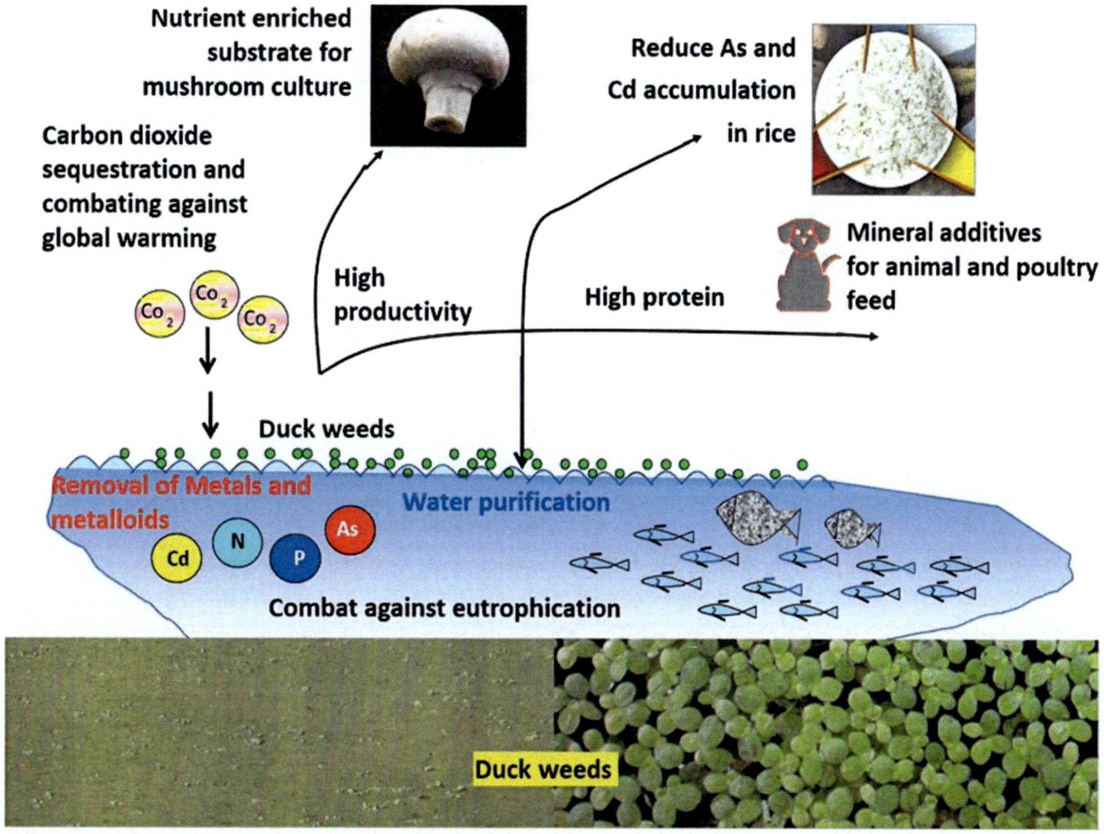

Fig. 12 Duckweed cultivation in wastewater ponds has several economic and environmental advantages. Duckweeds are rich in protein and act as a source of minerals for animal feed. Duckweeds filter cadmium and arsenic in wastewater. Above all, duckweeds combat eutrophication and carbon dioxide sequestration (Reproduced with permission from Prasad, 2015)

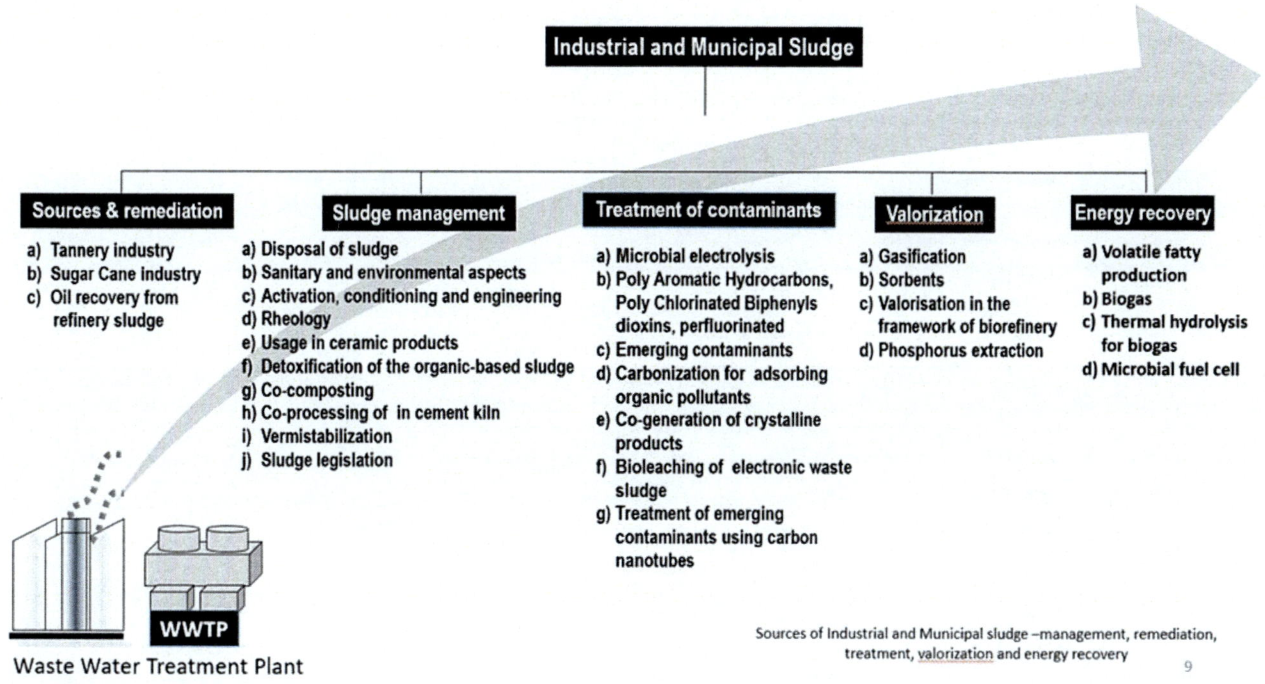

Fig. 13 Circular economy model for WWTPs (today, tomorrow, and in the future) with growing SMART Cities (Adopted from Neczaj & Grosser, 2018, Prasad et al., 2019)

Table 12 Notable natural coagulants used in treating wastewater from industries with their optimum parameters and results (Reproduced with permission from Owodunni & Ismail, 2021)

Coagulant	Form of application	Optimum conditions	Parameters and maximum removal efficiency
Agro-industrial wastewater (palm oil mill effluent)			
Cassia obtusifolia (seed gum)	Stock solution	Cd - 1.0 g/L pH - 3 Settling time - 45 min	TSS - 93.22 %
Chitosan	Stock solution (0.1 M HCl and distilled water)	Cd - 400 mg/L	COD – 68.31 % Turbidity – 98.35 % TSS – 98.95 % Color – 96 %
Moringa oleifera	Seed paste	Cd - 2000 mg/L	COD – 38.60 % Turbidity– 88.30 % TSS – 95.42 % Color – 90.15 % Oil & grease – 87.05 %
Zeolite	Powder	Cd - 1000 mg/L	COD – 61.89 % Turbidity – 91.98 % TSS – 96.82 %
Rice starch	Starch solution	Cd - 2 g/L pH - 3 settling time - 5 min slow stirring speed - 10 rpm	TSS - 84.1 %
Chitosan	Solution (1% acetic acid)	pH – 3–6 Cd– 0.08–8 g/L	TSS - 99 % Residue oil - 99 %
Moringa oleifera	Press cake (solid)	pH – 4–9 Cd – 0.5–6 g/L	TSS - 99.2 % COD - 52.5 %
Aquaculture industry wastewater			
Chitosan	Stock solution (0.1 M HCl)	Cd – 30 mg/L	Algae removal – 98 % Biomass recovery – 80 % TDS removal – 82 %
Sesamum indicum seeds extract	Stock solution	Cd – 0.4 g/L Settling time – 60 min pH - 2 Temperature – 303 K	COD – 38.60 %
Concrete plant wastewater			
Moringa oleifera + Alum	Powder	Cd - 20:80 (w/w) Mo and alum	Turbidity - <90 %
Domestic			
Oyster mushroom (*Pleurotus ostreatus*)	Solid	Cd - 600 mg/L Mixing rate – 150 rpm pH – 6–8	Turbidity – 84 % TSS – 90 %
Kaolin synthetic wastewater			
Jatropha curcas	Press cake (solid)	Cd - 120 mg/L pH – 3	Turbidity – 99.7 %
Roselle seeds	Suspension	Cd - 40 mg/L pH - 4	TSS - 93.13 %
Oil refinery wastewater			
Moringa oleifera	Stock solution	Cd – 70 mg/L	COD – 38.60 % Turbidity – 63.70 % TSS – 62.05 %
Moringa oleifera+ alum		Mo and alum – 2:1 Cd – Alum: 80 mg/L; Mo: 70 mg/L	COD – 50.41 % Turbidity – 86.14 % TSS – 81.52 %
Petroleum industry wastewater			
Cicer arietinum seed		Cd – 1.5 mg/L pH - 7	COD – 95.2 % Turbidity – 98.89 % Oil – 83.8 %
Eggplant seed	Filtrated solution	Cd – 2 mg/L pH - 7	COD – 92.18 % Turbidity – 99.42 % Oil – 81.8 %
Radish seed		Cd – 2 mg/L pH - 7	COD – 93.48 % Turbidity – 98.78 % Oil – 88.2 %
Raw river water			
Banana pith	Powder	Cd – 0.1 kg/m^3 pH - 4	TSS – 96.03 % COD – 54.3 % Turbidity – 98.5 %
Watermelon seed	Oil-free powder	Cd – 0.6 g/L	Turbidity – 87.9 %

Coagulant	Form of application	Optimum conditions	Parameters and maximum removal efficiency
Mixing speed – 120 rpm		pH – 7.5 Temperature – 25 °C	Color removal – 84.3 %
Watermelon seed	Oil-free powder	pH – 7 Cd – 0.1 g/L Speed – 100 rpm Time – 8 min	TDS – 39.3 % COD –79.5 % Turbidity – 98.5 % BOD – 91.5 %
Soap and Detergent industry wastewater			
Tamarind seed	Powder	Cd - 400 mg/L	Turbidity – 97.72 % COD – 39.55 %
Chitosan	–	Mass loading – 50 mm pH – 7 Speed – 150 rpm Time – 5 min	Color removal - 90 % COD – 83 %
Textile industry wastewater			
Chitosan	–	Cd - 25 g/L to 30 g/L	Color removal - 43 %
Maize seed *(Zea mays)*	Powder	Cd - 30 g/l	Color removal – 47.03 % COD – 68.82 %
Surjana seed	Powder	Cd - 25 g/l and 30 g/l	Color removal – 62.82 % COD – 74.11 %
Ocimum basilicum	Mucilage	Cd - 1.6 mg/L	Colour - 68.5 % COD - 61.6 %
Moringa Oleifera (Mo)	Powder	Cd – 60 mL	pH - 35 % Turbidity - 48 % TDS - 70 % TSS - 57 %
Tamarindus Indica (TI)	Powder	Cd – 60 mL	pH - 32 % Turbidity - 32 % TDS - 48 % TSS - 44 %
Strychnos potatorum (SP) (nirmali seeds)		SP (nirmali seeds): Cd – 5 g pH - 1.95 %	Turbidity -23.07 % TDS - 76.72 % TSS - 33.74 % COD - 45.7 % BOD - 51.21 %, Hardness - 9.94 %
Eirchorrnia crassipes (EC) (water hyacinth)	Pressed seedcake	Cd – 8 g pH - 2.37 %	Turbidity - 53.85 % TDS - 92.90 % TSS - 36.49 % COD - 50.80 % BOD - 51.21 % Hardness - 33.33 %
Paper mills industry wastewater			
Dolichas lablab;	Powder	Cd - 16 g	Turbidity - 37.45 %
Azadirachta Indica		Cd - 6.5 g	Turbidity - 63.01 %
Moringa Oleifera		Cd - 3 g	Turbidity - 31.47 %
Hibiscus Rosa Sinensis		Cd - 1 g	Turbidity - 12.95 %
Moringa oleifera seed	Powder	pH range – 6–8, 150 mg/L	Turbidity - 96.02 % COD – 97.28 %
Household laundry rinsing water			
Nirmali seed and Orange peel pectin	Pectin filterate	Cd mixture of Nirmali seed (6.4 g/L) and Orange peel pectin (1.6 mL/L) at 24 h	Turbidity- 89.5 %, TSS- 81.5 % COD- 56 %
Moringa oleifera seed	Powder	Cd - 120 mg/L	Turbidity – 83.63 % COD – 54.18 %
Pharmaceutical industry wastewater			
Date palm *(Phoenix dactylifera)*	Powder	Cd - 100 mg/L Settling time - 50 min. pH - 2	Color removal - 99.86 %
Dairy wastewater			
Chitosan		pH – 4 Cd – 10 mg/L	TDS – 57 % COD – 62 %
Concrete industry			
Lemon peels	Powder	Cd: 60 mg/L OpH range: 7.3–7.6	Turbidity- 87.5 % COD – 9.78 %
Neem leaves		Cd: 30 mg/L OpH range: 6.80–7.02	Turbidity- 88.75 % COD – 10.75 %

Coagulant	Form of application	Optimum conditions	Parameters and maximum removal efficiency
Tulsi leaves		Cd: 30 mg/L OpH range: 8.48–9.06.	Turbidity- 94.16 % COD – 40.38 %
Rambutan seeds		Cd: 20 mg/L OpH range: 7.43–8.68	Turbidity- 96.67 % COD – 46.18 %
Papaya seeds		Cd: 40 mg/L OpH range: 6.8–8.02	Turbidity- 83.75 % COD - 45.79 %

Cd: Coagulant dosage; OpH: Optimum pH; COD: Chemical oxygen demand; Mo: Moringa oleifera; TP: total phosphate; TSS: total suspended solids; TDS: total dissolved solids

Table 13 Merits and demerits of notable water purification techniques (reproduced with permission from Owodunni & Ismail, 2021)

Method	Means	Parameter	Advantages	Disadvantages
Ion exchange chromatography	Molecular charge: ion-exchange media (resin, liquids, membranes)	Choice of buffer (pH and ionic strength), simple, high cost of operation [Isoelectric point	Simple, easy to scale	High cost of operation
Chemical precipitation	Ammonium sulphate	Precipitation pH, relatively simple desalination is often required because of the high salt environment nature and concentration of ion	Relatively simple	Desalination is often required because of the high salt environment
Dialysis	Dialysis tubes	Molecular weight and size	Low energy consumption	Slow process
Ultrafiltration	Fractionation using membranes (molecular weights—5, 10, and 30 kDa)	Molecular weight and size, pressure of inert gas, stirring rate	A promising technique for purification	Fouling,
Lyophilization	Freeze drying	Freezing temperature	Longer shelf life, maintains coagulation efficiency	Quality overlap in molecular weights of natural coagulants and organic contaminants may result in the alteration of protein structure, Freezing and dehydration of the proteins may result in the alteration of protein structure and High running cost

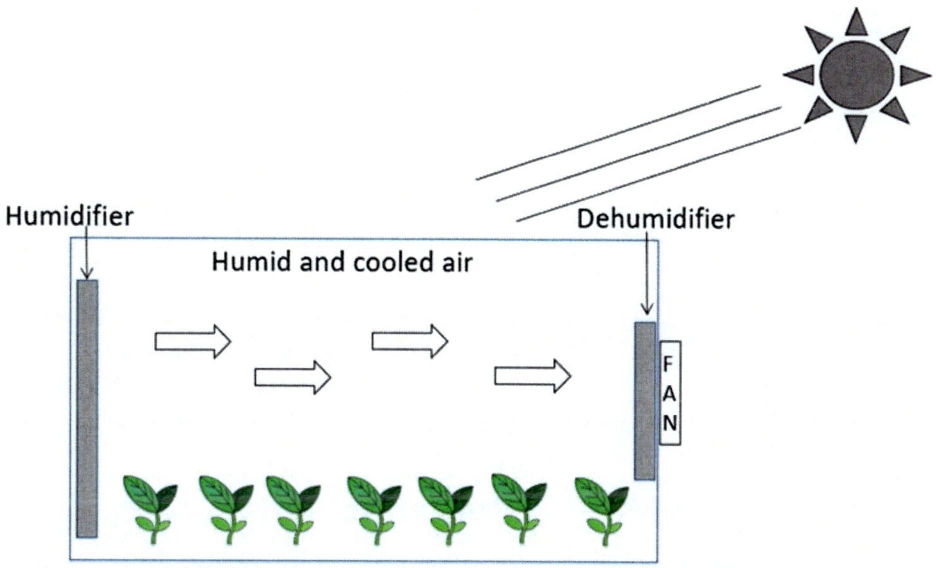

Fig. 14 Sea water greenhouse (Redrawn based on Parnian et al., 2020)

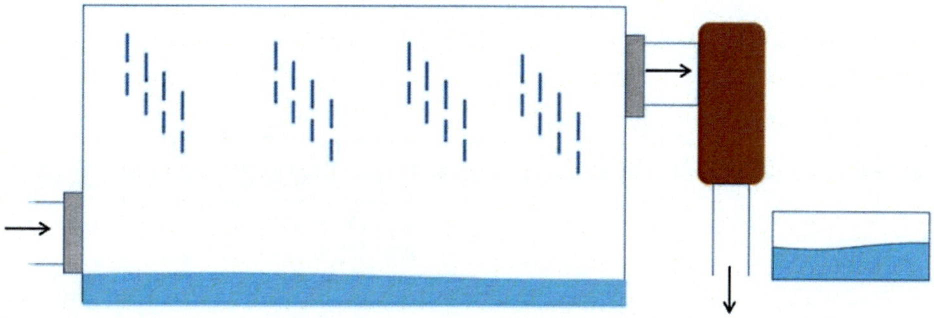

Fig. 15 Humidification and dehumidification within the seawater greenhouse. (Redrawn based on Parnian et al., 2020)

6 Challenges in the Hydrological Cycle

Water cycle is a global biogeochemical cycle. Water inflation and water insecurity in several parts of the globe, and in particular, Pan Africa, in the era of climate change, is alarming. In Sub-Saharan Africa, more than 400 million people lack access to safe drinking water. The price of water is rising at a rate higher than inflation, often costlier than food. Due to the unique geochemistry of Africa, high concentrations of fluoride, arsenic and chlorides render the groundwater unfit for drinking purposes without low-cost treatment. UN-SDG goal 6 clean water and sanitation to all by 2030. Huge gap between the current scenario of clean water availability and expectation.

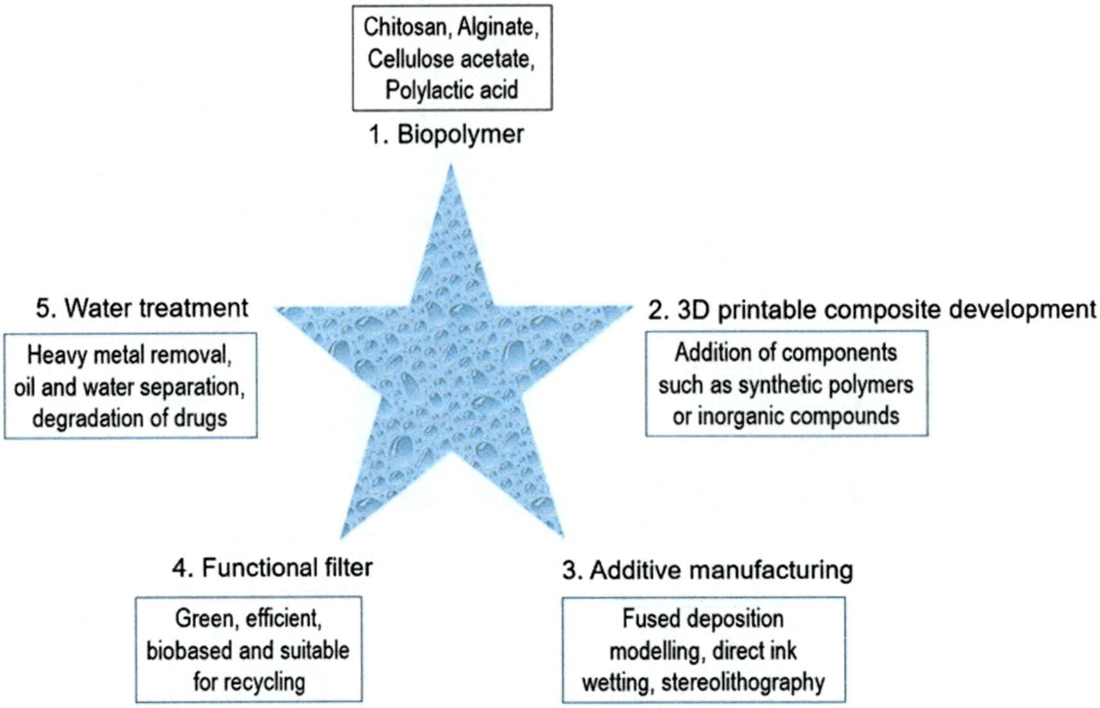

Fig. 16. 3D printable biopolymer-based materials for water treatment (Drawn based on Fijoł et al., 2022)

Acknowledgements Thanks are due to Elsevier for according permission to reproduce Figs. 6 and 7, Tables 1, 2, 7 and 8; Springer Nature for Figs. 11 and 12.

References

Achillas, C., & Bochtis, D. (2020). Toward a green, closed-loop, circular bioeconomy: Boosting the performance efficiency of circular business models. *Sustainability, 12*(23), 10142.

Adelodun, A. A., Olajire, T., Afolabi, N. O., Akinwumiju, A. S., Akinbobola, E., & Hassan, U. O. (2021). Phytoremediation potentials of *Eichhornia crassipes* for nutrients and organic pollutants from textile wastewater. *International Journal of Phytoremediation*, 1–9.

Ahmad, I., Abdullah, N., Koji, I., Yuzir, A., & Mohamad, S. E. (2021). Potential of Microalgae in bioremediation of wastewater. *Bulletin of Chemical Reaction Engineering & Catalysis, 16*(2), 413–429.

Ali, S. S., Al-Tohamy, R., Koutra, E., El-Naggar, A. H., Kornaros, M., & Sun, J. (2021). Valorizing lignin-like dyes and textile dyeing wastewater by a newly constructed lipid-producing and lignin modifying oleaginous yeast consortium valued for biodiesel and bioremediation. *Journal of Hazardous Materials, 403*, 123575.

Ammeri, R. W., Hassen, W., Hidri, Y., Di Rauso Simeone, G., & Hassen, A. (2021). Macrophyte and indigenous bacterial co-remediation process for pentachlorophenol removal from wastewater. *International Journal of Phytoremediation*, 1–12.

Arora, N., Patel, A., Sartaj, K., Pruthi, P. A., & Pruthi, V. (2016). Bioremediation of domestic and industrial wastewaters integrated with enhanced biodiesel production using novel oleaginous microalgae. *Environmental Science and Pollution Research, 23* (20), 20997.

Avellán, T., Nagabhatla, N., Jalan, I., & Liao, D. (2022). Integrating circularity to achieve sustainability: Examples of various wastewater treatment systems. In *Circular economy and sustainability* (pp. 21–35). Elsevier.

Azma, A., Kianfar, N., & Zhang, Y. (2021). Development on sustainable water recycle, reuse and management. In *3rd conference on environment, civil architecture urban development,* Valencia, Spain (pp 1–21).

Babatunde, A. O. (2022). Design concept and performance of a constructed wetland system engineered for the circular economy. In *Circular economy and sustainability* (pp. 199–214). Elsevier.

Balaei, B., Noy, I., Wilkinson, S., & Potangaroa, R. (2021). Economic factors affecting water supply resilience to disasters. *Socio-Economic Planning Science,* 76, 100961.

Behboudian, M., & Kerachian, R. (2021). Evaluating the resilience of water resources management scenarios using the evidential reasoning approach: The Zarrinehrud river basin experience. *Journal of Environmental Management, 284*, 112025.

Bender, M. V. (2021). Water for Bongo: Creative adaptation, resilience & Dar es Salaam's water supply. *Dædalus, 150* (4), 48–63.

Beyer, C. P., Gómez, S., Lara, G., Monsalve, J. P., Orellana, J., & Hurtado, C. F. (2021). Sarcocornianeei: A novel halophyte species for bioremediation of marine aquaculture wastewater and production diversification in integrated systems. *Aquaculture*, 736971.

Bricknell, I. R., Birkel, S. D., Brawley, S. H., Van Kirk, T., Hamlin, H. J., Capistrant-Fossa, K., Huguenard, K., Van Walsum, G. P., Liu, Z. L., Zhu, L. H., & Grebe, G. (2021). Resilience of cold water aquaculture: A review of likely scenarios as climate changes in the Gulf of Maine. *Reviews in Aquaculture, 13*(1), 460–503.

Cano-Aguilera, I., Haque, N., Morrison, G. M., Aguilera-Alvarado, A. F., Gutierrez, M., Gardea-Torresdey, J. L., & De La Rosa, G. (2005). Use of hydride generation-atomic absorption spectrometry to determine the effects of hard ions, iron salts and humic substances on arsenic sorption to sorghum biomass. *Microchemical Journal, 81* (1), 57–60.

Cardoso, L. G., Duarte, J. H., Costa, J. A. V., de Jesus Assis, D., Lemos, P. V. F., Druzian, J. I., de Souza, C. O., Nunes, I. L., &

Chinalia, F. A. (2021). Spirulina sp. as a Bioremediation agent for aquaculture wastewater: production of high added value compounds and estimation of theoretical biodiesel. *BioEnergy Research, 14*(1), 254–264.

Chen, J., Li, X., Jia, W., Shen, S., Deng, S., Ji, B., & Chang, J. (2021). Promotion of bioremediation performance in constructed wetland microcosms for acid mine drainage treatment by using organic substrates and supplementing domestic wastewater and plant litter broth. *Journal of Hazardous Materials, 404*, 124125.

Choudhary, K., Vivekanand, V., & Pareek, N. (2022). Recent advancements in microbial bioremediation of industrial effluents: challenges and future outlook. In *Microbial Biodegradation and Bioremediation* (pp. 293–303). Elsevier.

Das, S., Das, A., Mazumder, P. E. T., Paul, R., & Das, S. (2021). Lead phytoremediation potentials of four aquatic macrophytes under hydroponic cultivation. *International Journal of Phytoremediation*, 1–10.

de Freitas, F., Vendruscolo, S. J., Soares, M. A., Battirola, L. D., & de Andrade, R. L. T. (2021) Biomass of the macrophyte remedies and detoxifies Cd (II) and Pb (II) in aqueous solution. *Environmental Monitoring and Assessment, 193*(8), 1–14.

de Souza, D. S., Valadão, R. C., de Souza, E. R. P., Barbosa, M. I. M. J., & de Mendonça, H. V. (2021). Enhanced Arthrospira platensis biomass production combined with anaerobic cattle wastewater bioremediation. *Bioenergy Research*, 1–14.

Delattre, E., Techer, I., Reneaud, B., Verdoux, P., Laffont-Schwob, I., & Prohin, P. (2021). Chloride accumulation in aboveground biomass of three macrophytes (Phragmites Australis, Juncus Maritimus, and Typha Latifolia) depending on their growth stages and salinity exposure application for Cl-removal and phytodesalinization.

Dhankher, O. P., Li, Y., Rosen, B. P., Shi, J., Salt, D., Senecoff, J. F., Sashti, N. A., & Meagher, R. B. (2002). Engineering tolerance and hyperaccumulation of arsenic in plants by combining arsenate reductase and γ-glutamylcysteine synthetase expression. *Nature Biotechnology, 20* (11), 1140–1145.

Dong, D., Sun, H., Qi, Z., & Liu, X. (2021). Improving microbial bioremediation efficiency of intensive aquacultural wastewater based on bacterial pollutant metabolism kinetics analysis. *Chemosphere, 265*, 129151.

dos Santos, G. E. D. S., dos Santos Lins, P. V., de Magalhães Oliveira, L. M. T., da Silva, E. O., Anastopoulos, I., Erto, A., Giannakoudakis, D. A., de Almeida, A. R. F., da Silva Duarte, J. L., & Meili, L. (2021). Layered double hydroxides/biochar composites as adsorbents for water remediation applications: Recent trends and perspectives. *Journal of Cleaner Production, 284*, 124755.

El-Sheekh, M., Abdel-Daim, M. M., Okba, M., Gharib, S., Soliman, A., & El-Kassas, H. (2021). Green technology for bioremediation of the eutrophication phenomenon in aquatic ecosystems: a review. *African Journal of Aquatic Science*, 1–19.

Ferreira, A. D., Viana, D. G., Pires, F. R., Egreja Filho, F. B., Soares, M. B., de Carvalho, C. F. M., Bonomo, R., Cruz, L. B. S., & Nascimento, M. C. P. (2021). Arsenic phytoremediation in contaminated and flooded soil: Accumulation and translocation in two macrophytes. *Water, Air, & Soil Pollution, 232*(7), 1–13.

Fijoł, N., Aguilar-Sánchez, A., & Mathew, A. P. (2022). 3D-printable biopolymer-based materials for water treatment: A review. *Chemical Engineering Journal, 430*, 132964.

Fito, J., & Van Hulle, S. W. (2021). Wastewater reclamation and reuse potentials in agriculture: Towards environmental sustainability. *Environment, Development and Sustainability, 23*(3), 2949–2972.

Galizia, A., Mamo, J., Blandin, G., Verdaguer, M., Comas, J., Rodríguez-Roda, I., & Monclus, H. (2021). Advanced control system for reverse osmosis optimization in water reuse systems. *Desalination, 518*, 115284.

Géba, E., Rioult, D., Palluel, O., Dedourge-Geffard, O., Betoulle, S., Aubert, D., & Bigot-Clivot, A. (2021). Resilience of Dreissena polymorpha in wastewater effluent: Use as a bioremediation tool? *Journal of Environmental Management, 278,* 111513.

Guo, Q., Lei, C., Chen, W., Zhang, J., & Huang, B. (2021). Electric-field-driven nanoparticles produce dual-functional bipolar electrodes and nanoelectrolytic cells for water remediation. *Cell Reports Physical Science, 2*(1), 100299.

Haghnazar, H., Hudson-Edwards, K. A., Kumar, V., Pourakbar, M., Mahdavianpour, M., & Aghayani, E. (2021). Potentially toxic elements contamination in surface sediment and indigenous aquatic macrophytes of the Bahmanshir River, Iran: Appraisal of phytoremediation capability. *Chemosphere, 285*, 131446.

Hansen, H. K., Ribeiro, A., & Mateus, E. (2006). Biosorption of arsenic (V) with Lessonia nigrescens. *Minerals Engineering, 19*(5), 486–490.

Helmreich, B., & Horn, H. (2009). Opportunities in rainwater harvesting. *Desalination, 2009*(248), 118–124.

Hou, J., He, X., Zhang, S., Yu, J., Feng, M., & Li, X. (2021). Recent advances in cobalt-activated sulfate radical-based advanced oxidation processes for water remediation: A review. *Science of The Total Environment*, 145311.

Hua, Z. L., Li, X. Q., Zhang, J. Y., & Gu, L. (2021). Removal potential of multiple perfluoroalkyl acids (PFAAs) by submerged macrophytes in aquatic environments: tolerance of *Vallisneria natans* and PFAA removal in submerged macrophyte-microbiota systems. *Journal of Hazardous Materials*, 127695.

Huang, J. W., Poynton, C. Y., Kochian, L. V., & Elless, M. P. (2004). Phytofiltration of arsenic from drinking water using arsenic-hyperaccumulating ferns. *Environmental Science & Technology, 38*(12), 3412–3417.

Imani, M., Hasan, M. M., Bittencourt, L. F., McClymont, K., & Kapelan, Z. (2021). A novel machine learning application: Water quality resilience prediction model. *Science of the Total Environment, 768,* 144459.

Kadlec, R. H., & Knight, R. L. (1996). *Treatment Wetlands* (p. 893). Boca Raton: Lewis Publishers.

Kakwani, N. S., & Kalbar, P. P. (2020). Review of circular economy in urban water sector: Challenges and opportunities in India. *Journal of Environmental Management, 271*, 111010.

Kakwani, N. S., & Kalbar, P. P. (2022). Measuring urban water circularity: Development and implementation of a water circularity indicator. *Sustainable Production and Consumption,*31, 723–735.

Kamala, C. T., Chu, K. H., Chary, N. S., Pandey, P. K., Ramesh, S. L., Sastry, A. R. K., & Sekhar, K. C. (2005). Removal of arsenic (III) from aqueous solutions using fresh and immobilized plant biomass. *Water Research, 39*(13), 2815–2826.

Kataki, S., Chatterjee, S., Vairale, M. G., Dwivedi, S. K., Gupta, D. K. (2021). Constructed wetland, an eco-technology for wastewater treatment: A review on types of wastewater treated and components of the technology (macrophyte, biolfilm and substrate). *Journal of Environmental Management*, *283*, 111986.

Kaur Nagi, G., Chetry, R., Singh, N., Sinha, A., & Shinde, O. A. (2021). Bioremediation of coke plant wastewater from steel industry with mixed activated sludge–microalgal consortium in lab-scale semi-continuous mode. *Journal of Chemical Technology & Biotechnology, 96*(8), 2249–2256.

Khoo, K. S., Chia, W. Y., Chew, K. W., & Show, P. L. (2021). Microalgal-bacterial consortia as future prospect in wastewater bioremediation, environmental management and bioenergy production. *Indian Journal of Microbiology*, 1–8.

Krishna, R. S., Mishra, J., & Ighalo, J. O. (2020). Rising demand for rain water harvesting system in the world: A case study of Joda Town, India. *World Scientific News, 2020*(146), 47–59.

Kumar, V., & Dwivedi, S. K. (2021). Bioremediation mechanism and potential of copper by actively growing fungus Trichoderma lixii CR700 isolated from electroplating wastewater. *Journal of Environmental Management, 277,* 111370.

Kucukkaya, E., Kelesoglu, A., Gunaydin, H., Kilic, G. A., & Unver, U. (2021). Design of a passive rainwater harvesting system with green building approach. *International Journal of Sustainable Energy, 40* (2), 175–187.

Kumaresan, M., & Riyazuddin, P. (2005). Major ion chemistry of environmental samples around sub-urban of Chennai city. *Current Science, 91,* 1668–1677.

Kumari, P., Sharma, P., Srivastava, S., & Srivastava, M. M. (2006). Biosorption studies on shelled Moringa oleifera Lamarck seed powder: removal and recovery of arsenic from aqueous system. *International Journal of Mineral Processing, 78*(3), 131–139.

Lawan, M. S., & Surendran, S. S. (2021). Development of a community system for water reclamation from grey water in Gujba: a conceptual method. *International Journal of Environment and Waste Management, 27*(2), 211–223.

Ma, L. Q., Komar, K. M., Tu, C., Zhang, W., Cai, Y., & Kennelley, E. D. (2001). A fern that hyperaccumulates arsenic. *Nature, 409,* 579–579

Maksoud, M. A., Bedir, A. G., Bekhit, M., Abouelela, M. M., Fahim, R. A., Awed, A. S., Attia, S. Y., Kassem, S. M., Abd Elkodous, M., El-Sayyad, G. S., & Mohamed, S. G. (2021). MoS2-based nanocomposites: synthesis, structure, and applications in water remediation and energy storage: a review. *Environmental Chemistry Letters*, 1–37.

Manikandan, S., Subbaiya, R., Saravanan, M., Ponraj, M., Selvam, M., & Pugazhendhi, A. (2022). A critical review of advanced nanotechnology and hybrid membrane based water recycling, reuse, and wastewater treatment processes. *Chemosphere, 289,* 132867.

Mannina, G., Badalucco, L., Barbara, L., Cosenza, A., Di Trapani, D., Gallo, G., Laudicina, V. A., Marino, G., Muscarella, S. M., Presti, D., et al. (2021). Enhancing a transition to a circular economy in the water sector: The EU Project WIDER UPTAKE. *Water, 13*, 946.

Martínez-Gallardo, M. R., López, M. J., López-González, J. A., Jurado, M. M., Suárez-Estrella, F., Pérez-Murcia, M. D., Sáez, J. A., Moral, R., & Moreno, J. (2021). Microbial communities of the olive mill wastewater sludge stored in evaporation ponds: The resource for sustainable bioremediation. *Journal of Environmental Management, 279,* 111810.

Maryono, A., Fandeli, C., & Setyono, P. (2021, March). Rainwater harvesting as social capital for urban water supply: mitigation of floods and droughts. In IOP conference series: *Earth and environmental science* (Vol. 683, No. 1, p. 012144). IOP Publishing.

Meshram, S. G., Ilderomi, A. R., Sepehri, M., Jahanbakhshi, F., Kiani-Harchegani, M., Ghahramani, A., & Rodrigo-Comino, J. (2021). Impact of roof rain water harvesting of runoff capture and household consumption. *Environmental Science and Pollution Research, 28,* 49529–49540.

Moreno-Sader, K. A., Martínez-Consuegra, J., & González-Delgado, Á. D. (2021). An integrated biorefinery approach via material recycle/reuse networks for the extraction of value-added components from shrimp: Computer-aided simulation and environmental assessment. *Food and Bioproducts Processing, 127,* 443–453.

Mortaza, M., Abdullah, I., Shahabuddin, N. S., & Samat, H. A. (2021). Automated irrigation and rainwater harvest for sustainable community farming: a review. *International Journal of Agriculture, Forestry and Plantation, 12,* 199–205

Moula, A., Ellafi, A., Borgi, M. A., Rhimi, S., Akermi, N., Namsi, A., Chaieb, M., & Mekki, A. (2021). Enhanced Bioremediation of Heavy Metals from Phosphate Processing Wastewater Using the Indigenous Bacterium Serratia rubidaea NCTC12971. *Geomicrobiology Journal*, 1–10.

Mudhoo, A., Mohan, D., Pittman, C. U., Jr., Sharma, G., & Sillanpää, M. (2021). Adsorbents for real-scale water remediation: Gaps and the road forward. *Journal of Environmental Chemical Engineering, 9*(4), 105380.

Murugesan, G. S., Sathishkumar, M., & Swaminathan, K. (2006). Arsenic removal from groundwater by pretreated waste tea fungal biomass. *Bioresource Technology, 97*(3), 483–487.

Nahim-Granados, S., Martínez-Piernas, A. B., Rivas-Ibáñez, G., Plaza-Bolaños, P., Oller, I., Malato, S., Pérez, J. A. S., Agüera, A., & Polo-López, M. I. (2021). Solar processes and ozonation for fresh-cut wastewater reclamation and reuse: Assessment of chemical, microbiological and chlorosis risks of raw-eaten crops. *Water Research, 203,* 117532.

Neczaj, E., & Grosser, A. (2018). Circular economy in wastewater treatment plant–challenges and barriers. *Proceedings*, *2*, 614.

Ng, Y. S., & Chan, D. J. C. (2021). The enhancement of treatment capacity and the performance of phytoremediation system by fed batch and periodic harvesting. *RSC Advances, 11*(11), 6049–6059.

Ogwu, E. A., Igbax, S. I., & Ogwu, A. A. (2021). Assessing the viability of rainwater harvesting system for sustainable manufacturing in water stressed regions. In IOP conference series: *Earth and environmental science* (Vol. 665, No. 1, p. 012011). IOP Publishing.

Owodunni, A. A., & Ismail, S. (2021). Revolutionary technique for sustainable plant-based green coagulants in industrial wastewater treatment—A review. *Journal of Water Process Engineering, 42*, 102096.

Pardamean, M. A., Islamy, R., Hasan, V. E. R. Y. L., & Mutmainnah, N. U. R. U. L. (2021). Identification and physiological characteristics of potential indigenous bacteria as bioremediation agent in the wastewater of sugar factory. *Sains Malaysiana, 50*(2), 279–286.

Pari, L., Suardi, A., Stefanoni, W., Latterini, F., & Palmieri, N. (2021). 2021 Economic and environmental assessment of two different rain water harvesting systems for agriculture. *Sustainability, 13*, 3871.

Parnian, A., Furze, J. N., & Parnian, A. (2020). Water out of waste–solar-desalination-humidification-dehumidification auxiliaries processing extractive industrial operations. *Water Productivity Journal, 1*(1), 31–38.

Perera, K.R.S., & Yatawara, M. (2021). Phytoremediation of partially treated MSW leachate by selected free floating and emergent macrophytes in subsurface vertical flow constructed wetlands. *Environmental Technology & Innovation*, *24*, 101928.

Prasad, M. N. V. (2015). Phytoremediation crops and biofuels. In E. Lichtfouse (ed.), *Sustainable agriculture reviews* (vol 17, pp 159–261). Switzerland: Springer International Publishing. https://doi.org/10.1007/978-3-319-16742-8_7

Prasad, M. N. V., Favas, P. J. C., Meththika, V., Mohan, S. V. (2019). Industrialand municipal sludge emerging concerns and scope for resource recovery. ISBN:978–0.

Qiu, S., Yu, Z., Hu, Y., Chen, Z., Guo, J., Xia, W., & Ge, S. (2021). An evolved native microalgal consortium-snow system for the bioremediation of biogas and centrate wastewater: Start-up, optimization and stabilization. *Water Research, 196,* 117038.

Rai, A., Gowrishetty, K. K., Singh, S., Chakrabarty, J., Bhattacharya, P., & Dutta, S. (2021). Simultaneous bioremediation of cyanide, phenol, and ammoniacal-N from synthetic coke-oven wastewater using Bacillus sp. NITD 19. *Journal of Environmental Engineering, 147*(1), 04020143.

Rego, R. M., Kuriya, G., Kurkuri, M. D., & Kigga, M. (2021). MOF based engineered materials in water remediation: Recent trends. *Journal of Hazardous Materials, 403*, 123605.

Rögener, F. (2021). Upcoming challenges of water reclamation from unconventional sources. In Towards water secure societies: *Coping with water scarcity and quality challenges* (pp. 79–87).

Sáez, J. A., Pérez-Murcia, M. D., Vico, A., Martínez-Gallardo, M. R., Andreu-Rodríguez, F. J., López, M. J., Bustamante, M. A., Sanchez-Hernandez, J. C., Moreno, J., & Moral, R. (2021). Olive mill wastewater-evaporation ponds long term stored: Integrated assessment of in situ bioremediation strategies based on composting and vermicomposting. *Journal of Hazardous Materials, 402,* 123481.

Sauge-Merle, S., Cuiné, S., Carrier, P., Lecomte-Pradines, C., Luu, D. T., & Peltier, G. (2003). Enhanced toxic metal accumulation in engineered bacterial cells expressing Arabidopsisthaliana phytochelatin synthase. *Applied and Environmental Microbiology, 69* (1), 490–494.

Sebastian, A., Deepa, P., & Prasad, M. N. V. (2021). Azolla farming for sustainable environmental remediation. In *Handbook of assisted and amendment: Enhanced sustainable remediation technology* (pp. 517–533).

Seidel, H., Görsch, K., Amstätter, K., & Mattusch, J. (2005). Immobilization of arsenic in a tailings material by ferrous iron treatment. *Water Research, 39*(17), 4073–4082.

Seki, H., Suzuki, A., & Maruyama, H. (2005). Biosorption of chromium (VI) and arsenic (V) onto methylated yeast biomass. *Journal of Colloid and Interface Science, 281*(2), 261–266.

Sen Gupta, R., Padmavathy, N., & Bose, S. (2021). The journey of water remediation through biomimetic strategies: A mechanistic insight. *Advanced Sustainable Systems*, 2100213.

Seragadam, P., Rai, A., Ghanta, K. C., Srinivas, B., Lahiri, S. K., & Dutta, S. (2021). Bioremediation of hexavalent chromium from wastewater using bacteria-a green technology. *Biodegradation, 32* (4), 449–466.

Shammas, N. K., Wang, L. K., & Wang, M. H. S. (2021). Water reclamation and reuse for environmental conservation. In L. K. Wang, M. H. S. Wang, Y. T. Hung, N. K. Shammas (Eds.) *Environmental and natural resources engineering. Handbook of Environmental Engineering* (Vol. 19, pp. 53–71). Cham: Springer. https://doi.org/10.1007/978-3-030-54626-7_2

Sharma, P., Pandey, A. K., Kim, S. H., Singh, S. P., Chaturvedi, P., & Varjani, S. (2021). Critical review on microbial community during in-situ bioremediation of heavy metals from industrial wastewater. *Environmental Technology & Innovation, 24,* 101826.

Shi, Q., Chen, Z., Wei, F., Mao, Y., Xu, Q., Li, K., Lu, Y., & Hu, H. Y. (2022). Identification of surrogates for rapid monitoring of microbial inactivation by ozone for water reuse: A pilot-scale study. *Journal of Hazardous Materials, 424,* 127567.

Singh, A. K., Bilal, M., Iqbal, H. M., Meyer, A. S., & Raj, A. (2021). Bioremediation of lignin derivatives and phenolics in wastewater with lignin modifying enzymes: Status, opportunities and challenges. *Science of the Total Environment, 777,* 145988.

Souza-Chaves, B. M., Alhussaini, M. A., Felix, V., Presson, L. K., Betancourt, W. Q., Hickenbottom, K. L., & Achilli, A. (2022). Extending the life of water reuse reverse osmosis membranes using chlorination. *Journal of Membrane Science, 642,* 119897.

Smol, M. (2022). Circular economy approach in the water and wastewater sector. In *Circular economy and sustainability* (pp. 1–19). Elsevier.

Suganya, K., Kumar, J. E., Sebastian, S. P., Poornima, R., Kannan, B., Parameswari, E., & Kalaiselvi, P. (2022). *Exploring the phytoremediation potential of macrophytes for treating sewage effluent through constructed wetland technology (CWT) for sustainable agriculture in sustainable agriculture* (pp. 209–223). Springer.

Tossell, R. W., Binard, K. M., Rafferty, M.T. (2000). Phytoremediation as a replacement for pump-and-treat to manage flow of arsenic-bearing groundwater. In *Proceedings of the second international conference on remediation of chlorinated and recalcitrant compounds,* Monterey, California

Truchado, P., Garre, A., Gil, M. I., Simón-Andreu, P. J., Sánchez, G., & Allende, A. (2021). Monitoring of human enteric virus and coliphages throughout water reuse system of wastewater treatment plants to irrigation endpoint of leafy greens. Science of the Total Environment, 782, 146837.

Udume, O. A., Abu, G. O., & Stanley, H. O. (2021). Composting: A low-cost biotechnological approach to ameliorating macrophyte nuisance in fresh waters. *Journal of Agriculture and Ecology Research International*, 47–71.

Ummalyma, S.B., & Singh, A. (2021). Importance of algae and bacteria in the bioremediation of heavy metals from wastewater treatment plants. In *New trends in removal of heavy metals from industrial wastewater* (pp. 343–357). Elsevier.

Ummalyma, S. B., Sahoo, D., & Pandey, A. (2021). Resource recovery through bioremediation of wastewaters and waste carbon by microalgae: a circular bioeconomy approach. *Environmental Science and Pollution Research,* 1–20.

Wu, J. Y., Lay, C. H., Chen, C. C., Wu, S. Y., Zhou, D., & Abdula, P. M. (2021a). Textile wastewater bioremediation using immobilized *Chlorella* sp. Wu-G23 with continuous culture. *Clean Technologies and Environmental Policy, 23*(1), 153–161.

Wu, J. Y., Lay, C. H., Chia, S. R., Chew, K. W., Show, P. L., Hsieh, P. H., & Chen, C. C. (2021b). Economic potential of bioremediation using immobilized microalgae-based microbial fuel cells. *Clean Technologies and Environmental Policy*, 1–14.

Xie, W., Tang, P., Wu, Q., Chen, C., Song, Z., Li, T., Bai, Y., Lin, S., Tiraferri, A., & Liu, B. (2022). Solar-driven desalination and resource recovery of shale gas wastewater by on-site interfacial evaporation. *Chemical Engineering Journal, 428*, 132624.

Younas, F., Niazi, N. K., Bibi, I., Afzal, M., Hussain, K., Shahid, M., Aslam, Z., Bashir, S., Hussain, M. M., & Bundschuh, J. (2022). Constructed wetlands as a sustainable technology for wastewater treatment with emphasis on chromium-rich tannery wastewater. *Journal of Hazardous Materials, 422*, 126926.

Žvab, U., Kukulin, D. S., Fanetti, M., & Valant, M. (2021). Bioremediation of copper polluted wastewater-like nutrient media and simultaneous synthesis of stable copper nanoparticles by a viable green alga. *Journal of Water Process Engineering, 42,* 102123.

Zhang, X., & Liu, Y. (2022). Circular economy is game-changing municipal wastewater treatment technology towards energy and carbon neutrality. *Chemical Engineering Journal, 429*, 132114.

Zhao, M. M., Kou, J. B., Chen, Y. P., Xue, L. G., Fan, T. T., & Wang, S. M. (2021). Bioremediation of wastewater containing mercury using three newly isolated bacterial strains. *Journal of Cleaner Production, 299*, 126869.

Zhou, L., Wang, H., Zhang, Z., Zhang, J., Chen, H., Bi, X., Dai, X., Xia, S., Alvarez-Cohen, L., & Rittmann, B. E. (2021). Novel perspective for urban water resource management: 5R generation. *Frontiers of Environmental Science & Engineering, 15*(1), 1–13.

Monitoring of Circular Economy Implementation in Water Sector

Water Footprint of the business Sector and Circular Economy

Amirhossein Taghipour and Alexandros Stefanakis

Abstract

Water is vital in metal extraction and manufacturing. Grinding, flotation, gravity concentration, medium separation, and hydrometallurgical processes all use it. The kind of ore treated, the processing techniques, and the local climate all influence water use. Water is also required for dust suppression, cooling, equipment washing, and human consumption in mining operations. To drill for fossil fuels, for example, or to cool boilers and emission control systems in thermoelectric power plants, water is required throughout the entire life cycle of mining. In this study, a life cycle inventory is conducted to quantify water inputs and outputs, and a life cycle impact assessment to quantify possible environmental harm. Metals' water impacts are significant due to direct and indirect water use in mining and production, fuel extraction, and energy production, as well as sulfidic mine tailings disposal. To improve supply chain water efficiency, properly manage mining tailings, and boost the use of low-carbon and low-water energy sources and conversion technologies, the consumer electronics sector must work with suppliers to build long-term water management initiatives. Although strategies and the substitution or reduction of high-impact materials can help reduce water impacts, they may also entail sustainability trade-offs that must be evaluated and addressed in advance. To effectively connect design and material decisions to product consumption, longevity, and end-of-life management, supply chain coordination is required. Additional interventions, such as expanded producer responsibility policies, enhanced consumer participation, and increased supply chain transparency, will most likely be required to meet these aims.

Keywords

Water footprint • Circular economy • Mining • Sustainability

1 Introduction

Today, we discuss the research of manufacturing processes, such as carbon and water footprints, with the goal of increasing the sustainability of the industrial sector. Research on this emerging topic should focus on the circular economy, social and environmental responsibility, and sustainable governance (Nikolaou et al., 2021). Consumer electronics have a diverse array of components that require water to manufacture and may leak pollutants into drinking water, exacerbating freshwater scarcity and pollution. Due to the fact that the majority of consumer electronics research is focused on supply chain energy or carbon footprint, these water implications have not been adequately investigated. Scenarios are analyzed to determine the potential for improvement connected with product design changes such as alternative supply chains, material substitution, and recycled material utilization (Aivazidou & Tsolakis, 2019; Peña et al., 2021).

Water is the most valuable resource on the planet because it regulates ecosystems, protects the environment, and sustains life. While water covers more than 70% of the Earth's surface, the vast majority of it is salty water from the oceans and saline groundwater deposits, with less than 3% being freshwater (Singh et al., 2021). Glaciers, ice caps, and deep groundwater all contain freshwater that is inaccessible to

A. Taghipour
King Mongkut's Institute of Technology Ladkrabang, Bangkok, Thailand

A. Stefanakis (✉)
Laboratory of Environmental Engineering and Technology, School of Chemical and Environmental Engineering, Technical University of Crete, 73100 Chania, Greece
e-mail: astefanakis@tuc.gr

EURECA-PRO–The European University On Responsible Consumption and Production, 73100 Chania, Greece

M. Smol et al. (eds.), *Water in Circular Economy*, Advances in Science, Technology & Innovation,
https://doi.org/10.1007/978-3-031-18165-8_14

humans. Simultaneously, a little quantity remains in lakes, rivers, and aquifers, which are the primary sources of human water consumption yet account for only 0.26% of global freshwater resources (Agrawal et al., 2021).

On the other hand, overexploitation, pollution, and climate change have all had an effect on the water cycle in recent decades, resulting in scarcity challenges throughout the world (Stefanakis, 2020). Humans are consuming freshwater at an unsustainable rate for a growing population, and unless behaviors change to safeguard the planet's water resources, maintaining sufficient safe drinking water for everyone, including future generations, will become increasingly difficult. Water quality is a term that refers to the chemical, physical, and biological qualities of water that are frequently used to determine whether it is suitable for a certain use. "Proportion of bodies of water with good ambient water quality" provides information on the following major water quality impairments found in many parts of the world: oxygen (surface water), salinity (surface water and groundwater), nitrogen (surface water and groundwater), phosphorus (surface water), and acidification (surface water and groundwater).

Freshwater resources are becoming increasingly contaminated as a result of population development, urbanization, industrialization, agricultural expansion and intensification, and climate change. Water quality is predicted to deteriorate significantly further during the next few decades. Poor water quality contributes to ecological destabilization, increased production costs, reduced economic growth, and significant human health problems, such as waterborne diseases, which can cause hunger, sickness, and even death, despite their difficulty in detection (Weetman, 2020). As a result, long-term water quality management and monitoring are necessary, which includes safeguarding water sources against microbiological and chemical contamination and executing purification, disinfection, and anti-recontamination methods (Alkhuzaim et al., 2021).

While agriculture consumes 70% of all water worldwide, largely for irrigation, industry consumes 20% of all water worldwide, primarily for energy production, cooling/heating activities, and manufacturing (Giampietro & Funtowicz, 2020), from small-scale industries such as glass manufacturing (Gholipour et al., 2020), agro-industries (Gomes et al., 2018), and tanneries (Ramirez et al., 2019), to large-scale industrial activities such as oil and gas exploration (Stefanakis et al., 2018). Visualizing hidden water use behind objects enables a better understanding of freshwater's global character and quantifies the impact of consumption and trade on water resource use. This is an important aspect when estimating the water balance at a local or regional scale and the proper management route in a circular frame (Oral et al., 2021). Manufacturing, like any other business, consumes a significant amount of water, resulting in a significant environmental impact over the course of the product's life. Water is required for the bulk of beauty products' manufacturing processes, raw material growth, sanitation, and packaging. Additionally, clients must eat the goods. To preserve water and reduce its reliance on it, the industry's strategy and formulation and manufacturing methods must be rethought. Fortunately, increased customer awareness motivates businesses to create one-of-a-kind products (Brizga et al., 2020). Understanding and studying these aspects are keys to the sustainable use of water resources in a circular economy (Stefanakis & Nikolaou, 2021).

Consumer electronics now incorporate a diverse range of materials, including base metals like steel and aluminum, precious metals like gold and platinum, critical metals like indium and rare earth elements (REEs) (Aivazidou et al., 2018), and hazardous metals like lead and mercury. Several studies (Aivazidou, 2021) have been conducted. These materials are critical in achieving the desired form, finish, or functionality (Cámara-Creixell & Scheel-Mayenberger, 2019). For example, touch-enabled flat panel technology is reliant on rare materials such as indium, which is used in phones, tablets, and televisions (Mostert et al., 2021). Due to the high energy density supplied by minerals such as lithium and cobalt, lithium-ion batteries are frequently employed to power consumer gadgets (Al-Saidi et al., 2021). The demand for a varied range of materials is likely to increase and adapt as a result of shorter innovation cycles, shorter product lifecycles, and the introduction of new technologies (Rodias et al., 2020). While materials have reshaped the consumer electronics sector, rising demand has prompted new worries about sustainability (Esposito et al., 2020).

Literature on supply chain security, supply chain energy consumption, and supply chain carbon footprint has tackled water issues in a circumstantial manner, but has not examined the relationship between related sustainability issues and water use or pollution in detail. For example, electricity use is the largest contributor to carbon emissions in the electronics supply chain, as it consumes and degrades water resources through the combustion of fossil fuels. Additionally, mining waste, referred to as tailings, can leach toxic substances into water, increasing health problems and causing social and political unrest. As a result, it is critical to understand how the actions of the electronic material supply chain contribute to water contamination and to develop measures to mitigate these effects. Water is a critical component of the extraction and production of metals, especially considering circularity aspects. Grinding, flotation, gravity concentration, medium separation, and hydrometallurgical techniques are all examples of mining and refining operations that require it.

Water consumption in these processes is determined by a number of factors, including the mining site, the type of ore being processed, the processing protocols, and the local climate. Additionally, water is necessary for mining operations to perform auxiliary functions such as dust suppression, cooling, equipment washing, and human consumption. Moreover, mining is a high-energy operation that necessitates the use of water throughout the energy life cycle, from drilling and extracting fossil fuels to boilers, cooling towers, and emission control systems at thermoelectric power plants. For example, thermoelectric power plants account for around 40% of total water withdrawals in the United States, with 3% lost and not restored to surface water systems (Krikke, 2020; Zumegen, 2020).

Additionally, metal extraction and processing are typically conducted in water-stressed locations, straining local water supplies. China, a major producer of numerous consumer technology metals, has suffered severe water scarcity as a result of rapid industrialization over the preceding decade. Not only can mining operations deplete local watersheds, but they also have a detrimental effect on downstream water quality. Acid main drainage, the emission of heavy metals and processing chemicals, and mine tailings management all have an effect on surface and groundwater systems. Local populations frequently show opposition to mining initiatives as a result of these external constraints. Peru is one such South American country that is water-stressed. The Conga mine in Peru's Cajamarca region was supposed to produce gold and copper, two metals that are heavily utilized in consumer electronics, but it was forced to close owing to public outrage. To address these sustainability challenges and conserve water resources, it is necessary to gain a better understanding of how water-intensive businesses such as mining and consumer electronics manufacturing contribute to water quantity and quality risks in different locations.

A rising body of research has begun estimating the water footprint of consumer electronics in light of social, economic, and environmental concerns over water consequences. The life cycle volumetric water consumption of personal computers, printed circuit boards, lithium-ion batteries, and semiconductors has been estimated through case studies. Although material extraction accounts for approximately 10% of total volumetric water consumption, it may be disproportionately responsible for regional water stress impacts due to the fact that mining and refining processes are frequently conducted in water-stressed areas rather than in areas with abundant water. Additionally, compounds that consume little volumetric water may have a greater long-term degradative effect on water quality. By integrating the assessment of water usage and degradation consequences, it is possible to conduct a more holistic examination of consumer electronics materials and subsequently provide strategies to mitigate supply chain impacts. The life cycle assessment (LCA) method is particularly suited for this type of analysis because it provides a framework for examining cumulative resource inputs, emission releases, and associated environmental consequences across the whole material supply chain and/or product life cycle. These consequences include loss of freshwater resources and deterioration of water quality due to pollution discharges.

As a result, this chapter applies a life cycle-based water footprint technique to estimate the potential impact of metals used in the extraction and production of consumer electronics on freshwater quantity and quality. This chapter begins by examining the water effects of certain metals, then the metals in two example case study goods, and finally the electronics industry as a whole. The objective is to identify material "hotspots" and then examine scenarios involving supply chain adjustments, recycled content utilization, and material substitution. The study's findings are intended to assist product designers and manufacturers in decreasing consumer electronics' environmental impact toward a more circular management of resources.

2 Materials and Methods

The study used life cycle assessment (LCA) approaches to determine the water footprint of material supply chains in the electronics industry. These methodologies consist of four steps: defining the study's objective and scope, conducting a life cycle inventory to quantify water inputs and outputs, conducting a life cycle impact assessment to quantify the potential for environmental damage, and interpreting the results in light of the study objective. The following sections go into greater depth about each of these techniques. This systematic review was conducted in four basic steps. The objective was to do a search that was sufficiently broad to be exhaustive while remaining tightly focused on the topics at hand.

Following that, using inclusion and exclusion criteria, the titles and abstracts were evaluated for relevance to the review's topic. Articles that had no link to electronic device business were exclusively focused on wastewater treatment technologies.

3 Results

Sustainability was the first concept examined, followed by an interest in water, life cycle assessment, the mining industry, and water footprint.

The life cycle inventory was created in two stages: First, consumer electronic product bills of materials were analyzed to identify specific metals that are critical to the consumer electronics sector as a whole and are found in specific case study products; and second, production process data was compiled to model water inputs and emissions associated with the processing steps and upstream inputs required to produce all of the identified metals. Materials that are unique used bill of materials data to identify common materials in consumer electronics, which was then supplemented with information on the composition of complex components such as printed circuit boards, display units, and batteries. Base metals, precious metals, essential metals, rare earth elements, and hazardous metals were all found in consumer electronics. A few materials were excluded due to a lack of LCI data, including vanadium and several REEs (other than lanthanum, praseodymium, and neodymium).

Products derived from case studies: The smartphone and laptop were used as case studies to study how the material composition of a product affects its size. Multiple product models representing various brands and manufacturing years were utilized to determine the mass of certain materials using BOM data. Bulk materials such as steel, aluminum, and copper were included in BOMs, as well as complex components such as printed circuit boards, batteries, and displays.

Eco-invent data uses generic water flows that are assigned to various countries based on the best available information about the location of the unit processes. Only primary content and average worldwide production market mixtures were used to represent the baseline state of material production (no recycled content). While the LCI data utilized in this analysis include a broad variety of resource inputs and emission releases, it focuses on those that have a direct connection to the water impacts outlined in the section.

The flows quantified were the volume of freshwater consumed by each industrial operation, as well as the discharge of only those emissions to land, air, and water that could have an effect on water quality, as defined by the ISO 14046 standard. Assessment of the life-cycle impact Water consumption and quality impacts were initially quantified per kilogram of each metal produced, and then scaled to per product and per sector functional units based on the amount of each material contained in a smartphone or laptop, as well as the amount of each material used globally in the electronics industry. The next sections discuss in greater detail how to characterize these effects.

The impact of water scarcity: The water scarcity footprint (water-soluble fraction) was used to calculate the impact of water consumption. The water-soluble fraction is the product of the inventory flow and the water stress characterization variables.

Recovery of materials is crucial in a circular economy. The author's baseline results, which represent the worst-case scenario, incorporate the implications of primary metal production. Circular economy aims will require material recycling technologies even in high-purity applications such as electronics. Using open-loop allocation and cut-off approaches, we evaluated the utilization of recycled aluminum and gold recovered from e-waste. According to these principles, the principal user (i.e., the first life cycle) was responsible for material extraction and manufacture, while the second life cycle was responsible for collecting, recovering, and purifying a material from electronic garbage. The authors assessed the water scarcity impact of 1 kg of recycled material using industry-specific recovery rates. After that, the water-soluble fractions of gold and aluminum were estimated utilizing both primary and secondary material sources.

4 Discussion

According to this systematic review, the mining industry's negative environmental impacts manifest themselves in increased water consumption, carbon emissions, and energy footprints, particularly during the use phase of precious metal manufacturing, as well as market saturation in international mining markets and an increase in end-of-life metal waste. Although this is a relatively unexplored field in comparison to other global stressors that degrade water quality, such as nutrient enrichment, climate change, and land use changes, the author discovered increased research interest in the environmental effects of the mining industry. Apart from water consumption and management, the metal sector contributes to the depletion of other critical resources such as energy and soil. Because the phrase "steel industry" is used here to refer to both unfinished and finished items, such as slab and billet, it is reasonable to presume that it is also interchangeable when addressing the metal industry. The terms "metal industry" and "steel industry" refer to firms involved in the manufacture of finished steel slab, billet, and bar. Apart from the perpetual need to purge the "old" to make way for the "new," the shortened metal cycles create an ongoing void that must be filled with new ideas and designs. "Sustainability," "metal industry," and "water footprint" are three key topics that are gaining traction in the literature. Sustainability is a constant theme in the texts since it is the bedrock of environmental deterioration reduction. Below, the author highlighted a few of the most important research areas within this fast-growing science.

When individual materials are analyzed, data suggest that precious metals (rhodium, platinum, gold, and palladium) have the highest water scarcity footprint per kilogram, particularly when compared to base metals (steel, copper, and aluminum). While the absolute numbers of water-soluble fractions vary according to methodology, metals are all evaluated equally in terms of their impact on water shortage. Additionally, it was determined that the metals with the highest water-soluble fraction ratings are the most environmentally dangerous due to the pollutants released during metal extraction. Precious metals had the greatest per-kg impact on all three assessed water quality impacts: freshwater ecotoxicity, freshwater eutrophication, and aquatic acidification. Due to the fact that precious metals are typically present in trace amounts in ores, their extraction and refining need a greater amount of water and energy. These repercussions are expected to grow as demand for rare earth elements increases in electronics and resource needs for extraction and manufacturing processes increase as ore quality declines. Additionally, the underlying LCI data utilized to compute water use has an effect on the baseline water-soluble fraction values. Metal eco-invent approaches are commonly built using economic allocation algorithms to partition the effects of materials generated collaboratively. Due to the great economic value of precious metals, they are often accorded a higher level of impact, albeit this is an issue that warrants further exploration. However, depending on which reactions consumed the most water, the underlying processes that resulted in water-soluble fraction outcomes at the material level varied per metal.

For example, direct water consumption for mining and production operations accounts for 46% of the water-soluble fraction, whereas indirect water consumption associated with producing the energy utilized in mining and production accounts for 36% (Kumar & Murthy, 2011).

Throughout the metal supply chain, fuel and electricity are used in a variety of ways, including opening mining pits, crushing and grinding ores, refining metals, and basic plant operations. Upstream water impacts associated with the manufacturing of chemicals and materials used in mining and processing accounted for 16% of the total water-soluble fraction for gold, with the remaining 2% related to other activities.

Aluminum, on the other hand, is an energy-intensive material that accounts for 84% of its water-soluble fraction through indirect water consumption associated with energy generation, with the remaining 6% attributable to water consumed directly for mining and production activities, 6% to upstream water impacts of materials and chemicals, and 3% to other activities (Aguilera et al., 2015). Cobalt, on the other hand, is a critical metal with moderate water scarcity impacts, with 53% of the water-soluble fraction attributable to upstream water impacts associated with the production of chemicals and materials, 29% attributable to direct water consumed during mining and production, and the remaining fractions attributable to energy production (12%) and other activities (6%) (Escobar & Schäfer, 2009).

Pollutant emissions during the treatment of sulfidic mining tailings have been identified as the principal cause of freshwater ecotoxicity and eutrophication, according to process contribution analyses. The extraction and combustion of fossil fuels to generate the power required for mining activities have an effect on water quality. These upstream energy sources contribute to the acidification of all elements, as well as to the eutrophication and ecotoxicity of base metals such as copper, nickel, lead, and manganese. Acidification is also attributed to mine opening blasting, which releases nitrogen oxide, carbon monoxide, and ammonia into the atmosphere. Due to the fact that rock fragments generated during blasting are purged from the LCI database, the results indicate no effect on water quality, indicating a viable area for more inquiry and data collecting.

Other usual problems related to LCA models, LCI databases, and impact assessment approaches occur when measuring the water footprint of metal commodities. Volumetric fluxes and freshwater consumption, for example, differ by region. Due to a scarcity of publicly available data on water withdrawals and discharges, the eco-invent statistics do not adequately capture this variability for metal production processes. Water consumption is assumed to be constant per unit of material output in all places where metal is extracted under these conditions. Even within the same country, water stress varies by season and watershed, but this level of disaggregation is not yet recorded in LCI data. While such limits are unavoidable, given that they apply to all of the materials studied in this study, the related uncertainties are unlikely to have an effect on the fundamental material hotspot findings. Another big uncertainty that requires further investigation is whether material-focused discoveries can be applied to the water footprint of a whole electronic device's life cycle. Additional water inputs and various emissions that degrade water quality will be required for use and end-of-life management, but these are also subject to enormous variability that is beyond the scope of this study, such as evolving product lifetimes and product consumption and discard scenarios.

5 Conclusions

The findings indicate that direct water use in mining and production, indirect water use in fuel extraction and energy production, and sulfidic mine tailings disposal all contribute significantly to metals' water impacts, particularly precious metals, which were identified as materials of concern across all metrics evaluated. These findings emphasize the critical role of the consumer electronics industry in collaborating with suppliers to develop long-term water management programs that promote supply chain water efficiency, proper mine tailings management, and increased use of low-carbon and low-water energy sources and conversion technologies.

Additionally, if metals such as gold can be acquired from less water-stressed places, utilizing spatial heterogeneity in metal resources via diverse supply chains can reduce the water-soluble fraction by up to seventeen percent. Such data can be used to support policies that aim to mitigate water impacts holistically, such as supplier selection based on water intensity, local water stress, and potential trade-offs associated with supply chain interruptions. Transitioning to a circular economy model is also critical for mitigating material supply chain impacts and enhancing resilience to future demand for water and primary metals. Water-soluble fraction could be reduced by 19% or more by expanding the usage of recycled content components in new smartphones and laptops. Sustainable product design, as evidenced by these tactics and the substitution or reduction of high-impact materials, will be crucial for achieving lower water impacts; however, such solutions may also involve sustainability trade-offs that must be explored and addressed in advance. These solutions will require collaboration across the supply chain in order to effectively connect design and material choices to product consumption, longevity, and end-of-life management. Additional interventions, such as expanded producer responsibility policies to accelerate product recovery, increased consumer engagement to increase product collection, and increased supply chain transparency to fully calculate the water impacts of the growing electronics industry, will almost certainly be required to meet these targets.

References

Agrawal, R., Wankhede, V. A., Kumar, A., & Luthra, S. (2021). Analysing the roadblocks of circular economy adoption in the automobile sector: Reducing waste and environmental perspectives. *Business Strategy and the Environment, 30*(2), 1051–1066.

Aguilera, E., Guzmán Casado, G., Infante Amate, J., Soto Fernández, D., García Ruiz, R., Herrera, A., & González de Molina, M. (2015). Embodied energy in agricultural inputs. Incorporating a historical perspective.

Aivazidou, E. (2021). Water management in agriculture and industry: Challenges, trends, and opportunities. *Sustainability, 14*(1), 66.

Aivazidou, E., Tsolakis, N., Vlachos, D., & Iakovou, E. (2018). A water footprint management framework for supply chains under green market behaviour. *Journal of Cleaner Production, 197*, 592–606.

Aivazidou, E., & Tsolakis, N. (2019). Water footprint management in the fashion supply chain: A review of emerging trends and research challenges. *Water in Textiles and Fashion*, 77–94.

Alkhuzaim, L., Zhu, Q., & Sarkis, J. (2021). Evaluating emergy analysis at the nexus of circular economy and sustainable supply chain management. *Sustainable Production and Consumption, 25*, 413–424.

Al-, M., Das, P., & Saadaoui, I. (2021). Circular economy in basic supply: Framing the approach for the water and food sectors of the Gulf Cooperation Council countries. *Sustainable Production and Consumption, 27*, 1273–1285.

Brizga, J., Hubacek, K., & Feng, K. (2020). The unintended side effects of bioplastics: Carbon, land, and water footprints. *One Earth, 3*(1), 45–53.

Cámara-Creixell, J., & Scheel-Mayenberger, C. (2019). PetStar PET bottle-to-bottle recycling system, a zero-waste circular economy business model. In *Towards zero waste* (pp. 191–213). Springer, Cham.

Escobar, I. C., & Schäfer, A. (Eds.). (2009). Sustainable water for the future: Water recycling versus desalination. Elsevier.

Esposito, B., Sessa, M. R., Sica, D., & Malandrino, O. (2020). Towards circular economy in the agri-food sector. A systematic literature review. *Sustainability, 12*(18), 7401.

Gholipour, A., Zahabi, H., & Stefanakis, A. I. (2020). A novel pilot and full-scale constructed wetland study for glass industry wastewater treatment. *Chemosphere, 247*, 125966.

Giampietro, M., & Funtowicz, S. O. (2020). From elite folk science to the policy legend of the circular economy. *Environmental Science & Policy, 109*, 64–72.

Gomes, A. C., Silva, L., Albuquerque, A., Simões, R., & Stefanakis, A. I. (2018). Investigation of lab-scale horizontal subsurface flow constructed wetlands treating industrial cork boiling wastewater. *Chemosphere, 207*, 430–439.

Krikke, H. (2020). Value creation in a circular economy: an interdisciplinary approach. In *Decent work and economic growth* (pp. 1–15). Springer Nature: Cham, Switzerland.

Kumar, D., & Murthy, G. S. (2011). Impact of pretreatment and downstream processing technologies on economics and energy in cellulosic ethanol production. *Biotechnology for Biofuels, 4*(1), 1–19.

Mostert, C., Sameer, H., Glanz, D., & Bringezu, S. (2021). Climate and resource footprint assessment and visualization of recycled concrete for circular economy. *Resources, Conservation and Recycling, 174*, 105767.

Nikolaou, I. E., Jones, N., & Stefanakis, A. I. (2021). Circular economy and sustainability: The past, the present and the future directions. *Circular Economy and Sustainability, 1*, 1–20. https://doi.org/10.1007/s43615-021-00030-3

Oral, H. V., Radinja, M., Rizzo, A., Kearney, K., Andersen, T. R., Krzeminski, P., ... & Carvalho, P. N. (2021). Management of urban waters with nature-based solutions in circular Ccities—exemplified through seven urban circularity challenges. *Water, 13*(23), 3334. https://doi.org/10.3390/w13233334

Peña, C., Civit, B., Gallego-Schmid, A., Druckman, A., Pires, A. C., Weidema, B., & Motta, W. (2021). Using life cycle assessment to achieve a circular economy. *The International Journal of Life Cycle Assessment, 26*(2), 215–220.

Ramírez, S., Torrealba, G., Lameda-Cuicas, E., Molina-Quintero, L., Stefanakis, A. I., & Pire-Sierra, M. C. (2019). Investigation of pilot-scale constructed wetlands treating simulated pre-treated tannery wastewater under tropical climate. *Chemosphere, 234*, 496–504.

Rodias, E., Aivazidou, E., Achillas, C., Aidonis, D., & Bochtis, D. (2020). Water-energy-nutrients synergies in the agrifood sector: A circular economy framework. *Energies, 14*(1), 159.

Singh, S., Babbitt, C., Gaustad, G., Eckelman, M. J., Gregory, J., Ryen, E., & Seager, T. (2021). Thematic exploration of sectoral and cross-cutting challenges to circular economy implementation. *Clean Technologies and Environmental Policy, 23*(3), 915–936.

Stefanakis, A. I. (2020). Constructed wetlands for sustainable wastewater treatment in hot and arid climates: Opportunities, challenges and case studies in the middle east. *Water, 12*(6), 1665. https://doi.org/10.3390/w12061665

Stefanakis, A. I., & Nikolaou, I. (2021). *Circular economy and sustainability–management and policy* (Vol. I). Elsevier Publishing.

Stefanakis, A. I., Prigent, S., Breuer, R. (2018). Integrated produced water management in a desert oilfield using wetland technology and innovative reuse practices. In: A.I. Stefanakis (Ed.), *Constructed Wetlands for industrial wastewater treatment* (pp. 25–42). John Wiley & Sons Ltd, Chichester, UK.

Weetman, C. (2020). *A circular economy handbook: How to build a more resilient*. Kogan Page Publishers.

Zumegen, A. (2020). Environmental impacts of a local circular economy business model: a carbon and water footprint analysis for smartphone refurbishment.

Economic, Social and Environmental Dimensions in the Circular Economy (CE) Monitoring Framework in the Water and Wastewater Sector

Marzena Smol

Abstract

In the transformation process toward the circular economy (CE), sustainable and circular management of water resources is one of the strategic objectives. The CE requires the implementation of systematic changes in the whole value chain of water—from the collection of water from primary sources, through its use in commercial and industrial applications, to the management of the resulting waste, such as sewage or sewage sludge. To monitor the effectiveness of introducing changes during the transformation toward the CE, a monitoring framework was proposed in this paper. It contains a set of environmental, economic and social dimensions that are related to the water and wastewater sector. Those three dimensions are mutually dependent and related to each other. If a water-related company creates new green job places in a developing or emerging country, it affects positively the material standard of living (economic dimension) through knowledge and technology transfers. This also contributes to gaining better access to education and health for the local communities (social dimension). Due to this situation, income increases, and consumption also rises, so social and economic aspects are improved in the given region. The integration of analyzed three dimensions is a base of sustainability.

Keywords

Circular economy (CE) • Sustainable development (SD) • Monitoring • Water • Wastewater

M. Smol (✉)
Mineral and Energy Economy Research Institute of the Polish Academy of Sciences, Kraków, Poland
e-mail: smol@meeri.pl

1 Introduction

Water is the most important resource on earth, required for the life of all organisms and ecosystems (Ramm, 2021). Currently, the entire world is struggling with the progressive degradation of the environment and global warming (Kundzewicz et al., 2018). Climate change is increasing pressure on water bodies, and the impact of climate change on waters is expected to increase in the coming years due to an increased number of unexpected phenomena, such as floods and droughts, ocean acidification and rising sea levels. These changes mean that action needs to be taken more quickly across all regions of the world (Tabari, 2020). Cities, countries and regions are already adapting to the new situation, using more sustainable, nature-driven solutions to mitigate the effects of floods and to ensure that water is used in a smarter and more sustainable way to deal effectively with droughts (European Commission, 2019).

Water is a widely used resource both for municipal and industrial purposes (Smol, Adam & Preisner, 2020b). It is used for consumption (drinking water) and economic purposes. In industry, water resources are used as a heat carrier (cooling and heating); the mechanical energy of natural waters is converted in hydropower plants into other types of energy, e.g., electricity; energy in the form of steam is used to move steam turbines and steam piston engines (Frost & Hua, 2019). Water consumption in industrial production is variable (Chen et al., 2018). This is due to the specificity of individual types of production and the technology used. In practice, significant differences in water consumption are observed even under similar production conditions. In view of water shortages, in many industrialized regions there is a clear progress in introducing new water-saving production technologies. There is also significant progress in the implementation technologies that limit water consumption resulting from several initiatives for the protection of the environment (Gorman & Dzombak, 2018).

M. Smol et al. (eds.), *Water in Circular Economy*, Advances in Science, Technology & Innovation,
https://doi.org/10.1007/978-3-031-18165-8_15

To protect the water cycle, many implementation measures have been taken in recent decades. One of the most important was to include water aspects in the sustainable development goals (SDGs) (United Nations, 2015). The sustainable measures include the integration of economic, environmental and social aspects to protect the water resources for current and future generations. Moreover, to improve the circular water cycle in the economy, the water aspects have been also incorporated into the scope of the transformation process toward the circular economy (CE) model (Smol, Duda, et al., 2020c). The sustainable and circular management of water resources and waste generated in the water and wastewater sector is one of the strategic objectives of the CE model (Smol, Marcinek, et al., 2020d). The CE requires the implementation of systematic changes in the whole value chain of water—from the collection of water from primary sources, through its use in commercial and industrial applications, to the management of the resulting waste, such as sewage or sewage sludge (European Commission, 2015). To monitor the effectiveness of introducing changes during the transformation toward the CE, a monitoring framework should be proposed. The CE monitoring framework should contain a set of environmental, economic and social dimensions that are related to the water and wastewater sector. Therefore, this paper presents a proposal for economic, social and environmental dimensions in the CE monitoring framework in the water and wastewater sector.

2 Materials and Methods

The research was conducted in three steps. The research framework is presented in Fig. 1. The objective of the first step of the research was a clarification of the importance of water issues in the SD and the CE model. This step of research included a comprehensive literature review, using the desk research method. The review included the revision of the peer-reviewed scientific publications published in Elsevier, Springer, Multidisciplinary Digital Publishing Institute and Google Scholar. Moreover, the strategic and planning documents published on an international level (United Nations—UN, European Commission—EC) were also analyzed. Moreover, significant sources of data were specialist reports on the SD and the CE model. The keyword odes during the desk research were "water", "wastewater", "sustainable development", "SD", "circular economy", "CE", "monitoring", "measures" and "indicators". The results of this step of research are presented in Sects. 3 and 4. The second step of the research focused on the identification of the CE monitoring framework with a set of environmental, economic and social dimensions in the water and wastewater sector. The results of this step of research are presented in Sect. 5. The third step of research contained a discussion on the possibility of the use of the proposed CE monitoring framework at the macroeconomic level. The synthesis method, that is formulating generalizations based on recognized partial theorems, was used to interpret the obtained results. To compare the obtained results with other studies and authors, the desk research method was used. The findings, as well as future research directions, are described in Sect. 6.

3 Importance of Water in Sustainable Development (SD)

Water (H_2O) is an inorganic chemical compound that is one of the most abundant substances on the globe and the only one that occurs in nature in all three states of matter—solid

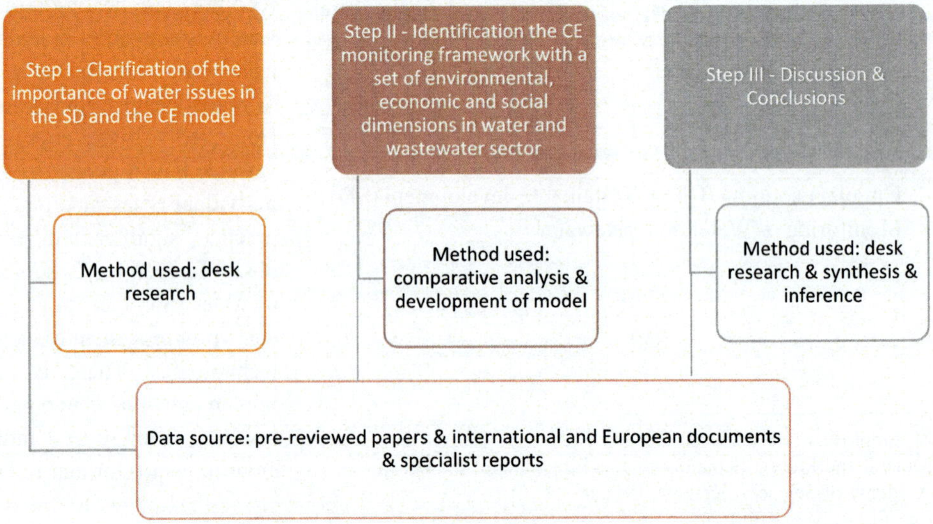

Fig. 1 The research framework

(as ice), liquid and gaseous (as water vapor). The water resources occur in the Earth's water shell (hydrosphere) in the form of surface waters (seas, oceans, rivers, lakes, marshes, wetlands, continental and mountain glaciers), groundwater (infiltration waters, condensation waters, juvenile waters, relict waters), in the atmosphere (water vapor in the troposphere, liquid water in the lower parts of the clouds and ice crystals in the upper parts of the clouds), the upper layer of the lithosphere, the pedosphere (water vapor in the soil cavities, soil solution, ground ice) and the biosphere (biological water constituting the basic component of all living organisms included in their cells and body fluids). Ensuring the security of water supply for consumer and industrial purposes requires improving practices in terms of both water abstraction and use, as well as putting it into circulation. In 2015, the United Nations (UN) launched 17 SDGs in the document "Transforming our World: Agenda for Sustainable Development—2030", which is an ambitious commitment to take action to reduce poverty in all its forms, ensure access to education, food and clean water, take action for equal opportunities, promote human rights, peace and stability in the world, protect the natural environment, mitigate climate change and access sustainable energy sources.

This document contains 17 SDGs and 169 related activities to be achieved by all parties—governments, international organizations, non-governmental organizations, the science and business sector, as well as citizens. They are concentrated around 5 areas: people, planet, prosperity, peace and partnership (5xP: people, planet, prosperity, peace, partnership). They constitute the basis of the 2030 Agenda for sustainable development in the world, taking into account the economic, social and environmental aspects. This plan also identifies the most important challenges of our time. It presents a vision of building a better world for the benefit of the inhabitants of the planet and its repair. The SDGs are a continuation of the Millennium Goals, and their elaboration was decided by the representatives of the UN member states during the United Nations Conference on Sustainable Development, which took place in June 2012 in Rio de Janeiro.

In the following SDGs, aspects of water issues were incorporated:

- SGD2—Zero hunger (end hunger, achieve food security and improved nutrition and promote sustainable agriculture)
- SDG6—Clean water and sanitation (ensure availability and sustainable management of water and sanitation for all)
- SGD7—Affordable and clean energy (ensure access to affordable, reliable, sustainable and modern energy for all)
- SGD14—Life below water (conserve and sustainably use the oceans, seas and marine resources for sustainable development), and others.

The inclusion of water aspects in the SDGs was the effect of the long-term collaboration between various groups of stakeholders involved in the implementation of the SD principles as policy makers, individual companies, non-governmental organizations (NGOs), research and education centers and society.

SDG 6: Ensure access to water and sanitation for all

An essential part of the sustainable world is clean, accessible water for all living organisms. The aim is to provide sufficient freshwater for everyone. However, due to ineffective economics or outdated infrastructure, millions of people (including kids) have no access to drinking water, which ultimately causes death. It is a consequence of inadequate water supply, hygiene and sanitation.

Poor water quality, water scarcity and inadequate sanitation have a negative impact on food security, livelihood choices and opportunities for education in third world countries and poor families across the world. Currently, more than 2 billion people are living in the regions where the risk of reduced access to freshwater occurs. The prediction showed that by 2050, at least one in four people will live in a country or region affected by recurring or chronic shortages of freshwater resources. Moreover, there is a high risk of drought in specific afflicts in the selected world's poorest countries, increasing malnutrition and hunger. In the last years, there is observed great progress regarding drinking sources and sanitation, which is an effect of the several actions of the stakeholders involved. Thanks to those efforts, up to 90% of the world's population now has access to improved sources of drinking water.

For the improvement of sanitation and access to drinking water, several investments in the management of freshwater ecosystems and sanitation facilities are needed. It includes the local level in several developing countries (Sub-Saharan Africa and Asia). The specific targets of SDG 6 are presented in Fig. 2 (United Nations, 2015).

SDG 14: Conserve and sustainably use the oceans, seas and marine resources

The world's oceans (their temperature, currents, chemistry and life) are the key components of all ecosystems, making the Earth habitable for humankind. Sea systems regulate rainwater, drinking water, climate, weather, coastlines, much of our food and even the oxygen in the air we breathe. Moreover, oceans and seas are a significant part of economic

Fig. 2 Targets of SDG 6 (United Nations, 2015)

6.1	By 2030, achieve universal and equitable access to safe and affordable drinking water for all
6.2	By 2030, achieve access to adequate and equitable sanitation and hygiene for all and end open defecation, paying special attention to the needs of women and girls and those in vulnerable situations
6.3	By 2030, improve water quality by reducing pollution, eliminating dumping and minimizing release of hazardous chemicals and materials, halving the proportion of untreated wastewater and substantially increasing recycling and safe reuse globally
6.4	By 2030, substantially increase water-use efficiency across all sectors and ensure sustainable withdrawals and supply of freshwater to address water scarcity and substantially reduce the number of people suffering from water scarcity
6.5	By 2030, implement integrated water resources management at all levels, including through transboundary cooperation as appropriate
6.6	By 2020, protect and restore water-related ecosystems, including mountains, forests, wetlands, rivers, aquifers and lakes
6.A	By 2030, expand international cooperation and capacity-building support to developing countries in water- and sanitation-related activities and programmes, including water harvesting, desalination, water efficiency, wastewater treatment, recycling and reuse technologies
6.B	Support and strengthen the participation of local communities in improving water and sanitation management

activities as they have been vital conduits for trade and transportation. Therefore, the sustainable management of this essential global resource is a key feature of a sustainable future. Unfortunately, the continuous deterioration of coastal waters contributes to pollution, while ocean acidification has an adversarial effect on the functioning of ecosystems and biodiversity. There is also a negative impact on small-scale fisheries observed. The UN strongly underlined that marine protected areas need to be effectively managed and well-resourced. Moreover, dedicated regulations have to be proposed to reduce marine pollution, overfishing and ocean acidification. The specific targets of SDG 6 are presented in Fig. 3 (United Nations, 2015).

In general, from the SD perspective, sustainable water management has to be improved in various regions and countries in the world. Sustainable water management is based on the usage of water in a way that could meet current, ecological, social and economic needs without compromising the ability to meet the needs of the future generations (Tahir, 2018). Further efforts and initiatives are planned in this area in those three dimensions—environmental, economic and social.

4 Importance of Water in the Circular Economy (CE) Model

The circular economy is defined as "a regenerative growth model that gives back to the planet more than it takes, advance towards keeping its resource consumption within planetary boundaries, and therefore strive to reduce its consumption footprint and double its circular material use rate in the coming decade" (European Commission, 2020). The importance of sustainable management of water and wastewater was mentioned by the European Commission as one of the elements in the transition toward the CE model in Europe. The first EC communication on the CE is strongly connected with waste management as it focuses on the "zero waste" strategy. From this perspective, the water and wastewater sector is a significant source of raw materials contained in water-based wastes. Therefore, scope of circular approaches in water-based waste management was proposed in the EC documents toward the CE. It was underlined that sustainable resource management is based on interactions within water, waste, energy and raw materials. Therefore, it is recommended to integrate activities in terms of water—as a resource, and waste—as a source of raw materials and energy. In these areas, further innovation developments and their implementations are expected. The aim is to improve water management should, but also management of water-based wastes (wastewater, sewage sludge, sewage sludge ash and others) (European Commission, 2014). Those waste streams are an important source of energy and biogenic raw materials (European Commission, 2013).

In the second CE communication on the CE model that has been published in 2015, 54 actions for the CE implementation were proposed (European Commission, 2015). They were presented as the first CE Action Plan for the European Union. There were several proposed initiatives for the water and wastewater sector. Due to water scarcity

Fig. 3 Targets of SDG 14 (United Nations, 2015)

14.1 By 2025, prevent and significantly reduce marine pollution of all kinds, in particular from land-based activities, including marine debris and nutrient pollution

14.2 By 2020, sustainably manage and protect marine and coastal ecosystems to avoid significant adverse impacts, including by strengthening their resilience, and take action for their restoration in order to achieve healthy and productive oceans

14.3 Minimize and address the impacts of ocean acidification, including through enhanced scientific cooperation at all levels

14.4 By 2020, effectively regulate harvesting and end overfishing, illegal, unreported and unregulated fishing and destructive fishing practices and implement science-based management plans, in order to restore fish stocks in the shortest time feasible, at least to levels that can produce maximum sustainable yield as determined by their biological characteristics

14.5 By 2020, conserve at least 10 per cent of coastal and marine areas, consistent with national and international law and based on the best available scientific information

14.6 By 2020, prohibit certain forms of fisheries subsidies which contribute to overcapacity and overfishing, eliminate subsidies that contribute to illegal, unreported and unregulated fishing and refrain from introducing new such subsidies, recognizing that appropriate and effective special and differential treatment for developing and least developed countries should be an integral part of the World Trade Organization fisheries subsidies negotiation

14.7 By 2030, increase the economic benefits to Small Island developing States and least developed countries from the sustainable use of marine resources, including through sustainable management of fisheries, aquaculture and tourism

14.A Increase scientific knowledge, develop research capacity and transfer marine technology, taking into account the Intergovernmental Oceanographic Commission Criteria and Guidelines on the Transfer of Marine Technology, in order to improve ocean health and to enhance the contribution of marine biodiversity to the development of developing countries, in particular small island developing States and least developed countries

14.B Provide access for small-scale artisanal fishers to marine resources and markets

14.C Enhance the conservation and sustainable use of oceans and their resources by implementing international law as reflected in UNCLOS, which provides the legal framework for the conservation and sustainable use of oceans and their resources, as recalled in paragraph 158 of The Future We Want

having worsened in some parts of the European Union in previous decades, acceleration of the market for secondary raw materials and water reuse was proposed. They include water-efficiency measures, reuse of treated water and wastewater in cost-effective and safe conditions. These water management options were mentioned as valuable but still under-used means of the methods of increasing water supply and alleviating pressure on over-exploited water resources on the continent. Special attention was paid to water reuse in agriculture due to its contribution to biogenic raw material recycling by substitution of mineral fertilizers. Several actions to promote the reuse of treated water and wastewater (legislation on minimum requirements for reused water) were proposed. The following water-related proposals were included in this communication, in the area of the market for secondary raw materials:

- 2017: Proposed legislation setting minimum requirements for reused water for irrigation and groundwater recharge;
- 2016–2017: Promotion of safe and cost-effective water reuse, including guidance on the integration of water reuse in water planning and management, inclusion of best practices in relevant Best available techniques Reference document (BREFs), and support to innovation

(through the European Innovation Partnership and Horizon 2020) and investments (European Commission, 2015).

In the second CE Action Plan from 2020, water aspects were incorporated in the section dedicated to "Food, water and nutrients". Special attention was paid to the importance of water reuse. The EC proposed to develop a new Water Reuse Regulation that should encourage circular approaches to water reuse in agriculture. The EC declared to facilitate water reuse and efficiency, including in industrial processes. Furthermore, the important part of the water sector is the proposition to develop an Integrated Nutrient Management Plan in the following years, with a view to ensuring more sustainable application of nutrients (including phosphorus, nitrogen and others) and stimulating the markets for recovered nutrients. The EC also mentioned considering reviewing directives on wastewater treatment and sewage sludge. One of the tasks of the EC will be more strictly monitoring and supporting the implementation of the requirements that are indicated in the Drinking Water Directive to make drinkable tap water accessible in public places (European Commission, 2020).

The CE that has its foundation in the sustainability of water use, renewable materials are used where possible, water is used sustainably, energy is provided from renewable sources in a sustainable manner, natural systems are preserved and enhanced, and waste and negative impacts are designed out. Materials, water, products and components are instead managed in loops, maintaining them at their highest possible intrinsic value (Tahir, 2018). Application of the CE principles can help achieve this, but this requires mapping the interactions of the water cycle, how it is used, and where within the river basin and urban water cycle value they can be extracted and new enterprises established. Moreover, the relationship between CE principles and water systems management needs to be identified, and disseminated among individuals, organizations and companies in order to integrate their efforts in this area. The water-circular economy relationship was presented in the White Paper "Water and Circular Economy" developed by the Ellen MacArthur Foundation in 2018 (Tahir, 2018). This foundation is one of the most important actors in the transition process to the CE. It published a lot of reports and documents on CE presenting the main CE approaches, CE models and White Papers dedicated to specific sectors or areas. The most important assumptions of the Water and Circular Economy White Paper are presented in Fig. 4. They include the design of waste externalities, keeping the resources in use, regeneration of natural capital by the reduction of consumption and non-consuming use of water.

As it was mentioned above, water is fundamental to meeting the basic needs of all living things. This service to life (including human consumption) cannot be satisfied by substituting water with other measures or materials. However, beyond this water is used and offers value in a number of different ways. These dimensions of water use were clustered in the White Paper on the CE implementation in the water sector into the three themes:

- Service,
- Energy and
- Carrier.

They have been indicated in Fig. 5. The dimensions of water also map against the Food-Water-Energy Nexus, e.g. Nutrients (Service), reducing consumptive uses (service) and generating energy (hydropower/biomass). Whether these can be harnessed and value generated and optimized, and

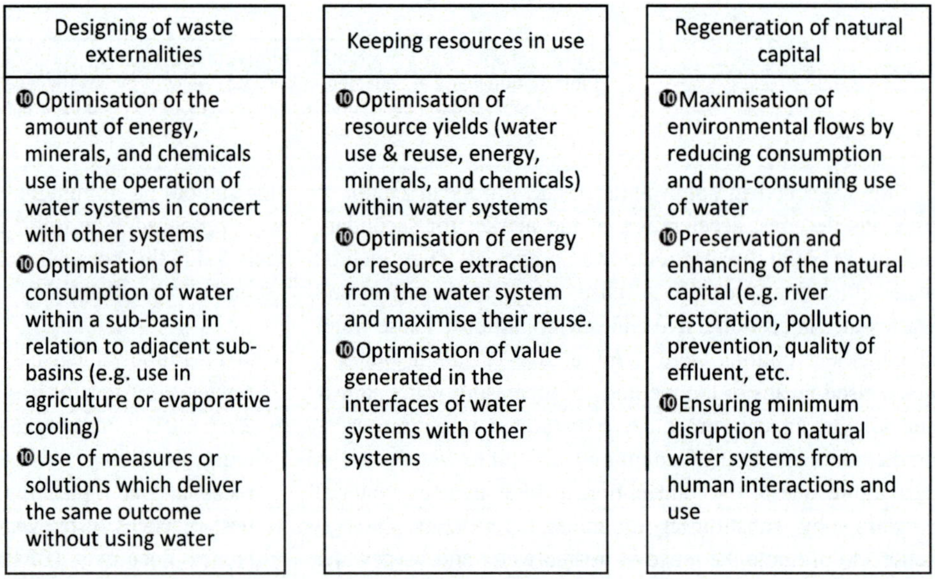

Fig. 4 Relationship between CE principles and water systems management, own based on (Tahir, 2018)

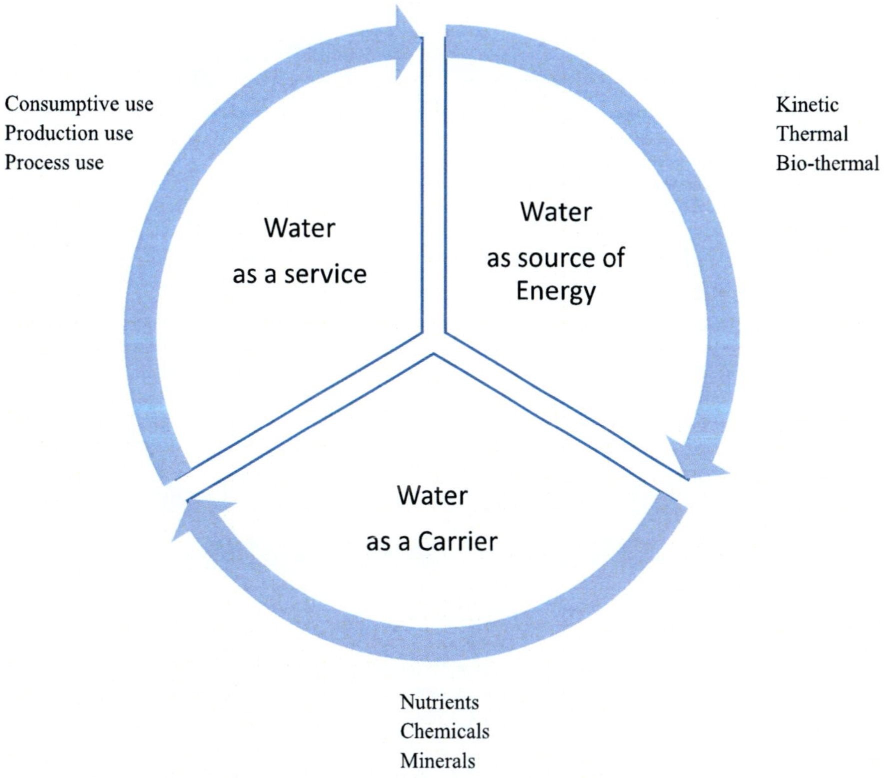

Fig. 5 Research approaches in the CE management of water resources, own based on (Tahir, 2018)

appropriately balanced between nature and human needs will vary from basin to basin, societal development level and the use category (Agricultural, Municipal, Industrial, and Environmental) (Tahir, 2018).

The dimensions of water as carrier, energy source and service provider are not exclusive, and confluences exist across them. For example, in cities wastewater conveys human/organic wastes while at the same time it can be harnessed as a source of thermal Energy. This confluence is central to the opportunity that CE offers the water sector.

Currently, the issues of transforming the water and wastewater sector to the CE model are also one of the key works of the International Water Association (IWA). In 2016, the IWA proposed a framework that may support the water utility leaders in the identification of opportunities that can boost their progress along the pathway to the CE. The following approaches have been identified:

- integrated urban resource management, connecting to stakeholders beyond traditional boundaries (urban and basin),
- leadership innovation and
- new business models.

The IWA framework Water Utility Pathways in a Circular Economy aims to demystify what systems and integrated approach to sustainability might mean for water, sanitation and wastewater service providers, assisting them to identify new operational and business models that help them prepare now for the circular economy of the future. The general idea assumes that the water cycle needs to be managed from catchment to consumer, back to catchment, and the transition to the CE model needs to consider both the consumption and production of resources across this entire value chain. It also must create both connections and synergies within the water cycle for more efficient water management and connections outside the sector.

For the water sector—all institutions and individuals responsible for managing the water cycle—transitioning to a CE presents an opportunity to accelerate and scale up recent scientific and technological advances that support greater efficiency in the sector. Water, sanitation and wastewater utilities ("water" utilities) can become engines for the CE. Water utilities have an opportunity to start to see water as a medium of valuable resources and more significantly have an important role to play as resource stewards. The IWA provided three main research areas in the form of the Utilities Pathways:

- Water Pathway,
- Materials Pathway and
- Energy Pathway,

which are indicated in Fig. 6 (International Water Association, 2016).

In summary, water plays a critical role in the world's transition from the linear consumption of resources toward

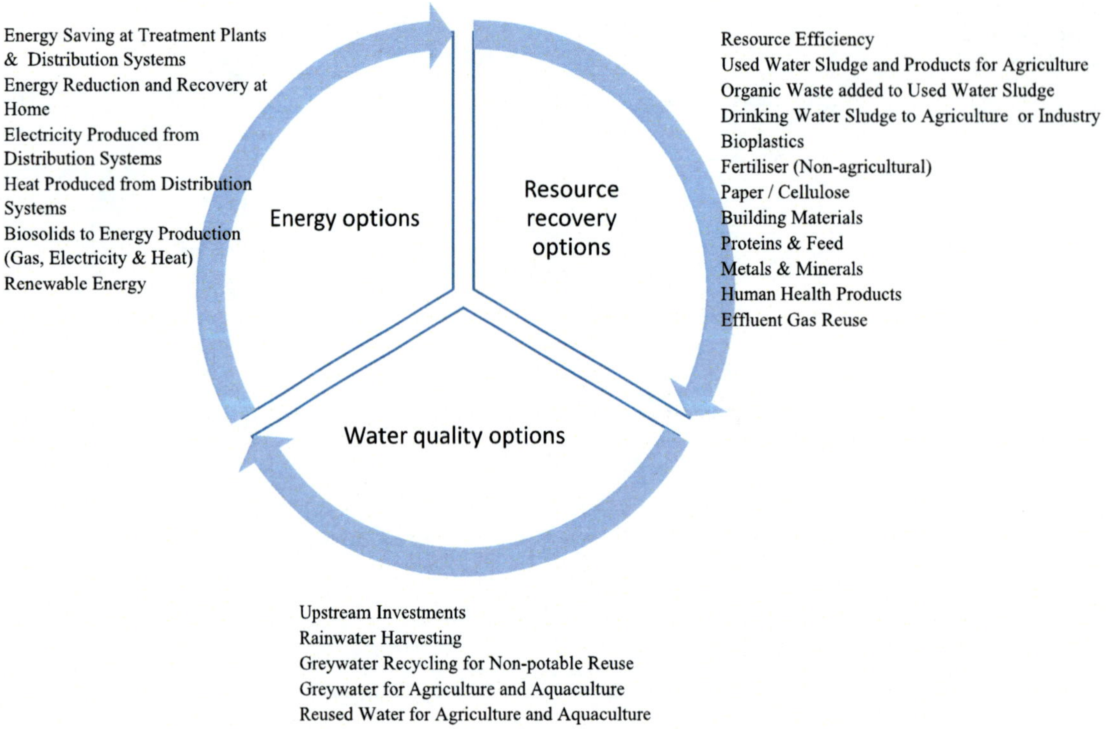

Fig. 6 Research pathways in the CE management of water resources, own based on (International Water Association, 2016)

their circular use. Water issues are incorporated in the CE model due to water being a carrier of materials and energy. Therefore, it is important to develop and implement technologies focused on water reuse or recovery of resources and energy in water and sewage management in accordance with the "nutrients-energy-water" paradigm ("NEW").

5 CE Monitoring Framework in Water and Wastewater Sector—Environmental, Economic and Social Dimensions

The CE model aims for more sustainable management of raw materials from primary and secondary sources. In this context, the CE model should integrate the three aspects of the SD—environmental, social and economic, into all CE activities. These three aspects of the SD can be also illustrated as three dimensions of the SDGs that are closely interlinked and mutually dependent. If one factor affects a particular dimension, it can be expected that this may have also positive or negative impact on the other dimensions. These can be measured by selected indicators, the identification of which is often discussed by scientists and specialists due to certain aspects being difficult to quantify. In general, those three dimensions overlap to some extent and have common points of intersection. The interdependencies within the sustainability triangle are presented in Fig. 7.

In the project "Monitoring of water and sewage management in the context of the implementation of the circular economy assumptions" (MonGOS), the assumptions for the CE model in the water and wastewater sector were proposed and published in (Smol et al., 2020b). This CE model framework was developed based on "xR" models in waste management (Huang et al., 2018; Marino & Pariso, 2020; Vaverková et al., 2019), as well as the European waste management hierarchy (European Union, 2018). The proposed CE model framework in the water and wastewater sector has been classified into selected groups of activities that fit into the following assumptions of the CE:

- *"reduction*—prevent wastewater generation in the first place by the reduction of water usage and pollution reduction at source;
- *reclamation (removal)*—application of effective technologies for the prevention of inclusion of hazardous pollutants into wastewater and removal of pollutants from water and sewage;
- *reuse*—reuse of wastewater as an alternative source of water supply (non-potable usage);
- *recycling*—recovery or reclamation of water from wastewater for potable usage;
- *recovery*—recovery of resources as nutrients and energy from water-based waste;

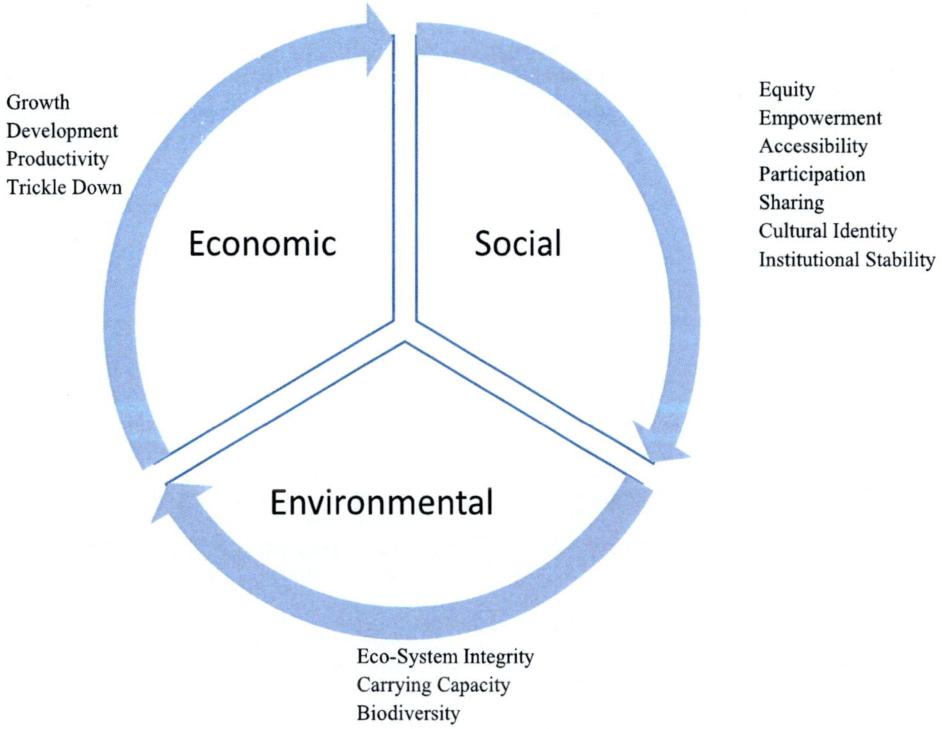

Fig. 7 Three dimensions of SDGs for sustainable development, own based on (Basiago, 1998)

- *rethink*—rethinking how to use resources to create a sustainable economy, which is free of waste and emissions."

There is also landfilling of water-based wastes. The concept of the CE model framework in the water and wastewater sector is presented in Table 1.

Economic dimension in the CE monitoring framework

Economic sustainability is defined as a system of production that satisfies present consumption levels without compromising future needs (Basiago, 1998). In this context, the CE monitoring framework should seek the economic sustainability of the whole economic system itself. In the water and wastewater sector, economic sustainability can be achieved by several activities that involve the use of water and water-based wastes. The presented CE monitoring framework includes various aspects of water and waste management. Economic sustainability can be achieved by the reduction of costs resulting from reduced water consumption (by introducing technologies that optimize the use of water) as well as by reducing the amount of water used in households (by acting little as short showers instead of bathing in a bathtub, turning off the water tap while brushing teeth, and others). Introducing infrastructure in the proposed elements of the CE monitoring framework—water recovery, raw materials recovery, energy recovery, highly effective wastewater and water treatment—also creates investment costs. However, in the long-term, they will be compensated by savings revenues from reclaimed water (for drinking and non-drinking purposes), energy and raw materials (e.g., fertilizers). The economic dimension seeks to ensure the economic efficiency of all units and individuals. It aims to ensure the greatest possible benefit from the resources to as many users as possible within the limits of the available opportunities and water resources. The implementation of the CE model in the water and wastewater sector should lead to economic growth, economic independence of the city/region/company, job creation, generating returns through integrated water-related projects, lowering the operating costs of households and enterprises, and minimizing the need to use raw materials.

Social dimension in the CE monitoring framework

Social sustainability implies a system of social organization that alleviates poverty (Basiago, 1998). In the context of the water and wastewater sector, this social dimension is mainly dedicated to public access to potable water in various regions of the world. The concept of the CE monitoring framework in the water and wastewater sector includes several activities that are dedicated to achieving this goal by reducing unsustainable water consumption per person. It is required to change behaviors in the individual households and increase the knowledge of people. Implementation of the CE solutions can also positively affect the number of green job places, which also goes hand in hand with economic aspects.

Table 1 Proposal for three dimensions of SDGs for sustainable development in the CE monitoring framework in the water and wastewater sector

Element of the CE monitoring framework	SD dimension		
	Economic	Social	Environmental
Reduction	• Costs of water consumption (industry; households) • Investment in infrastructure—water usage reduction • Income from reduced water consumption	• Water consumption per person • Household water-saving activities	• Water footprint • Water savings in the industry • Water savings in the household
Reclamation (removal)	• Cost of water/wastewater treatment • Investment in infrastructure—water/wastewater treatment • Income from treated water/wastewater sales (circulation)	• Access to clean water • Access to toxic-free food from the aquatic environment • Exposure to pollutants from the aquatic environment	• Clean aquatic environment • Eutrophication • Toxicity to aquatic organisms
Reuse	• Cost of water/wastewater reuse (non-potable usage) Investment in infrastructure—water/wastewater reuse	• Access to recirculated wastewater	• Water footprint • Water savings in the industry • Water savings in the agriculture
Recycling	• Cost of water/wastewater recycling (potable usage) • Investment in infrastructure—water/wastewater recycling	• Access to recirculated water	• Water footprint • Water savings in the household
Recovery	• Cost of energy recovery • Cost of materials (incl. nutrients) recovery Investment in infrastructure—energy/raw materials recovery	• Access to recovered energy • Access to recovered raw materials (incl. nutrients)	• Limited pressure on primary resource deposits • Limited pressure on fossil fuel deposits
Rethink	• Costs of training for specialists • Costs of training for society	• Increasing ecological awareness in the field of water and wastewater management • Changing consumer behavior	• Protection of water resources

Moreover, the implementation of activities for CE included in the monitoring framework, such as the recovery of water, raw materials and energy, allows access to these resources by society in various regions of the world. Social sustainability can be also achieved through access to toxic-free food coming from the aquatic environment and reduced exposure to pollutants from the aquatic environment. The social dimension is one of the elements of building social equality all over the world. It is important to create a system in which everyone has an equal right to access water resources to the extent that guarantees a dignified life, irrespective of the economic situation. The transformation toward the CE model should create local development opportunities for all social groups and sectors, security, access to infrastructure and green areas for various stakeholder groups at all stages of life.

Environmental dimension in the CE monitoring framework

Environmental sustainability requires maintaining natural capital as both a provider of economic inputs called "sources" and an absorber called "sinks" of economic outputs called "waste" poverty (Basiago, 1998). Environmental sustainability should be an integral part of the CE monitoring framework in the water and wastewater sector due to the fact that it concerns the management of the most important resource necessary for the survival of present and future generations—water. One of the main goals is to reduce water use. Moreover, the highly effective technologies for the removal of pollutants from water and wastewater could contribute to a clean aquatic environment, and reduce eutrophication and toxicity to aquatic organisms. The activities related to the recovery of raw materials, and energy could positively affect the unsittable pressure on natural deposits. The main goal is to improve the water footprint of all the inhabitants of the planet and to protect the earth's water resources. The environmental dimension strives for an ecological balance where ecosystems are to be treated like other water users, so their rights to access the necessary resources to the extent that they exist and remain sustainable must be guaranteed on an equal footing with people's right to use resources. In terms of the impact on the natural environment, the implementation of the CE solutions should contribute to the improvement of water quality, reduction of non-renewable energy and water consumption, material recycling and infrastructure recycling. It is worth noticing that environmental protection requirements, including the SD and the CE, impose the use of new technologies that are safe for this environment.

6 Discussion and Conclusions

The water and wastewater sector is currently facing many challenges arising from the implementation of the CE, which is a political priority in the EU (Smol, 2019), China (Geng et al., 2012), Japan (Ogunmakinde, 2019), Canada (Lederer, 2021) and other countries in the world (Despeisse et al., 2015). This sector is an important element in building the CE as waste such as wastewater and sewage sludge are valuable sources of raw materials and energy—in the CE model, they are transformed into resources (Smol, Adam & Anton Kugler, 2020a). To measure progress toward a circular economy, in 2018 the EC proposed a set of 10 CE indicators (European Commission, 2018), none of which directly refer to the water and wastewater sector. This is a big gap in determining the direction of transformation toward the CE in the area of water and wastewater. The current paper presents a proposal for the integration of three dimensions of the SDGs—economic, social and environmental into the CE monitoring framework, that has been developed based on the European waste management hierarchy and xR management methods. For the research areas presented in Table 1, a set of specific indicators to monitor the possibilities of implementing the CE in sector will be further proposed. Those indicators should monitor the implementation of integrated water and wastewater management options in the process of transformation toward the CE, such as.

- protection and conservation of water resources (natural water balance) (Halkos & Matsiori, 2014),
- water security, including access to water supply and sufficient water quality (Bogardi et al., 2012),
- reduced water consumption (Orlowsky et al., 2014),
- water and wastewater reuse to reduce the problem of water scarcity (Ramm, 2021),
- increasing evaporation in cities to improve the microclimate (Qiu et al., 2013),
- promoting the multiple, sustainable use of water (Carvalho et al., 2019),
- implementation of local separation solutions in the construction law contaminated water (double sewage system) and providing it use (Misiołek & Głowkowski, 2017), and
- increasing awareness of society on the importance of water resources (Wei et al., 2017).

To transform toward the CE model in the water and wastewater sector, many changes are required, in the organization, technologies improvement and development, financing structure, law restrictions, and also in society (European Commission, 2014). The CE has its foundation in the sustainability of water use, usage of renewable materials where possible and providing energy from renewable sources in a sustainable manner. It creates sustainable natural systems that are preserved and enhanced, and waste and negative impacts are designed out. Materials, water, products and components are instead managed in loops, maintaining them at their highest possible intrinsic value (Tahir, 2018). The application of the CE principles can help achieve this, but this will require mapping the interactions of the water cycle, how it is used, and where within the river basin and urban water cycle value they can be extracted and new enterprises established. Therefore, it is important to monitor the CE implementation in all sectors of the economy, including water and wastewater aspects.

The proposed CE monitoring framework underlies the importance of integration of sustainability dimensions with the CE activities in the waste and wastewater sector. It should be pointed out that the individual dimensions are mutually dependent and related to each other. If a water-related company creates new green job places in a developing or emerging country, it affects positively the material standard of living (economic dimension) through knowledge and technology transfers. This also contributes to gain better access to education and health for the local communities (social dimension). Due to this situation, income increases, consumption also rises, so social and economic aspects are improved in the given region. However, in the short to medium term, increased consumption and production could lead to an increase in CO_2 emissions. However, in the longer term, higher incomes allow the local communities to invest in more environment-friendly and sustainable goods, such as more efficient water treatment systems. In addition, the implementation of innovative technologies by foreign companies could contribute to making local production more resource-efficient (environmental dimension).

The important factors for further CE implementation in the water and wastewater sector are political will and commitment, social capital and adequate financing, taking into account the context of long-term planning and opportunities for return on investment in infrastructure, as well as comprehensive monitoring and evaluation of the effects of activities in the political, social and implementation fields (Krauze & Wagner, 2014).

Acknowledgements This work was funded by the National Agency for Academic Exchange (NAWA) as a part of the project "*Monitoring of water and sewage management in the context of the implementation of the circular economy assumptions*" (MonGOS; mon-gos.eu) within the International Academic Partnerships Programme. Part of the work was funded by the Subsidy of the Division of the Biogenic Raw Materials in the Mineral and Energy Economy Research Institute, Polish Academy of Sciences. The funders had no role in the design of the study; in the collection, analyses or interpretation of data; in the writing of the manuscript, or in the decision to publish the results.

NAWA POLISH NATIONAL AGENCY FOR ACADEMIC EXCHANGE

Conflicts of Interest The author declares no conflict of interest.

Financial or non-financial interests The author declares no relevant financial or non-financial interests to disclose.

Data Availability Statement The datasets generated during and/or analyzed during the current study are available from the corresponding author on reasonable request.

Ethics Approval Not applicable.

References

Basiago, A. D. (1998). Economic, social, and environmental sustainability in development theory and urban planning practice. *The Environmentalist, 19*(2), 145–161. https://doi.org/10.1023/A:1006697118620

Bogardi, J. J., et al. (2012). Water security for a planet under pressure: Interconnected challenges of a changing world call for sustainable solutions. *Current Opinion in Environmental Sustainability, 4*(1), 35–43. https://doi.org/10.1016/j.cosust.2011.12.002

Carvalho, P., Spataru, C., & Bleischwitz, R. (2019). Integration of water and energy planning to promote sustainability. *Journal of Sustainable Development of Energy, Water and Environment Systems, 7*(2), 229–252. https://doi.org/10.13044/j.sdewes.d6.0246

Chen, Y., et al. (2018). Multi-criteria design of shale-gas-water supply chains and production systems towards optimal life cycle economics and greenhouse gas emissions under uncertainty. *Computers and Chemical Engineering, 109*, 216–235. https://doi.org/10.1016/j.compchemeng.2017.11.014

Despeisse, M., et al. (2015). Towards a circular economy for end-of-life vehicles: A comparative study UK–Japan. In *Procedia CIRP* (vol 29, pp. 668–673). Elsevier B.V. https://doi.org/10.1016/j.procir.2015.02.122

European Commission. (2013). Consultative communication on the sustainable use of phosphorus. In COM(2013) 517 final, (p. 19). https://doi.org/10.1017/CBO9781107415324.004

European Commission. (2015). Communication from the Commission. Closing the loop–An EU action plan for the Circular Economy', p. (COM no. 614, 2015).

European Commission. (2018). A monitoring framework for the circular economy, p. (COM no. 29, 2018). http://ec.europa.eu/environment/circular-economy/index_en.htm

European Commission. (2019). Communication from the Commission: The European Green Deal', p. (COM no. 640, 2019).

European Commission (2020) Communication from the Commission. Circular Economy Action Plan for a cleaner and more competitive Europe, p. (COM no. 98, 2020). https://ec.europa.eu/environment/circular-economy/pdf/new_circular_economy_action_plan.pdf

European Commission. (2014). Communication from the commission–towards a circular economy: A zero waste programme for Europe, p. (COM no. 398, 2014). https://ec.europa.eu/environment/circular-economy/pdf/circular-economy-communication.pdf

European Union. (2018). Directive 2018/851 amending Directive 2008/98/EC on waste Framework. *Official Journal of the European Union*, (1907), L-150/109–140.

Frost, K., Hua, I. (2019). Quantifying spatiotemporal impacts of the interaction of water scarcity and water use by the global semiconductor manufacturing industry. In Water Resources and Industry, (vol 22, p. 100115). Elsevier B.V. https://doi.org/10.1016/j.wri.2019.100115

Geng, Y., et al. (2012). Towards a national circular economy indicator system in China: An evaluation and critical analysis. *Journal of Cleaner Production, 23*(1), 216–224.

Gorman, M. R., & Dzombak, D. A. (2018). 'A review of sustainable mining and resource management: Transitioning from the life cycle of the mine to the life cycle of the mineral. *Resources, Conservation and Recycling, 137*, 281–291. https://doi.org/10.1016/j.resconrec.2018.06.001

Halkos, G., & Matsiori, S. (2014). Exploring social attitude and willingness to pay for water resources conservation. *Journal of Behavioral and Experimental Economics, 49*, 54–62. https://doi.org/10.1016/j.socec.2014.02.006

Huang, B., et al. (2018). Construction and demolition waste management in China through the 3R principle. *Resources, Conservation and Recycling, 129*, 36–44. https://doi.org/10.1016/j.resconrec.2017.09.029

International Water Association. (2016). Water utility pathways in a circular economy. https://iwa-network.org/water-utility-pathways-cir

Krauze, K., & Wagner, I. (2014). Woda w przestrzeni miejskiej a zintegorwane zarządzanie maistem. *Zrównowazony Rozwój Zastosowania, 5*, 95–112.

Kundzewicz, Z. W., et al. (2018). Assessment of climate change and associated impact on selected sectors in Poland. *Acta Geophysica, 66*(6), 1509–1523. https://doi.org/10.1007/s11600-018-0220-4

Lederer, M. (2021). Climate change politics in Canada and the EU—from carbon democracy to a green deal? *Canadian Journal of European and Russian Studies, 14*(2), 9–28. https://doi.org/10.22215/cjers.v14i2.2763

Marino, A., & Pariso, P. (2020). Comparing European countries' performances in the transition towards the Circular Economy. *The Science of the Total Environment, 729*, 138142. https://doi.org/10.1016/j.scitotenv.2020.138142

Misiołek, A., & Głowkowski, W. (2017). Ecological safety of Silesia Metropolis. Perspectives, tasks and threats. *Scientific Journals of the Higher School of Occupational Safety Management in Katowice, 1* (13), 95–110.

Ogunmakinde, O. E. (2019). A review of circular economy development models in China, Germany and Japan. *Recycling, 4*(3), 27. https://doi.org/10.3390/recycling4030027

Orlowsky, B., et al. (2014). Todays virtual water consumption and trade under future water scarcity. *Environmental Research Letters. IOP Publishing, 9*(7), 1–10. https://doi.org/10.1088/1748-9326/9/7/074007

Qiu, G. Y., et al. (2013). Effects of evapotranspiration on mitigation of urban temperature by vegetation and urban agriculture. *Journal of Integrative Agriculture Central Iron and Steel Research Institute, 12*(8), 1307–1315. https://doi.org/10.1016/S2095-3119(13)60543-2

Ramm, K. (2021). 'Considerations related to the application of the EU water reuse regulation to the production of snow from reclaimed water. *Circular Economy and Sustainability* https://doi.org/10.1007/s43615-021-00075-4

Smol, M. (2019). The importance of sustainable phosphorus management in the circular economy (CE) model: The Polish case study. *Journal of Material Cycles and Waste Management, 21*(2), 227–238. https://doi.org/10.1007/s10163-018-0794-6

Smol, M., Adam, C., & Anton Kugler, S. (2020a). Inventory of Polish municipal sewage sludge ash (SSA)–Mass flows, chemical composition, and phosphorus recovery potential. *Waste Management, 116*, 31–39. https://doi.org/10.1016/j.wasman.2020.07.042

Smol, M., Adam, C., Preisner, M. (2020b). Circular economy model framework in the European water and wastewater sector. *Journal of Material Cycles and Waste Management, 22*(3). https://doi.org/10.1007/s10163-019-00960-z

Smol, M., Duda, J., et al. (2020c). Transformation towards circular economy (CE) in municipal waste management system: Model solutions for Poland. *Sustainability (Switzerland)*, 12(11). https://doi.org/10.3390/su12114561

Smol, M., Marcinek, P., et al. (2020d). Importance of sustainable mineral resource management in implementing the circular economy (CE) model and the European green deal strategy. *Resources, 9* (5). https://doi.org/10.3390/RESOURCES9050055.

Tabari, H. (2020). Climate change impact on flood and extreme precipitation increases with water availability. *Scientific Reports, 10* (1), 13768. https://doi.org/10.1038/s41598-020-70816-2

Tahir, S., et al. (2018). White paper. Water and circular economy. Ellen MacArthur Foundation Report.

United Nations. (2015). Resolution adopted by the General Assembly on 25 September 2015. Transforming our world: the 2030 Agenda for Sustainable Development. United Nations General Assembly, p. A/RES/70/1. https://doi.org/10.1163/157180910X12665776638740

Vaverková, M. D., et al. (2019). Environmental consequences and the role of illegal waste dumps and their impact on land degradation. *Land Use Policy, 89*, 104234. https://doi.org/10.1016/j.landusepol.2019.104234

Wei, J., Wei, Y., Western, A. (2017). Evolution of the societal value of water resources for economic development versus environmental sustainability in Australia from 1843 to 2011. *Global Environmental Change Elsevier Ltd., 42*, 82–92. https://doi.org/10.1016/j.gloenvcha.2016.12.005

Circular Economy Indicators and Measures in the Water and Wastewater Sector—Case Study

Monika Zubrowska-Sudol and Agnieszka Bisak

Abstract

The objective of this paper was to create a list of indicators for tracking the transformation of water and wastewater enterprises towards a circular economy (CE). The study was performed based on the case study of the Municipal Water Supply and Sewerage Company in Warsaw, a joint stock company. The applied methodology for the analytical and conceptual work included a detailed analysis of statutory activities implemented at the company, and a review of the benchmarking indicators and the enterprise's documentation for the identification of indicators that could be used in the system of monitoring of water and wastewater enterprises in terms of their transformation towards circular economy. Particular attention was paid to the undertaken activities related to minimising the use of primary resources (particularly water), renewable energy production, waste management and possibilities of resource recovery. The possibility of the analysis of the implementation of an information-educational strategy aimed at increasing social awareness in the scope of pro-CE activities was also considered, as well as expanding the skills of staff in the CE area. The proposed monitoring system includes 29 indicators divided into the following six groups: Resources Used, Resources Recovery, Energy, Emissions, Waste and Other (economic and social indicators).

Keywords

Circular economy indicators • Water and wastewater enterprises • Environmental indicators • Economic indicators • Social indicators

1 Introduction

The basis of the innovative thinking developing in the political, economic and business fields to replace the linear economy model with a circular economy (CE) one is the conservation of natural resources while minimising their use and maximising the reuse of materials in circulation (European Commission, 2015).

According to the programme called "Closing the loop—An EU action plan for the Circular Economy" adopted by the European Commission in December 2015, the primary goal is to create a low-emission, resource-efficient, innovative and globally competitive European economy (European Commission Comunicat 2015). This action plan has forced a change in the operations of all industry sectors that use natural resources, which, due to the high speed of economic development in recent decades, are close to depletion. One of the key natural resources that is depleting fast is water. It is an important raw material not only due to the economic development in question, but also because of the fact that it supports the basic human vital functions. An important recipient of this raw material is the water supply and wastewater sector. This is due to the strategic goals it implements, e.g., collection, treatment and supply of water to consumers.

. Thanks to the opportunity to better manage this raw material important to human life by introducing solutions to limit the amount of water taken from the primary sources and the implementation of water recovery from treated wastewater, the water and wastewater sector is a potent player able to implement strategic actions that are part of the

M. Zubrowska-Sudol (✉)
Warsaw University of Technology, Warsaw, Poland
e-mail: monika.sudol@pw.edu.pl

M. Zubrowska-Sudol · A. Bisak
Municipal Water Supply and Sewerage Company in Warsaw Joint Stock Company, Warsaw, Poland
e-mail: A.Bisak@mpwik.com.pl

M. Smol et al. (eds.), *Water in Circular Economy*, Advances in Science, Technology & Innovation,
https://doi.org/10.1007/978-3-031-18165-8_16

circular economy model (CE). However, no activities performed by the entities from this sector are limited to those related to better water management (Tran et al., 2017). In this field, there is also the potential for investments concerning the recovery of valuable raw materials or achieving complete self-sufficiency in terms of energy and heat by using renewable sources (Raheem et al., 2018).

In Poland, water and wastewater management lies largely with public entities. Water and wastewater infrastructure managed by these is classified as the so-called critical infrastructure. Therefore, construction facilities, equipment, plants and services used/provided by these entities are considered strategic in terms of ensuring the security of the State and its citizens. The status given to the water and wastewater infrastructure obliges the operators to keep prescribed safety procedures. The water and wastewater management sector should be considered of key importance not only in terms of the operation of the economy as such, but also in terms of satisfying the basic needs of the state's residents. Water and wastewater management is a crucial element of environmental protection, as well as of the health and safety of society (Kuśmierz, 2020).

On the basis of the guidelines included in the EU circular economy action plan, certain areas have been pointed out in which the operations of the water and wastewater sector are important for strengthening or implementing CE assumptions in them. These areas are closely connected to the statutory activities performed by entities operating within this sector. The key areas must include water resources management, waste management and related raw materials recovery, plus the area related to electricity generation from renewable sources. In the field of technological/technical innovation, the market is already providing a wide array of solutions that meet the challenges of a circular economy. Furthermore, the issue of resource and energy recovery in water and wastewater facilities has been a leading trend in research and development for several years (Raheem et al., 2018; Yu et al., 2021). A challenge for water and wastewater enterprises that manage critical infrastructure is to implement a system of monitoring their transformation towards a circular economy. The system would enable ongoing evaluation of actions taken/investments made by these enterprises and making the proper corrections to achieve the set objectives, thus, a system that would contribute to the implementation and development of the CE model in such enterprises.

To fill this gap, steps were taken to create a list of indicators for tracking the transformation of such enterprises towards CE. They were based on the analysis of the operations of Municipal Water Supply and Sewerage Company in Warsaw, a joint stock company. The results are presented in this paper.

2 Materials and Methods

a. Brief characteristics of the company

The Municipal Water Supply and Sewerage Company in Warsaw, a joint stock company, deals with (on behalf of the municipality) collective water supply as well as wastewater disposal and treatment. The annual amount of water supplied to the residents of the Warsaw metropolitan area is 125 mln m^3. The basic source of water for the Warsaw M.A. are the Vistula and Zegrze Lakes, which cover approximately 99% of the demand. The remaining 1% comes from local underground water intakes in Wawer and Wesoła districts. Another equivalent statutory operation of the Company is the discharge of municipal and industrial wastewater as well as infiltration and rainwater to the sewerage system, their subsequent treatment and discharge to the receiver. The wastewater treatment process takes place in four facilities: "Czajka", "Południe", "Pruszków" and "Dębe" WWTPs. The annual amount of wastewater collected from the residents and treated in the WWTPs operated by MPWiK Warszawa S.A. is approximately 170 million m^3. The product of wastewater treatment is sewage sludge. The sludge is treated in the process line. An important stage of the line is the methane fermentation process, which produces biogas of approximately 65% methane content. The gas captured in the separate fermentation chambers is combusted in power generators. This solution enables the Company to generate approximately 45,200 MWh of electricity a year to cover for its auxiliary power. In the process of biogas combustion, the intermediate product is generated, i.e., heat, also used for the Company's auxiliaries. The sludge stream which remains after the fermentation process undergoes dewatering and is later disposed in the Sludge Thermal Utilisation Station. The Company generates approximately 150 thousand Mg by dry weight of dewatered, digested sludge a year. The product of combustion is ash which is later transferred to external parties for recovery. The strategic objective of MPWiK in the Capital City of Warsaw is to prevent environmental pollution, reduce water losses, energy consumption and carbon dioxide emissions to the atmosphere. In addition, the Company conducts a number of environmental campaigns to increase the environmental awareness of its recipients. An example of such activities is the campaign called "Warszawska Kranówka" (Warsaw tap water) campaign to encourage using tap water instead of bottled water. It includes the construction of rainwater tanks to reduce the amount of water taken from primary sources for watering city plants and creating compost heaps using biodegradable organic fraction generated in the Company's own plants to produce high-quality fertiliser, i.e., compost (MPWiK w m.st. & Warszawa S.A., 2021a).

In addition, the Company has already started and plans to implement projects which qualify as investments in line with the circular economy model. These investments will enable the recovery of raw materials (e.g., phosphorus) from wastewater and will also make the Company more self-sufficient in terms of energy due to, i.a., the construction of photovoltaic farms or the implementation of the co-digestion process. The latter process will be conducted in partnership with the producers of waste characterised by high methane generation potential, which will not only allow keeping the waste in circulation, but will also provide an example of industrial symbiosis, so crucial for the implementation of the circular economy model. The implementation of the internal benchmarking for monitoring the Company's transformation towards a circular economy described in Chap. 3 is crucial not only because of the assessment of the current degree of circularity of MPWiK in the Capital City of Warsaw, but also due to the need to have a tool that will enable its users to set further directions for future investments in accordance with CE principles (MPWiK w m.st. & Warszawa S. A., 2020a).

b. Description of the research framework

Works on the preparation of the list of indicators for tracking the transformation towards CE of MPWiK in the Capital City of Warsaw commenced in 2020.

The adopted methodology included two analytical and conceptual steps. The starting point was to identify the activities realised/executed in the enterprise that meet the CE paradigms, and search for solutions to increase the circularity of the enterprise. The key element of this step was the application of the entrepreneurial discovery process with the participation of working groups representing the entirety of the enterprise's activity. Particular attention was paid to measures undertaken in the aforementioned sector related to minimising the use of primary resources and their protection (particularly water), renewable energy production, waste management and possibilities of resource recovery. The possibility of the analysis of the entire life cycle of products/services was considered, as well as an information-educational strategy aimed at increasing social awareness in the scope of pro-CE activities.

The next step was to analyse indicators applied to assess the circularity of the European Union (Eurostat, 2021) and the CE transformation indicators for the economic activities in Poland suggested in the "GOSPOSTRATEG—Oto-GOZ" project initiated by the Mineral and Energy Economy Research Institute of the Polish Academy of Sciences (Nowaczek et al., 2020a, 2020b) with regard to the characteristics of the water and wastewater industry. In parallel, there was a review of the enterprise's documentation (environmental reports, final administrative decisions, internal guidelines and documents, socio-economic reports), and benchmarking developed by the Chamber of Commerce Polish Waterworks to identify indicators that could be used in the system of monitoring of water and wastewater enterprises in terms of their transformation towards a circular economy. This step involved analysing 256 indicators.

The results of both steps were applied to creating a list of indicators that could be used by the company to evaluate CE implementation (MPWIK w m.st. & Warszawa S. A. 2021b).

3 Results and Discussion

Proposal of indicators for monitoring the implementation of circular economy principles in the Municipal Water Supply and Sewerage Company in Warsaw, a joint stock company.

The proposed monitoring system includes 29 indicators divided into the following six groups: Resources Used, Resources Recovery, Energy, Emissions, Waste and Other (Table 1). It is proposed to determine the indicators for calendar years.

Group I. Resources Used.

The first group of indicators is related to the consumption of primary resources, which should be limited in line with CE rules. Since water is the principal resource taken from the environment in water and wastewater enterprises, the proposed monitoring system offers an indicator to enable tracking reasonable management of this resource:

$$R_u^{W_{RES}} = \frac{W_t}{W_s} * 100[\%] \quad \text{(Based on internal benchmarking of the company (MPWiK w m. st.\& Warszawa 2020b))} \tag{1}$$

where

$R_u^{W_{RES}}$—use of water resources [%].
W_s—water sold by the company during the year [m^3].
W_t—water taken from primary sources during the year [m^3].

The indicator is determined against the consumers' actual water demand specified as the amount of water sold by the company during the year.

Reductions desirable from the perspective of CE can be achieved by reducing water losses in intake, treatment and supply processes, as well as by reducing the amount of tap water used for process purposes in the company, e.g., washing of equipment, rinsing filters, preparation of chemical solutions and operation of water scrubbers. Such activities facilitate achieving a situation where the amount of water taken from primary sources is adjusted to the consumers'

Table 1 List of indicators for monitoring the implementation of circular economy principles in the Municipal Water Supply and Sewerage Company in Warsaw, a joint stock company

Groups of indicators	No	Name of Indicator	Unit	Expected direction of changes	References
Environmental indicators	Resources used				
	1	*Use of water resources*	%	↓	Based on internal benchmarking of the company (MPWiK w m.st. & Warszawa, 2020b)
	Resources recovery				
	2	*Recycling degree of water recovered from used water and/or wastewater*	[%]	↑	–
	3	*Water recycling including methods of recovered water reuse*	[%]	↑	–
	4	*Recovery of critical raw materials (X_c)*	[Mg X_c] [Mg X_c/y]	↑	Based on Eurostat (Eurostat: Circular Economy Indicators: Secondary Raw Materials)
	5	*Recovery of other raw materials (X)*	[Mg X] % (Mg (X) /y)]	↑	Based on Eurostat (Eurostat: Circular Economy Indicators: Secondary Raw Materials)
	Energy				
	6	*Total energy consumption in the company*	[GWh]	↓	Based on internal benchmarking of the company (MPWiK w m.st. & Warszawa S.A. 2020b)
	7	*Energy consumption of the water supply system*	[GWh] [kWh/m^3]	↓	– Chamber of Commerce Polish Waterworks benchmarking (Izba Gospodarcza "Wodociagi Polskie" 2019)
	8	*Energy consumption of the wastewater collection and treatment system*	[GWh] [kWh/m^3] [kWh/PE]	↓	– Chamber of Commerce Polish Waterworks benchmarking (Izba Gospodarcza "Wodociagi Polskie" 2019) Based on the Chamber of Commerce Polish Waterworks benchmarking (Izba Gospodarcza "Wodociagi Polskie" 2019)
	9	Share of renewable energy in the company's total consumption of electricity	[%]	↑	Nowaczek et al. (2020a)
	10	*Share of renewable energy consumed for the water supply processes in the company's total consumption of electricity*	[%]	↑	–
	11	*Share of renewable energy consumed for the processes of wastewater collection and treatment and sludge management in the company's total consumption of electricity*	[%]	↑	–
	12	*Share of renewable energy in the total energy consumption by RES type*	[%]	↑	–
	13	*Share of renewable energy consumed for the water supply processes in the total energy consumption by RES type*	[%]	↑	–
	14	*Share of renewable energy consumed in the wastewater collection and treatment, as well as sludge management processes in the total energy consumption by RES type*	[%]	↑	–
	Waste				
	15	*Amount of particular types of waste (e.g., by European Waste Codes) generated in the company, consid ering the waste characteristics:* • *Hazardous waste* • *Neutral and non-hazardous waste*	[Mg]	↓	Based on Nowaczek et al. (2020b)
	16	*Waste management hierarchy:* • *Amount of recycled waste* • *Amount of recovered waste* • *Amount of disposed waste, divided into amounts of waste disposed of within the company and by transferring to third-party Contractors*	[%] [%] [%] [%] [%]	↑↑ ↑ ↓	Nowaczek et al. (2020b)
	17	*Substances/materials/equipment transferred for reuse (which, due to their transfer possibility, i.e., reuse, are not considered as waste)*	[Mg]	↑	–
	Emissions				
	18	*Total emissions of particular greenhouse gases*	[eq kg CO_2]	↓	Nowaczek et al. (2020a)

(continued)

Table 1 (continued)

Groups of indicators	No	Name of Indicator	Unit	Expected direction of changes	References
Economic indicators	Other				
	19	*Number of public procurement projects performed pursuant to the procurement procedure under the PPL Act, with CE principles included in the selection criteria*	[pcs] [%]	↑	Based on Nowaczek et al. (2020a)
	20	*Number of public procurement projects performed pursuant to the procurement procedure under the regulations, with CE principles included in the selection criteria*	[pcs] [%]	↑	Based on Nowaczek et al. (2020a)
	21	Charges or fees paid for the use of the environment	[PLN]	↓	Based on Nowaczek et al. (2020b)
	22	Charges or fees paid for the use of the environment, considering the increase (over the years) of the unit costs indicated in the operating schedules for these charges/fees	[PLN]	↓	Based on Nowaczek et al. (2020b)
	23	Share of costs incurred for the use of the environment in the total costs of the company	[%]	↓	Nowaczek et al. (2020b)
Social and economic indicators	24	Social initiatives related to CE	[PLN] [%]	↑	Based on Dygas and Generowicz (2020)
	25	Number of e-services in the Company • *External* • *Internal*	[pcs]	↑	Based on Nowaczek et al. (2020a)
	26	Availability of e-services	[%]	↑	–
	27	Number of instances of industrial symbiosis for waste use/management or other CE-related operations	[pcs]	↑	Based on Nowaczek et al. (2020b)
	28	Number of individuals trained in the field of CE	[ind.]	↑	Nowaczek et al. (2020b)
	29	Share of jobs related to CE operations versus the total employment in the company	[%]	↑	Nowaczek et al. (2020a)

demand. Here, it is worth mentioning that the data from the national benchmarking conducted by the Polish Waterworks Chamber of Commerce show that water losses versus water injected into the water supply network between 2018 and 2019 amounted to 12.82% and 11.07% respectively for large companies, i.e., entities providing services to more than 100,000 inhabitants. The above data show a decrease in this indicator, which reflects the expected direction of changes. However, it must be said that this indicator does not enable the monitoring of water losses along the whole chain of the water supply system. The indicator does not consider the losses in water intake processes and during its treatment in Water Purification Station Plants. Thus, it does not offer a comprehensive overview of the significant issue, i.e., the economical management of this raw material.

A reduction in $R_u^{W_{RES}}$ can also be achieved by replacing primary water resources with secondary ones, i.e., treated wastewater. The solutions to reuse water recycled from wastewater for drinking are developed largely in countries with limited primary freshwater resources (Lee & Tan, 2016).

Group II. Resource Recovery

The second group of indicators enables the monitoring of activities related to resource recovery. In this case, indicators related to the recovery of water as well as critical and other resources were distinguished.

In a water and wastewater enterprise, water can be recovered from the flow of treated wastewater and from the flows of used water from operational processes related to water treatment and distribution. The application of technological processes for water recovery is closely connected with the purpose of its reuse; thus, in the proposed monitoring system the term "reclaimed water" was defined as a total flow of water coming from treated wastewater and used water in water supply systems, directed to reuse.

The formula below was proposed to determine the degree of water recycling:

$$D_{W_{TS}}^{W_{RECD}} = \frac{W_{RECD}}{W_{TS}} * 100\ [\%] \quad (2)$$

where

$D_{W_{TS}}^{W_{RECD}}$—recycling degree of water recovered from treated wastewater and used water in water supply systems.

W_{RECD}—water recovered and diverted to reuse during the year [m^3].

W_{TS}—total amount of treated wastewater and used water in water supply systems used during the year [m^3].

Recovered water can be used for various purposes (X_{RECD}): for the company's auxiliaries, irrigation of green areas including agricultural applications, as a resource for a number of industries, for domestic use, for aquifer supply or for the preparation of potable water. For the purpose of tracking the reuse of recovered water, the following indicator is proposed:

$$D_{X_{RECD}}^{W_{RECD_X}} = \frac{W_{RECD_X}}{W_{TS}} \cdot 100\ [\%] \tag{3}$$

where

$D_{X_{RECD}}^{W_{RECD_X}}$—water recycling including methods of recovered water reuse.
W_{RECD_X}—water recovered and reused for the purpose of X_{RECD}.

Apart from water recovery, water and wastewater enterprises have great potential to perform processes which enable the recovery of resources, including critical ones, which can be reused within the economy. In this context, they become a source of secondary resources alternative to primary ones. This particularly concerns wastewater treatment plants, which have recently been considered biorefineries using primarily waste streams (Żubrowska-Sudoł, 2022). In order to determine the role of these operations with regard to the national demand for a given resource and its management globally, it is proposed to determine the weight of the individual resources recovered during the year, while classifying them as critical and other resources. In addition, it is worth showing the amount of recovered raw materials as relative values. In this case, the reference unit depends on the place of recovery (WWTP, WTS) and on the characteristics of the technological process applied. As one of these values, the amount of a given resource in the flow(s) coming into the company is proposed. Using such an indicator, it will be possible to evaluate recovery efficiency.

Group III. Energy

Another group of indicators concerns energy management. It enables a two-tier tracking of energy intensity and the use of renewable energy. The first tier includes the company as a whole, while the second one consists of two separate systems related to the statutory objectives of water and wastewater enterprises, i.e., the water supply system (water intake, treatment and supply to receivers) and the wastewater collection and treatment system (wastewater collection, treatment, as well as sludge treatment).

Indicators T_{Energy}, WSS_{Energy} and $WWCTS_{Energy}$ (nos. 6, 7 and 8 in Table 1) show the absolute energy consumption during the year for the whole company, the water supply system and the wastewater collection and treatment systems, respectively.

Indicator $E_{W_t}^{WSS}$ (no. 7 in Table 1) stands for the relative energy consumption in the water supply system calculated with the following formula:

$$E_{W_t}^{WSS} = \frac{WSS_{Energy}}{W_t}\left[\frac{kWh}{m^3}\right] \tag{4}$$

(Izba Gospodarcza ''Wodociagi Polskie'' 2019)

where

$E_{W_t}^{WSS}$—energy consumption in the water supply system in relation to the amount of water taken from the primary source $[\frac{kWh}{m^3}]$.
WSS_{Energy}—energy consumption in the water supply system [kWh].

Indicators $E_{WW_T}^{WWCTS_{WW_T}}$ and $E_{PE}^{WWCTS_{PE}}$ (nos. 8 in Table 1) are the relative energy consumption in wastewater collection and treatment systems:

$$E_{WW_T}^{WWCTS_{WW_T}} = \frac{WWCTS_{Energy}}{WW_T}\left[\frac{kWh}{m^3}\right] \tag{5}$$

(Izba Gospodarcza ''Wodociagi Polskie'' 2019)

$$E_{PE}^{WWCTSPE} = \frac{WWCTS_{Energy}}{PE}\left[\frac{kWh}{PE}\right] \tag{6}$$

(Based on Izba Gospodarcza ''Wodociagi Polskie'' 2019)

where

$E_{WW_T}^{WWCTS_{WW_T}}$—energy consumption of the wastewater collection and treatment system in relation to the amount of treated wastewater $[\frac{kWh}{m^3}]$
$E_{PE}^{WWCTS_{PE}}$—energy consumption of the wastewater collection and treatment system in relation to the People Equivalent (PE) $[\frac{kWh}{PE}]$
$WWCTS_{Energy}$—energy consumption in the wastewater collection and treatment system [kWh].
WW_T—wastewater treated during the year [m^3].
PE—population equivalent.

The other indicators included in group III enable the evaluation of renewable energy use. In this case, the total energy consumption in the enterprise was applied as the unit of reference

$$E_{RES_x} = \frac{RES_x}{T_{Energy}} \cdot 100[\%] \tag{7}$$

where
E_{RES_X}—renewable energy share in the company's total consumption of electricity, water supply processes consumption and wastewater collection and treatment system.

RES_x—amount of energy from RES for the company's total consumption of electricity, water supply processes consumption and wastewater collection and treatment system [kWh].
x = enterprise, water supply system (WSS), wastewater collection and treatment system (WWCTS).

In addition, it is proposed to monitor renewable energy share in the total energy consumption by RES type:

$$E_{RES_X^Y} = \frac{RES_X^Y}{T_{Energy}} \cdot 100[\%] \quad (8)$$

where

$E_{RES_x^y}$—renewable energy share in the company's total consumption of electricity, water supply d treatment syprocesses consumption and wastewater collection anstem.
RES_x^y—amount of energy from a type of RES for the company's total consumption of electricity, water supply processes consumption and wastewater collection and treatment system.
y—type of RES.

Group IV. Waste

The fourth group of indicators concerns properly performed waste management, which, in line with CE guidelines, should be aimed at reducing the generated waste flow by, i. a., the implementation of the so-called zero-waste technologies, reuse of waste in production processes or the reduction of the generated waste flow by applying the process of resource recovery from the waste in question. Still, the activity of utmost significance is to minimise the amount of generated waste and to drive towards the "zero waste" model. In order to bring this proposal to life, it is important to perform appropriate tasks in each area of the company's operations. This proposal should be addressed at the very stage of concept development and be one of the key criteria for the selection of a given solution at every stage of the investment process (concept/design/execution), and at any further stages including the operation of technical facilities, as well as administrative or promotional activities.

In this group, an indicator was proposed to monitor the flow of substances/materials/equipment to be directed for reuse, and therefore they will not increase the waste flow generated in the company (indicator no. 17 in Table 1). In situations when waste generation cannot be avoided, it is important to keep the correct waste management hierarchy (Waste Management Act of December 14, 2012). Pursuant to the quoted legal act, the principal waste management method is to take appropriate steps aimed at preparing waste for reuse. Less valuable solutions for properly executed waste management (in relation to the ones described above) are activities based on waste recycling or recovery using methods other than recycling or, finally, its disposal. For the proper monitoring of the transition of the current waste management towards the circular economy model, the key operation is to monitor the weight of generated waste, grouped into categories of hazardous, neutral and non-hazardous (indicator no. 15 in Table 1). The expected result of the change in the value of this indicator in the compared intervals, e.g., within a unit of one year, is a decrease in the amount of generated waste. In addition, it is important to monitor the percentage of various waste management methods versus the total amount of waste generated in the company, divided into hazardous and non-hazardous categories (indicator no. 16 in Table 1):

$$I_z^{WAS} = \frac{WAS_z}{\Sigma WAS_A} \cdot 100[\%] \quad \text{(Nowaczek et al. (2020b))} \quad (9)$$

I_z^{WAS}—Percentage of waste subject to the analysed management method (z) versus the total waste amount generated in a given accounting year [%].
$\sum WAS_A$ —total amount of waste flow generated in a given accounting year [Mg].
WAS_z—weight of waste undergoing the analysed management method (z) [Mg].
z—waste management method: recycling, recovery or disposal.

Group V. Emission

It is believed that the CE model can significantly contribute to achieving climate neutrality. In the new EU action plan concerning circular economy for a cleaner and more competitive Europe, the EC has indicated the need to increase synergies between the closed cycle and the reduction of greenhouse gas emissions. In view of the above, European and national CE-related documents underline the significance of steps to promote the reduction of greenhouse gas emissions. Implementing circular economy principles in the activities of enterprises should enable them to reduce the emissions of these gases and thus contribute to limiting negative climate change. It is important to introduce a correct indicator for monitoring in terms of emissions and all actions taken by enterprises, including new investments planned.

The indicator proposed in the monitoring system, presented here, designed to measure the emission of the particular greenhouse gases calculated as CO_2 equivalent (indicator no. 18 in Table 1), will enable the monitoring of the changes in the values of gas emissions introduced into the atmosphere.

The value of this indicator should show a downward trend in successive comparative periods. However, it should be mentioned that the implementation of new installations and processes into the existing process system, as well as the execution of measures towards transforming the unorganised emissions (not monitored to date) into organised ones, e.g., by hermetisation and deodorisation of previously open facilities, may indicate an annual increase in the value of this indicator despite the fact that the implemented investments will prove beneficial for the environment. Therefore, the indicator for emission monitoring is auxiliary and should always be considered in relation to the performed investment operations.

Group VI. Other (economic and social indicators)

The monitoring of actions taken by water and wastewater enterprises, aimed at the transformation towards CE, apart from purely environmental aspects, should consider economic as well as social aspects, particularly in the areas of public procurement, provided services (including e-services), employment, and promotional and educational initiatives. In order to monitor the transformation in the above areas, 11 indicators were proposed, included in Group VI Other.

The first of the indicators in this group are key to monitoring one of the significant EU objectives, i.e., actions aimed at linking the criteria for contractor/supplier selection in the announced public procurement procedures with the requirements of a circular economy (i.e., repairability, sustainability and recyclability). In order to meet this objective, comprehensive works are being performed in the EU, including the introduction of ecolabels or applicable certificates to confirm the degree of circularity. They are designed, i.a., to facilitate the definition of CE-compatible criteria in public procurement. The monitoring of public procurement procedures conducted by the water and wastewater sector entities, which are granted based on the selection criteria in line with the principles of the CE model, will enable the CE paradigms to be reflected in the whole value chain. In order to monitor the above, two types of indicators were proposed (indicators nos. 19 and 20 in Table 1) respectively for public procurement procedures to which the provisions of the Public Procurement Law (PPL) apply (Public Procurement Law of September 11, 2019) and procedures to which these provisions do not apply. The above division results from a different level of restrictions, especially legal ones, for conducting each type of these procedures, as well as from the fact that it is easy to introduce additional bid evaluation criteria to procedures to which the provisions of the Public Procurement Law do not apply. For each of the indicators listed in items 19 and 20 of Table 1, two reference units were proposed, i.e., a relative unit shown as a percentage of procedures in which the criterion of bid selection based on CE paradigms was applied versus the total number of PP procedures conducted in the company, and an absolute unit shown as a number of procedures in which the criteria based on CE principles were applied at the bid selection stage. The formula for the calculation of the absolute value of each of these indicators is presented below

$$I_u^{PPCE} = \frac{PP_u^{CE}}{\Sigma PPu} \cdot 100[\%] \text{ (Based on Nowaczek et al. (2020a))} \tag{10}$$

$I_u^{PP_{CE}}$—percentage of PP conducted according to one of the selected procedures (u) including CE principles in the bid selection criteria [%].

PP_u^{CE} —number of PP including CE principles in bid selection criteria conducted according to one of the selected procedures [pcs].

$\sum PP_u$ —total number of PP conducted according to one of the selected procedures [pcs].

u—type of PP procedure, i.e., tender procedures to which the provisions of the Public Procurement Law (1) apply and tender procedures below the threshold amount specified in the PPL, conducted on the basis of the internal regulations of the company (2).

Another economic aspect significant for monitoring the transformation towards CE is the amount of charges/fees for the use of the environment, e.g., for operations related to water intake, amount of treated wastewater discharged to the receiver, stormwater discharges, rainwater discharged to the receiver or emission pollutants to the atmosphere. The expected result of implementing solutions/technologies/ investments that meet the principles of the circular economy model is the gradual reduction of these costs. In order to evaluate the cost changes, three indicators were proposed, presented in items 21, 22 and 23 of Table 1. The first two indicators enable the monitoring of the absolute value of the costs incurred for the use of the environment. The difference between these indicators is the introduction of an applicable coefficient for the indicator in item 21, designed to eliminate the impact of the increase in unit costs indicated in the charge/fee operating schedules on the actual amounts of the incurred charges/fees. For the calculation of this coefficient, the following formula was proposed:

$$F_{ENVy}^{ENVb} = \frac{U_{cy}^{ENV}}{U_{cb}^{ENV}} [-] \tag{11}$$

$F_{ENV_y}^{ENV_b}$—coefficient to eliminate the impact of the increase in unit costs indicated in the operating schedules for charges/fees on the interpretation of changes in the value of indicator no. 21 in Table 1: fees incurred for the use of the environment [–];

$U_{C_y^{ENV}}$—unit cost indicated in the operating schedule for environmental charge in the analysed calendar year (y) [PLN];

$U_{C_b^{ENV}}$—unit cost indicated in the operating schedule for environmental charges/fees in the reference year (b) [PLN];
y—the calendar year under consideration for which the indicators for monitoring the transformation of the enterprise towards CE are calculated;
b—base year taken as a reference year for the values of the individual indicators to monitor the progress towards CE (the implementation year of the monitoring of the company's transformation towards CE).

The introduction of a correct coefficient is particularly significant for the interpretation of the results. Failure to implement such a coefficient may indicate an increase in the costs of environmental charges/fees in spite of the actual reduction of emissions to the environment due to the executed steps for reduction.

The third indicator in the group of indicators concerning incurred environmental charges/fees is the one that measures the share of costs for environmental charges/fees versus the total costs incurred by the company within its operations:

$$I_{\Sigma Cy}^{C_y^{ENV}} = \frac{C_y^{ENV}}{\Sigma Cy} * 100[\%] \quad (12)$$

(Based on Nowaczek et al. (2020b)

$I_{\sum C_y}^{C_y^{ENV}}$—share of costs incurred for the use of the environment versus the total costs of the company in the reporting year (y) [%].
C_y^{ENV}—value of charges/fees for the use of the environment in the reporting year (y) [PLN].
$\sum C_y$—value of total costs incurred by the company in the reporting year (y) [PLN].
y—the reporting year.

Other indicators included in this group concern the area of education provided by enterprises in the water and wastewater sector. As the players of this sector operate in the public area, their actions and social campaigns can influence local communities, e.g., promote pro-CE behaviour. In order to monitor the actions in the field of active social campaigns related to CE, indicator no. 24 in Table 1 was developed, showing the amount of costs incurred on such initiatives. For the above-mentioned indicator, two methods of monitoring the applied values were proposed. The first is to present the total costs incurred on promotional campaigns concerning pro-CE behaviour, in PLN. The second method is to calculate the percentage (according to formula no. 13) of the costs incurred on promotional campaigns concerning pro-CE behaviour versus the total costs of the company for social campaigns:

$$I_{SC}^{C_{SC}^{CE}} = \frac{C_{SC}^{CE}}{\Sigma C_{SC}} \cdot 100[\%] \quad (13)$$

(Based on Dygas and Generowicz (2020))

$I_{sc}^{C_{SC}^{CE}}$—share of costs for social campaigns to promote behaviour consistent with CE paradigms versus the total costs of the company for social campaigns in the reporting year [%].
C_{SC}^{CE}—costs incurred for social campaigns to promote behaviour consistent with CE paradigms in the reporting year [PLN].
$\sum C_{sc}$—total costs incurred by the company for social campaigns in the reporting year [PLN].

Still, the above educational value should not be directed only outwards of the company. It is also significant to take steps to increase the awareness and knowledge of the staff of the entity in the field of EU policy concerning the transition from a linear to a circular economy model. For this purpose, specific training should be organised for employees on CE issues, and the share of positions related to the performance of tasks from the field of the circular economy model should be increased. To assess progress in these areas, indicator nos. 28 and 29 in Table 1 were proposed.

In addition, the primary challenge of the new policy aimed at reaching circularity is to keep all the links of the production chain, i.e., products, materials, raw materials and waste, in circulation. In order to meet the above proposal, a strategic activity is to look for proper business partners and conclude cooperation agreements that enable both parties to achieve increased industrial circularity. Such a type of cooperation is called industrial symbiosis and is an important factor in EU proposals. To monitor progress in this area, indicator no. 27 (Table 1) was proposed.

In many papers on CE, key actions include increasing digitalisation of enterprises. Actions in this respect lead to a large reduction in the use of printing consumables, e.g., paper, toner and ink, and therefore reduce the carbon footprint. An additional benefit of increasing the degree of digitalisation of business entities is their growing innovation and better competitive position in the market. In order to monitor the progress of digitalisation in water and wastewater enterprises, two indicators, nos. 25 and 26 in Table 1, were proposed. The purpose of the first indicator is to show the availability of the so-called e-services, both for the company's stakeholders and staff. The second indicator enables monitoring the time of system availability in the period under consideration:

$$I_T^{E-SERVICE} = \frac{TE - SERVICE}{8760} * 100[\%] \quad (14)$$

(Based on Nowaczek et al. (2020a))

$I_T^{E-SERVICE}$—availability of e-services during the year [%].
$T_{E-SERVICE}$—time of e-services availability during the year [h].
8760—total number of hours during the year [h].

3.1 Future Perspective

Works on the herein-described development of the system of monitoring the transformation of the Municipal Water Supply and Sewerage Company in Warsaw, a joint stock company, towards CE were conducted in the period from July 2020 to August 2021. In March 2022, based on a resolution of the company's Supervisory Board, the system was implemented in the activity of the enterprise (MPWIK w m. st. & Warszawa S.A., 2022).

Due to the dynamic development in the area of CE expected in the upcoming years, the monitoring system implemented in the enterprise is of open character. It was presumed to be annually verified/modified over the period 2022–2025. An important element of such modifications will be the analysis of current literature reports in the scope of indicators proposed for tracking the transformation of the water and wastewater sector towards CE. It should be emphasised that while scientific papers regarding the measurement of circularity in the water and wastewater sector were scarce in 2020 (when works on the discussed monitoring system commenced), the issue has become increasingly popular among researchers in recent years (Kakwani & Kalbar, 2022; Preisner et al., 2022; Smol & Koneczna, 2021).

4 Conclusions

A set of 29 indicators is proposed to monitor the transition of the Municipal Water Supply and Sewerage Company in Warsaw, a joint stock company, towards CE. The proposed monitoring system includes indicators to measure the use of water from primary sources (as the main resource taken from the environment by water and wastewater enterprises), the use of renewed water (i.e., a secondary resource, an alternative to the primary one), the amount of recovered raw materials, including critical ones, the energy intensity of the company, the use of energy from renewable sources, the amount of waste generated, greenhouse gas emissions, as well as a number of economic and social indicators to measure the costs incurred on CE-related actions, digitisation of the company, pro-CE social and educational initiatives, and finally staff qualifications in the field of CE.

Further analytical and conceptual work will focus on specifying circularity degrees. In the area of monitoring the circular economy within a company, it is also important to identify key indicators that would enable a utilitarian evaluation of an entity's circularity against the background of both a given industry and broader economic operations.

References

Dygas, P., & Generowicz, N. (2020). Identyfikacja wskaźników z zakresu gospodarki o obiegu zamkniętym raportowanych przez firmy krajowe i zagraniczne z branży energetycznej; Wskaźniki monitorowania gospodarki o obiegu zamkniętym (Identification of circular economy indicators reported by domestic and foreign companies from the energy sector; Circular economy monitoring indicators), Wydawnictwo IGSMiE PAN.

European Commission, (2015). Comunication from the Commission. Closing the loop -An EU Action Plan for the Circular Economy, COM 614; Brussels, Belgium: Europen Commission.

Eurostat: Circular Economy Indicators: Secendary Raw Materials https://ec.europa.eu/eurostat/cache/metadata/en/cei_srm030_esmsip2.htm

Gospodarcza, I. "Wodociagi Polskie", (2019). Benchmarking: Wyniki przedsiębiorstw wodociągowo-kanalizacyjnych w Polsce za rok 2018 (Benchmarking: Results of Water Suplies and Sewerage Companies in Poland for 2018).

Kakwani, N. S., & Kalbar, P. P. (2022). Measuring urban water circularity: Development and implementation of a water circularity indicator. *Sustainable Production and Consumption 31* 723–735

Kuśmierz, A., Ramm, K., Lutowska, A., Gołębiewski, A. (2020). Charakterystyka sektora gospodarki wodno-ściekowej i jego potrzeb w zakresie kompetencji (Characteristics of the water and sewage management sector and its competence needs), Competence Council in the Water and Sewage Management and Reclamation Sector, Poland).

Lee, H., & Tan, T. P. (2016). Singapore's experience with reclaimed water: NEWater. *International Journal of Water Resources Development 32* 4 611–621.

MPWiK w m.st., & Warszawa, S. A. (2020a). Dokument wewnętrzny: Rekomendacje dotyczące transformacji działalności Miejskiego Przedsiębiorstwa Wodociągów i Kanalizacji m.st. Warszawa S.A. w kierunku gospodarki o obiegu zamkniętym (Integral document: Recommendations for the transformation of the activities of the Municipal Water Supply and Sewerage Company in Warsaw, joint stock company towards a circular economy).

MPWiK w m.st., & Warszawa S. A. (2020b). Dokument wewnętrzny: Księga Zintegrowanego Systemu Zarządzania (Integral document: Book of the Integrated Management System)

MPWiK w m.st., & Warszawa, S. A. (2021a). Raport Odpowiedzialności Społecznej 2020 (Social Responsibility Raport for 2020).

MPWiK w m.st., & Warszawa, S. A. (2022). Dokument wewnętrzny: Uchwała nr 110/2022 z dnia 22 marca 2022 r. Zarządu Miejskiego Przedsiębiorstwa Wodociągów i Kanalizacji w m.st. Warszawie S. A. w sprawie akceptacji i przyjęcia do realizacji rekomendowanych w dokumencie pt. „Mapa drogowa transformacji Spółki w kierunku Gospodarki Obiegu Zamkniętego" działań ukierunkowanych na zwiększenie cyrkularności Przedsiębiorstwa (Internal document: Resolution No. 110/2022 of 22 March 2022 of the Management Board of the MPWiK w m.st. Warszawie S.A on the acceptance and implementation of the recommendations from the document entitled "Roadmap for the transformation of the Company towards the Circular Economy" of activities aimed at increasing the circularity of the Company.).

MPWIK w. m.st., & Warszawa, S. A. (2021b). System monitoringu transformacji MPWIK w m. st. Warszawie S.A. w kierunku GOZ. Mapa drogowa transformacji MPWIK w m. st. Warszawie S.A. w kierunku GOZ (System for monitoring the transformation of

MPWIK w m.st. Warszawa S.A towards circular economy . Roadmap for transformation MPWIK w m.st. Warszawa S.A towards circural economy").

Nowaczek, A., Kulczycka, J., & Bączyk, A. (2020a). Propozycja wskaźników pomiaru transformacji gospodarki polskiej w kierunku GOZ; Wskaźniki monitorowania gospodarki o obiegu zamkniętym (Proposal of indicators for measuring the transition of the Polish economy towards CE; Indicators for monitoring circular economy), Wydawnictwo Instytutu Gospodarki Surowcami Mineralnymi i Energią PAN.

Nowaczek, A., Kulczycka, J., & Bączyk, A. (2020b). Postulowane mierniki monitorowania transformacji w kierunku gospodarki o obiegu zamkniętym; Wskaźniki monitorowania gospodarki o obiegu zamkniętym (Proposed metrics for monitoring the transition towards circular economy; Indicators for monitoring circular economy), Wydawnictwo Instytutu Gospodarki Surowcami Mineralnymi i Energią PAN.

Ustawa z dnia 14 grudnia.2012 r. o odpadach, Dz. U.2021 poz. 779, 784, 1648, 2151 (Waste Management Act of December 14, 2012 Journal of Laws of 2021, items 779, 784, 1648, 2151).

Preisner, M., Smol M., Horttanainen, M., Deviatkin, I., Havukainen, J., Klavins, M., Ozola-Davidane, R., Kruopienė, J., Szatkowska, B., Appels, L., Houtmeyers, S., & Roosalu, K. (2022). Indicators for resource recovery monitoring within the circular economy model implementation in the wastewater sector. *Journal of Environmental Management 304*, 114261.

Raheem, A., Sikarwar, V. S., He, J., Dastyar, W., Dionysiou, D.D., Wang, W., & Zhao, R. M. (2018). Opportunities and challenges in sustainable treatment and resource reuse of sewage sludge: A review. *Chemical Engineering Journal, 337* 616–641

Regulation (EC) No 2150/2002 of the European Parliament and of the Council of 25 November 2002 on waste statistics.

Smol, M., & Koneczna, R. (2021) Economic indicators in water and wastewater sector contributing to a circular economy (CE) *Resources, 10* 129.

Tran, Q. K. Jassby, D., & Schwabe, K. A. (2017). The implications of drought and water conservation on the reuse of municipal wastewater: Recognizing impacts and identifying mitigation possibilities. *Water Research, 124* 472–481.

Ustawa z dnia 11 września 2019 r., & Prawo zamówień publicznych, Dz. U. (2019) poz. 2019 (Public Procurement Law of September 11, 2019. Journal of Laws of 2019, item 2019).

Yu, B., Luo, J., Xie, H., Yang, H., Chen, S., Liu, J., Zhang, R., & Li, Y.-Y. (2021). Species, fractions, and characterization of phosphorus in sewage sludge: A critical review from the perspective of recovery. *Science of the Total Environment, 786* 147–437.

Żubrowska-Sudoł, (2022). Oczyszczalnia ścieków jako element biogospodarki o obiegu zamkniętym (Wastewater treatment plant as an element of circular bioeconomy), Oficyna Wydawnicza Politechniki Warszawskiej.

Business Opportunities in the Water and Wastewater Sector Through the ReSOLVE Framework

Marzena Smol, Alfonso Mejia, and Marina Howarth

Abstract

Circular economy (CE) is a regenerative system where the value of resources is maintained in the economy and the production of waste is minimised. It refers to the whole value chain of a key resource that is used in the economy for both municipal and industrial applications, such as water. Given that water is a key resource, it should be managed in the most circular manner, especially in regions of water scarcity. Sustainable and circular water management are critical tasks for the water and wastewater sector, given their prominent role in assigning economic value to water. Therefore, there is a strong need to further implementation of business opportunities in this sector to maximise the use of water potential, as well as to accelerate the transformation process towards a CE model. This paper presents an overview of the CE business opportunities for different actor's bodies in the water and wastewater sector. Examples of good practices were divided into six action areas for businesses and countries; transforming from a linear to a CE model using the Regenerate, Share, Optimise, Loop, Virtualise and Exchange (ReSOLVE) framework. There are several possibilities for CE implementation in the analysed sector, including recovering renewable energy and nutrients from waste generated in water and wastewater treatment plants; providing water supply systems in cities and rural areas; optimising the amount of energy, minerals and chemicals in water systems operations; reusing water for agricultural and industrial purposes; adding digitisation operations in water treatment stations and wastewater treatment plants; and/or replacing industrial devices with more energy-saving options. The proposed solutions should contribute to creating a balance between economic, social and environmental dimensions that are the bases of sustainable solutions.

Keywords

Circular economy (CE) • Water • Wastewater • Business • Circular business models

1 Introduction

A circular economy (CE) is a regenerative system where the value of resources is maintained as long as possible in the economy and the production of waste is minimised (European Commission, 2020). The CE is indicated as a main economic model that supports the implementation of Green Deal strategies in various regions of the world, such as the European Union (EU) (European Commission, 2019), United Kingdom (HM Govenment, 2018), United States (US) (US Government, 2018), Canada (MacArthur et al., 2020), and South Korea (Chowdhury, 2021). In the green transformation, more attention is paid to the creation of added value in the whole chain, based on the CE approach, and thus new business models are needed in individual companies and sectors (Nikolaou et al., 2021). These new models—Circular Business Models (CBMs)—create business opportunities for individual actors involved in wastewater management around the world.

The literature on CBMs has grown rapidly in the last decade and contains a variety of different typologies (Iyer-Raniga, 2019; Nikolaou et al., 2021; Nikolaou & Tsagarakis, 2021). These typologies have important differences in the level of granularity and the classification approach

M. Smol (✉)
Mineral and Energy Economy Research Institute of the Polish Academy of Sciences, Kraków, Poland
e-mail: smol@meeri.pl

M. Smol · A. Mejia · M. Howarth
Department of Civil and Environmental Engineering, The Pennsylvania State University, State College, USA
e-mail: aim127@psu.edu

M. Howarth
e-mail: mch263@psu.edu

M. Smol et al. (eds.), *Water in Circular Economy*, Advances in Science, Technology & Innovation,
https://doi.org/10.1007/978-3-031-18165-8_17

taken. The CBMs refer to the business models in which the material loop is closed (Mentink, 2014). They can also focus on a value chain perspective that structures business models into circular design, optimal use and value recovery types (Kumar et al., 2022). IMSA (van Renswoude et al., 2019) underlined short loops, long loops, cascades and pure cycles in their models while Lewandowski (2016) focused on regeneration, sharing, optimisation, or looping methods (Lewandowski, 2016). The specific areas of CBMs implicit in all of these typologies overlap significantly, but are often given different names, depending on the model. The concept of CBMs focuses on promoting the reorganisation of current value creation architectures and supply chains to create a sustainable system of production and consumption (Hofmann, 2019). Therefore CBM configurations should contribute to a more holistic and radical change in current business logics than approaches that achieve incremental resource efficiency improvements (Nußholz, 2017).

The transformation towards a CE relies on the whole value chain of a key resource that is used in the economy for both municipal and industrial applications—water (Smol et al., 2020). Since water is one of the Earth's key resources, it should be managed in the most circular manner possible, especially in regions of water scarcity and vulnerability (Židonienė & Kruopienė, 2015). Sustainable and circular water management is one of the main tasks for the water and wastewater sector, where water and water-related resources (as nutrients and energy) have an economic value. There is, therefore, a strong need to develop CBMs in this sector that maximise the use of water potential, as well as accelerate the transformation process towards the CE model. This paper presents an overview of the CE business opportunities for different bodies involved in the water and wastewater sector, and provides examples of key benefits from CBM implementation.

2 Materials and Methods

This research presents CBMs that could be used in the water and wastewater sector, within the specific areas of the ReSOLVE framework developed by the Ellen MacArthur Foundation and McKinsey in 2015 (Ellen MacArthur Foundation, 2015). The structure of the study contains a literature review of the CE approach within the context of ReSOLVE's six business actions and an inventory of the CBMs that could be used in each of these actions is presented. Results are supported by the identification of key benefits in CBM implementations. Our research framework is shown in Fig. 1, while a description of the method used is presented below.

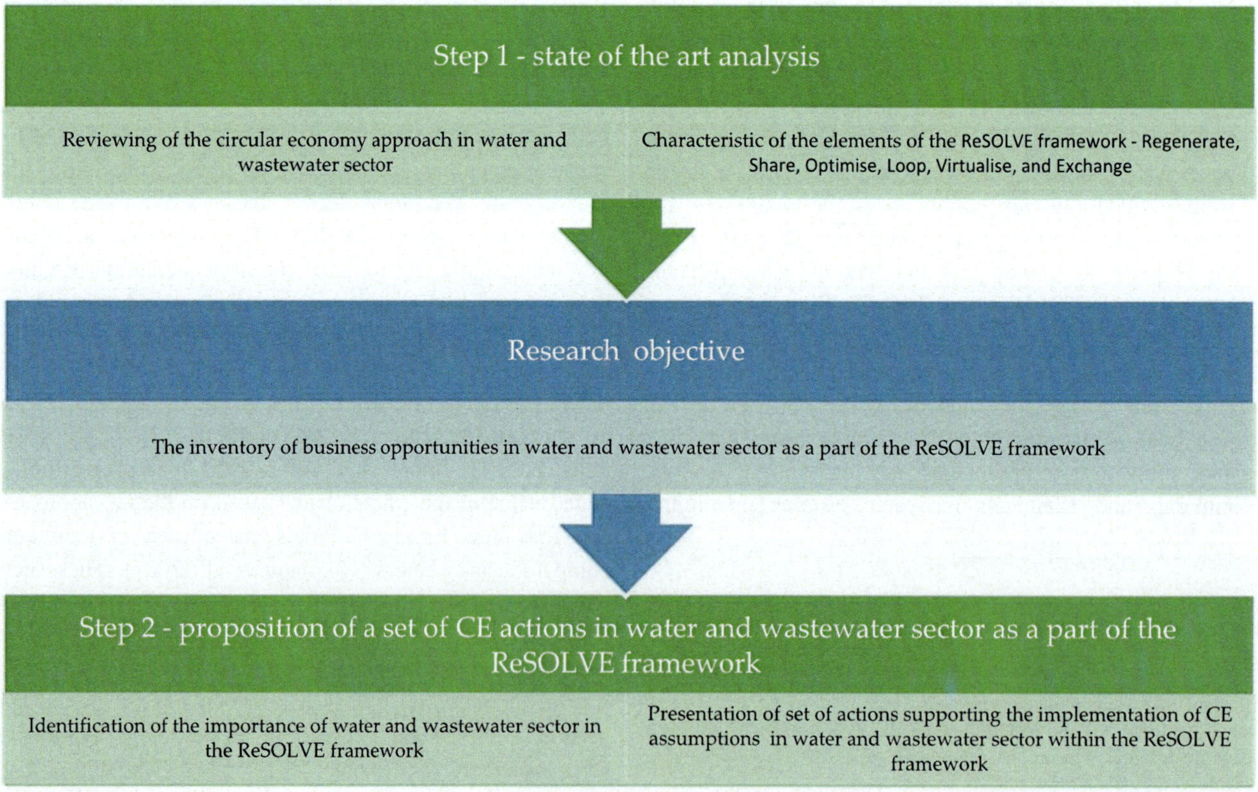

Fig. 1 Research framework

Table 1 Characteristics of the ReSOLVE model

Action	Description	Examples
Regenerate	Shifting to renewable energy and materials; reclaiming, retaining and regenerating health of ecosystems; and returning recovered biological resources to the biosphere	• Shift to renewable energy and materials • Reclaim, retain and restore health of ecosystems • Return recovered biological resources to the biosphere
Share	Keeping the product loop speed low and maximisation of the utilisation of products by sharing them among users, reusing them throughout their technical lifetime, and prolonging their life through maintenance, repair, and design for durability	• Share assets (e.g. cars, rooms, appliances) • Reuse/secondhand • Prolong life through maintenance, design for durability, upgradability, etc.
Optimise	Increasing the performance/efficiency of a product and removal of the waste in production and the supply chain	• Increase performance/efficiency of product • Remove waste in production and supply chain • Leverage big data, automation, remote sensing and steering
Loop	Keeping components and materials in closed loops, and prioritisation inner loops. In the case of the finite materials, it means remanufacturing products or components, and as a last resort recycling materials	• Remanufacture products or components • Recycle materials • Digest anaerobically • Extract biochemicals from organic waste
Virtualise	Delivering and using the utility virtually	• Books, music, travel, online shopping, autonomous vehicles, etc
Exchange	Replacing the old materials with advanced non-renewable materials and applying of the new technologies	• Replace old with advanced non-renewable materials • Apply new technologies (e.g. 3D printing) • Choose new product/service (e.g. multimodal transport)

Source Own based on (Ellen MacArthur Foundation, 2015)

In the first step of research, a comprehensive literature review was conducted with the use of desk research method (Smol, 2021b). The analysis included characteristics of the ReSOLVE framework with revisions to specify for CE business opportunities in the water and wastewater sector. The following sources of data were analysed: national and international documents on the CE model, thematic reports on the CE, peer-reviewed papers published (available in Elsevier, Springer, Google Scholar, Multidisciplinary Digital Publishing Institute). The following keywords were used during searching the relevant papers: "circular economy", "CE", "business model", "CE business model", "Circular Business Model", "CBM", "water", "wastewater", "recycling".

In the second step of research, examples of specific solutions supporting the transformation towards the CE model in water and wastewater were identified and inventoried using synthesis (combining distinguished factors) and induction (deriving conclusions from the premises that are their individual cases). The examples of CE solutions were identified and inventoried for all areas of circularity indicated in the ReSOLVE framework: Regenerate, Share, Optimise, Loop, Virtualise and Exchange (Ellen MacArthur Foundation, 2015). For each presented area, good practices were listed and key benefits were provided based on the implementation of CE assumptions in the water and wastewater sector. The basic description of the selected elements of the ReSOLVE framework is provided in Table 1 and the inventory of the examples of CE solutions in the water and wastewater sector within the ReSOLVE framework is presented in Fig. 3.

3 Results

This section provides a description of the results of the paper, including characteristics of the ReSOLVE framework, clarification of the importance of the water and wastewater sector in this framework, and identification of the CBMs that could be used in the water and wastewater sector to foster CE implementation.

3.1 Characteristics of the ReSOLVE Framework

The ReSOLVE framework was proposed by the Ellen MacArthur Foundation and McKinsey in 2015. The ReSOLVE framework includes six action areas for businesses and countries in their transformation from a linear model to the CE model: Regenerate, Share, Optimise, Loop, Virtualise and Exchange. It offers businesses and countries a tool for generating circular strategies and growth initiatives. These action areas can be used as a template for deepening the implementation of CE principles in various branches of the economy, including the water and wastewater sector. The current work focuses on inventory of CBMs and good practises that can be implemented in water and wastewater sector, which have been grouped in the core areas indicated in this ReSOLVE framework. A set of six business actions in the analysed framework refer to three key principles of the CE model:

(i) preserve and enhance natural capital by controlling finite stocks and balancing renewable resource flows;
(ii) optimise resource yields by circulating products, components and materials at the highest utility at all times in both technical and biological cycles;
(iii) foster system effectiveness by revealing and removing negative externalities, such as water, air, soil and noise pollution; toxins; congestion; climate change and negative health effects related to resource use.

The characteristics of individual actions in the ReSOLVE framework are presented in Table 1. For each action, examples of good practises are provided which present a major circular business opportunity enabled by the technology revolution. In various ways, these actions increase the utilisation of physical assets, prolong their life and shift raw material use from finite to renewable sources. Moreover, each action reinforces and accelerates the performance of the other actions (Ellen MacArthur Foundation, 2015).

The ReSOLVE framework was proposed for businesses and government bodies to support their decisions on CE implementation in specific sectors of the economy (Ellen MacArthur Foundation, 2015). This model framework was an inspiration to adopt its assumptions to different areas of the economy (Cattelan Nobre & Tavares, 2020), as business operation (Lewandowski, 2016), building and construction industry (Iyer-Raniga, 2019), municipal waste management (Sl et al., 2020), service-oriented technology companies (Heyes et al., 2018) and others. This confirms the great importance of the ReSOLVE framework for the transformation process toward CE model in the economy.

3.2 Importance of Water and Wastewater Sector in the ReSOLVE Framework

To build the case for business action and potential policy interventions in the water and wastewater sector, the economic impact of each prioritised CE opportunity needs to be assessed. One of the reports of the Ellen MacArthur Foundation describes an important tool to conduct this impact quantification for individual opportunities. The ambition level for the quantification action must be considered before the specific action is initiated. The understanding of the impact has key importance to help mobilise business and build the base for policy intervention. Moreover, the related (potentially high) resource demands also need to be balanced with those of other steps from the proposed toolkit (Ellen MacArthur Foundation, 2015). Figure 2 presents the indicative prioritisation of ReSOLVE action areas for the water sector in Europe. This confirms the importance of this sector in the transformation process from a linear to a CE model. In each action group in the ReSOLVE framework, there are different importance levels indicated for water-related aspects termed here as "water supply, waste and remediation":

Regenerate—high,
Share—low,
Optimise—low,
Loop—middle,
Virtualise—middle,
Exchange—low.

However, this prioritisation model, which is based on the economic and resource impacts of different action areas, does not present the final places where potential can be found. It was proposed in 2015, and since then, there has been significant progress in the area of water, water-related waste and remediation.

The CE model provides various value creation mechanisms that aim to decouple economic activities from the consumption of finite resources (Smol & Kulczycka, 2019). In general, consumption in a full CE can only occur in effective bio-cycles, while in other cases use replaces consumption. In a bio-cycle, the raw materials should be regenerated or recovered, while in a technical cycle they

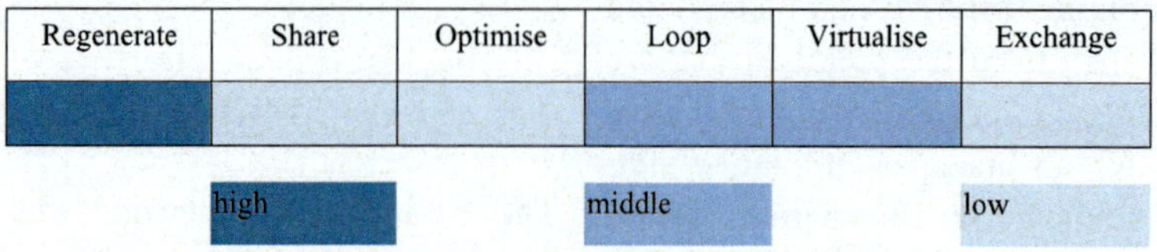

Fig. 2 Indicative prioritisation of ReSOLVE action areas water supply, waste and remediation sector in Europe. *Source* own based on (Ellen MacArthur Foundation, 2015)

should be restored. The aim of the bio-cycle is to regenerate disordered materials, without or despite human intervention (Nika et al., 2020). In the technical cycle, the CE-related technological solutions and CE business models focus on maximisation of the value extracted from finite resources, including also various waste streams in industrial sectors (Kakwani & Kalbar, 2020). Water is related to both those cycles. It should be underlined that the transformational change in the use of new business models in the water sector is necessary to maximise the recirculation of value from water cycles in the bio-cycles and technological cycles, at all scales (city, river basin, industrial plants, individual and municipal buildings) to increase the efficiency in usage of water resources. Moreover, new CE business models in the CE could contribute to prevent further degradation of individual environmental components (water, soil, air, etc.). The adaptation of the CE approach, based on the following three principles of the CE: (i) designing out waste externalities; (ii) keeping resources in use; (iii) and regenerating natural resources; includes tremendous opportunities for water-related businesses, central and regional governments, and cities to minimise various waste streams and create greater value from industrial activities and agriculture practices, while regenerating the environment (Tahir et al., 2018).

The CE model aims to increase long-term resilience of the economy and creation of innovation. The current economic systems, including the water sector, put emphasis on the importance of a CE, through the opportunities they can create in business and economics, as well as through providing environmental and social benefits. The CE promotes changes in the resource use practiced in linear economic models and creates a positive reinforcing development cycle in all sectors (Ellen MacArthur Foundation, 2015). The ReSOLVE framework through a water context focuses on the importance of the CE model and its foundation in sustainability of water management, including sustainable use of water, use of renewable materials where possible, energy provision from renewable sources in a sustainable manner, preservation of natural systems and removal of waste and negative impacts in designs. All raw materials, water, resources, products and components should be managed in loops, keeping them at their highest possible intrinsic value (Tahir et al., 2018).

3.3 Examples of CE Implementation in Water and Wastewater Sector Within the ReSOLVE Framework

CE and resilience principles create an opportunity to tackle the water and wastewater challenges in cities and rural areas by providing a transformative and systemic approach to water supply, sanitation services and wastewater treatment in more a sustainable, efficient, inclusive and resilient manner. Despite these basic obligations arising from the activities of the water and wastewater sector, various business opportunities can be created in this sector. Examples of the CE model application to water systems using the ReSOLVE action areas are presented in Fig. 3. Some examples can be classified in more than one element of the ReSOLVE framework, for example, energy recovery from wastewater, sludges and other water-related wastes can be indicated in the *regenerate* and *loop* action areas.

The *Regenerate* action area focuses on renewable energy and materials and returning recovered biological resources to the biosphere. An example for business opportunity in this area is biosolids use for energy production, in the form of gas, electricity or heat (Roy et al., 2011). Durdević et al. (2019) studied recovery of renewable energy from sewage sludges generated in wastewater treatment plants in Croatia. Their research indicates that incineration of sewage sludge generates substantially lower greenhouse gas emissions (GHGs) than energy production from fossil fuels. Specifically, for the same amount of energy, the analysed sewage sludge emitted 58% less than natural gas, and 80% less than hard coal or fuel oil (Durdević et al., 2019). Li et al. (2017) studied anaerobic digestion as a promising method for energy recovery from sewage sludge. Their study compared mesophilic and thermophilic anaerobic digestion (CAD and TAD), anaerobic digestion with thermal hydrolysis pre-treatment (THPAD) and mesophilic and thermophilic high-solids anaerobic digestion (HSAD and THSAD). The results showed that in the case of high-organic-content sludge, thermal processes such as THSAD and TAD contributed the least to environmental impact, while THSAD and THPAD showed the best economic performance. In particular, the environmental and economic assessments showed that THSAD (compare with CAD) has 44% lower environmental impact with 118% higher net present value (NPV) for a assumed project (treatment capability of 100 t dry solids per day). However, for low-organic-content sludge, high-solids processes like THSAD and HSAD are much better than the others mainly owing to their less consumption of thermal energy. The usage of this kind of feed sludge, method THSAD could bring 40% less environmental burden with 31% more NPV than method CAD (Li et al., 2017).

There is also possibility to obtain renewable energy from water. Gopal et al. (2013) identified five major groups of renewable energy source water pumping systems: solar thermal water pumping systems, solar photovoltaic water pumping systems, biomass water pumping systems, wind energy water pumping systems and hybrid renewable energy water pumping systems (Gopal et al., 2013). In the case of photovoltaic panels, their costs are higher than the other

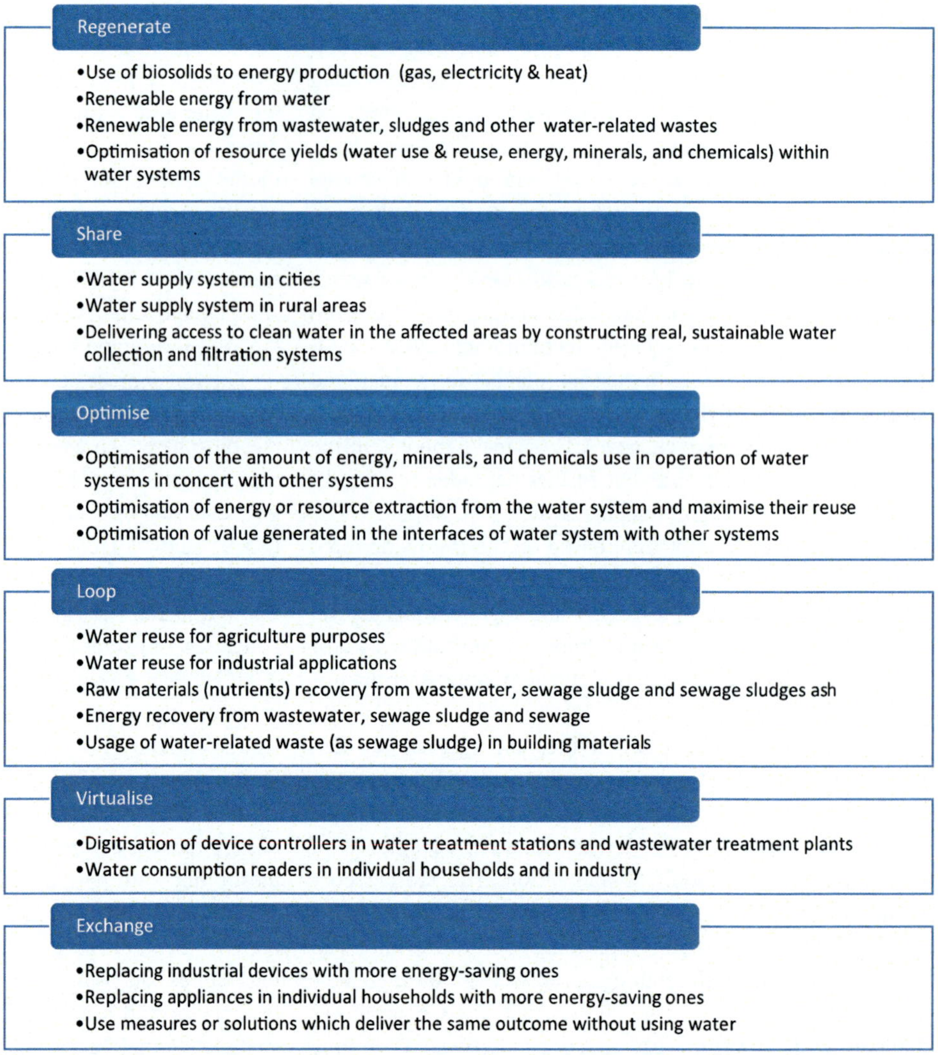

Fig. 3 Examples of the CE application in water systems in the ReSOLVE action areas. *Source* own based on (International Water Association, 2016; Tahir et al., 2018)

energy generating devices (eg. diesel engines and electrical motors), however, this solution does not require running costs during their lifetime, except for the maintenance cost (which are high for grid connected and diesel-powered water pumping systems). Therefore, after the payback period, the solar thermal water pumping systems can run without significant running cost and can be classified as the most feasible option in comparison to conventional grid connected and diesel powered systems for the rural areas facing a shortage of electricity (Meah et al., 2008).

The *Share* option in the ReSOLVE framework focuses on keeping the product loop speed low while maximising the use of products by sharing them among different users. In the water and wastewater sector, the services provided by city and village managers can be treated as an element of the sharing economy. This includes providing water supply systems in cities and rural areas. It is important to note that the contribution of the water industry to energy consumption and related GHG emissions is relatively low compared with other sectors of the economy (as transport of production). However, in long-term view, it can escalate energy use and contribute to further climate change which consequently affects the sustainability of available water resources (Vakilifard et al., 2018). Therefore, sustainable solutions for water supply systems are very much needed. Moreover, it also contains delivering access to clean water in the affected areas by constructing real, sustainable water collection and filtration systems.

The *Optimise* action area is dedicated to increasing the performance or efficiency of a product and removing any waste in production and the supply chain. In the water and wastewater sector, business opportunities can be found in this option from the reduced costs of energy, minerals and chemicals as a consequence of optimisation of their use in operation of water systems in concert with other systems. Mala-Jetmarova et al. (2017) conducted a detailed review of operational optimisation of water distribution systems. Works that include optimisation of the operation of water distribution systems mainly focus on optimal pump operation to minimise pumping costs as well as optimal water quality management to meet the limits indicated in standards for drinking water. The operational optimisation of water distribution systems include optimal pump operation, valve control and system operation for water quality purposes of both urban drinking and regional multi-quality water distribution systems (Mala-Jetmarova et al., 2017). The whole operation of the water system can be expressed with a generic mathematical optimisation model, in which the cases where water demand exceeds water availability could be taken into account by assigning priorities to the users (Kondili et al., 2010). *Optimise* can also be dedicated to optimisation of energy or resource extraction from the water system and maximise their reuse and optimisation of value generated in the interfaces of water system with other systems (International Water Association, 2016).

Loop focuses on keeping products, materials and resources in closed loops, and prioritising inner loops. In the case of the water and wastewater sector it can be dedicated to water reuse for agricultural purposes or industrial applications, raw materials (nutrients) and energy recovery from wastewater, sewage sludge and sewage sludge ash, and usage of water-related waste (as sewage sludge) in construction and other sectors (Smol et al., 2015). Water reuse is a recommended option in the CE model. Water reuse is related to three main sources: urban wastewater, rainwater and greywater (wastewater from household baths and showers, and kitchen sinks, hand basins—without food waste). One of the good examples for it is already implemented technology for the uses of reclaimed water on green areas in ski resort producing artificial snow (Kasina, Poland). This solution is not popular in Europe, while in Australia and the USA more and more installations for the production of artificial snow through water reuse are being developed (Ramm, 2021). One of the most important benefits of water and wastewater use in agriculture is the decrease in pressure on freshwater sources (Jaramillo & Restrepo, 2017). There is also the possibility to recover nutrients from waste generated in the water sector. In Poland, the theoretical phosphorus recovery potential from the sewage sludges as from plants in Lodz, Cracow, Gdansk, Gdynia, Szczecin, and Kielce was estimated at 1613.8 Mg, of which 33.9% is bioavailable (Smol et al., 2020a), and thus could be recovered by thermochemical treatment of ashes (Smol et al., 2020b). It is worth to notice that the economic feasibility assessments of raw materials recovery and water-reuse projects should quantitatively evaluate economic, environmental and resource availability (Molinos-Senante et al., 2011). Individual households can also play an important role in *closing loop* in the ReSOLVE framework. An example of this could be to store runoff from the roof in tanks to later use for watering greenery within the building. If the ground and water conditions allow it, the stored water can be use rained in gardens, ponds, green roofs, green walls, infiltration ditches, boxes or drainage chambers and many other available systems that by local management of rainwater will relieve sewage systems that would have to additionally accept runoff from the roofs of these buildings and the surfaces sealed within it (Burszta-Adamiak & Spychalski, 2021).

Virtualise means delivering and using the utility virtually. In the water and wastewater sector, it can be associated with digitisation of device controllers in water treatment stations and wastewater treatment plants. Attaran (2021) indicated that fifth-generation (5G) networks can be used in several sectors. For example, drones that could fly over a field of crops, using sensors on the ground, to sort, pick, feed and water individual plants. Moreover, 5G networks enable factory robots to communicate their position task, allowing them to do more tasks wirelessly and efficiently (Attaran, 2021). In general, digitalisation to support circularity is more and more often used in the urban metabolism. It provides urban managers, planners and administrators, as well as policy makers the useful tool for identifying, controlling and assessing various data concerning the flows of social, environmental and economic resources, including those that are connected with water supply systems and wastewater treatment services. Fixed and mobile digital infrastructures play a crucial role here, including for example real-time monitoring stations, augmented reality, virtual sharing platforms, smart grids, social media dashboards, and the like in the development and strengthening of the efficiency and quality of the circularity of resources, including water (D'Amico et al., 2022). In the *virtualise* option, the households and industry plants can be equipped with water consumption readers, which can contribute to a more effective monitoring of the level of water consumption, as well as later—to introduce improvements in this area, so as to reduce the pressure on natural water resources and prevent non-controlled water leakage (or excessive use) from the system.

Exchange focuses on replacing old materials and products with advanced non-renewable materials and applying the use of new technologies. An example of this option could be replacing industrial devices with more energy-saving ones. Individual households, environmental attitudes and behaviour of the inhabitants can play a key role (Smol et al., 2018). There are several examples of good ways to exchange the current non-CE practises for circular ones, such as installing composting toilets; installing water-saving showerheads, shower timers and low-flow faucet aerators; and exchanging for a high-efficiency washing machine (Smol, 2021a). These energy saving and water-saving actions should take place both in the home and at the workplace (Wells et al., 2016). There is also a possibility to replace appliances in individual households with more energy-saving ones and use measures or solutions which deliver the same outcome without using water.

4 Conclusions

In this paper, the examples of CE implementation in water and wastewater sector within the ReSOLVE framework were presented. The analysed examples of good practices were divided into six action areas for businesses and countries transforming from linear to the CE model: Regenerate, Share, Optimise, Loop, Virtualise and Exchange. There are several possibilities to implement CE in analysed sector, including recovery of renewable energy and nutrients from waste generated in water and wastewater treatment plants, providing water supply systems in cities and rural areas, optimisation of the amount of energy, minerals and chemicals use in operation of water systems, water reuse for agricultural and industrial purpose, digitisation of device controllers in water treatment stations and wastewater treatment plants, or replacing industrial devices with more energy-saving ones. It is worth to notice that the implementation of the proposed solutions should contribute to create the balance between economic, social and environmental aspects, that is a base for sustainability. Therefore, economic feasibility assessments of proposed projects should takes into account also social, environmental and resource availability. The current economic systems, including water sector, put emphasis on the CE matters, due to they can create a series of business and economic opportunities, next to environmental and social benefits.

Acknowledgements This research was funded by the project "CircuWater" co-financed by the Kosciuszko Foundation—The American Centre of Polish Culture under the Exchange Program to the US and subsidy of the Division of Biogenic Raw Materials at the Mineral and Energy Economy Research Institute of the Polish Academy of Sciences.

The funders had no role in the design of the study; in the collection, analyses, or interpretation of data; in the writing of the manuscript, or in the decision to publish the results.

Conflicts of Interest
Author declares no conflict of interest.

Financial or Non-financial Interests
Author declares no relevant financial or non-financial interests to disclose.

Data Availability Statement
The datasets generated during and/or analysed during the current study are available from the corresponding author on reasonable request.

Ethics Approval Not applicable.

References

Attaran, M. (2021). The impact of 5G on the evolution of intelligent automation and industry digitization. *Journal of Ambient Intelligence and Humanized Computing*. Berlin: Springer. https://doi.org/10.1007/s12652-020-02521-x.

Burszta-Adamiak, E., & Spychalski, P. (2021). Water savings and reduction of costs through the use of a dual water supply system in a sports facility. *Sustainable Cities and Society, 66*, 102620. https://doi.org/10.1016/j.scs.2020.102620

Cattelan Nobre, G., & Tavares, E. (2020). Assessing the role of big data and the internet of things on the transition to circular economy: Part I - An extension of the ReSOLVE framework proposal through a literature review. In *Technology review* (pp. 19–31).

Chowdhury, S. (2021). South Korea's Green New Deal in the year of transition, UNDP Seoul Policy Centre, https://www.undp.org/blog/south-koreas-green-new-d.

European Commission, (2019). Communication from the Commission: The European Green Deal, (COM no. 640, 2019).

D'Amico, G., et al. (2022). Digitalisation driven urban metabolism circularity: A review and analysis of circular city initiatives. *Land Use Policy. Elsevier Ltd, 112*, 105819. https://doi.org/10.1016/j.landusepol.2021.105819

Durdević, D., Blecich, P., & Juri´, Ž. (2019). Energy recovery from sewage sludge: The case case study of Croatia. Energies, *12*(10) (1927). https://doi.org/10.3390/en12101927.

Ellen MacArthur Foundation, (2015). Delivering the circular economy a toolkit for policymakers executive summary.

European Commission (2020) Communication from the Commission. Circular Economy Action Plan for a cleaner and more competitive Europe, (COM no. 98, 2020). Available at: https://ec.europa.eu/environment/circular-economy/pdf/new_circular_economy_action_plan.pdf.

Gopal, C., et al. (2013). 'Renewable energy source water pumping systems - A literature review', renewable and sustainable energy reviews. *Elsevier, 25*, 351–370. https://doi.org/10.1016/j.rser.2013.04.012

HM Government, (2018). Our waste, our resources: A strategy for England. www.gov.uk/government/publications/resources-and-w.

Heyes, G., et al. (2018). 'Developing and implementing circular economy business models in service-oriented technology companies. Journal of Cleaner Production. *Elsevier Ltd, 177*, 621–632. https://doi.org/10.1016/j.jclepro.2017.12.168

Hofmann, F. (2019). Circular business models: Business approach as driver or obstructer of sustainability transitions? *Journal of Cleaner Production, 224*, 361–374.

International Water Association, (2016). Water utility pathways in a circular economy. https://iwa-network.org/water-utility-pathways-cir.

Iyer-Raniga, U. (2019). Using the ReSOLVE framework for circularity in the building and construction industry in emerging markets. *IOP Conference Series: Earth and Environmental Science, 294*, 012002. https://doi.org/10.1088/1755-1315/294/1/012002

Jaramillo, M. F., & Restrepo, I. (2017). Wastewater reuse in agriculture: A review about its limitations and benefits.

Sustainability (switzerland), 9(10), 1734. https://doi.org/10.3390/su9101734

Kakwani, N. S., & Kalbar, P. P. (2020). Review of Circular Economy in urban water sector: Challenges and opportunities in India. *Journal of Environmental Management. Elsevier Ltd, 271*, 111010. https://doi.org/10.1016/j.jenvman.2020.111010

Kondili, E., Kaldellis, J. K., & Papapostolou, C. (2010). A novel systemic approach to water resources optimisation in areas with limited water resources. Desalination. Elsevier B.V., *250*(1), 297–301. https://doi.org/10.1016/j.desal.2009.09.046.

Kumar, A., Mangla, S. K., & Kumar, P. (2022). An integrated literature review on sustainablefood supply chains: Exploring research themes and future directions. *Science of The Total Environment*, 153411.

Lewandowski, M. (2016). Designing the business models for circular economy-towards the conceptual framework. *Sustainability (switzerland), 8*(1), 43. https://doi.org/10.3390/su8010043

Li, H., et al. (2017). Environmental and economic life cycle assessment of energy recovery from sewage sludge through different anaerobic digestion pathways. *Energy, 126*, 649–657.

MacArthur, J. L., et al. (2020). Canada's Green New Deal: Forging the socio-political foundations of climate resilient infrastructure? *Energy Research and Social Science. Elsevier, 65*(February), 101442. https://doi.org/10.1016/j.erss.2020.101442

Mala-Jetmarova, H., Sultanova, N., & Savic, D. (2017). Lost in optimisation of water distribution systems? A literature review of system operation. Environmental Modelling and Software. *Elsevier Ltd, 93*, 209–254. https://doi.org/10.1016/j.envsoft.2017.02.009

Meah, K., Ula, S., & Barrett, S. (2008). Solar photovoltaic water pumping-opportunities and challenges. *Renewable and Sustainable Energy Reviews, 12*(4), 1162–1175. https://doi.org/10.1016/j.rser.2006.10.020

Mentink, B. (2014). Circular business model innovation: A process framework and a tool for business model innovation in a circular economy, Delft University of Technology.

Molinos-Senante, M., Hernández-Sancho, F., & Sala-Garrido, R. (2011). Cost-benefit analysis of water-reuse projects for environmental purposes: A case study for Spanish wastewater treatment plants. *Journal of Environmental Management. Elsevier Ltd, 92* (12), 3091–3097. https://doi.org/10.1016/j.jenvman.2011.07.023

Nika, C. E., et al. (2020). Water cycle and circular economy: developing a circularity assessment framework for complex water systems. *Water Research. Elsevier Ltd, 187*, 116423. https://doi.org/10.1016/j.watres.2020.116423

Nikolaou, I. E., Jones, N., & Stefanakis, A. (2021). Circular economy and sustainability: The past, the present and the future directions, circular economy and sustainability. *Circular Economy and Sustainability, 1*(2), 783–783. https://doi.org/10.1007/s43615-021-00054-9

Nußholz, J. L. K. (2017). Circular business models: Defining a concept and framing an emerging research field. *Sustainability (switzerland), 9*(10), 1810. https://doi.org/10.3390/su9101810

Ramm, K. (2021). Considerations related to the application of the EU water reuse regulation to the production of snow from reclaimed water, circular economy and sustainability. In *Circular economy and sustainability*. https://doi.org/10.1007/s43615-021-00075-4.

Roy, M. M., et al. (2011). Review of biosolids management options and co-incineration of a biosolid-derived fuel. *Waste Management, 31* (11), 2228–2235.

Smol, M. (2021a). Circular economy approach in the water and wastewater sector. *Circular Economy and Sustainability - Environmental Engineering, 2*, 1–19.

Smol, M. (2021b). *Inventory and Comparison of Performance Indicators in Circular Economy Roadmaps of the European Countries*. Springer International Publishing. https://doi.org/10.1007/s43615-021-00127-9

Smol, M., & Kulczycka, J. (2019). Towards innovations development in the European raw material sector by evolution of the knowledge triangle. *Resources Policy, 62*, 453–462. https://doi.org/10.1016/j.resourpol.2019.04.006

Smol, M., et al. (2018). Public awareness of circular economy in southern Poland: Case of the Malopolska region. *Journal of Cleaner Production, 197*, 1035–1045. https://doi.org/10.1016/j.jclepro.2018.06.100

Smol, M., Adam, C., & Kugler, S. A. (2020a). Inventory of Polish municipal sewage sludge ash (SSA) - mass flows, chemical composition, and phosphorus recovery potential. *Waste Management, 116*, 31–39.

Smol, M., Adam, C., & Preisner, M. (2020c). Circular economy model framework in the European water and wastewater sector. Journal of Material Cycles and Waste Management. *Springer Japan, 22*(3), 682–697. https://doi.org/10.1007/s10163-019-00960-z

Smol, M. et al. (2015). The possible use of sewage sludge ash (SSA) in the construction industry as a way towards a circular economy. *Journal of Cleaner Production, 95*. https://doi.org/10.1016/j.jclepro.2015.02.051

Smol, M., Adam, C., & Kugler, S. A. (2020b) Thermochemical treatment of Sewage Sludge Ash (SSA)-potential and perspective in Poland. *Energies, 13*(20). https://doi.org/10.3390/en13205461.

Smol, M. et al. (2020). Transformation towards circular economy (CE) in municipal waste management system: Model solutions for Poland. *Sustainability (Switzerland), 12*(11). https://doi.org/10.3390/su12114561.

Tahir, S. et al. (2018). White paper. Water and circular economy. Ellen MacArthur Foundation Report.

US Government (2018). Resolution. Recognizing the duty of the Federal Government to create a Green New Deal, p. H. RES. 109. https://doi.org/10.5553/tcc/221195072021011001005.

Vakilifard, N., et al. (2018). The role of water-energy nexus in optimising water supply systems – Review of techniques and approaches. Renewable and Sustainable Energy Reviews. *Elsevier Ltd, 82*, 1424–1432. https://doi.org/10.1016/j.rser.2017.05.125

Van Renswoude, D. R., Visser, I., Raijmakers, M. E., Tsang, T., & Johnson, S. P. (2019). Real-world scene perception in infants: What factors guide attentionallocation?. *Infancy, 24*(5), 693–717.

Wells, V. K., et al. (2016). The role of generativity and attitudes on employees home and workplace water and energy saving behaviours. Tourism Management. *Elsevier Ltd, 56*, 63–74. https://doi.org/10.1016/j.tourman.2016.03.027

Židonienė, S., & Kruopienė, J. (2015). Life cycle assessment in environmental impact assessments of industrial projects: Towards the improvement. *Journal of Cleaner Production, 106*, 533–540.

Circular Economy Implementation in the Water Sector: Wastewater Treatment Plant Tychy–Urbanowice Case Study

Monika Zubrowska-Sudol and Zbigniew Gieleciak

Abstract

The paper presents the roadmap of the wastewater treatment plant (WWTP) Tychy–Urbanowice from an object producing electricity sufficient to cover approximately 20% of the plant's own needs, to one where production of the medium exceeds its needs. In order to intensify the production of biogas, constituting fuel for electricity and heat production, it used codigestion of sewage sludge with biodegradable waste sourced from external suppliers. The use of renewable energy was not limited to own needs, but combined with the needs of the regional community in the scope of implementation of the concept of the 'Symbio-Tychy' smart city. The concept is based on the cooperation between water and wastewater, waste and energy sector, and on the philosophy of the knowledge triangle (Ecology, Economy and Education). In 2021, the amount of electricity produced by cogenerating units was 13 955 MWh, whereas the demand of the WWTP and the water park of Tychy was at a level 10 530 MWh. The energy self-sufficiency indicator for these objects, therefore, exceeded 130%. An important element that contributed to the development of WWTP in Tychy was the adopted management model that next to technological growth also prioritises meeting environmental and social norms.

Keywords

Biodegradable waste • Renewable energy • Sewage sludge • Industrial symbiosis • Management model

M. Zubrowska-Sudol (✉)
Warsaw University of Technology, Warsaw, Poland
e-mail: monika.sudol@pw.edu.pl

Z. Gieleciak
Regional Centre of Water and Sewage Management JSC Tychy, Tychym, Poland
e-mail: z.gieleciak@rcgw.pl

1 Introduction

The implementation of circular economy principles in wastewater treatment plants (WWTP) entails a change in the current wastewater treatment paradigm. It ceases to only involve removal of pollutants. The technological processes applied in this type of object should permit the recovery of energy, heat and valuable resources. Due to the popularisation of the idea of resource efficient and low-emission circular economy, but most importantly, due to the acknowledgement of the necessity to apply the model in all sectors of the economy in the years to come, one of the most popular areas of research and development (R&D) regarding wastewater treatment plants is research on the recovery of resources and energy. A lot of attention is paid to the issue of water recovery from wastewater and its reuse. More and more objects implement technologies permitting phosphorus recovery. The possibility of recovery of many other compounds has also been documented, offering substitution for resources obtained from primary sources, among others:

- dissolved organic compounds used in wastewater treatment plants as an alternative source of organic carbon for supporting nutrient removal from wastewater (Zubrowska-Sudol, 2018);
- nitrogen and phosphorus used as a component of fertilisers (Larriba et al., 2020);
- poly-β-hydroxyalkanoates constituting raw material in the production of bioplastics (Larriba et al., 2020);
- alginates applicable in the textile, construction and paper industry (Nieuwenhuijzen et al., 2017);
- cellulose used as insulation material, a binding component in asphalt and in the paper industry (Nieuwenhuijzen et al., 2017);
- proteins applicable as a component of fodders (Hwang et al., 2008);

M. Smol et al. (eds.), *Water in Circular Economy*, Advances in Science, Technology & Innovation,
https://doi.org/10.1007/978-3-031-18165-8_18

- enzymes constituting an addition to detergents and biocatalysts in extraction reactions in many industry sectors (Raheem et al., 2018);
- rare earth elements necessary for high-tech development (used among others to produce electric engines, catalytic converters, permanent magnets, optical filters, luminophores, glass, ceramics, polishing powders, dyes, fertilisers or insulation fibres) (Raheem et al., 2018).

The analysis of the current state of implementation of technologies meeting the challenges of CE reveals that wastewater treatment plants are dominated by solutions aimed at energy recovery. One of the examples of such a situation is the already long-established anaerobic digestion process as one of the methods of stabilisation of sewage sludge. In that process, in anaerobic conditions, organic compounds are transformed into biogas, composed among others of methane. Owing to the current trends, we perceive the process not only as a method of stabilising sewage sludge but also as a solution allowing for electricity and heat recovery (with the application of cogeneration), and additionally cold with the application of trigeneration. Moreover, the potential of anaerobic digestion involves the production of not only methane but also compounds such as biomethanol, biohydrogen or biomethane (Farsi, 2021; Iqbal et al., 2022). The latter can be supplied to gas networks or serve as fuel in transport. The driving force of the development of energy recovery in wastewater treatment plants is the possibility of obtaining independence from an external supply and creating an opportunity for building up the position of bioheat and power plants. An important benefit is providing for industrial biocoenoses with producers of organic waste that can be codigested with sewage sludge to obtain production of electricity exceeding own needs of the wastewater treatment plant. An example of a good practice in this field is the WWTP Tychy–Urbanowice (Poland) with 261 265 population equivalent, managed by the Regional Centre of Water and Sewage Management JSC (RCWSM JSC). In order to intensify biogas production, it used codigestion of sewage sludge with biodegradable waste sourced from external suppliers (among others dairy waste and waste from food industry). The use of renewable energy was not limited to own needs, but combined with the needs of the regional community in the scope of implementation of the concept of the 'Symbio-Tychy' smart city. The concept is based on the cooperation between water and wastewater, waste and energy sector, and on the philosophy of the knowledge triangle (Ecology, Economy and Education). The smart city concept determined all decisions by RCSM JSC regarding the development and functioning of the wastewater treatment plant. The company identifies factors affecting the plant's technological progress, as well as obstacles and problems restricting it. Based on that, concepts are delivered on an ongoing basis regarding development mechanisms and tools applicable in the stimulation and control of the processes occurring in the wastewater treatment plant. Pursuant to the principles adopted by the company, technological progress cannot only generate technological growth but also needs to meet environmental and social norms. Therefore, the regulation model implemented by the company develops individual career paths of employees and promotes learning, innovativeness and creativity, among others owing to the adopted employee evaluation system (based on internal documents of the company).

The paper adopted an approach based on a case study to demonstrate the potential efficiency of the implementation of circular economy solutions in wastewater treatment plants and intensification of renewable energy production. The presented example of good practices in this aspect is the roadmap of WWTP Tychy–Urbanowice from an object producing electricity sufficient to cover approximately 20% of the plant's own needs, to one where production of the medium exceeds its needs more than twice. Next to technological issues, the paper also discusses selected management and social aspects.

2 Implementation of Circular Economy in WWTP Tychy

The implementation of the codigestion process in the Tychy–Urbanowice wastewater treatment plant, and therefore the development of management of produced heat and electricity was described based on operation data from the object. The data come from the period 2006–2021 (archive of WWTP Tychy–Urbanowice).

2.1 Heat and Power Recovery

2.1.1 Biogas Production—from Mono- to Codigestion

Anaerobic digestion is one of the processes applied in sludge processing in large wastewater treatment plants. In the years 2006–2008, WWTP Tychy–Urbanowice conducted monodigestion of primary sludge, constituting waste generated in primary tanks as a result of sedimentation of suspended solids, and excess sludge generated in the processes of biological wastewater treatment. The process was conducted in two anaerobic digesters with a volume of 2200 m^3 each. The amount of the produced biogas was in a range of $1.18*10^6$–$1.59*10^6$ m^3/year, and the mean percent content of methane in biogas was approximately 61% (Fig. 1). After the modernisation of the object completed in 2009,

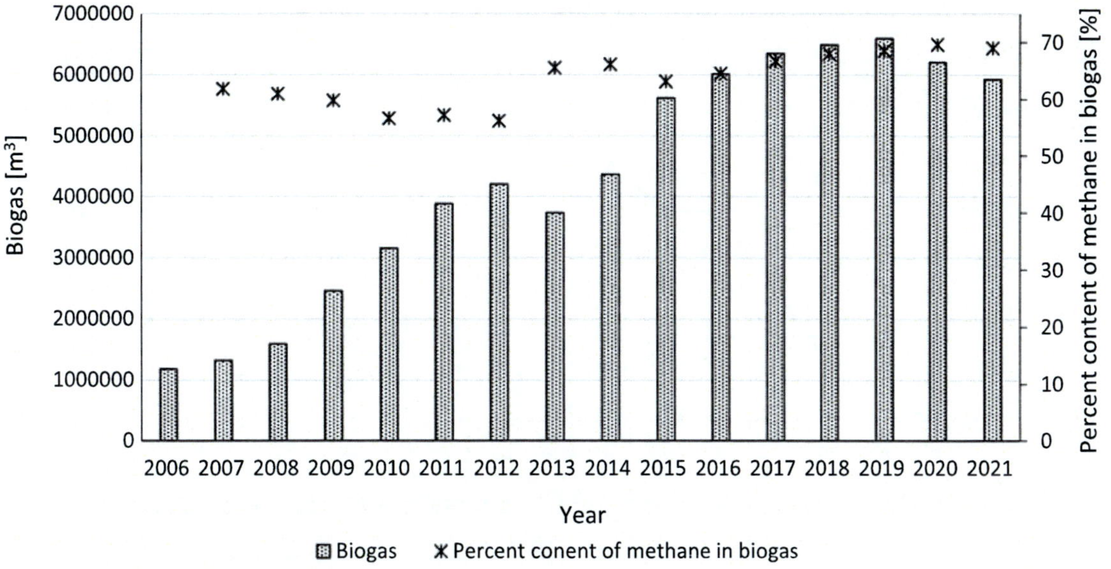

Fig. 1 Biogas production in wastewater treatment plant Tychy–Urbanowice (years 2006–2008: monodigestion of sewage sludge; years 2009–2021: codigestion of sewage sludge and biodegradable industrial waste)

co-substrates started being dosed to the biogas plant, composed of two anaerobic digesters with a volume of 5500 m^3 each. The development of codigestion began with whey and other dairy wastes, with year-to-year expansion of the portfolio of wastes accepted and stabilised at the wastewater treatment plant (the wastewater treatment plant currently accepts 22 types of waste classified in the European Waste Codes). According to data presented in Fig. 1, these activities led to obtaining a substantial increase in the amount of produced biogas. Already in 2009, the amount was 2.07 times higher than the value recorded in 2006 and successively increased, reaching values from 4.76 to 5.59 times higher beginning in 2015. In 2021, WWTP Tychy–Urbanowice produced $5.93*10^6$ m^3 of biogas. The mean percent methane content was 68.98%. The successive increase in biogas production is related to the implementation of the policy of approaching waste as raw material. The analysis of available biodegradable wastes and continuous research permitted the improvement of the codigestion process. Approximately 100 thousand m^3 of sludge and 95 thousand MG of biodegradable waste were dosed to the digestion process annually.

The implementation of codigestion entailed the necessity to expand the infrastructure. Among others, storage tanks for biodegradable waste accepted from external suppliers were constructed, with a total volume of 5410 m^3, as well as a veterinary waste pasteurisation station with an efficiency of 144 m^3/d, and an additional tank for storing biogas with a volume of 6370 m^3.

Next to the expansion of the technical infrastructure, the key element of the transition from mono- to codigestion was and is industrial development research. The wastewater treatment plant conducts research both at the laboratory and pilot scale. Substrates are selected at the laboratory from among particular organic wastes after the analysis of their methane potential and susceptibility to biodegradation. WWTP Tychy–Urbanowice also includes a pilot research station composed of two systems of two-stage anaerobic digestion and two single-stage systems. The station is fully automated, and the working volume of the research chambers is 110 and 220 l, respectively. Research conducted at the station aims at optimising anaerobic digestion parameters (among others temperature and retention time), intensification of biogas production and modelling of processes conducted at the technological scale.

2.1.2 Combined Heat and Power Production and Their Utilisation

Based on the data shown in Fig. 2, in the period when monodigestion was conducted, electricity production at a level of 2662 MWh/year was obtained. Two CHP units with a power of 345 kW each were installed in the wastewater treatment plant. The produced energy was utilised for own needs of the wastewater treatment plant, covering 45% of the object's demand for the medium. The implementation of codigestion did not only allow for obtaining energetic self-sufficiency but also for electricity production exceeding the medium consumption in the object. In 2015, the energy production reached 8397 MWh/year which at the plant's demand at a level of 5645 MWh/year corresponded to an energy self-sufficiency index of 148.7%. In order to utilise this substantial energy surplus, the construction of a water

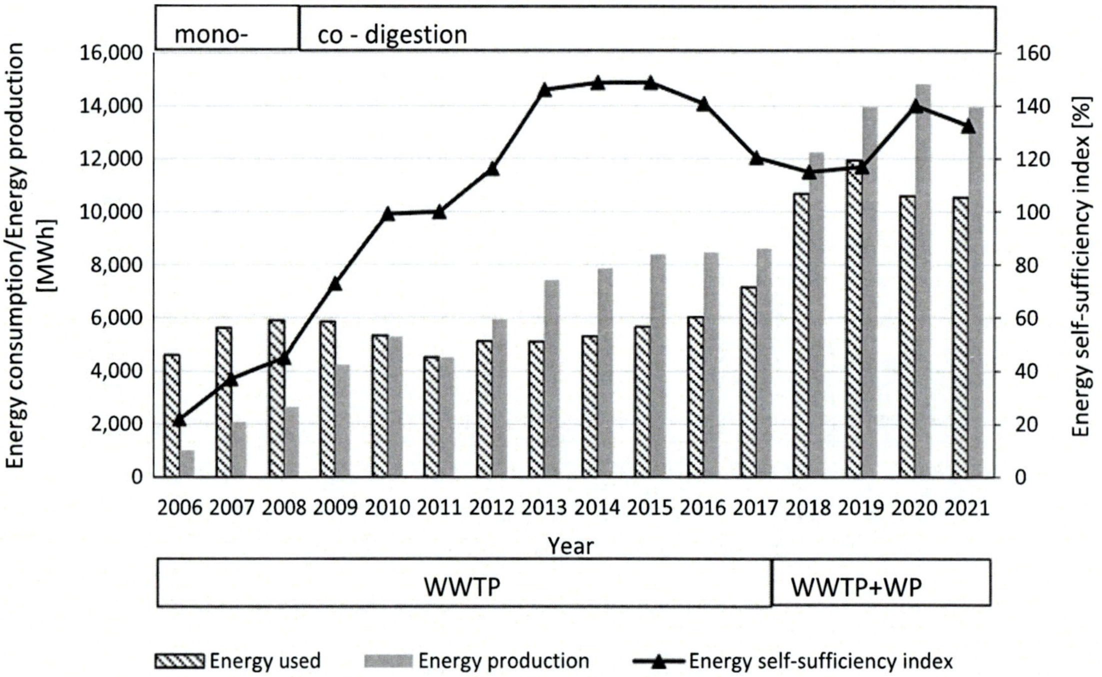

Fig. 2 Electricity consumption and production (from 2018 the values illustrate total electricity production and consumption for both objects, i.e. wastewater treatment plant and water park, supplied with biogas from anaerobic digestion)

park (WP) commenced, supplied with energy from biogas produced in the codigestion process at the WWTP. The investment also involved the construction of a gas pipeline with a length of 6.2 km connecting both objects, and an additional codigestion installation at the WP with a total active power of 1200 kW. In the years 2018–2021, when both objects were in operation, the amount of produced electricity reached 12,263–14,827 MWh/year. This allowed for complete covering of the energy needs of both the wastewater treatment plant and the water park, and generated an energy surplus over the needs reaching from 15 to 40%. By referring the total amount of produced electricity to only own needs of the wastewater treatment plant, an energy self-sufficiency index of 172–210% is obtained. These values suggest that the amount of produced electricity was from 1.7 to more than 2 times higher than the consumption of the medium at the wastewater treatment plant. It is worth emphasising that the Tychy–Urbanowice wastewater treatment plant achieves the highest value of the energy self-sufficiency index among Polish wastewater treatment plants. Other objects that considerably increased their energy self-sufficiency index after the implementation of codigestion of sewage sludge with organic waste include the wastewater treatment plant in Chorzów (Kalemba, 2018) and Swarzewo (Cytawa, 2018). In the former case, the value of the analysed index increased from approximately 40% in 2014 to more than 70% in 2016. In the latter case, the plant reached energy self-sufficiency. Examples from outside the territory of Poland can be found in papers by Mattioli et al. (2017) and Macintosh et al. (2019). According to data presented by Mattioli et al. (2017), in the Rovereto wastewater treatment plant (Italy), the implementation of the municipal waste fraction as a cosubstrate will allow for obtaining 85% coverage of the energy needs of the object, whereas in the case of monodigestion, the amount of produced electricity covered 50% of demand for electricity. According to Macintosh et al. (2019), the effect of codigestion of sewage sludge with food waste in the Grüneck wastewater treatment plant (Germany) was an increase in coverage of the energy needs of the object by 16%. The cited data distinguish Tychy–Urbanowice wastewater treatment plant among both Polish and European objects.

The CHP installation also provided for heat production that considerably exceeded the demand of both WWTP and WP. For example, in the years 2015–2017, when the biogas plant supplied only WWTP, heat production averaged 39,190 GJ/year at the average demand of the wastewater treatment plant at a level of 21,973 GJ/year. Data illustrating heat production and consumption at WWTP and WP for the last two years are presented in Table 1. It should be emphasised that at the water park, heat is produced not only from biogas but also from heat recovery from rinse waters or cold production.

It is worth emphasising that the produced electricity and heat are not only utilised for own purposes of the wastewater treatment plant and water park (the latter object in operation

Table 1 Heat consumption and production at the wastewater treatment plant and at the water park

Year	Wastewater treatment plant [GJ]		Water park [GJ]		
	Heat consumption	Heat production from biogas	Heat production	Heat production from biogas	Heat recovery from other sources
2020	28,487	35,774	22,697	29,075	2200
2021	28,217	34,219	23,778	27,681	2158

Table 2 Electricity sold to the electricity network and heat sold to the heating network

Year	Electricity [MWh]	Heat [GJ]
2015	2932	–
2016	2715	–
2017	1846	–
2018	2636	1397
2019	2917	905
2020	4975	5098
2021	4568	4157

since 2018). The generated production surpluses are sold to the electricity and heating network (Table 2).

2.1.3 Emission Reduction

As a renewable energy source, biogas is one of the basic components of low-emission economy. The use of this renewable fuel contributes to a reduction of CO_2 emission to the environment in comparison to the use of fossil fuels.

From the beginning of 2007 to the end of 2021, the installation at the WWTP and WP produced 121 470 MWh of electricity. The production of an equivalent amount of energy would require the consumption of approximately 50 thousand tonnes of hard coal by a conventional power plant. Considering heat also produced in cogeneration, the amount of the resource would increase to 77 thousand tonnes of coal.

Because pursuant to the emission allowance trading scheme of the Community and IPCC, CO_2 emission from biomass combustion is not included in total emission from fuel combustion, the index equals zero. Due to this, since 2006, when RCGW S.A. commenced energy production from biogas, the avoided CO_2 emission reached 92 thousand Mg.

2.2 Water Reuse

The activity of the Regional Centre of Water and Sewage Management in Tychy described above, corresponding with the assumptions of circular economy, is primarily related to the production of green energy from sludge and biodegradable waste. An equally important area of activity in accordance with the regulations of CE is the recovery and management of raw materials produced in processes conducted at the Wastewater Treatment Plant Tychy–Urbanowice. An example is the use of treated sand obtained at the mechanical stage of wastewater treatment for own purposes, or the use of treated wastewater. In the analysed object, treated wastewater is used as technological water for cleaning screens, cooling the heat distribution centre, rinsing sand, dilution of polyelectrolyte used in the processes of sludge thickening and dehydration, spraying biofilters and washing tanks. Technological water also found application in the installation for the removal of carbon dioxide from biogas. The biogas treatment station is composed of a system of devices and pipelines, including compressor units where biogas pressure is increased, absorbers where biogas comes in contact with sprayed technological water, demisters, scrubbers, heat exchangers and reductors ensuring appropriate pressure on the installation outlet. The purpose of the installation is an increase in methane content in biogas to approximately 75%.

The demand for water for technological purposes is not substantial in comparison to the total amount of treater wastewater, but it permits the elimination of consumption of tap water for these purposes. For example, in 2021, the amount of treated wastewater used for technological purposes reached 241 846 m^3 (2.16% of the total volume of treated wastewater).

2.3 Management Model

An important element that contributed to the implementation of circular economy in WWTP Tychy was the adopted management model that next to technological growth also prioritises meeting environmental and social norms.

Regional Centre of Water and Sewage Management in Tychy was the first company in the Polish municipal sector with a management system implemented and certified pursuant to norm PN-EN ISO 50001:2012 'Energy management system'. The instrument contributes to a more efficient functioning of the company in terms of organisation and brings quantifiable economic benefits. The aforementioned norm was integrated with management systems already functioning in the company, including among others: environmental management norm ISO 14001, Ecomanagement and Audit System in the Community EMAS, and Business Social Responsibility Strategy pursuant to Norm SA 8000. Owing to continuous measurements and analyses, based on specified criteria and energy parameters, as well as cyclical inspections and energy audits, the implemented management instruments permit continuous monitoring and assessment of the implementation of planned energy goals and tasks. One of the expected effects of the adopted management model is the improvement of the energy safety of the city of Tychy.

Pursuant to the rules adopted by the company, the technological progress cannot only generate technological growth but must also meet environmental and social norms. Therefore, RCGW S.A. each time measures results by means of a three-tier account of profits: economic, social and environmental. Due to the adopted knowledge triangle approach (Ecology, Economy and Education), the provided services are sustainable and efficient, transparent, safe and socially adaptable. Particular attention is paid to education of society. An example of such activities is the Water Academy at the Water Park. The 32 stations prepared at the Academy teach about water as a source of life, energy and transport.

The aforementioned knowledge triangle approach is the key element of the SymbioTychy concept developed by the company. It aims at building an Ecocity focusing on the protection of the environment and its resources with the commercial application of knowledge on knowledge.

3 Conclusions

Successful and well-planned reuse of resources can help achieve sustainability. WWTP in Tychy-Urbanowice is an example of good practice in this field. The major achievements of the object include co-fermentation process implementation and biogas production intensification. The management model based on the three pillars of sustainable development was developed to exploit the potential of WWTP in Tychy for the recovery of energy and resources. Moreover, the R&D section is still being developed to find more sustainable and effective solutions for both treatment and resource recovery. The case of WWTP in Tychy–Urbanowice clearly shows that the wastewater treatment plant has great potential for the implementation of circular economy solutions.

References

Archive of WWTP Tychy-Urbanowice, Operation data from the period 2006–2021.

Cytawam, S. (2018). Możliwość poprawy efektywności energetycznej oczyszczalni ścieków poprzez wspólną fermentację osadu i różnych odpadów organicznych, (Possibility of improvement of the energy efficiency of a wastewater treatment plant through codigestion of sludge and different organic wastes), Materiały Konferencyjne, 2. Konferencja Efektywność energetyczna w przedsiębiorstwach wod-kan, Chorzów/Tychy, 20–22 listopada 2018, 63–71 (Conference materials, 2nd Conference Energy efficiency in water supply and sewage companies).

Farsi, M. (2021). Biomass conversion to biomethanol. In *Advances in bioenergy and microfluidic applications* (pp. 231–252).

Hwang, J., Zhang, L., Seo, S., Lee, Y.-W., & Jahng, D. (2008). Protein recovery from excess sludge for its use as animal feed. *Bioresource Technology, 99*, 8949–8954.

Iqbal, K., Saxena, A., Pande, P., Tiwari, A., Joshi, N.C., Varma, A., Mishra, A. (2022). Microalgae-bacterial granular consortium: Striding towards sustainable production of biohydrogen coupled with wastewater treatment. *Bioresource Technology, 354*, 127203.

Kalemba, K. (2019). Współfermentacja jako proces zwiększający niezależność energetyczną oczyszczalni ścieków (Codigestion as a process improving energy self-sufficiency of wastewater treatment plants), Materiały Konferencyjne, 13. Konferencja Metody Zagospodarowania osadów ściekowych, Władysławowo, 16–18 września 2019, 127–132 (Conference materials, 13th Conference Methods of management of sewage sludge).

Kledyński, Z., Bogdan, A., Jackiewicz-Rek, W., Lelicińska-Serafin, K., Machowska, A., Manczarski, P., Rolewicz-Kalińska, A., Rucińska, J., Szczygielski, T., Walczak, J., Wojtkowska, M., & Żubrowska--Sudoł, M. (2020). Condition of circular economy in Poland. *Archives of Civil Engineering, 66*(3), 37–80.

Larriba, O., Rovira, E., Juznic-Zonta, Z., Guisasola, A., & Baeza, J. A. (2020). Evaluation of the integration of P recovery, polyhydroxyalkanoate production and short cut nitrogen removal in a mainstream wastewater treatment process. *Water Research, 172*, 115474.

Macintosh, C., Astals, S., Sembera, C., Ertl, A., Drewes, J. E., Jensen, P. D., & Koch, K. (2019). Successful strategies for increasing energy self-sufficiency at Grüneck wastewater treatment plant in Germany by food waste co-digestion and improved aeration. *Applied Energy, 242*, 797–808.

Mattioli, A., Gatti, G. B., Mattuzzi, G. P., Cecchi, F., & Bolzonella, D. (2017). Co-digestion of the organic faraction of municipal solid waste and sludge improves the energy balance of wastewater treatment plants: Rovereto case study. *Renewable Energy, 113*, 980–988.

Nieuwenhuijzen, A., Lewis, R., Bergsma, G., Uiterlinde, C. (2017). Life cycle assessment of resource recovery from primary and waste activated sludge. In *Proceedings IWA specialist conference on sludge management sludge tech*.

Raheem, A., Sikarwar, V. S., He, J., Dastyar, W., Dionysiou, D. D., Wang, W., & Zhao, R. M. (2018). Opportunities and challenges in sustainable treatment and resource reuse of sewage sludge: a review. *Chemical Engineering Journal, 337*, 616–641.

Zubrowska-Sudol, M. (2018). Carbon source recovery from excess sludge by mechanical disintegration for biological denitrification. *Water Science and Technology, 77*(7), 1942–1950.

Wastewater-Based Circular Economy Operations in East Kolkata Wetlands (the Largest Ramasar Site in South Asia), India

Jayanta Kumar Biswas and Majeti Narasimha Vara Prasad

Abstract

Wastewater generated from domestic and industrial sources contains a cocktail of countless beneficial and harmful chemicals. Conventional wastewater treatment plants used for removing toxic metals and nutrients from wastewater are not considered a viable option in developing countries because of their energy and capital-intensive nature. Nevertheless, wastewater treatment should not let nutrients wasted down the drains and cause eutrophication and environmental deterioration of the receiving water bodies. Rather it is desired to accomplish recovery and recycling of nutrients and water to bioregenerative use by closing the loop through circular economy approach towards sustainable development. Ecotechnology has evolved as an economical ecofriendly holistic approach to wastewater remediation, for fulfilling human needs while protecting environment from pollution, by harnessing and subtly manipulating ecosystem components and natural forces to leverage their beneficial effects. The East Kolkata Wetland (EKW), a Ramsar site located at the east of Kolkata city and composed of salt marshes, agricultural fields, sewage-fed fish farms and settling ponds, exemplifies an eco-engineered natural wastewater treatment system, receiving daily up to 65% of Kolkata's sewage and reusing it for producing fish and vegetables by local communities. the present book chapter portrays the genesis and the state-of-the-art of ecotechnological practices spearheaded by wastewater-fed aquaculture (WWFA) sustaining in the East Kolkata Wetlands (EKW), its present problems and future scope for improved circular economic approaches to converting 'waste into wealth' in meeting the win–win goals of environmental protection and livelihood security. However, some of these activities pose potential risk for the environment and human health due to presence of toxic heavy metals, and pathogenic microorganisms, which warrants comprehensive ecotoxicologicl risk assessment studies.

Keywords

Wastewater treatment • Ecotechnology • Bioregenerative practices • Resource recovery and recycling • Circular economy • Ecotoxicological risk assessment

1 Introduction

Rapid urbanization has been putting mounting pressure on the world's fragile and dwindling freshwater resources and overloaded sanitation systems, leading to pervasive environmental degradation. Wastewater generated from domestic and industrial sources contains a cocktail of countless useful and harmful chemicals. Wastewater in any form raw, diluted or treated is a misplaced or misused resource and unrecognized wealth of increasing global importance, particularly in urban and peri-urban agriculture. With proper management, wastewater use can contribute immensely to livelihood support, food security and enhanced environment quality. While the code 'reduce' is the 'Holy Grail' of wastewater management, the code 'recycle' transforms 'waste into wealth', 'garbage into gold' or 'trash into treasure'. Thus, waste(water) may be a 'necessary evil' but waste is not waste until we waste it. But in most of the developing countries little wastewater treatment takes place in urban areas, and two-thirds of urban wastewater is discharged without any treatment into water bodies due to lack of

J. K. Biswas (✉)
Enviromicrobiology, Ecotoxicology and Ecotechnology Research Laboratory (3E-MicroToxTech Lab), Department of Ecological Studies, International Centre for Ecological Engineering, University of Kalyani, Kalyani, West Bengal Nadia-741235, India
e-mail: jkbiswas@klyuniv.ac.in

M. N. V. Prasad
School of Life Sciences, University of Hyderabad (an Institution of Eminence), Hyderabad, Telangana 500046, India

M. Smol et al. (eds.), *Water in Circular Economy*, Advances in Science, Technology & Innovation,
https://doi.org/10.1007/978-3-031-18165-8_19

conventional wastewater treatment plants (WWTPs). Such WWTPs used for removing toxic metals and nutrients from wastewater are not considered a viable option in developing countries because of their energy and capital-intensive nature. Besides, they demand high maintenance costs, sophisticated equipment, technological knowhow and technical skills. Nevertheless, wastewater treatment should not let nutrients wasted down the drains and cause eutrophication and environmental deterioration of the receiving water bodies. Rather recovery and recycling of nutrients and water is desirable for bioregenerative use by closing the loop following circular economy approach towards sustainable development. This calls for the development of low-cost, sustainable ecotechnologies for wastewater treatment that safeguard natural environment and human health. Waste stabilization ponds have drawn much interest as an ecotechnological tool for reducing nutrient and pathogen loads and wastewater remediation. There is an emerging need for a holistic natural ecosystem-based approach for the treatment, reclamation and recycling of both nutrients and water in an energy efficient and cost-effective manner by harnessing and subtly manipulating ecosystem components and natural forces to leverage their beneficial effects, while providing employment generation, livelihood security and low-cost food for the poor. Based on the existing body of the knowledge (Table 1) this book chapter portrays the genesis and the state-of-the-art of wastewater-fed farming systems sustaining in the East Kolkata Wetlands (EKW), its present problems and future scope for improved circular economic approaches in meeting the win–win goals of environmental protection and livelihood security.

2 Kolkata as an Ecologically Subsidized City

Kolkata is blessed with two special ecological systems: the river Ganges (Hooghly river) on the west and East Kolkata Wetlands (EKW) (Fig. 1) on the east towards which the city is inclined to use it as a *robust ecological infrastructure, huge sink, inimitable natural kidney, WAR (waste as resource) field or a WRR (waste recycling region)*. EKW exists in the eastern part of Kolkata City and bordering Salt Lake and Rajarhat, extending over districts of North 24-Parganas and South-24 Parganas in the State of West Bengal, India. It is located between latitudes 20°25’ –20°35’ N and longitudes 88°20’–88°35’ E, spreading over an area of 12,500 hectares and comprising 254 sewage-fed fish farms (Kundu et al., 2008). The fish farms consist of units of various sizes from large holdings locally called *bheris* and relatively smaller ones called *jheels* due to their trench-like elongated shapes. But all these fish farms generally have similar types of produce, farming practice and distribution system. EKW is a classic example of waste recovery for over a century through a mechanism traditionally developed by the local people and traditional fishermen. This resource recovery system is the largest of its kind in the world

Table 1. Selected studies on ecotechnological practices, problems and prospects of East Kolkata Wetlands (EKW), the largest Ramsar site in South Asia. The table is not exhaustive

Salient aspects of studies	References
(A) Practices	
i.Wastewater fed aqua- & agriculture in EKW	Ghosh, (1985; 1988; 1990; 1999)
ii. Wastewater treatment and recycling: Converting waste land to waste-not land	Furedy, (1987)
iii. Social aspects of human excreta reuse: implications for aquaculture	Furedy, (1988)
iv. EKW as the city sewage treatment system	Majumdar, (2008)
v. Traditional practices at EKW for wastewater management	Raychaudhuri et al., (2008)
vi. Developing waterlogged areas for urban fishery and a waterfront recreation project	Ghosh and Sen, (1992)
vii. Land reclamation and reuse of waste water	Roy et al., (2021)
viii. Traditional management practices for aquaculture in EKW	Saha, (2014)
ix. Holistic approach to EKW management	Bhattacharyya, S., (2021)
x. Biodiversity, traditional practices and sustainability issues of EKW	Bhattacharya et al., 2012
xi. Aquaculture in EKW: soil-water relationship	Paul, (2011a), (2011b)
xii. Temporal variations in bio-indices of *Oreochromis niloticus* in EKW	Kumar et al., (2021)
xiii. Water hyacinth-mediated wastewater phytoremediation in EKW ponds	Chaudhuri et al., (2008)
xiv. Bacterial diversity based water quality assessment of sewage-fed fish farms	Tyagi et al., (2021)
xv. Seasonal changes in zooplankton and macro-fauna populations in EKW fish ponds	Saha et al., (2017)

(continued)

Table 1. (continued)

Salient aspects of studies	References
xvi. Modeling phosphorus dynamics in EKW	Mandal et al., (2015)
xvii. Waste water utilization in EKW: From local practice to sustainable option	Ghosh, D., (2008)
xviii. Relating peri-urban poor and ecology	Pakrashi, (2021)
xix. Dynamics of peri-urban agriculture	Ghosh, R., (2022)
(B) Problems	
i. Wastewater-fed aquaculture in the East Kolkata Wetlands, India: anachronism or archetype for resilient ecocultures?	Bunting et al., (2010)
ii. Traditional aquaculture practice at EKW: the safety assessment	Raychaudhuri et al., (2008)
iii. Urban expansion and evaluating urbanogenic interventions in EKW	Basu and Bhattacharjee, (2022)
iv. Smart global megacities spatial strategies for Kolkata Metropolitan Region	Bedi, (2022)
v. Assessment of the bioaccumulation pattern of Pb, Cd, Cr and Hg in edible fishes of EKW	Dutta et al., (2022)
vi. Bioaccumulation of heavy metals in muscle tissue of fishes from selected aquaculture ponds in east Kolkata wetlands	Kumar et al., (2011)
vii. Bioaccumulation of metals in the EKW	Biswas and Santra, (2000); Chattopadhyay, (2002)
viii. Trace metal distribution in tissues of cichlids (*Oreochromis niloticus* and *O. mossambicus*) grown in wastewater-fed fishponds in EKW	Chatterjee et al., (2006)
ix. Assessment of total mercury level in fish in sewage-fed aquaculture	Bhattacharyya et al., (2010)
x. Fate of metals in fish under variable sewage input in fish ponds	Maiti and Banerjee, (2012)
xi. Impact of chromium from tannery effluents on EKW ecosystem	Chattopadhyay, 2000
xii. Toxicity and profitability of rice cultivation under wastewater irrigation	Mukherjee et al., (2013)
xiii. Threat of land subsidence to EKW	Sahu and Sikdar, (2011)
xiv. Framework for building urban resilience to climate change for sustainable urbanism	Kumar et al., (2022)
xv. Effect of climate change induced acidification on Pb in EKW	Dutta et al., 2021
xvi. Threat to EKW avi-fauna (*Synoicus chinensis*) due to development activities	Naskar et al., (2022)
xvii. Stress responses of fish cultured in EKW	Das et al., (2021)
xviii. Effect of enhanced salinity	Bhattacharyya et al., (2021)
xix. Effect of biological invasion by *Pterygoplichthys* spp. in wastewater-fed fish farms of EKW	Hussan et al., (2021)
xx. Threats to fish diversity of EKW and conservation needs	Hussan, (2016)
(C) Prospects	
i. EKW for environmental protection	Bandyopadhyay, 2011
ii. Removal of chromium (VI) by *Bacillus subtilis* isolated from EKW	Basu et al., (2014)
iii. Carbon sequestration by horticultural plants in East Calcutta Wetlands ecosystem, a Ramsar Site in India	Adhikari et al., (2021)
iv. Microbial genetic resource mapping of EKW	Chaudhuri. And Thakur, (2006)

(Ghosh, 1999). The resources recovered from city sewage are used in three kinds of economic activities such as wastewater-fed fish culture, vegetable farming on a garbage substrate (garbage farms) and paddy cultivation outside the core area using pond effluent, in the order of present power, priority and potential. Since the sewage-fed fish culture is the 'heart' of the entire waste recycling region, it has been described in detail.

The EKW system is the glorious example of a unique low-cost, ecological municipal waste management system, the largest and possibly the oldest integrated resource recovery system in the world supporting a combination of agricultural and aquacultural practices (Ghosh, 1990, 1999; Edwards, 2000; Ray Chaudhuri et al., 2008). According to Article 8 of the Ramsar Bureau lists under the Ramsar Convention, the EKW has been designated as Ramsar Site (No. 1208), i.e. wetland of international importance on 19 August 2002, owing to their outstanding contribution in rendering vast array of ecosystem goods and services of overarching importance, such as wastewater treatment, sustainable

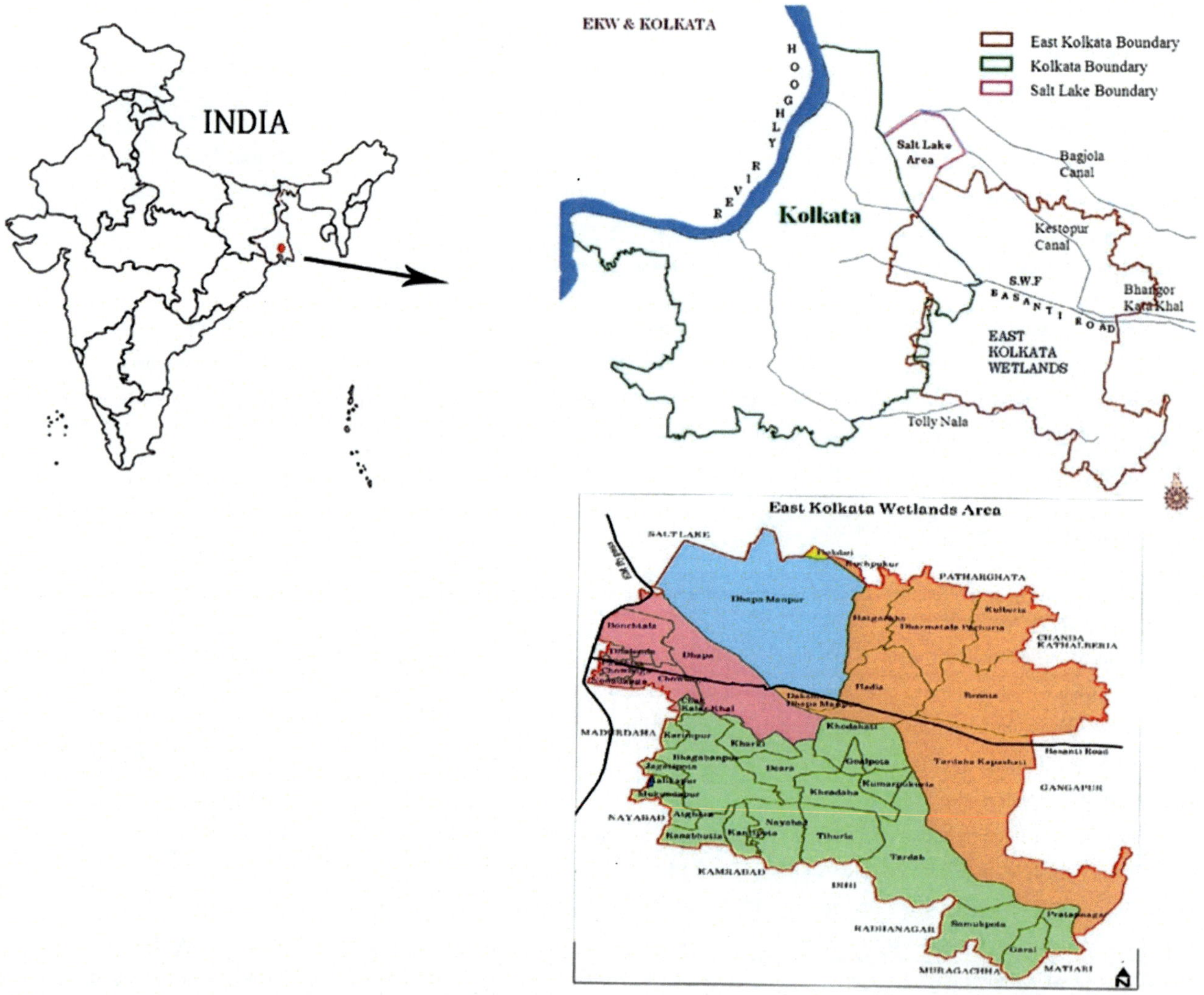

Fig. 1 Geographical location of the East Kolkata Wetlands (EKW), the largest Ramsar site in South Asia. Different wetland areas are clearly shown in inbox

nutrient recovery in the form of valorized biomass, fish protein and vegetables production, water recovery and integration with terrestrial food production ecosystems, particularly wastewater management and environmental protection (Majumdar, 2008; Sarkar et al., 2009; Bunting et al., 2010; Bhattacharya et al., 2012a, 2012b; Bhattacharjee, 2022; EKWMA/WISA, 2021; Mundoli et al., 2022; Kumar et al., 2022). Over 20,000 economically poor families rely on different forms of biological products for livelihood. The EKWs are spatially spread to offer trinity of usages: sewage-fed fish farming (3898.7 ha), sewage-fed agricultural activities (4959.86 ha) and solid waste garbage disposal (602.78 ha), while a total of 1326.52 ha area is used for urban and rural settlements (Sengupta, 2018). Total wastewater holding capacity of the fisheries is > 35,000 million liters.

It serves as a *natural kidney* for Kolkata, receiving 250 million gallons of anthropogenic wastewater (both municipal and industrial) daily and ameliorates it by natural processes with a daily capacity to remove around 237 kg of BOD (Chattopadhyay et al., 2002). Several canals and wastewater channels bring up sewage from domestic and municipal sources from different parts of the city to EKWs. A total of 254 domestic and municipal sewage-fed fish farms are reported to be used in aquaculture practices (Chandra et al., 2020; Sengupta, 2018).

Moreover, approximately 13,000 tons of fishes/year was produced and exported to different parts of the country from this sewage-fed wetland. About 50 fish species such as *Cirrhinus mrigala* (Mrigal), *Cirrhinus reba* (reba), *Labeo bata* (Bata), *Labeo rohita* (Rohu) and *Labeo catla* (Catla) were reported from these sewage-fed fishery farms (Hussan, 2016). The garbage farms produce approximately 150 tonnes of fresh vegetables daily, while the rice paddies produce approximately 16,000 tonnes of rice annually.

3 Ecological History of EKW: Genesis of Sewage-Fed Farming

The EKW system is originally part of the Gangetic delta. The land formation in this region was influenced through centuries by the Ganges River system where upland and tidal depositions continued to create land. Before the establishment of Calcutta (now called Kolkata), the saltwater lake created here was a backwater swamp and the spill basin of once mighty river Bidyadhari (Furedy & Ghosh, 1984; Ghosh, 1990, 1999; Ghosh & Sen, 1987, 1992). The tidal channels of this region joined further down with a deltaic channel of the Sundarbans, through which salt water flowed into the water bodies (Stewart, 1836). Due to tidal interplay, these water bodies were saline in nature, hence the name Salt Lake, where traditional captured fishery was practiced by the local people. Rapid silt deposition in the rivers which started since the end of the fourteenth century eventually led to the River Bidyadhari becoming completely defunct by the end of nineteenth century (Ghosh, 1999). The entire spill basin was transformed into stagnant water bodies due to combined effects of the sudden change in the course of the Ganges towards east caused by neo-tectonic movement and loss of the tidal subsidy. The spill area and the channel beds further shrunk due to increasing anthropogenic activities including brackish water fisheries which led to the death of the tidal creek and genesis of the vast saltwater marsh to the eastern side of Kolkata between the levee of the Hooghly on the west and the Kulti Gong on the east (Ghosh & Sen, 1987). Before 1830, these low-lying regions were traditionally utilized for capture fishery for the cultivation of brackish water fishes such as Bhetki (*Lates calcarifer*), Parse (*Mugil parsia*), Bhangar (*Mugil tade*) and Prawn (*Macrobrachum rosenbergii*) (Ghosh & Sen, 1987). These were originally saline in nature, which gave rise to its name as Salt Lake. Subsequently, local fishermen tried to utilize city sewage flowing towards eastern side through canals as source of water and nutrient which became successful in 1929–30.

Kolkata city has a natural gradient towards east. As the city grew, the authorities decided to drain the city into the salt takes forming the world's largest wastewater ecosystem. Huge composite effluent mixed city sewage (around 50,000 $m^3 d^{-1}$) from Kolkata is discharged in the raw state, without any pre-treatment via nine drainage pumping stations (DPS). The bheries are connected to a network of canals/channels (locally called *khals*) running from west to east, that are segregated as primary, secondary and tertiary channels based on their functional features. There are two main parallel canals close to each other that run up to the Bantala lock gates, namely, the Dry Weather Flow Canal (DWF) that starts at the Topsia pumping station and the Strong Weather Flow Canal (SWF) that starts at the Ballygunj pumping station The city sewage from the south-central part of the city is collected and discharged from higher level into the Dry Weather Flow (DWF) canal at Topsia pumping station which flows towards the eastern part of the City and reaches Bantala, wherein a number of other canals from central and northern part of the city join DWF. By operating the sluice gates at Bantala, the sewage is diverted into the Fishery Feeding Channel (FFC) which draws sewage into EKW with a network of sewage-carrying canals. From different sewage canals wastewater enters the fisheries either via gravity, siphoning or pumping. Besides large volume of municipal wastewater, these canals also carry effluents from at least 538 tanneries in three prominent agglomerates and approximately 5500 other small-scale industrial establishments such as rubber industries, electroplating industries, pigment manufacturing units, potteries and battery manufacturing plants. The city sewage mixed with industrial effluents enters into the fisheries through a bamboo screen and gets stored there for about a month or so and thereafter gets cleaned through several well-established steps and methodology (Sarkar et al., 2009; Vicziany et al., 2017). Thus with the diversion of city sewage from Kolkata city and influx of storm water into this area along with decaying and death of the river Bidyadhari there was a gradual decrease in salinity in the aquatic environment which created a new window for fish culture and livelihood support (Ghosh, 2008; Bunting & Little, 2005; Mukherjee, et al., 2021a; Nadella & Sen, 2021; Pakrashi, 2021; Roy et al., 2021).

4 Ecotechnology of Wastewater-Fed Fish Farming

Raw sewage from Kolkata City directly flows into the fish ponds which accomplish a wide range of physical, chemical and biological processes to grow fish using the raw sewage as nutrient, and ultimately improving the quality of the effluents from these fisheries which is disposed into Bidyadhari River at Kulti Ganj, about 33 kms away from Kolkata City. The degree of fish pond fertilisation by wastewater depends on the wastewater characteristics, weather conditions, physical features of the pond and fish species, which are best judged with the traditional knowledge. The wastewater fishponds act as *solar reactors* and accomplish most of their biochemical reactions with the help of solar energy. The basins of such ponds are shallow which allows full vertical circulation of water to the surface where algal blooms occur. This is highly favourable for photosynthesis, as this depth gives a better ratio between pond volume and pond surface than a deeper pond. This helps to create sufficient oxygenation, which, in turn, can reduce BOD and fecal coliform load (EKWMA, 2011). The photosynthetic activity in the pond is one of the reasons for the purification of sewage. The biological remediation process is carried on by algae and bacteria which transform sewage into useful nutrients for fish and animals as well as maintain dynamic balance of zooplankton and macro-fauna populations (Saha et al., 2017; Sifa, 1997). The mix of algae and bacteria is also more efficient at reducing coliform bacteria than conventional sewage treatment plants. The biochemical reactions occurring in the wastewater ponds include photosynthesis, aerobic oxidation, organic acid formation, methanogenesis, nitrification and denitrification (Biswas & Rana, 2014; Bhattacharya et al., 2008). Macrophytes (e.g. water hyacynth *Eichhornia crassipes*) is capable of stabilizing the biological oxygen demand levels and removing nitrogen and heavy metals (Bhattacharya et al., 2008; Chaudhuri et al., 2008). Reduction of biochemical oxygen demand (BOD) takes place due to the unique functional accomplishment of algae-bacteria symbiosis where energy is drawn from algal photosynthesis. The BOD of the sewage effluent becomes drastically reduced in sewage-fed fish ponds. The effluents from these fisheries are 99% free from BOD and are being used for irrigating the paddy fields downstream. It has been estimated that EKW removes around 237 kg of BOD per day acting as a natural kidney for Kolkata city (Bhattacharya et al., 2012a, 2012b). Sometimes a train of different waste stabilization and maturation ponds is used. For example, Mudiali farm receives 35 million L/day of untreated wastewater which is drawn into an anaerobic pond (stabilization pond), passes into the facultative pond, and then into a 425 m macrophyte-channel having *Eichhornia* sp., *Phragmites* sp., *Pistia* sp. and *Lemna* spp. which are used to remove nutrients, heavy metals, suspended solids, oil and grease (Jana, 1998).

The removal efficiency of nutrients, contaminants and pathogens may vary for each pond, but they are connected in a series with varying numbers. The cumulative removal efficiency of these integrated ponds is considered for wastewater treatment which removes nutrients (N, P and K), organic loading (BOD) and contaminants by microbial (metabolic processes of heterotrophic bacteria and autotrophic algae), chemical, physio-chemical and hydrodynamic processes (Biswas & Rana, 2014; Paul, 2011a, 2011b). *Aerobic waste stabilization ponds* are shallow ponds which maximize light penetration and the growth of algae through photosynthetic activity. Anaerobic waste stabilization ponds require no dissolved oxygen for microbial activity as the anaerobic and facultative organisms use oxygen from compounds like nitrates and sulphides as their terminal electron and hydrogen acceptors, and give end products like methane and carbon dioxide. *Facultative waste stabilization ponds* are partly aerobic and partly anaerobic. They support algal growth along with the growth of facultative microorganisms. In *maturation ponds* fish are cultured. The algae in the pond serve as food of the cultured fish. In a pond ecosystem, the presence of both bacteria and algae in symbiotic action is essential to convert the complex organic substrate contained in the wastewater to simpler inorganic end products like the nitrates and phosphates whose uptake leads to the production of more bacteria and algae, some of which flow out with the effluent (Biswas & Rana, 2014). In the ponds, two mechanisms are involved in removing phosphorus from the wastewater received. (i) phosphorus uptake by the algae that comprises only about 1% of total P and (ii) precipitation of phosphorus, for example, as calcium hydroxyapetite as the major mechanism. Nitrogen removal mechanisms are (i) loss with algae in effluent, (ii) loss through seepage and (iii) denitrification.

Chemical characterization of sewage used for fish culture showed a distinct gradient, with a progressive rise in the values of pH and DO, bur reduction in carbon. A considerable reduction of organic loading or BOD of the wastewater is a prerequisite for such wastewater-fed system. Dilution with freshwater or proper loading of wastewater provides congenial environment and aerobic conditions for fish culture. Retention of wastewater in the ponds before the initial stocking of fish for a considerable residence time allows bacterial population to decompose and metabolize the organic matters of the sewage (Ghosh, 2008). However, the rate of biodegradation at EKW has been found to be very high. Wastewater becomes stabilized in EKW within 10 days of retention (Sarkar et al., 2009). Lime is applied to the ponds at a rate of 400-500kh/ha during the winter months when the ponds dry up. Lime is also applied occasionally, depending on the water quality. Lime addition before the release of raw sewage into the fisheries helps to reduce the pathogen/fecal coliform load by 96–99% (Ray Chaudhuri et al., 2012). No supplementary feeding or pond fertilisation is performed during the culture period. No aeration device is installed in these fish ponds, but regular boating and netting operations are performed at night, which helps to augment the DO level.

5 Culture Practices

Composite fish culture is performed using different fish species which utilize different ecological niches on pond ecosystem. The *polyculture* system practiced selecting species from the following: Indian major carp: Rohu (*Labeo rohita*), Catla (*Catla catla*), Mrigal (*Cirrihinus mrigala*); Indian minor carp: Bata (*Labeo bata*); exotic variety: Silver carp (*Hypophthalmichthys molitrix*), common Carp (*Cyprinus carpió*), Grass Carp (*Cenopharyngodon idella*), Tilapia (*Oreochromis nilotica* and *Tilapia mosambica*). For example, Mudialy fish farm entails nine species while Captain bheri has eight species. Apart from these cultured varieties (except Tilapia), some other varieties including forage fishes are also available occasionally in the *bheris*. They are Punti (*Puntius japónica*), Sole (*Charma striatus*), Lata (*Channa punctatus*), Chyang (*Channa gachua*), Singi (*Heteropneustes fossillis*), Magur (*Ciarias batrachus*), Fouli (*Notopterus notopterus*), Pungus (*Pangasius pangasius*), etc. Stocking density varies depending on seed availability, stocking time and culture system. The annual stocking density is high in Mudialy farm (200–300 × 10^3/ha/yr) while in Captain bheri it is 150 × 10^3/ha/yr. The introduction of fishes follows a short food chain which is a profitable policy. Mo specific stocking ratio among fishes is maintained, but fish farmers use greater proportion of more tolerant and benthivorous fish like tilapia, common carp and mrigal compared to others. The culture period varies—one year for Mudialy and 3–4 years for Captain bheri. When required the ponds are completely dried up for reducing organic load, pollutants and toxicants.

Judicious selection of fish species in *polyculture* is a key factor for effective utilization of sewage in fish culture. Omnivorous and bottom dwelling fishes consume directly the organic detritus of sewage-fed ponds, and thereby help to remove obnoxious gases from pond bottom and save the bottom from being anaerobic through bioturbation. The suitable species are Indian carp, Israeli carp, silver carp, bighead carp, grass carp, common carp, hybrid buffalo, cat fish, largemouth bass, tilapia, freshwater prawn, etc. (Ghosh, 1985). Traditionally, aquaculture in Kolkata was dominated by the production of Indian major carps (IMCs), specifically a polyculture of rohu (*Labeo rohita*), catla (*Catla catla*) and mrigal (*Cirrhinus mrigala*). However, there has been a trend towards culturing tilapia (*Oreochromis niloticus*), usually as part of a polyculture with IMCs (Kumar et al., 2021). Tilapia is an ideal species for rearing because it can tolerate relatively low dissolved oxygen concentration, wide variation in salinity and high ammonia concentration. Carp are sensitive to low DO and high ammonia levels, and require larger water area for growth. Tilapia culture is associated with several advantages: self-recruitment in ponds, which is absent for the IMCs; demanding less investment in stocking; rapid growth; and reportedly good disease resistance. Their omnivorous feeding behaviour also means they exploit various feeding niches, and the ability to feed efficiently on phytoplankton makes them highly productive in systems receiving significant fertiliser inputs, but where supplementary feeding is limited. In general, fish catch starts after 2–3 months daily or weekly interval at their fingerling stage (100–200 g). In Mudialy fish farm, the average monthly fish catch is 400 kg/ha (Jana, 1998).

6 Garbage Farming

Another system of waste recycling in the EKW is the use of garbage and wastewater as manure and water for irrigation in paddy and vegetable farming. The old solid waste dumping grounds (called *Dhapa area*) have been converted into cultivable lands using solid municipal garbage, and sludge filled soils are used for agriculture (Furedy, 1987, 1988). Between 2,000 and 3,000 metric tonnes of diverse solid wastes are being deposited daily in the wetland areas. Garbage farming yields nearly 370,650 kg ha^{-1} vegetables, which amounts to 150 metric tonnes of vegetables daily. The paddy fields can produce 16,000 tonnes of winter paddy varieties cultivated during post-monsoon period annually. Different varieties of green leafy vegetables cultivated in these areas are *Amaranthus caudatus, Amaranthus blithum, Spinacia oleracea* etc. (Ghosh, 2008; Adhikari et al., 2021).

7 Traditional Knowledge: Cornerstone of Sewage-Fed Fish Farming

The knowledge that has emerged provides an alternate and wise use of wetlands. Sewage-fed fish farms were introduced in 1930. Overtime, local fishermen have innovated, developed, upgraded and perfected the wise use of wetlands and the self-sufficient farming process based on traditional skill, enterprise and innovation. In the process, the city developed a *vernacular wastewater management system* through nature-based remediation and resource recycling (Ghosh, 1990, 1999; Saha, 2014). The fish of EKW are grown by fishermen in large ponds fed with the city's sewage which is regarded as '*liquid gold*'. The sewage-fed fish farms are able to produce two to four-fold greater yields than conventional fish farms. Sewage produces edible fish as a result of the application of the fish farmers' traditional knowledge of how to purify the sewage water and harness biomass. The first purification process is done by solar radiation and the shallow depth (50–150 cm) of the bheries. The sunlight supports photosynthesis and promotes growth of planktons which nourish the fish while the shallowness allows the pond waters to easily circulate from bottom to top in response to wind currents, bioturbation by fish and physical agitation by the fishing boats, and the deliberate stirring of the mud by the fishermen, promoting oxygenation.

Pisciculture, agriculture and horticulture are common practices in which local peoples of EKW area are occupationally involved. They perform wastewater-fed fish culture, garbage farming irrigating agro-horticultural fields with wastewater and application of sludge (derived from wastewater carrying canals) in agricultural fields. These creative and sustainable practices are generated, developed, carried and transmitted by the traditional community over generation after generation. These wetlands have been preserved and nurtured by them. To the traditional fishers, such fish farms are more than waterbody and fish crop. They are their community's heritage and future. Three quarter of the working population of the adjoining areas depends on fish farming, agriculture and horticulture for their survival. They focus on the value of the belief-knowledge-practice complex (the *Kosmos—Corpus—Praxis* complex) in relation to wetland system and its biodiversity. To them the *Kosmos* (i.e. Nature) is the productive source and core of (agri)cultural identity; the *Corpus* is the repository of ecological knowledge (local, collective, diachronic and holistic) and the *Praxis* comprise human-Nature symbiosis, subsistence and principle of diversity of resources and practices entailing optimum utilization of land; recycling of materials, energy and wastes; multiple use and product diversification; integration of different practices.

The manner of managing the seasonal weather cycles by fish farmers shows how human intervention contributes to water purification. There are five phases of fish farming: pond preparation, primary fertilisation, fish stocking, secondary fertilisation and fish harvesting. At an interval of 3–4 years, in January, the ponds are subjected to drying in rotation through natural evaporation; the dikes and boundaries are repaired and the bhery land is utilized to grow vegetables. Following harvesting the vegetables, the fishers dig silt traps around the edges of the fish ponds. When the fish are harvested at the end of the season, the fishers drag their nets across the ponds so that the silt is scraped together and stored in the silt traps, providing a ready supply of soil at the edges of the ponds to repair the earthen banks separating the bheries. Just before the release of the sewage into the bheries as a means of primary stage of fertilisation, the ground tilling is performed and lime treatment is done to reduce the acidity of the pond waters, making them more suitable for the fish culture (Sarkar et al., 2009). The bottom sediments of the bheries are stirred to increase oxygenation for enhancing microbial activities in the mud. The water purification process is manifested by the changing appearance of the ponds. After the sewage flows in, the water turns black, but the rapid growth of algae changes the colour to green in three or four days. After 7 days, algae growth reaches its peak to mark algal blooms. The water becomes crystal clear after the fisherfolks remove the algae. Then it is allowed to breather for about 25 days before the fish are introduced (Sarkar et al., 2009). In mid-March, the pond water quality is judged by releasing a fish sample. Through visual observation the fishermen can judge if the waters are ready for larger batches of fish. If the test fish survives, fish of various varieties are stocked in roughly four batches over a period of months, for example, silver carp in July and ordinary carp in December. The decision is made by the fishers exclusively on the basis of their past experience, without scientific measurement of any water quality parameter. With stocking of additional fish additional quantum of sewage is introduced. The dosing of this secondary fertilisation depends on the stocking density, i.e. the number of fish and the size of the pond. The precautionary principle for managing the additional sewage and plankton growth is not to deplete the dissolved oxygen level too low to sustain the fish (Ghosh, 1988). A delicate natural balance is required to be maintained between the oxygen demand of the microorganisms and that of the fish. All stages of managing the bheries and the fish are based on the fishermen's traditional knowledge and experience to guarantee the precise sustainability conditions for the fish. Fish harvesting happens during different periods of the year depending on the fish age: older fish are harvested in May–July, more recently-introduced fish in August–October, and silver carp are removed in December. The manner in which the fish farmers manage their bheries presents an excellent example of the power of traditional knowledge. The fish farmers' main tools are observation, experimentation and caution. This is especially important while deciding when to transfer the fish through their various breeding stages (Dey & Banerjee, 2013). The culturing fish varieties in the bheries represents *polyculture* developed by local, innovative farmers taking due consideration of exotic fish varieties along with the indigenous fish varieties and selecting different fishes which occupy different micro-niches in the ponds—top, middle and bottom feeders. The fishers also make proper use of the land when it is not inundated with water by growing vegetables such as radishes and leafy greens exemplifying an example of the *intercropping system* practiced in traditional farming. The fishermen also apply their knowledge of the habits and habitats of plants and animals to improve the productivity of the bheries. Exotic water hyacinths (*Eichhornia crassipes*) are used to line the edge of the ponds in the form of a 3–4 m wide band controlled by a unique kind of fencing system (Ghosh, 1988). During the hottest months, the hyacinths are gathered at one end of the ponds to provide shade and shelter for the fish. The wire or bamboo fencing holds the hyacinths in place, allows the fish to eat the roots, and contributes to the stability of the mud banks. The hyacinths also remove heavy metals present in the water and serve as windbreaks, preventing the erosion of the. Traditional knowledge practices also include the use of natural pesticides. Many insects and slugs are attracted to the sewage, but numbers are kept under control by scattering the seeds of the mahuya tree and the neem tree to the pond soils when they are dry.

8 Contraints, Concerns, Threats and Challenges

Urbanization, industrial pollution, siltation, weed infestation and the changed land use patterns are continuously adversely affecting the ecosystem health of the EKW (Basu & Bhattacharjee, 2022; Bedi, 2022; Bera et al., 2021; Naskar et al., 2022). The current and potential contraints, concerns, threats and challenges (Table 1) comprise the following: (i) unsustainable use of groundwater posing a risk of land subsidence; (ii) progressive shift in the land use pattern leading to a gradual dominance of agriculture ($\sim$40%) and reduction in fish farms area which results in shrinkage of its capacity to recycle wastes and attenuate floods (Ghosh, 2008; Sengupta, 2018); (iii) increasing siltation in the canals and fish ponds reducing the quantity of sewage flowing to the fisheries and making the fish ponds much shallower, which impact fish yield; (iv) pressure from profit-intensive groups to develop areas for residential and industrial purposes, which has negative implications for groundwater recharge and flood; (v) improper and unscientific management of water regimes (through maintenance of sluice gates, running the pumping system, regulating the storm weather flow and the dry weather flow channels of the drainage system), technology integration and weak marketing, post marketing and value addition opportunities; (vi) contamination of the groundwater and water insecurity due to long term contaminated wastewater-fed fish culture practice (Sikdar et al., 2001a, 2001b; Sahu & Sikdar, 2008; Sikdar & Sahu, 2009; Mukherjee et al., 2021a, 2021b); (vii) contaminants loading and nutrient impoundment due to industrial and municipal wastes (Mandal et al., 2015); (viii) lack of sewage water due to inefficiency in management practices; (ix) high input costs; (x) lack of government initiatives; (xi) conflicts over tenurial rights (e.g. owner-worker conflicts); (xii) microbial load (e.g. *E.coli* and *S. aureus*) in organs of fish grown in sewage-fed fish ponds (Tyagi et al., 2021); (xiii) threats to fish diversity due to invasive species (Hussan, 2016; Hussan et al., 2021); (xiv) cascade effects of climate change (Bhattacharyya et al., 2021; Kumar et al., 2022). (xv) Increasing concentrations of heavy metals in the sewage used in the wetlands *is the most critical health concern.* Unregulated discharge of contaminated industrial effluents poses great threat to the environment and human health. Plants and animal species including fishes of the EKW can bioaccumulate the heavy metals such as lead, chromium, zinc and mercury from the contaminated wastewater and industrial effluents creating severe health hazards (Chatterjee et al., 2006; Das et al., 2021, 2022). Many researchers (Chatterjee et al., 2007; Chattopadhyay et al., 2002; Ahmed et al., 2019; Biswas & Santra, 2000; Bhattacharyya et al., 2010; Kumar et al., 2010, 2011; Aich et al., 2012; Mukherjee et al., 2013; Nath and Bhowmik, 2013; Dutta et al., 2016, 2017; Dutta et al., 2017, 2022; Biswas Roy et al., 2022; Dutta et al., 2022a, 2022b; Maiti and Banerjee, 2012) reported that the use of sewage for growing vegetables and fishes resulted in contamination and toxicity of heavy metals (lead, cadmium, copper, chromium, mercury, zinc, etc.) in those consumable biomass grown in the Dhapa area (Table 1). The highest bio-concentration factor was observed in *T. mossambicus* followed by *C. mrigela* and *L. rohita* (Bhattacharyya et al., 2010). Mercury concentration in *T. mossambicus* in all season and *C. mrigela* in pre monsoon season had cross permissible level for human consumption (0.23–0.47 µg/g of wet weight of fish/day) recommended by World Health Organization (WHO, 1990) and Prevention of Food Adulteration Act, India (PFA Act 1954, amended in 2002). Major sources of Cr in the composite wastewater are the tanneries and electroplating industries (Basu et al., 2014; Chattopadhyay et al., 2002). Cauliflower, radish and amaranth grown in these farmlands accumulated significant amount of metals. Different parts of these plants contained different amounts of metals. In cauliflower, maximum accumulation of Cr, Zn and Mn was observed in the stem while maximum accumulation of Pb, Fe and Mg occurred in root material. Copper accumulated in leaf material. In radish, on the other hand, maximum accumulation of Cr, Cu and Mg was observed in root while Zn, Mn, Fe and Pb accumulated in leaves.

For rejuvenation and restoration of drainage structures and EKW the following measures need to be adopted: (i) for proper maintenance and operation of the wastewater recycling process in the EKW, all drainage channels and distributaries need to be brought under a comprehensive action plan; (ii) desiltation can improve water circulation and flushing; (iii) improved waste segregation, proper and safe handling of wastes, and sustainable waste management strategies can help to mitigate pollution in the landfill areas; (iv) pollution control and installation of common effluent treatment plants for in situ mitigation of release of untreated contaminanted effluents from industries in general and nearby tanneries in particular, for the polluting industries; (v) cultivation of hyperaccumulators of the toxic contaminants in the contaminated zones for their removal and remediation; (vi) imposing restrictions on use of insecticides, pesticides and other harmful chemicals in fish culture; (vii) Monitoring fish health; (viii) comprehensive ecotoxicologicl risk assessment; (vix) down-to-earth mitigation and adaptation strategies for combating climate change.

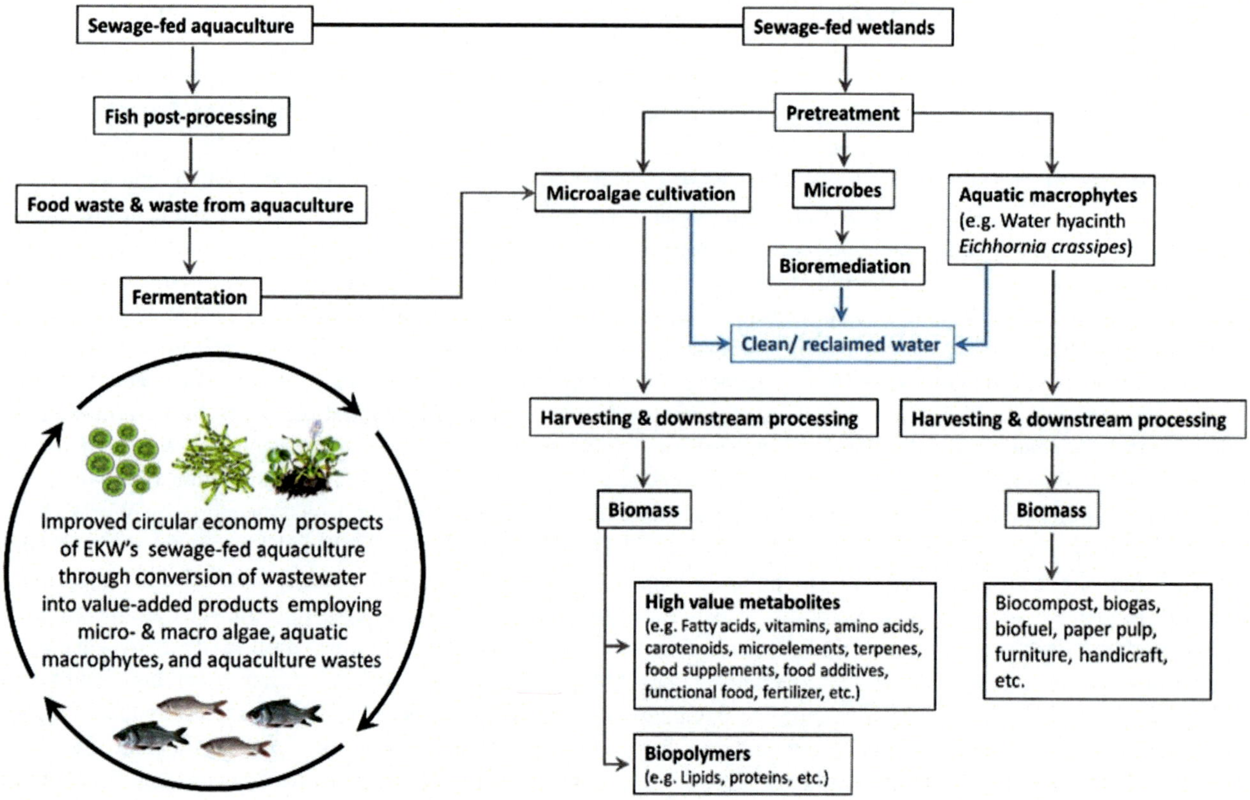

Fig. 2 Future scope for improved circular economy prospects of EKW's sewage-fed farming through conversion of wastewater into value-added products and metabolites employing micro- and macroalgae, aquatic macrophytes and aquaculture wastes

9 Prospect of Bioprospecting in EKW Following Circular Economic Approach

Living systems infrastructure has the potential to serve the needs of many while providing a portfolio of options that, through their flexibility, pave the way for sustainable cities. A detailed biodiversity and genetic database of the wetlands should be made, which, in turn, can give the scope for using varieties of resistant and tolerant genes for research purposes. Isolation and culture of microbial strains and microbial genetic mapping will be beneficial for implementing bioremediation strategies (Chaudhuri & Thakur, 2006). The EKW model can be improved based on circular economy (CE) approach inclusive of its diversification for converting wastewater into value-added products employing micro- and macroalgae, aquatic macrophytes and aquaculture wastes (Regueiro et al., 2021). Applications of biotechnological and microbiological methods are essential in this regard. It will explore creative designs that could offer in the long run the potential to enhance profitability and sustainability through the valorization of by-products and side streams (Fig. 2). This concept includes technologies from recirculation to the implementation of (i) fish pond sludge for biogas production, co-incineration or fertiliser generation; (ii) algal biomass harvesting and downstream processing to produce high-value metabolites such as fatty acids, vitamins, amino acids, carotenoids, microelements, terpenes, food supplements, food additives, functional food, fertiliser, etc.); (iii) macrophytic biomass harvesting and downstream processing to produce biocompost, biogas, biofuel, paper pulp, furniture, handicraft, etc. Since CE should be restorative and regenerative by design the biological loops pursue the recycling of bio-based materials for the same manufacturing processes but also for new possible applications.

10 Conclusions

When cities are often considered as fabricated, heterotrophic parasites on the global landscape, use of nature-based blue-green living infrastructure in urban areas can add value to land use and environmental planning and management. Our challenge is to make cities more autotrophic, more self-nourishing, and aid in the evolution of urban agri/aquaculture and other green businesses in the underutilized regions and/or so-called wastelands of urban and peri-urban environments. Circular economy offers the solution to such pressing problems. EKW is a bright example of an ecological marvel, 'soft' ecosystemic infrastructure and a high-performing, ecologically engineered and managed ecosystem tightly bound to urban fabric. It is the largest ensemble of sewage-fed fish ponds in the world which is enriched in biodiversity and genetic resources. It furnishes not only the material manifestations of a unique living infrastructure, but also the underlying networks of energy, material, capital, social ties, and influences that collectively steer the development trajectories of the cities like Kolkata and its suburbs. The sewage-fed aquaculture has great potential for converting city's refuse into resource being treated in wetland biome, producing low-cost fish for the urban poor, low-cost wastewater treatment, and mitigating environmental pollution. This has developed over a century to treat urban wastewater, and convert wastewater into marketable fish and vegetables, create employment, and support the livelihood of local population and economy. The agriculture and fisheries activities in the EKW provide the city with two outstanding urban facilities: treatment of the city's sewage, and nutrient recovery from which an abundant supply of fish and vegetables is the end product. The integrated resource recovery system provided by the agriculture and fisheries activities, making use of city waste, provides a comprehensive answer to the city's sanitation and resource recovery needs.

The uniqueness of the EKW is that it is the world's largest wastewater ecosystem created to sustain successive resource recovery practices in the form of pisciculture, agriculture, and horticulture (vegetable farming) on a garbage substrate as a wonder of the waste land. EKW is unique as a Ramsar site for the wise use of sewage water, as a model for multi-use resource recovery system linked with agri-aquacultural activities for production of fish and vegetables.

The historical evolution of the bheries in the EKW provides evidence of how traditional knowledge led to setting one of the rarest examples of environmental protection and development management where complex ecological processes have been adopted and adapted by the local farmers for mastering the resource recovery activities which has drawn much international attention as a model system for the reuse of urban wastewater and resource recovery. These sewage-fed fisheries act simultaneously for the purification process like removal of heavy metals, coliform load reduction, and fish and vegetable production at a commercial scale. In EKW the sewage-fed fish farms facilitate a wide range of physical, biological and chemical processes which help improve the quality of the water. Acting as the *kidney of the city* EKW turns 'bad' water into 'good' water closing the loop in the urban environment. It has the potential to develop further as a low-cost eco-sanitation technology recognizing environmental and economic benefits for the urban areas. It shows how *welfare ecology* ensures the protection of environmental system and nature automatically produces welfare dividends to the people depending on it.

The EKW demonstrates a high degree of resilience attributed to the scale of operation, adaptive production strategies that optimize resource use while minimizing risks, self-organization among stakeholders, and timely legal instruments and institutional interventions to preserve the natural character. But wastewater-fed aquaculture is in decline. The introduction of externalizing technologies, erosion of social capital and wearing of traditional ecological wisdom threaten to undermine this resilience. However, the procedure being applied in EKW raised several scientific concerns and questions about food safety from public health perspective. The accumulation of toxic heavy metals in fishes and vegetables rings alarm bell with respect to ecological health and integrity of these wetlands.

EKW resource recovery system is apparently time tested and has so far not created any perceptible epidemiological issue, although many researchers apprehend that such long-term wastewater-fed fisheries may contaminate the groundwater in near future if the 'business as usual' continues. For promoting continued resilience of the EKW system in a constantly evolving situation, an objective and holistic approach is required. The cardinal guiding principles are maintenance of the unique features of the ecosystem, sustainable use of its resources and economic development for the wetland community. Wise use of wetlands would bring massive returns in long term with respect to development of green economy, in parallel with mainstream socio-economic development. The cost-effectiveness, community participation and natural processes involved in this reliable traditional technology system of municipal wastewater management which has changed the grammar of conventional wastewater treatment need to be replicated elsewhere all over the world, particularly in the developing countries as an ecofriendly alternative to cost and energy intensive conventional wastewater treatment systems for bringing economic gains and ensuring ecological balance.

References

Adhikari, S., Pal, S., Barik, A., Chakraborty, S., & Mukhopadhyay, S. K. (2021). Carbon sequestration by horticultural plants in East Calcutta Wetlands ecosystem, a Ramsar Site in India. *International Journal of Chemical and Environmental Sciences, 2*(2), 18–27.

Ahmed, A. S., Sultana, S., Habib, A., Ullah, H., Musa, N., Hossain, M. B., Rahman, M. M., & Sarker, M. S. I. (2019). Bioaccumulation of heavy metals in some commercially important fishes from a tropical river estuary suggests higher potential health risk in children than adults. *PLoS ONE, 14*(10), 0219336. https://doi.org/10.1371/journal.pone.0219336.

Aich, A., Chakraborty, A., Sudarshan, M., Chattopadhyay, B., & Mukhopadhyay, S. K. (2012). Study of trace metals in Indian major carp species from wastewater-fed fishponds of East Calcutta Wetlands. *Aquaculture Research, 43*(1), 53–65.

Basu, S., Dasgupta, M., & Chakraborty, B. (2014). Removal of chromium (VI) by *Bacillus subtilis* isolated from East Calcutta wetlands, West Bengal, India. *International Journal of Bioscience, Biochemistry and Bioinformatics, 4*(1), 7.

Basu, A., & Bhattacharjee, A. (2022). Urban expansion around Kolkata: evaluating urbanogenic interventions in New Town, Rajarhat. In *Anthropogeomorphology* (pp. 571–603). Springer, Cham.

Bedi, P. (2022). Smart global megacities spatial strategies for Kolkata metropolitan region. In *Smart global megacities* (pp. 383–407). Singapore: Springer.

Bera, B., Bhattacharjee, S., Shit, P. K., Sengupta, N., & Saha, S. (2021). Anthropogenic stress on a Ramsar site, India: Study towards rapid transformation of the health of aquatic environment. *Environmental Challenges, 4*, 100158.

Bhattacharjee, C. (2022). Solid waste disposal on inland water wetlands: A sustainable restoration of Wetland ecosystems. In *Handbook of research on monitoring and evaluating the ecological health of wetlands* (pp. 171–181). IGI Global.

Bhattacharya, S., Ganguli, A., Bose, S., & Mukhopadhyay, A. (2012a). Biodiversity, traditional practices and sustainability issues of East Kolkata Wetlands: A significance Ramsar site of West Bengal, (India). *Research & Reviews in BioSciences, 6*(11), 340–347.

Bhattacharya, S., Ganguli, A., Bose, S., & Mukhopadhyay, A. (2012b). Biodiversity, traditional practices and sustainability issues of East Kolkata Wetlands: A significance Ramsar site of West Bengal, (India). *Research & Reviews in BioSciences, 6*(11):340-347.

Bhattacharyya, S., Chaudhuri, P., Dutta, S., & Santra, S. C. (2010). Assessment of total mercury level in fish collected from East Calcutta Wetlands and Titagarh sewage fed aquaculture in West Bengal, India. *Bulletin of Environmental Contamination and Toxicology, 84*(5), 618–622.

Bhattacharyya, S., Chanda, A., Hazra, S., & Choudhury, S. B. (2021). Effect of enhanced salinity on the chlorophyll-a concentration and partial pressure of CO_2 of a sewage-fed freshwater aquaculture pond: A microcosm experiment. *Chemistry and Ecology, 37*(9–10), 840–849.

Bhattacharyya, A., Sen, S., Roy, P. K., & Mazumdar, A. (2008). A critical study on status of East Kolkata Wetlands with special emphasis on water birds as bio-indicator. In *Proceedings of Taal2007: The 12th World Lake Conference*, Vol. 1561 (p. 1570).

Biswas, J. K., & Rana, S. (2014). Treatment wetlands as ecotechnological tools for regenerative reclamation of wastewater: Experiences from working with Kalyani model. *Journal of Clean Energy Technologies, 2*(1), 23–27.

Biswas, J. K., & Santra, S. C. (2000). Heavy metal levels in marketable vegetables and fishes in Calcutta Metropolitan Area, India. In B. B. Jana, R. D. Banerjee, B. Guterstam, & J. Heeb (Eds.), *Waste Recycling and Resource Management in the Developing World: Ecological Engineering Approach* (pp. 371–376). University of Kalyani, India and International Ecological Engineering Society.

Biswas Roy, M., Nag, S., Halder, S., & Kumar Roy, P. (2022). Assessment of wetland potential and bibliometric review: A critical analysis of the Ramsar sites of India. *Bulletin of the National Research Centre, 46*(1), 1–11.

Bunting, S. W., & Little, D. C. (2005). The emergence of urban aquaculture in Europe. In C. A. B. I. Publishing (Ed.), *Urban Aquaculture* (pp. 119–135). Wallingford.

Bunting, S. W., Pretty, J., & Edwards, P. (2010). Wastewater-fed aquaculture in the East Kolkata Wetlands, India: Anachronism or archetype for resilient ecocultures? *Reviews in Aquaculture, 2*(3), 138–153

Chandra, K., Raghunathan, C., & Mao, A. A. (2020). Biodiversity profile of East Kolkata wetlands Zool. *Survey of India*, 326.

Chatterjee, S., Chattopadhyay, B., & Mukhopadhyay, S. K. (2006). Trace metal distribution in tissues of cichlids [*Oreochromis niloticus* and *O. mossambicus*] collected from wastewater-fed fishponds in East Calcutta wetlands, a ramber site. *Acta Ichthyologica et Piscatoria, 2*(36): 119–125.

Chatterjee, S., Chattopadhyay, B., Mukhopadhyay, S.K., Mohanta, B., Sudarshan, M. & Chakraborty, A. (2007). East Calcutta wetlands as a sink of industrial heavy metals: A PIXE study. *International Journal of PIXE, 17*(03n04): 129–142.

Chattopadhyay, B., Chatterjee, A., & Mukhopadhyay, S. K. (2002). Bioaccumulation of metals in the East Calcutta wetland ecosystem. *Aquatic Ecosystem Health & Management, 5*(2), 191–203.

Chaudhuri, S. R., & Thakur, A. R. (2006). Microbial genetic resource mapping of East Calcutta wetlands. *Current Science, 25*, 212–217.

Chaudhuri, S. R., Salodkar, S., Sudarshan, M., Mukherjee, I., & Thakur, A. R. (2008). Role of water hyacinth-mediated phytoremediation in waste water purification at east Calcutta wetland. *Environmental Sciences, 5*(1), 53–62.

Chaudhuri, S. R., Mukherjee, I., Ghosh, D., & Thakur, A. R. (2012). East Kolkata wetland: A multifunctional niche of international importance. *Journal of Biological Sciences, 12*(2), 84–95.

Das, A., Nagesh, T., Das, S. K., & Abraham, T. J. (2021). Stress responses of Indian major carps cultured in the East Kolkata Wetland, West Bengal India. *Aquatic Research, 4*(4), 351–362.

Das, S.K., Nagesh, T.S., Das, A., Abraham, T.J., & Vishwanath, T.S. (2022). February. stress mitigating and growth-enhancing effect of dietary Vitamin E in indian major carps cultured in East Kolkata wetlands, India. In *Proceedings of the zoological society* (pp. 1–13). India: Springer.

Dey, D., & Banerjee, S. (2013). Ecosystem and livelihood support: The story of East Kolkata wetlands. *Environment and Urbanization Asia, 4*(2), 325–337.

Dutta, J., Saha, A., & Mitra, A. (2016). Impact of acidification on heavy metal levels, in a bheri of East Kolkata Wetland (EKW), a ramsar site in the Indian sub-continent. *International Journal of Advanced Research in Biological Sciences, 3*, 154–159.

Dutta, J., Choudhary, G., & Mitra, A. (2017). Bioaccumulation of toxic heavy metals in the edible fishes of eastern Kolkata wetlands (EKW), the designated Ramsar Site of West Bengal India. *International Journal of Aquaculture and Fishery Sciences, 3*(1), 18–21.

Dutta, J., Zaman, S., Thakur, T. K., Kaushik, S., Mitra, A., Singh, P., Kumar, R., Zuan, A. T. K., Samdani, M. S., Alharbi, S. A., & Datta, R. (2022a). Assessment of the bioaccumulation pattern of Pb, Cd, Cr and Hg in edible fishes of East kolkata Wetlands India. *Saudi Journal of Biological Sciences, 29*(2), 758–766.

Dutta, S., Basak, R., & Chakraborty, S. (2022b). Isolation and determination of chromium (VI) and nickel (II) resistant siderophore producing *Bacillus cereus* (sp) and analysis of the protein expression using bioinformatics tools. *Journal of Microbiology, Biotechnology and Food Sciences*, e3611-e3611.

Edwards, P. (2000). Wastewater-fed aquaculture: state of the art. Waste recycling and resource management in the developing world. *University of Kalyani, India and International Ecological Engineering Society, Switzerland*, 37–49.

EKWMA (East Kolkata Wetland Management Authority). (2011). Conservation and management plans of east Kolkata wetlands, India: Report to be submitted to the Ministry of Environment and Forests, Government of India.

EKWMA, WISA. (2021). East Kolkata wetlands: Management action plan 2021–2026. East Kolkata wetlands management authority and wetlands International South Asia. Kolkata, India.

Furedy, C., & Ghosh, D. (1984). Resource-conserving traditions and waste disposal: The garbage farms and sewage-fed fisheries of Calcutta. *Conservation & Recycling, 7*(2–4), 159–165.

Furedy, C. (1987). From waste land to waste-not land: The role of the salt lakes, East Calcutta, in Waste Treatment and Recycling, 1845–1930. Sinha, ed.

Furedy, C., 1988. Social aspects of human excreta reuse: implications for aquacultural projects in Asia. In Wastewater-fed Aquaculture. In *Proceedings of the International Seminar on Wastewater Reclamation and Reuse for Aquaculture, Calcutta* (pp. 251–66).

Ghosh, D., & Sen, S. (1987). Ecological history of Calcutta's wetland conversion. *Environmental Conservation, 14*(3), 219–226.

Ghosh, D., & Sen, S. (1992). Developing waterlogged areas for urban fishery and a waterfront recreation project. *Ambio, 21*(2), 185–186.

Ghosh, D. (1985). Dhapa Report : From Disposal Ground to WAR (Waste-As-Resource) Field, East Calcutta Wetland Project, Department of Fisheries, Government of West Bengal, Calcutta, India, pp.1–32.

Ghosh, D. (1988). Wastewater fed aquaculture in the wetlands of Calcutta: An overview. In *Proceedings of the international seminar on wastewater reclamation and reuse for aquaculture, 6–9 December 1988, Calcutta, India.*

Ghosh, D. (1990). Wastewater-fed aquaculture in the wetlands of Calcutta–an overview. Wastewater-fed Aquaculture. Bangkok: Environmental Sanitation Information Centre, Asian Institute of Technology, pp.49–56.

Ghosh, D. (1999). Wastewater utilisation in East Calcutta wetlands. UWEP Occasional Paper, WASTE, Nieuwehaven, the Netherlands. pp.1–20.

Ghosh, D. (2008). Waste water utilization in East Calcutta Wetlands from local practice to sustainable option. *eSS Occasional Papers, 1* (1): 36–56.

Ghosh, R. (2022). Not a 'peripheral'connection: dynamics of peri-urban agriculture in Kolkata Metropolitan Area, West Bengal. In *Sustainable Urbanism in Developing Countries* (pp. 277–290). CRC Press.

Hussan, A. (2016). Threats to fish diversity of East Kolkata Wetlands and conservation needs. *Aquaculture times, 2*(6), 10–15.

Hussan, A. J. M. A. L., Mandal, R. N., Hoque, F., Sundaray, J. K., Das, A., Chakrabarti, P. P., Adhikari, S., Udit, U. K., Choudhury, G., & Pillai, B. R. (2021). Strategies to control invasion of Sailfin Armoured Catfish, Pterygoplichthys spp. in wastewater-fed aquaculture bheries of East Kolkata Wetland, India with suggestion of a modified barrier based on the biological and behavioural characteristics. *International Journal of Aquatic Biology, 9*(3): 187–199.

Jana, B. B. (1998). Sewage-fed aquaculture: The Calcutta model. *Ecological Engineering, 11*(1–4), 73–85.

Kumar, B., Senthil Kumar, K., Priya, M., Mukhopadhyay, D., & Shah, [illegible]. (2010). Distribution, partitioning, bioaccumulation of trace elements in water, sediment and fish from sewage fed fish ponds in eastern Kolkata India. *Toxicological & Environment Chemistry, 92*(2), [illegible]43–260.

Kumar, B., Mukherjee, D. P., Kumar, S., Mishra, M., Prakash, D., Singh, S. K., & Sharma, C. S. (2011). Bioaccumulation of heavy metals in muscle tissue of fishes from selected aquaculture ponds in east Kolkata wetlands. *Annals of Biological Research, 2*(5), 125–134.

Kumar, A., Das, S. K., & Bhakta, D. (2021). *Temporal variations in bio-indices of Oreochromis niloticus (Linnaeus, 1758) in East Kolkata wetlands.*

Kumar, S., Banerji, H., & Ghosh, S. (2022). Reflections from East Kolkata Wetlands: A conceptual framework for building urban resilience to climate change for sustainable urbanism in developing countries. In *Sustainable Urbanism in Developing Countries* (pp. 343–364). CRC Press.

Kundu, N., Pal, M., & Saha, S. (2008). East Kolkata wetlands: a resource recovery system through productive activities. In *Proceedings of Taal 2007: the 12th world lake conference.*

Maiti, P., & Banerjee, S. (2012). Fate of metals in fish under variable sewage input in fish ponds. *International Journal of Scientific and Research Publications, 2*(6), 1–13.

Majumdar, A. (2008). The East Calcutta (Kolkota) wetlands: The city sewage treatment system of Kolkota, a city of 14 million. *Wetland Science and Practice, 25*(1), 21–23.

Mandal, S., Goswami, A. R., Mukhopadhyay, S. K., & Ray, S. (2015). Simulation model of phosphorus dynamics of an eutrophic impoundment–East Calcutta Wetlands, a Ramsar site in India. *Ecological Modelling, 306*, 226–239.

Mukherjee, V., Das, A., Akhand, A., & Gupta, G. (2013). Toxicity and profitability of rice cultivation under wastewater irrigation: The case of the East Calcutta Wetlands. *Ecological Economics, 93*, 292–300.

Mukherjee, J., Morera, R., Guerrin, J., & Véron, R. (2021a). Histories of urban Deltascapes: A comparison of Arles and Kolkata. *Global Environment, 14*(3), 505–534.

Mukherjee, S., Sikdar, P. K., Pal, S., & Schütt, B. (2021b). Assessment of environmental water security of an Asian deltaic megacity and its peri-urban wetland areas. *Sustainability, 13*(5), 2772.

Mundoli, S., Dechamma, C.S., Auddy, M., Sanfui, A., & Nagendra, H. (2022). A new imagination for waste and water in India's peri-urban interface. In *Water security, conflict and cooperation in Peri-urban South Asia* (pp. 27–43). Cham: Springer.

Nadella, A., & Sen, D. (2021). Evolution of the urban wastewater bio-treatment and reuse system of East Kolkata Wetlands, India: An appraisal. *Wetlands, 41*(8), 1–18.

Naskar, A., Alam, I., Majumder, A., & Maheswaran, G. (2022). Recently resighted population of Blue-breasted Quail (*Synoicus chinensis*) in and around East Kolkata Wetland is under threat due to development activities. *Records of the Zoological Survey of India, 121*(4), 527–535.

Nath, S., & Bhoumik, M. (2013). Levels of toxic metals in edible fish from a wetland of India. *International Journal of Environmental Sciences, 3*(5), 1509–1515.

Pakrashi, H. (2021). The peri-urban poor and ecology in the megacity of Kolkata. In Urbanization in the global South (pp. 242–254). India: Routledge.

Paul, A. (2011a). Aquaculture in artificially developed wetlands in urban areas: Relationship between soil and surface water in landscape ecology. *Journal of Institute of Town Planners, India, 8*, 50–52.

Paul, A. (2011b). Aquaculture in artificially developed wetlands in urban areas: An application of the bivariate relationship between soil and surface water in landscape ecology. *Journal of Environmental Science & Engineering, 53*(1), 135–136.

Ray Chaudhuri, S., Mishra, M., Nandy, P., & Thakur, A. R. (2008). Waste management: A case study of ongoing traditional practices at East Calcutta Wetland. *American Journal of Agricultural and Biological Sciences, 3*(1), 315–320.

Ray Chaudhuri, S., Mukherjee, I., Ghosh, D., & Thakur, A. R. (2012). East Kolkata wetland: A multifunctional niche of international importance. *Journal of Biological Sciences, 12*(2), 84–95.

Regueiro, L., Newton, R., Soula, M., Mendez, D., Kok, B., Little, D. C., Pastres, R., Johansen, J., & Ferreira, M. (2021). Opportunities and limitations for the introduction of circular economy principles in EU aquaculture based on the regulatory framework. *Journal of Industrial Ecology*. https://doi.org/10.1111/jiec.13188.

Roy, P.K., Roy, M.B., Chatterjee, S., Halder, S., Nag, S., & Majumder, A. (2021). Land reclamation and reuse of waste water on the backdrop of urban sprawl of Kolkata metropolitan: a case study of

East Kolkata, Wetland, India. In *Modern Cartography Series*,Vol. 10 (pp. 557–579). Academic Press.

Saha, S. (2014). Traditional management practices for aquaculture in east Kolkata wetlands West Bengal. *Traditional Knowledge, 31*, 31

Saha, S., Saha, T., & Basu, P. (2017). Seasonal changes in zooplankton and macro-fauna populations of the east Calcutta wetland fish ponds. *Proceedings of the Zoological Society, 70*(2): 156–164.

Sahu, P., & Sikdar, P. K. (2008). Hydrochemical framework of the aquifer in and around East Kolkata Wetlands, West Bengal India. *Environmental Geology, 55*(4), 823–835.

Sahu, P., & Sikdar, P. K. (2011). Threat of land subsidence in and around Kolkata city and east Kolkata wetlands, West Bengal India. *Journal of Earth System Science, 120*(3), 435–446.

Sarkar, S., Ghosh, P. B., Mukherjee, K., Sil, A. K., & Saha, T. (2009). Sewage treatment in a single pond system at East Kolkata Wetland India. *Water Science and Technology, 60*, 9.

Sengupta, K. (2018). Shrinkage of East Kolkata Wetlands and its effect. *International Journal of Academic Research and Development, 3* (1), 57–67.

Sifa, L. (1997). *Aquaculture and its role in ecological wastewater treatment* (pp. 38–49). CRC Lewis Publishers.

Sikdar, P. K., & Sahu, P. (2009). Understanding wetland sub-surface hydrology using geologic and isotopic signatures. *Hydrology and Earth System Sciences, 13*(7), 1313–1323.

Sikdar, P. K., Sarkar, S. S., & Palchoudhury, S. (2001a). Geochemical evolution of groundwater in the quaternary aquifer of Calcutta and Howrah India. *Journal of Asian Earth Sciences, 19*(5), 579–594.

Sikdar, P. K., Sarkar, S. S., & Palchoudhury, S. (2001b). Geochemical evolution of groundwater in the quaternary aquifer of Calcutta and Howrah India. *Journal of Asian Earth Sciences, 19*(5), 579–594.

Stewart, D. (1836). Report on the project of the salt water lakes reclamation and irrigation company limited, pp.1–7 in Selections from the records of the Bengal government, containing papers from 1865 to 1904, Government of Bengal, Calcutta, India, pp.v + 404.

Tyagi, I., Tyagi, K., Bhutiani, R., Chandra, K., & Kumar, V. (2021). Bacterial diversity assessment of world's largest sewage-fed fish farms with special reference to water quality: A Ramsar site. *Environmental Science and Pollution Research, 28*(31), 42372–42386.

Vicziany, M., Chattopadhyay, D., & Bhattacharyya, S. (2017). Food from sewage: Fish from the east Kolkata wetlands and the limits of traditional knowledge. *South Asia: Journal of South Asian Studies, 40*(3), 619–644.